EC Competition Law

Second Edition

D. G. GOYDER

Solicitor: Visiting Professor in Law at King's College London and the University of Essex

CLARENDON PRESS · OXFORD

Oxford University Press, Walton Street, Oxford OX2 6DP
Oxford New York Toronto
Delhi Bombay Calcutta Madras Karachi
Kuala Lumpur Singapore Hong Kong Tokyo
Nairobi Dar es Salaam Cape Town
Melbourne Auckland Madrid
and associated companies in
Berlin Ibadan

Published in the United States
by Oxford University Press Inc., New York

First Edition 1988
Second Edition 1993, Reprinted (with corrections) 1993

British Library Cataloguing in Publication Data
(Data available)
ISBN 0–19–825727–9
ISBN 0–19–825728–7 (Pbk)

Library of Congress Cataloging in Publication Data
EC competition law / D. G. Goyder.—2nd ed.
(Oxford European Community law series)
Rev. ed. of: EEC competition law. 1988.
Includes bibliographical references and index.
1. Antitrust law—European Economic Community countries.
2. Restraint of trade—European Economic Community countries.
3. Competition—European Economic Community countries.
I. Goyder, D. G. EEC competition law. II. Title. III. Series
KJE6456.G69 1992 343.4.072—dc20 [344.0372] 92 19005
ISBN 0–19–825727–9
ISBN 0–19–825728–7 (Pbk)

Printed in Great Britain by
Biddles Ltd
Guildford & King's Lynn

OXFORD EUROPEAN COMMUNITY LAW SERIES

General Editor: F. G. Jacobs
Advocate-General, The Court of Justice
of the European Communities

EC COMPETITION LAW

OXFORD EUROPEAN COMMUNITY LAW SERIES

The aim of this series is to publish important and original studies of the various branches of European Community Law. Each work will provide a clear, concise, and critical exposition of the law in its social, economic, and political context, at a level which will interest the advanced student, the practitioner, the academic, and government and Community officials.

Dedicated to
Sir Arthur fforde
(1900–1985)
Lawyer, Headmaster, and Poet

General Editor's Foreword to First Edition

This is the first of a new series of books on European Community law. The general aims of the Oxford Series, which will deal in a systematic way with different areas of Community law, are set out in the fly-leaf of this book. The aims are perhaps ambitious, but the need is undeniable. Despite the appearance in recent years of both general student texts and specialized practitioners' works, there are very few works containing informed and readable discussion in depth of the main branches of Community law. The few books in this category have been individually designed and, not forming part of an overall plan, often overlap. They rarely appeal to both student and practitioner alike. Yet both the novelty and the importance of Community law demand a treatment which will set out the underlying principles of the subject and the practical effect of its rules.

The need can only increase with the approach of the European internal market, which will have a decisive impact on the life and work of everyone in the Community, and of many outside. But the need exists already, since the Community is a Community of law, and its law already affects profoundly the life and commerce of its citizens.

Nowhere is this more true than in the field covered by this first volume. Competition law is a subject of the highest practical significance to all those doing business in, or with, the Community. A sophisticated and principled body of law has developed over the past thirty years. It affects a vast range of commercial activity. It is implemented by the Commission of the European Communities, endowed for this purpose with a panoply of powers unparalleled in other areas of Community law, or for the most part in national law. It has to be applied by the courts of the Member States. And it has to be observed by all those subject to it. It has the greatest impact on business life in the obvious form that it renders anti-competitive agreements and practices void, exposes firms to intensive investigations and enquiries, and may give rise to very substantial fines. Seen more constructively, it gives firms the opportunity to compete fairly and the opportunity to compete for the advantages of the European market; it protects the consumer; and it compels a more efficient use of economic resources.

The study of this branch of law is both a necessity and a subject of great intrinsic interest. But its study requires more than a knowledge of legal rules; it requires a grasp of the underlying economic principles and of the nature of competition policy. To write a good book on the subject, of course, requires much more: not least, a combination of academic and practical expertise, and great clarity and skill in the exposition of a technical and complex subject.

The author of this book is admirably qualified for the task, and the resulting work has many virtues which the reader will appreciate. The reader may be struck, in particular, by the perception, and clarity of exposition, of the interconnections between law and economics; by the historical perspective which enables the current law to be properly understood; and not least by the freshness of approach to old problems. An additional original feature is the concern to look at the subject, not only from the viewpoint of the lawyer, but from the perspective of those who administer the system.

This book augurs well, then, for the future of the series, which in general is designed for the advanced student, the practitioner, and government and Community officials, but the book seems destined for a wider audience, both in the UK and beyond. In the UK the timing of its appearance is topical, in view of the current proposals to re-shape major areas of domestic competition law in line with Community law. Similar reforms of domestic law have been introduced, or are projected, in other Member States. Such reforms make particularly relevant the historical perspective of this book, which confronts the fundamental issues of competition law from first principles. On the Community level, the subject is of constantly increasing importance, with new developments in the field of mergers, in air transport and elsewhere.

Interest in this book will certainly not be confined to specialists, or to those with particular concerns in the interaction of law and economics and in competition policy, but will extend to many involved in the life and business of the European Community.

Francis Jacobs

Kings College, London
26 April 1988

Author's Preface to First Edition

The actual process of writing a book is necessarily solitary: in spite of Voltaire's belief that 'the happiest of all lives is a busy solitude', there are moments when one could wish for a more gregarious task. Fortunately, however, there are other stages (both at the beginning and the end of the process) which cannot be completed without the help of many others, and that has been especially true on this occasion.

Before writing a general account of the European Community's competition law, it seemed to me important to interview a number of the officials of Directorate-General IV of the Commission, which is responsible for the administration of Community competition policy, particularly as some of the more senior among them would shortly be reaching retirement. I am grateful to all those officials who allowed me to share their memories of the pioneer days of DG IV, as well as their views of past and present events and policy issues. Some of them also later read and commented on individual chapters. As those who helped me in this way were numerous, I have elected not to name them individually, with but two exceptions. The first is Aurelio Pappalardo, now retired from the Commission, who made it possible for me to spend time in DG IV and conduct these interviews; the second is Karen Williams, formerly a colleague at the University of Essex, who after her move to Brussels to join DG IV continued to give me enthusiastic encouragement and practical help on many aspects of the work.

I was also fortunate to be able to spend time in Brussels in the offices of Cleary, Gottlieb, Steen & Hamilton, where Don Holley and his colleagues made me welcome, allowed me the use of their excellent library, and kept me going with copious cups of black coffee. I am also grateful to Don for reading the drafts of several chapters and for his wise comments upon them. Professor Arved Deringer allowed me on a visit to Cologne to benefit from his unrivalled knowledge of the historical roots both of the Treaties of Paris and Rome, and also of the circumstances in which Regulation 17 was enacted in 1962. The Leverhulme Trust awarded me a research grant to help meet part of the travelling and other expenses involved in the writing of the book, and I am pleased to acknowledge their help.

I must also express my thanks to Byron and Eleanor Fox, Barry Hawk, Sir Alan Neale, Noel Ing, Philippa Watson, Francis B. Jacobs (of the European Parliament's Secretariat), and Chris Docksey, all of whom gave me help, in a variety of different ways, during the writing of the book.

Most of this writing was done in the Library of the University of Essex at Colchester where Peter Luther ensured that I had all the relevant authorities. I must also express my appreciation to my colleagues, both in Ipswich and at the University Law Department, for their tolerance of my preoccupation with this project at times when they must have wished it was less time-consuming.

All the typing of the manuscript has been done by my former secretary, Elaine Golledge, whom I was fortunately able to persuade to return from premature domestic retirement; she and her husband Brian have tolerated the presence of a word processor in their dining room for eighteen months, which has enabled her to produce a series of consistently accurate typescripts to a succession of increasingly unreasonable deadlines.

My relationship with the Oxford University Press has been pleasant, and I would like to thank Richard Hart for ensuring a smooth production process. My final expressions of thanks must, however, be directed towards three people in particular; to my Consulting Editor, Professor Francis Jacobs of London University, for cheerful and admirably constructive advice throughout; to my daughter Joanna, Barrister, for her help, especially with French language material; and finally to my wife Jean, without whose unfailing support this book could certainly not have been written.

This book is dedicated, with the approval of Lady fforde, to Sir Arthur fforde, who died in 1985. After a distinguished career as a City of London solicitor and wartime civil servant, he came as headmaster to Rugby School from 1948 to 1957. Many of us who were in the school during that time remember him with particular respect and affection. I am glad to have this opportunity to acknowledge my own debt of gratitude to him; his inquiring and subtle mind would, I like to think, have enabled him to enjoy the intricacies of EC competition law.

I have finally to issue the customary disclaimers. Though I have received help from many sources, responsibility for the contents of the book and for any errors within it are exclusively mine. Neither the contents nor the errors are to be attributed in any way to any of the persons or organizations referred to in this Preface, nor to any other body with which I am, or have been, associated.

The law is stated as at 31 March 1988.

D.G.G.

20–26 Museum Street
Ipswich, Suffolk

April 1988

Author's Preface to Second Edition

Less than four years have passed since the first edition of this book was published. In the intervening period, however, many important new developments have occurred. The Merger Regulation has been brought into force and already a number of important decisions rendered under it. The Court of First Instance has produced its first (and impressive) judgments, and the European Court itself has considerably expanded the jurisprudence of Article 86 as well as giving a number of other important judgments in the field of competition law. Several Member States (though not yet the United Kingdom) have adopted new national laws incorporating the basic principles of Articles 85 and 86. The Maastricht Treaties amending the Treaty of Rome in important respects were signed in December 1991 though ratification by Member States and the European Parliament remains to be completed.

With the formation likely from the beginning of 1993 of the European Economic Area and with some EFTA members in any case seeking full membership in the Community, the detailed principles of Community competition law are likely in the near future to become applicable in an even wider area of Europe. In the long run the recent far-reaching changes in Eastern Europe may also have extremely wide consequences for the administration of competition policy in the Community. All these developments mean that the subject of Community competition law has become increasingly important in the United Kingdom.

The first part of the book, being largely historical, remains virtually unchanged. The second part has required substantial revision and expansion in those chapters dealing with Article 86 and with mergers, whilst minor revisions have been necessary in other chapters to take account of new group exemption regulations and recent case-law. For reasons of space the chapter on State Aids contained in the first edition has been taken out. With the inevitable inclusion of so much new material I felt a temptation to remove references to some of the early cases whose importance has inevitably diminished with the passage of time. Nevertheless, with but few exceptions, I have retained them because of my belief that they help to provide an understanding of the historical development of this complex subject, which remains of essential importance to both lawyers and law students. The final part of the book dealing with 'Conclusions and Prospects' has been generally updated and expanded in the light both of these many changes since the publication of the first edition and the likelihood of further change over the next few years.

In preparing this second edition I have again been helped by a number of the officials of DG IV in Brussels for whose time and trouble I am grateful. I have also benefited once more from being able to spend time in the Brussels office of Cleary, Gottlieb, Steen, & Hamilton (where excellent cups of black coffee were still to be found) and where Don Holley was kind enough to comment on some of the new material.

The law is stated as at 1 March 1992. The important decision of the Court of First Instance in the *Italian Flat Glass* cases was delivered a few days after this date and is noted in a footnote in Chapter 19, though it has not been possible to incorporate comment on the decision into the text. One or two other additional cases have also been briefly mentioned in footnotes.

D.G.G.

King's College
Strand, London

March 1992

Contents

Table of Cases from the European Court of Justice and Court of First Instance (1990 onwards)

Table of European Commission Decisions

Table of Cases from Other Jurisdictions

United Kingdom

United States

United Kingdom Monopolies and Mergers Commission Reports

Commission Notices

Table of European Community Treaties

Table of European Community Secondary Legislation

Directives

Regulations

Table of International Legislation

France

Germany

Netherlands

United Kingdom

United States

Chronological Table (1951-1991)[1]

Year	Date	Chronological table of events	European Community Council and Commission Regulations	Commission: leading decisions	European Court of Justice: leading decisions
1951	18 Apr.	Signature by France, Germany, Italy, and Benelux countries ('the Six') of Treaty of Paris establishing European Coal and Steel Community (ECSC).			
1953	10 Feb.	Common Market in coal comes into force.			
	1 May	Common Market in steel comes into force.			
1957	25 Mar.	Signature of Treaty of Rome establishing European Economic Community by the Six (and also establishing EURATOM).			
1958	1 Jan.	Treaty of Rome comes into force.			
1961	7 Sept.	European Parliament report (the Deringer Report) on the draft of Reg. 17 published.			
1962	13 Mar.	First Council Reg. 17 implementing Articles 85 and 86 of the Treaty of Rome comes into force.	COUNCIL. Reg. 17/62 Commission's powers and procedures in competition matters.		*Bosch/van Rijn*
	11 May		COMMISSION Reg. 27/26 Notification arrangements.		
	30 July		COUNCIL Reg. 26/62 applying certain competition rules to agriculture.		
	1 Nov.	Deadline for notification of multilateral agreements under Reg. 17/62 (approx. 800 lodged).			
	26 Nov.		COUNCIL Reg. 141/62 retrospectively from 13 Mar. 1962 excluding transport from scope of Reg. 17/62.		

[1] For reasons of space, some abbreviations have been used and names of decisions are occasionally given in either abbreviated or 'popular' form. Names of cases may change when taken on appeal to the Court from the Commission.

Year	Date	Chronological table of events	European Community Council and Commission Regulations	Commission: leading decisions	European Court of Justice: leading decisions
1962	24 Dec.	'The Christmas Message' issued by DG IV; Notices on patent licensing and commercial agents.			
1963	31 Jan.	Deadline for registration of bilateral agreements under Reg. 17/62 (approx. 34,000 lodged).			
	24 July	First informal decision by DG IV prohibiting agreement under Article 85(1), the 'Convention Faience' recommendations.			
	25 July		COMMISSION Reg. 99/63 governing procedures of hearings of individual cases by the Commission.		
1964	13 May	First publication of Notice of proposed decision (the 'Convention Faience' case) under Article 19(3) of Reg. 17/62.		*Grosfillex–Fillistorf; Grundig-Consten; Bendix–Mertens & Straet: Nicholas–Vitapro; Dutch Engineers and Contractors Assoc.*	
1965	2 Mar.	First Council Reg. authorizing the granting of group exemptions by the Commission.	COUNCIL Reg. 19/65 conferring power on Commission to issue group exemptions for Exclusive Distribution and Exclusive Purchasing and Industrial Property Rights licences	*DRU-Blondel Hummel–Isbecque; Maison Jallatte–Voss*	
	8 Apr.	Signature of Treaty on amalgamation of executives of ECSC, EC, and EURATOM.			
1966	13 July	First European Country Justice decision on Art. 85, (*Grundig–Costen/Commission*)		*Grundig–Costen/Commission; Italy/Council Commission; STM/Maschinenbau Ulm*	

Year	Date	Chronological table of events	European Community Council and Commission Regulations	Commission: leading decisions	European Court of Justice: leading decisions
1967	1 May	First Commission Reg. granting group exemption comes into force.	COMMISSION Reg. 67/67 group exemptions for specified forms of Exclusive Distribution and Exclusive Purchasing Agreements.	*Transocean Marine Paint* (no. 1)	*Brasserie de Haecht* (no. 1)
	1 July	Effective date for amalgamation of executives of ECSC, EC, and EURATOM.			
1968	1 July		COUNCIL Reg. 1017/68 rules of competition applied by Council to rail, road, and inland waterway transport.	*Eurogypsum: ACEC/Berliet: Cobelaz*	*Parke, Davis/ Probel*
	29 July	Commission issues Notice concerning co-operation between undertakings.			
1969	16 July	First fines imposed by Commission for a substantive violation of Article 85 of the Treaty of Rome (the *Quinine Cartel*).		*Quinine Cartel: Jaz/Peter; Dyestuffs; Christiani–Nielsen*	*Walt Wilhelm/ BKA; Völk/Vervaecke*
1970	27 May	Commission issues first Notice concerning Agreements of Minor Importance.		*Omega: Kodak: ASBL; German Ceramic Tiles*	*Quinine Cartel*
1971	2 June	First formal decision of Commission on Article 86 (*GEMA*).		*GEMA; Continental Can;*	*Sirena/Eda: DGG/Metro;*
	7 June	European Parliament first requests the Commission to prepare an Annual Report on Competition Policy.		*VCH: Burroughs–Del planque; Burroughs–Geha: Henkel–Colgate*	*Béguelin/GL Import*
	20 Dec.		COUNCIL Reg. 2821/71 conferring power on Commission to issue group exemptions for Specialization Agreements; Reg. 2822/71 relieving some Specialization Agreements from the requirements of notification by amending Reg. 17/62.		

Year	Date	Chronological table of events	European Community Council and Commission Regulations	Commission: leading decisions	European Court of Justice: leading decisions
1972		First Annual Report on Competition Policy issued.	COMMISSION Reg. 2591/72 extending Reg. 67/67 for a further ten years. Reg. 2779/72 group exemption for Specilization Agreements	*Saviem/MAN; Raymond/ Nagoya; Davidson Rubber; Cimbel; Pittsburgh Corning; Commercial Solvents; WEA–Filipacchi; GISA*	*Dyestuffs*
	19 Dec.		COUNCIL Reg. 2743/ 72 introducing minor amendments to Reg. 2821/71.		
1973	1 Jan.	UK, Ireland, and Denmark enter the Community.		*Sugar Cartel; Kali–Salz; Prym–Beka; Transocean Marine Paint* (no. 2)	*Continental Can; Brasserie de Haecht* (no. 2);
1974			COMMISSION Reg. 2988/74 bringing in rules relating to limitation of actions from 1 Jan. 1975.	*GM Continental; Europair–Durodyne; SHV/Chevron; Belgian Wallpapers; FRUBO; Franco-Japanese Ballbearings; BMW*	*BRT/SABAM; Commercial Solvents; Van Zuylen/Hag; Centrafarm Sterling and Winthrop; French Merchant Seamen*
1975	1 Jan.			*Bayer/Gist–Brocades; KEWA/United Processors; United Brands; IFTRA Rules; AOIP/Beyrard; Bomée Stichting*	*Sugar Cartel; GM Continental; Belgian Wallpapers; FRUBO Kali–Salz*
	15 Dec.	European Patent Convention signed.			
1976				*Vitamins (Hoffmann); Pabst–Richarz/ BNIA; Junghans; Gerofabriek; Reuter/BASF; Miller International*	*Fonderies Roubaix; EMI/CBS; Terrapin/ Terranova*
1977			COMMISSION Reg. 2903/77 amending and extending Reg. 2779/72 on Specialization Agreements.	*ABG; Hugin/Liptons; Cobelpa/VNP; Campari; GEC/Weir; Vacuum Interrupters* (no. 1) *De Laval/Stork;*	*Concordia; Metro/SABA* (no. 1); *Inno/ATAB*

Year	Date	Chronological table of events	European Community Council and Commission Regulations	Commission: leading decisions	European Court of Justice: leading decisions
1977	29 Dec.	Amended Notice concerning Agreements of Minor Importance.		*Spices; Distillers; BMW Belgium*	
1978	7 Feb.	Publication by Commission of new draft Reg. amending 67/67.		*Fedetab; Maize Seed; Black Powder; Zanussi Guarantee*	*Miller International; United Brands; BP' Commission; Hoffman–la Roche/ Centrafarm Centrafarm/American Home Products*
1979	3 Mar.	Publication by Commission of new draft Regs. containing group exemption for patent licences.		*Floral Sales; Atka/BP Kemi; Rennet; Pioneer; Beecham, Parke Davies*	*Vitamins (Hoffman); Hugin/Lipton; BMW Belgium; Greenwich/ SACEM*
1980				*Italian Cast Glass; Hennessy–Henkell*	*Fedetab; Perfumes; Camera Care; Distillers; Coditel* (no. 1)
1981	1 Jan.	Greece enters the Community.		*Anseau–Navewa; VBVB–VBBB; Hasselblad; Michelin*	*Coöperative Stremsel; Salonia/ Poidomani; Züchner/ Bayerische Vereinsbank*
1982	1 Sept.	Post of Hearing Officer for DG IV created		*AROW/BNIC; Amersham/ Buchler; AEG/Telefunken; SSI; Rolled Zinc (CRAM); Toltecs/Dorcet*	*Maize Seed (Nungesser); Polydor/ Harlequin; Keurkoop/Nancy Kean; Coditel* (no. 2)
1983	1 Jan.	New Specialization Agreement group exemption comes into force	COMMISSION Reg. 3604/82 amended group exemption for Specialization Agreements replacing Reg. 2779/72, and Regs. 1983/83 and 1984/83 replacing Reg. 67/67 and providing group exemptions for Exclusive Distribution and Exclusive Purchasing Agreements, lwith special provision for beer and petroleum products.	*Remia–Nutricia; Vimpoltu; Ford; Saba* (no. 2); *Rockwell/Iveco; Carbon Gas Technology; Int. Energy Authority; Windsurfing International*	*Anseau–Navewa; GVL; Michelin; Pioneer; AEG/Telefunken; Demo-Studio Schmidt; Kerpen & Kerpen*

Year	Date	Chronological table of events	European Community Council and Commission Regulations	Commission: leading decisions	European Court of Justice: leading decisions
1983	1 July	New Exclusive Distribution and Exclusive Purchase group exemption Regs. come into force.			
1984				*Synthetic Fibres; Zinc Cartel; Peroxygen; Woodpulp; Aluminium; IBM Personal Computers; BPCL/ICI; Carlsberg/Grund Met*	*VBVB–VBBB; Hasselblad; Ford of Europe (interim measures); Rolled Zinc (CRAM); Hydrotherm*
1985	1 Jan.	Commission Reg. granting group exemption for patent licences comes into force.	COMMISSION Reg. 2349/84 group exemption for patent licences.	*Grundig; Villeroy–Boch; Olympic Airways; BP/Kellogg; ECS/AKZO; Ivoclar*	*British Telecom; CBEM–CLT; Toltecs; Ford of Europe (selective distribution system); Leclerc (Books); Leclerc (Petrol); BNIC/Clair (Cognac); Binon; ETA Fabriques; Pharmon* v. *Hoechst; Remia–Nutricia; SSI*
	1 Mar.	Specialization and R & D group exemptions Regs. comes into force.	COMMISSION Reg. 417/85 Specialization Agreements group exemption replacing 3604/82.		
	1 July	Automotive Dealers Distribution Agreements group exemptions Regs. come into force	COMMISSION Reg. 418/85 R & D Group Agreement exemption. COMMISSION Reg. 123/85 Automotive Dealers group exemption.		
	5 Aug.		COMMISSION Reg. 2526/85 amending Reg. 27/62 by introducing a new form A/B for notification purposes.		
1986	1 Jan.	Spain and Portugal enter the Community.		*Polypropylene; Optical Fibres; Meldoc; Fatty Acids; Pronuptia; Yves Rocher; Belasco; Boussois/ Interpane Mitchell–Cotts/ Sofiltra*	*Nouvelles Frontières; Pronuptia; Windsurfing; International; Metro/SABA* (no. 2); *British Leyland*
	17 and 28 Feb.	Signature by all Member States of the Single European Act (amending certain provisions of the Treaty of Rome).			
	9 Sept.	Commission Notice on Agreements of Minor Importance issued.			
	22 Dec.		COUNCIL Reg. 4056/ 86 applying competition rules to maritime transport.		

Year	Date	Chronological table of events	European Community Council and Commission Regulations	Commission: leading decisions	European Court of Justice: leading decisions
1987	1 July	Single European Act comes into force		*Sandoz: Hilti*	*Philip Morris* (merger)
	1 July	Council Reg. on applying competition rules to maritime transport comes into effect.			
1988	1 Jan.		COUNCIL Regs. 3975–3976/87 applying competition rules to air transport.		
				Bayer/BP Chemicals; British Dental Trade Assoc.; VBA; BSC/Napier Brown;Delta Chemie;BPB Industries; PVC/LDPE Cartels; Magill;Net Book Agreement; Continental/Michelin; ServiceMaster; Italian Flat Glass	*Allen & Hanbury/Generics; Erauw–Jacquery; Bodson; Van Eycke/Aspe; Woodpulp* (Jurisdiction *only*) *Novasam/Alsatel; Renault/Maxicar; Volvo/Veng*
	30 Nov.	New Commission Regulations adopted for franchising and know-how licensing.			
1989	1 Feb.	Commission Reg. granting group exemption for franchise agreements comes into force	COMMISSION Reg. 4087/88 group exemption for franchise agreements	*Dutch Banks; Dutch Courier Services; Italian Fire Insurance;*	*Ahmed Saeed Ottung/Klee Belasco; SACEM/Lucazeau; Hoechst/Dow/Solvay*
	1 Apr.	Commission Reg. granting group exemption for know-how licensing agreements comes into force.	COMMISSION Reg. 556/89 group exemption for know-how licensing agreements.	*APB; Bayo-n-ox; Welded Steel Mesh; UIP*	
	1 Sept.	Establishment of Court of First Instance (CFI) in Luxembourg.			
	21 Dec.	New Council Regulation adopted on control of concentrations (mergers).			
1990	14 Aug.	Commission notices on concentration/co-operation joint ventures and on ancillary restrictions published.		*Alcatel/ANT; Elopak/Metal Box (Odin); Moosehead/ Whitbread Soda Ash (Solvay–ICI)*	*Sandoz; Tetrapak (CFI); Hag* (no. 2)
	21 Sept.	Council Regulation on control of mergers, and also Commission Regulation on merger procedures comes into force.	COUNCIL Reg. 4064/89 on control of mergers. COMMISSION Reg. 2367/90 on merger procedures.		

Year	Date	Chronological table of events	European Community Council and Commission Regulations	Commission: leading decisions	European Court of Justice: leading decisions
1991	Sept.	First concentration subject to Reg. 4064/89 declared incompatible with Common Market (Aerospatiale–Alenia)		*Screensport/EBU; Gosmé/Martell; Vichy; Ecosystems; Aerospatiale–Alenia* (merger)	*Delimitis/ Henninger Bräu; ECS/AKZO; Magill (CFI)*

Abbreviations

CMLR	Common Market Law Reports
CML Rev	*Common Market Law Review*
ECLR	*European Competition Law Review*
ECR	European Court Reports
EIPR	*European Intellectual Property Review*
EL Rev	*European Law Review*
ICLQ	*International and Comparative Law Quarterly*
IIC	*International Review of Industrial Property and Copyright Law*
JO	*Journal Officiel (des Communautés Européennes)*
MLR	*Modern Law Review*
MMC	Monopolies and Mergers Commission
OFT	Office of Fair Trading
OJ	*Office Journal of the European Communities*
US	United States Supreme Court Reports
YEL	*Yearbook of European Law*

PART I

The Early Years

Contemplation of the actual is succeeded at once by reflection on its implications and possibilities, so men say 'if this is what we have, we can and shall make something different'. We see the process going on all the time. For instance, in politics, governments are always telling us how they will change our world for the better, but when they have brought in their Acts of Parliament, they are very surprised by the consequences, and begin to ask anxiously what exactly is happening, why this new world is behaving in such a strange and unexpected way: and a new crowd of observers set to work to examine the new world, to see what it really is, and to write books and articles about it.

Joyce Cary, *Art and Reality*
(Cambridge University Press, 1958)

1
An Introduction

The aim of the Oxford Series on European Community Law is to provide a group of books dealing with specific aspects of Community law, each of which offers a clear, concise, and critical exposition of law in its social, political and economic context. This work is concerned with the subject of competition law whose relevance and influence has increased in recent years. It is hoped, therefore, that it will be read not only by legal practitioners and students of competition law but equally by administrators and civil servants within the Community and by businessmen and others interested in acquiring a general understanding of the origins, current status, and outlook for this subject.

Competition law, however, is a complex and often highly technical subject, which does not lend itself to easy summary or concise clarification. To these inherent complexities must be added the further difficulties caused by the fact that the institutions of the European Community operate necessarily through a number of languages and that the process of translation itself brings the risk of blurring or misunderstanding the point at issue. My objective, therefore, has been to make easier the understanding of the substantive law by explaining its historical origins and early development, before going on to illustrate the development of the main areas of substantive law arising both from the decisions and policy of the Commission itself and from the many decisions of the European Court of Justice. I have also provided a chronological table to give guidance to the sequence of major cases, Council and Commission Regulations, and other notable landmarks over the last forty years.

It is not the purpose of this series to produce encyclopaedias, and for this reason I have not attempted to refer to (let alone describe) every decided case on individual topics. Nevertheless I have tried to make at least a brief reference to any case that can fairly be considered to have significance in the gradual development of the relevant principles of substantive law. I have no doubt from my own experience that the study of EC competition law gains substantially from comparison with other systems of competition law and in particular by reference to the wealth of cases found in, and from academic and professional commentary available on, the United States antitrust[1] law. The book, however, is not a study in

[1] 'Antitrust' is often used as a convenient portmanteau expression which covers US statute, case-law, and administrative processes relating to monopoly and competition. The phrase

comparative law, though the reader will discover that, on occasions, I found it difficult to resist the temptation to draw comparisons which have normally been confined to the footnotes. It is moreover a book on substantive, rather than procedural, issues although inevitably reference has to be made on occasion to important procedural questions, particularly in Parts I and III.

Competition law is a subject of great practical importance. It involves the establishment and development of legal principles and policies for the benefit of the public interest, enforced mainly (but not exclusively) through the work of public authorities; but these principles and policies are applied to a wide range of private agreements and arrangements which commercial undertakings have made for themselves, or with each other, on the basis of existing rights under contract and property law. Whilst the doctrine of restraint of trade has been familiar in the common law for several centuries,[2] the approach of the English Courts has largely been that the basis for refusal to enforce the individual restrictive clauses in contractual documents was that they appeared unreasonable in respect of the parties involved, either as employer and employee, or as vendor and purchaser of a business, rather than because they offended against the interests of third parties let alone those of the State or the public interest generally. Today, however, the justification, both of the national and Community competition law in interfering in such contractual and property rights derives far less from concern for the interests of the parties and far more from what is claimed to be the economic interests of the nation or Community as a whole. Moreover, whereas when the Treaty of Rome was first signed, there was no Member State which had a competition law of a comprehensive nature and proven effectiveness, today all twelve Member States have some form of competition law even though in some cases still limited in both scope and implementation.[3]

The fact that European competition law (by which of course is meant competition law that is European rather than merely national in its application) is now of great practical importance because of its influence upon the everyday conduct of business and industry within the Community of the Twelve provides, of course, by itself no valid reason why students should choose to study it, especially as it is demonstrably a subject provid-

derives from the technical device adopted in the late 19th cent. for vesting shareholdings (held by a number of enterprises) in individual trustees, who then issued trust certificates to those enterprises for whose benefit the pooling of interests had taken place, normally to facilitate the operation of cartels and other restrictive arrangements.

[2] The history of restraint of trade is fully dealt with in J. D. Heydon, *The Restraint of Trade Doctrine* (London: Butterworths, 1971).

[3] For a contemporary review of the competition laws of the Six in the early days of the Community, see S. A. Riesenfeld, 'Antitrust Laws in the E.E.C.', 50 *California Law Review* 459 (1966).

ing but few simple answers; indeed on the contrary, it appears still to raise many difficult and unsolved questions. Nevertheless, it is a subject eminently worth learning both as part of the general study of the legal framework of the EC, and also because of the importance of its substantive rules for both government and business within the geographical area of the European Community. It needs to be seen, however, as part of the wider sphere of EC law so that the relative importance of both the Treaty of Rome, and of the institutions of the Community, can be fully appreciated in their application to many other fields of law as well as to the field of competition law.[4] For freedom of competition is only one of the principles underlying the Common Market and at times may have to yield priority to the requirements of other objectives.

The structure of the book requires some explanation. Part I comprises a historical introduction to the subject. Chapter 2 discusses briefly the subject of modern competition by considering the elements of competition and an account of its relevance to the European Community. We then begin our review in Chapter 3 of the early origins of the competition law of the Community by considering the Treaty of Paris which established the European Coal and Steel Community (ECSC) in 1951.

These events themselves lead directly to the negotiations in 1955 and 1956 that culminated in the signature of the Treaty of Rome on 25 March 1957. Reference will be made to the influence of the Spaak Report upon whose firm foundations these final negotiations leading to the adoption of Articles 85 to 94 of the Treaty of Rome were based (dealing with various aspects of the subject of monopoly and competition) and to other influences of importance in establishing the final form of this part of the Treaty.

We next consider in Chapter 4 the early history of how the Commission began its task of implementing the Treaty's provisions on such matters, from its early days through to the eventual development of the key procedural regulation known as Regulation 17 of 1962. It is upon this Regulation that the entire subsequent operations of the Directorate-General of the Commission responsible for competition policy, known invariably as DG IV, have been and still remain based. In view of its subsequent importance to DG IV and to the Community as a whole, detailed consideration is given at this stage to the powers which the Regulation confers and to the influence which its adoption has had on later developments in substantive law. In particular, the early emphasis placed by DG

[4] For the purposes of this book, it is assumed that the reader has at least an outline knowledge of the main provisions of the Treaty of Rome and of the institutions of the Community. The Bibliography at p. 516 includes a number of books to which reference can be made if this assumption proves unfounded, together with some comments on the other sources for obtaining information about developments in EC competition law.

IV on reviewing agreements concerned with exclusive distribution leads on finally in Chapter 5 to a description of the origins of the first group exemption.[5]

Part II, which comprises some three-quarters of the book, is an account of the present substantive competition law of the European Community, set in the context of its development over the last thirty-five years. Chapters 6 and 7 analyse key elements in Article 85(1), and Chapter 8 deals briefly with the effect of Article 85(2) and in some detail with the conditions for exemption provided by Article 85(3). Chapter 9 considers some of the major case-law of the last thirty years under Article 85, dealing particularly with horizontal agreements relating to price-fixing, market-sharing, and other 'defensive' applications of the Article including 'crisis cartels'. Chapter 10 then deals with some types of agreements in which the justification for the inclusion of some restraints on competition may be more easily established, including agreements on joint research and development and specialization between undertakings, whilst Chapter 11 considers the application of the Article to joint ventures.

Chapters 12 to 14 deal with vertical agreements relating mainly to the distribution of goods. Once general principles relating to exclusive distribution arrangements had been evolved, it became possible for DG IV to turn its attention also to the special problems raised by particular goods and services with unique characteristics, whether of technical complexity or luxury image; the relevant case-law and ensuing group exemptions applicable for such products are explained. Reference is also made to the practice of distributing goods and services by franchising, and of efforts made to control the preference of some manufacturers to retain or impose resale price maintenance upon distributors of their products. The practice of price discrimination is dealt with at this point, although it is considered also in the treatment of Article 86 in Chapter 18.

Chapters 15 and 16 deal with another subject of major importance to which the Commission has devoted considerable attention, namely the basis upon which licences for patents and other forms of intellectual property may be approved notwithstanding that they combine territorial exclusivity with other restrictions upon the licensee; and also with the question of the reconciliation of the rights of holders in industrial property (including patents, trade marks, and copyrights) given by national law of Member States with the overriding needs of the Community for the free movement of goods, as provided through Articles 30 to 36 of the Treaty.

Chapters 17 to 19 deal with Article 86, which prohibits an abuse of

[5] Reg. no. 67/67. Throughout the book, the phrase 'group exemption' has been used in preference to 'block exemption', as being closer to the terminology of the original language of Article 85 (French 'catégorie' and Italian 'categoria'; German 'Gruppen' and Dutch 'groep').

dominant market position by undertakings within the Common Market or which have effect within the Market. Although fewer cases have been dealt with by the Commission and Court under this Article than under Article 85, those that have been considered are of major significance both in terms of their effect upon commercial practice and also because of the significant size of many of the companies involved. Chapter 19 discusses the relationship of Article 86 to Article 85 which has now been considered in a number of cases. Chapter 20 provides an account of the development of merger control within the Community, culminating in the adoption of Regulation 4064/89 which for the first time established control by the Commission over large-scale mergers with a Community dimension. This is followed by a discussion in Chapter 21 of some of the cases decided under both Articles 85 and 86 during the last thirty years involving trade associations, which have figured prominently in cases decided by the Commission and the Court.

The final three chapters of Part II attempt to stand back from detailed case-law relating to individual substantive topics and to review the operation of the competition laws of the Community as a system, both in co-operation and in conflict with a variety of national institutions, administrative, legislative, and judicial, whose operation has considerable effect on the workings and enforceability of Community competition policy. Chapter 22 thus deals with the problem of the co-ordination of the efforts of DG IV with those of national Courts and legislatures and examines the extent to which Member States, through these institutions, show themselves willing and able to support the policies of the Commission itself. Chapter 23 deals with the extent to which Member State involvement in commercial activities can relieve the involved undertakings from compliance with Articles 85 and 86. Finally in this Part, Chapter 24 considers the role of competition policy in the context of wider trading relationships outside the Community, including the problem of extraterritorial jurisdiction.

Part III completes the book by a review of developments since 1958, analysing the achievements of the Commission in general and DG IV in particular, with special attention to their successes and failures over these years. Consideration is here given to the major influence of the decisions of the European Court of Justice over the period and also to the role in the area of competition policy of the other institutions of the Community. The final chapter suggests some ways in which the Commission and DG IV may have to adapt in the future in order to be able to deal with some of their existing problems as well as with new challenges arising from the probable extension of the Community in the near future.

2

The Nature and Importance of Competition Policy

The competition policy of the European Economic Community is the underlying theme of virtually every chapter in this book. It may, therefore, be advisable to consider at this early stage what is meant in this context both by the expression 'competition' and also by 'competition policy'. This chapter will attempt no more than a brief account of what is a highly complex topic. It may, however, prove easier for those first meeting the subject to become familiar with some of its basic tenets before starting to grapple with its application to European competition law. Those familiar with the subject can, of course, pass on straight away to the following chapters.

The Treaty of Rome itself refers to 'competition' as a concept in both Articles 3(*f*) and 85. but perhaps wisely offers no definition. It is, however, advisable to have at least a general understanding of how the word is used, as well as important to realize that Article 85 can itself have no application unless the agreement, decision, or practice concerned has either an objective or effect of restricting competition.

Competition then is the relationship that exists among any number of undertakings which find themselves selling goods or services of the same kind at the same time to an identifiable group of customers. Each undertaking having made a commercial decision to place its goods or services on the market, utilizing its production and distribution facilities, will by this act have of necessity brought itself into a relationship of potential contention and rivalry with the other undertakings in that same geographical market, whose limits may be a single shopping precinct, a city, a region, a country, a group of countries, or the entire European Community, or even now in some cases, the entire world. Whilst competition law tends to focus more on the activities of, and relationship between, sellers, it logically applies equally to the like activities and relationships between buyers, also finding themselves in contention in particular product and geographic markets. Moreover, it applies not only to the markets for tangible goods and objects but to the fast growing variety of commercial and personal services available and necessary for the smooth operation of a developed modern economy. There are important differences that exist between the operation of competition in economic theory and its practical workings in real markets. Nevertheless, in spite of the imperfections of the workings of competition in real markets, the presence of competition is normally considered by economists to confer real advantages. Among these the following are normally referred to:

(*a*) the part that it has played in allocating resources in the direction preferred by consumers (this is generally described as allocative efficiency and has the merit that it reduces the risk that goods or services will be produced which are not really wanted or at least are not wanted at the price at which they are being offered);
(*b*) the constant process of dynamic adjustment to the changes in consumer preferences which are continually taking place and an incentive for producers to invest in research and development and to innovate;
(*c*) the keeping of such producers, and all sellers in the market, under continual pressure to keep costs, and therefore prices, down for fear of losing custom to other sellers who may be finding ways to attract business either by general price cuts or by special discounts to favoured buyers.

Now the economist illustrates the workings of the competitive process by reference to two models which are designed to show, on certain assumptions, what would be the rational behaviour of sellers under two extreme sets of conditions. These models are respectively perfect monopoly and perfect competition: in the real world it is extremely rare to find either, but they represent the two ends of the spectrum between which markets in the real world are found. Perfect monopoly is a situation where the provider of goods or services has 100 per cent control of the market and where there are no close or even similar products or services to which buyers can turn.[1] It is normally only possible as the result of State intervention when a public service or utility is provided either directly by a Government department or agency, or through a nationalized body such as British Rail or the Post Office, though even in these cases competition may exist with other forms of State provided goods and services, e.g. roads and inland waterways in the former case, private delivery couriers in the latter. Private monopolies too are rare in their pure form, controlling 100 per cent of a given market, unless perhaps the control is over a unique physical feature or product simply not available elsewhere.[2]

[1] In technical terms, such a market might be described as 'not contestable'.

[2] In US antitrust law, this would often be referred to as a 'bottleneck monopoly'. The classic example is the case of *United States* v. *Terminal Railroad Association of St. Louis* 224 US 383 (1912), where the Association controlled the only rail approach from the west into St Louis and was ordered by the Supreme Court to make it available to all railroads which desired to serve St Louis, on non-discriminatory terms. See A. Neale and D. Goyder, *The Antitrust Laws of the United States*, 3rd edn. (Cambridge University Press, 1980), 128–34. An example from the EC case-law is found in the case of *FRUBO* v. *Commission* Case 71/74 [1975] ECR 563: 2 CMLR 123, where the scarce facility was the fruit and vegetable market at Rotterdam, for whose use the controlling trade associations imposed restrictions on importers which the Court of Justice ruled were in violation of Article 85(1).

The opposite of pure monopoly is pure (or atomistic) competition, which is also very rare. For such a condition to exist, the number of competitors must be so great that the market share of each is tiny and none has, therefore, sufficient influence to alter price levels or the balance of supply and demand by its own efforts alone. In such a market, undertakings could, in theory at least, remain totally uninterested in the actions or reactions of their rivals since the market works simply to respond to the 'invisible hand' whose operations were first explained to us by Adam Smith.[3] The most likely type of market to approach this extreme would be one consisting of a large number of agricultural producers of the same crop, none of whom has a large enough percentage of total production to be able to influence market conditions or prices. In practice, the Common Agricultural Policy of the European Community usually prevents this kind of market from coming into existence.

Far more important for us, however, are the other two principal forms of market, oligopoly and workable competition. Oligopoly is the form of market where some degree of competition remains but where there is still a mere handful of competitive undertakings (probably not less than two nor more than eight) and the nature of the rivalry between them is substantially affected by this fact. Workable competition, on the other hand, is to be found when the number of rivals in the market is smaller than would be needed for perfect competition, but where total numbers involved and other relevant circumstances mean that there is a sharper degree and different tempo of mutual reaction than can be found in the oligopolistic situation. It can, however, on occasions be quite difficult to distinguish between oligopolistic and workably competitive markets, as the borderline is indistinct. Some oligopolies are notable for distinct elements of competition existing among their members, whilst some forms of what appear to be workably competitive markets necessarily are significantly affected by oligopolistic tendencies. The vast majority of relevant markets that will concern us within the Community, however, fall within one or the other of these last two categories; throughout the remainder of this book references to 'competition' will normally mean 'workable competition', not excluding markets with recognizable features of oligopoly.

Workable competition itself, however, is not easily defined. Its operation may be quite easy to understand if the product is homogeneous, possibly offered for sale in identical form by all competitors within a market, or where product differentiation that is claimed by suppliers to exist is not objectively based, as possibly with the differences between various brands of petrol or oil placed on the market by a number of competing oil companies. Other product differences may be tangible or objective, attributable

[3] In his famous work *The Wealth of Nations*, published in 1776.

to different physical or chemical formulations or other differences between the relevant ingredients and components of which competing products are composed. If the rival products being marketed are substantially identical without any real or ostensible differentiation being perceived by buyers, competition is likely to take place largely on the basis of price and to a lesser extent in terms of level of service or conditions of sale.

In many markets, however, competition is of a more complex kind since the product, whilst generally appropriate for the same end uses, has certain distinct characteristics, perhaps unique to itself and not shared by its rivals. Potential customers are asked to value and to assess any one of these heterogeneous products as better suited to their needs than any other brand. In this category would fall such popular consumer products as cars, cameras, and television sets. Here the competitive struggle between undertakings is waged on a number of fronts.[4] At one time, price cuts or secret discounts may be the chosen weapon whilst at other times it may be better warranties or service or changes in product specification or an increase in the range of models or varieties offered backed up by widespread advertising and promotion. For some consumer products the choice of particular kinds of distribution systems may be thought necessary to promote interbrand competition.[5] With capital goods and those other products of a non-capital nature which are sold by one industrial concern to another rather than to individuals, the competitive struggle is usually waged at a more objective level. For buyers here will usually have more information about the product which they wish to purchase and will be less susceptible to the blandishments of the advertising skill or promotional ingenuity of the seller. In the case of consumer goods, however, advertising and sales promotion and the establishment of consumer recognition for the seller's brand may be all important, and changing market shares far more influenced by the skill of this promotion than the inherent quality of the product and its pricing, at least in the short term.

An important characteristic of consumer goods markets, involving the distribution of products through retail outlets, is whether manufacturers of such goods can compete fully with each other on an interbrand basis. They can of course choose to sell direct to the public through wholly owned distribution channels, that is on a vertically integrated basis. In many cases, however, they may prefer to sell at arm's length to independent dealers or distributors.

If each manufacturer can obtain access to sufficient outlets of the required quality in appropriate locations then even if these vertical agree-

[4] For a full-scale analysis of the competitive process from the viewpoint of the undertaking itself, see M. E. Porter, *Competitive Strategy* (Harvard University Press, 1980) and *Competitive Advantage* (Harvard University Press, 1985).

[5] Examples of these are exclusive and selective distribution systems. See Chs. 12 and 13.

ments contain some degree of restriction e.g. by requiring exclusivity, interbrand competition should not be seriously affected. On the other hand, if sufficient outlets are not available to all potential manufacturers then their consequent 'foreclosure' from the market may through the reduction in competition be damaging to consumer and manufacturer alike.

Foreclosure can also occur if, although there are an adequate number of outlets available, manufacturers so arrange matters that only a limited proportion of them can obtain the goods that consumers actually require. This situation may nevertheless be sought to be justified by manufacturers on the grounds that too many dealer appointments may deter the best outlets from adequate investment in premises, equipment, staff, and promotional activities. The reason will be the fear that those who make a minimal investment or enter the market at a later stage will inherit as 'free riders' the benefit of the larger capital outlay which the original entrants have made. Similar arguments can be made by licensees in the context of the terms upon which intellectual property rights are made available.

It is against this background that the need for competition policy arises, to deal with both horizontal and vertical agreements. Horizontal agreements are those which are made between undertakings in competition with each other (actually or potentially) at the same level, for example between manufacturers or between wholesalers. Vertical agreements by contrast are those entered into between undertakings whose relationship is complementary, such as manufacturer and distributor or licensor and licensee.

In almost any markets left without government intervention or regulation, it is likely that the competitive process itself will ensure that sooner or later participants in them will gain a degree of market power. The effect of gaining such market power may in turn enable such undertakings to exercise a measure of strategic power over prices and thus increase their profit margins. This increase in profitability may allow them, whether by taking advantage of economies of scale or by pursuing successful research and development, to gain still further advantage over their competitors. Moreover, the obtaining of market power in this way places the holder in a stronger position to meet with adverse economic trading conditions, including general economic recession as well as temporary falls in demand. In such circumstances, the likely demise or weakening of some of its competitors may still further strengthen its position.

All these consequences, however, are in themselves often the normal result of a successful effort by an undertaking to compete and are not of themselves necessarily undesirable. Competition policy seeks to protect the process of competition, not simply to retain competitors in existence. This remains the case, even though paradoxically the health and integrity of the competitive process is to some extent dependent on the numbers and individual strengths of the undertakings in the relevant market.

Nevertheless, it is the function of competition policy to seek to prevent the unfair acquisition of market power by individual undertakings, without at the same time becoming too overprotective of rivals. The extent, however, to which such policy should seek to intervene in the way in which business conducts its affairs is a central issue of controversy between economists. The traditional school of economists would recommend a twofold form of intervention. The first, which is that taken under Section 1 of the Sherman Act of the United States, the Restrictive Trade Practices Act, and under Article 85 of the Treaty of Rome, would seek to prevent restraints of competition by way of agreements or concerted practices between undertakings, rendering them unenforceable and imposing civil or criminal penalties on those who participate in them. The second method would aim to prevent the undue acquisition of power by undertakings through merger or by monopolization of markets or through abuse of dominant positions by exclusionary anti-competitive or predatory practices. These are the targets of action taken in the United States under Section 2 of the Sherman Act and Section 7 of the Clayton Act, in the United Kingdom under the monopolies and mergers provisions of the Fair Trading Act and the anti-competitive practices provisions of the Competition Act, and under Article 86 of the Treaty of Rome dealing with abuse of dominant position.

The aims of the European Community in this field are well summarized in a publication of the Organization for Economic Co-operation and Development:

> Competition policy has as its central economic goal the preservation and promotion of the competitive process, a process which encourages efficiency in the production and allocation of goods and services, and over time, through its effects on innovation and adjustment to technological change, a dynamic process of sustained economic growth. In conditions of effective competition, rivals have equal opportunities to compete for business on the basis and quality of their outputs, and resource deployment follows market success in meeting consumers' demand at the lowest possible cost.[6]

From the viewpoint of the Community, a further important element flowing from a competitive system is the opportunities which it may give to small and medium-sized undertakings within the Market, which are regarded by both the Commission and European Parliament as potentially the main providers of new employment for the future, as well as giving the individual the opportunity of starting his or her own business.

[6] *Competition and Trade Policies: Their Interaction* (Paris: OECD, 1984), para. 232.

These advantages, however, to be expected from a competitive system have been well understood for many years and can be found analysed in detail in the pages of economists of different schools. From the viewpoint of the framers of the Treaty, however, the need for an active competition policy to encourage competitive markets and discourage practices restraining trade between Member States arose not from economic theory but in the main from political necessity. If the Common Market was to be, as its name implies, a market where economic progress would come from the efforts of independent undertakings, large and small, from any of the Member States, then the individual territorial markets of Member States had to be made open and kept available to them. If this were prevented, then to a great extent the other important political and economic aims of the Treaty would be rendered abortive.

The aim of the integration of national markets, therefore, became a necessity, and not just an optional extra, as part of the Community competition policy.[7] It is the working out in the individual cases of its application that forms much of the early history of the policy which we now have to examine. This emphasis on integration markets, moreover, will be found to play an especially important part in a number of the individual areas of the law, namely abuse of dominant position, price discrimination, the exercise of intellectual property rights, and in the consideration of various distribution systems. Indeed, on occasions, it seems to have been given priority by DG IV even over the encouragement and protection of the competitive process itself, though in many cases the twin aims of market integration and enhancement of competition can be simultaneously pursued.

[7] By comparison, in the USA the first Federal antitrust law, the Sherman Act 1890, operated on the assumption that its principal commercial and industrial markets were organized on at least an interstate, if not a nationally integrated, basis to which Federal rather than State law would necessarily apply under the US constitution.

3

The Origins of European Competition Law

1. The Treaty of Paris and the Formation of the European Coal and Steel Community

The twin pillars of the competition law of the European Community are Articles 85 and 86 of the Treaty of Rome. Community competition law consists mainly, therefore, of an examination of how these two Articles have been applied in many different situations and to a great variety of agreements and practices. In order to understand the ideas that underlie the drafting of these particular Articles, the events which led up to the signing of the Treaty of Rome on 25 March 1957 have first to be considered.

The history of Western Europe during the period from the end of the Second World War up to the actual signature of the Treaty shows a recurring pattern of events.[1] An individual statesman puts forward an idea for European co-operation within a particular area of activity. Then follows acceptance by European nations of that idea as a broad political aim, with later development of that aim itself into a detailed international treaty. Upon the basis of this treaty permanent institutions are then formed in order to implement the broad political aim now established, but which themselves lack the authority or political will to ensure that the co-operation itself develops into any kind of supranational integration. The lack of any real progress in this direction leads other politicians then to propose new ideas, in the hope that these, in spite of the disappointing experience of the past, will have the result of bringing about more than just co-operation between individual European nations.

The first of these European institutions was the Organization for European Economic Co-operation[2] established in 1948, primarily to allocate Marshall Aid, which was the inspired idea of General George Marshall, the then United States Secretary of State. The object of this plan

[1] For a detailed account of the negotiations leading up to the Treaty of Paris 1951 and the Treaty of Rome 1957, see R. Mayne, *The Recovery of Europe* (London: Weidenfeld & Nicolson, 1970), esp. chs. 8–10, and D. Swann, *The Economics of the Common Market*, 5th edn. (London: Penguin Books, 1984), esp. ch. 1.

[2] This body changed its name in 1960, when joined by the USA and Canada, to that of the Organization for Economic Co-operation and Development (OECD). Among other responsibilities it numbers the International Energy Authority (IEA) which was to feature in an important decision of the Commission referred to in Ch. 8.

was the economic rehabilitation of Europe by the provision of much needed financial assistance, not only for its immediate needs of food, fuel, and other necessities, but also the capital investment required for the modernization and re-establishing of much of the industrial capacity destroyed, or allowed to deteriorate, in the course of the Second World War and the depressed days immediately following. Whilst discussions for the implementation of a customs union were held within the framework of OEEC in the late 1940s, the essential political will to convert this organization into one with any elements of supranational integration was lacking. OEEC went on in the following years to play an important role in the liberalization of trade within Western Europe, but its constitution remained that of a permanent conference of sovereign States without any supranational powers. It lacked, therefore, the ability to effect any major changes leading to integration of the economies of the countries of Western Europe.[3]

The second major European institution, created soon afterwards, was the Council of Europe which, however, also lacked any supranational power. Its founders were chiefly Georges Bidault, French Foreign Minister, and Paul-Henri Spaak, the Belgian Foreign Minister, both of whom had attended the famous Congress of Europe at The Hague in May 1948. The original project agreed upon at that Congress had been of a European legislative Assembly to which some sovereign powers, even if of only a limited nature, could be transferred at a future date. British opposition ensured, however, that the Assembly would have a purely consultative rather than a legislative role. The Statute of the Council was signed in May 1949 and the Assembly first met in August that year under the presidency of Paul-Henri Spaak. Its Committee of Ministers could operate only on the basis of unanimity and its Assembly (composed of members of national parliaments) had no legislative powers.

To belittle the achievements of the Council over subsequent decades would be unjust. It was responsible in 1950 and afterwards for the signature by the great majority of European States of the European Convention for the Protection of Human Rights, whose influence has risen steadily over the intervening years. Over one hundred other conventions have been drafted and negotiated by the staff of the Council, with the support of the Assembly and the Committee of Ministers, covering a wide range of topics including narcotics, terrorism, data protection, enforcement of foreign judgments, family law, and a variety of other social issues. These conventions have had a significant influence on the adoption of related national legislation in the twenty-one member countries participating in

[3] Richard Mayne records in *The Recovery of Europe* at p. 127 that during the period 1947 to 1950 when France urged that OEEC should be given greater powers it was the UK which repeatedly opposed all such moves.

the Council. The Council, however, had no influence in the direction of greater co-operation on purely economic and industrial issues. Progress, therefore, towards the economic or political integration of Western Europe during the period prior to 1950 through the creation of new international institutions was limited.

In the meantime, however, important developments had been taking place in Germany. The country had at the end of the Second World War in 1945 been divided up by the Potsdam Agreement into four zones, the Eastern Zone being placed under the control of the USSR and the three Western Zones under the United States, the United Kingdom, and France respectively. As no agreement could be reached between the occupying powers as to the basis upon which German industry should be reorganized, each power gradually introduced its own legislative and administrative measures in respect of its Zone. The original policy objectives behind these measures had been largely negative (to prevent Germany from again becoming a threat to peace) and restrictive, the 'decartelization' of the existing structures of industry in the various sectors (coal, iron and steel, chemicals, plastics, heavy engineering, banking, etc.). Some of the largest 'konzerns' were split up; for example IG Farben (the huge chemicals combine) was split into three parts which became Bayer, Hoechst, and BASF respectively, and considerable reorganization took place in some other sectors. As democracy gradually returned, the more positive elements in the programme of the Western powers were given more prominence, in an attempt to ensure that the German economy recovered to a sufficient extent to provide an adequate standard of living for its people. The punitive element in the Allied programmes of reorganization gradually faded, and emphasis switched to the establishment of the most effective and efficient means of ensuring that the German economy could be stabilized and so linked on a non-discriminatory basis with the undertakings of France and other Western European nations. Coal and steel were the industries where the need of this reform was of the greatest urgency.

It was at this point that a fresh impetus to change was provided by Jean Monnet, well-known French administrator and civil servant, who throughout his life provided a copious flow of original ideas of economic co-operation between nations both in war and peace.[4] His idea was the bold one of creating a common market for iron, steel, and coal in Western Europe by the removal of all customs duties, tariffs, quotas, and other market restrictions; these rules would be administered by an independent High Authority endowed with supranational powers to which all participating countries would be subject. These ideas were keenly supported and elaborated by Robert Schuman, then French Foreign Minister, and in

[4] Mayne, ibid. 177. See also Mayne's translation of the Memoirs of Monnet (London: Collins, 1978).

April 1950 the detailed proposals for the implementation of Monnet's original ideas were put forward as the Schuman Plan.

Not surprisingly, in view of its past negative attitude to the development of other European institutions, the United Kingdom again refused to contemplate participating in any body which was to have powers which could derogate from those of the United Kingdom Parliament. Notwithstanding the fact that the United Kingdom would not join them,[5] the Six (France, Germany, Italy, The Netherlands, Belgium, and Luxembourg) decided to proceed with the Schuman Plan influenced by both the economic and political advantages that the creation of a body with supranational powers could bring to industries so essential to the development of the industrial strength of Western Europe. On 18 April 1951, the Treaty of Paris was signed creating a new legal entity with a status in international law, the European Coal and Steel Community.

The Treaty of Paris was a document of one hundred clauses, three annexes, and a number of protocols. Its duration was stated to be fifty years from the date of its entry into force.[6] The objectives of the Treaty were set out in Articles 1 and 2 which define the task of the Community as 'to contribute, in harmony with the general economy of the Member States and through the establishment of a Common Market . . . to economic expansion, growth of employment and a rising standard of living in the Member States'. Article 3 set out the duties of the institutions of the Community as being to ensure orderly supplies of materials in the Common Market, promote the expansion and modernization of production and the rational uses of the raw materials available within the Community, and the promotion of international trade in those products. Article 4 continued by listing the following practices as incompatible with the new market and, therefore, to be 'abolished and prohibited', namely:

(*a*) import and export duties, or charges having equivalent effect, and quantitative restrictions on the movement of products;
(*b*) measures which discriminate between producers, purchasers, or consumers relating to prices, delivery terms, or transport rates and conditions, or which interfere with the purchaser's free choice of supplier;
(*c*) subsidies or aids granted by States; and
(*d*) restrictive practices 'which tend toward the sharing or exploiting of markets'.

Article 5 continued the reference to protection of competition contained in Article 4(*d*) with a positive obligation placed upon the Community to

[5] At a later date the UK did enter into a Treaty of Association with the ECSC, but on terms which did not involve the yielding of any parliamentary sovereignty to the new High Authority.

[6] Article 97.

'ensure the establishment, maintenance and observance of normal competitive conditions . . .'. Clearly, therefore, an early requirement for the new Community would be that its institutions should take responsibility for identifying any anti-competitive practices and eliminating them. These institutions would include the executive body to be known as the High Authority, which had primary responsibility for carrying out the objectives of the Community.[7] Other relevant institutions created by the Treaty were an Assembly, a Special Council of Ministers. and significantly a Court of Justice required to ensure the implementation and application of the Treaty, to provide interpretation of its terms, and to give rulings binding both on the High Authority and Member States as well as on undertakings within the Community.

At this stage it would be appropriate to ask the principal reasons for according so much importance to the value of pro-competitive measures within the framework of the Treaty. The chief element was of course that the absence of, or at least the major imperfections of, competition readily visible in the markets covered by the 1951 Treaty of Paris were felt to be major weaknesses in industries that needed the spur of rivalry in order to achieve a higher level of performance. Beyond that element, however, lay the known example of the United States, an economy that had apparently reached its position of pre-eminent industrial strength by its reliance over many decades on free competition, the principle that provided the mainspring of antitrust. If the Coal and Steel Community were to avoid the need for a form of supranational control, involving management on a completely centralized basis, then allowing to its undertakings a fuller freedom of exposure to the forces of competition was unavoidable. This conclusion was accepted even though at the same time the individual Member States had little tangible evidence from their own national experience to support a wholesale move away from the existing norm of an industry regulated by governmental and private agencies towards the uncertainties of a freer and less regulated market.[8]

Nevertheless, the structure of the Treaty of Paris retained certain elements of central control; so long as supplies of coal or steel were in reasonable balance, competitive forces could operate, but if a serious imbalance arose between supply and demand, the Treaty permitted the

[7] The functions of the High Authority are set out in Articles 8 to 19 of the Treaty. The functions of the Assembly (Articles 20 to 25), the Council (Articles 26 to 30), and the Court of Justice (Articles 31 to 45) followed immediately after in the Treaty.

[8] H. Smit and P. Herzog, *The Law of the EEC: A Commentary on the Treaty*, 6 vols. (New York: Matthew Bender, 1976). vol. ii., ch. 3, para. 13. On the other hand, the Freiburg School, favouring the social market economy as the only means of combining individual liberty with social justice, had substantial influence on the policies of the Christian Democrats under Ludwig Erhard at this time, which in turn led to the enactment of the German Law against Restraints of Competition with effect from Jan. 1958.

High Authority to take measures to bring them back into closer accord. This would necessarily involve interference with the operation of the market. Moreover, the High Authority had jurisdiction under the terms of the Treaty to deal with restraints on competition if these occurred purely within an individual Member State, not merely with those that affected trade between two or more States within the ECSC.[9]

Articles 65 to 66, therefore, provided detailed legislative backing for the enforcement of Article 4(*d*). Article 65 contained several provisions relating to the preservation of a competitive structure in the Community. First, paragraph (1) of Article 65 stated that agreements between undertakings, decisions by associations of undertakings, and concerted practices were prohibited if they tended directly or indirectly to prevent, restrict, or distort 'normal competition' within the Common Market. Although the prohibition applied to all agreements, decisions, and concerted practices, the attention of the High Authority was drawn in particular to those which tended (*a*) to fix or determine prices, (*b*) to restrict or control production, technical development, or investment, or (*c*) to share markets, products, customers, or sources of supply.

Having laid down the basic rule in paragraph (1), paragraph (2) proceeded with some qualifications; specialization agreements or joint buying or selling agencies could be authorized by the High Authority if it found that the agreement (*a*) would make for substantial improvement in production or distribution, (*b*) was also essential to achieve such results and is not more restrictive than necessary for that purpose, and (*c*) was not liable to give undertakings power to determine prices or restrict production or marketing of a substantial part of the products within the Common Market or shield them against effective competition within the Market. Other agreements found to be 'strictly analogous' and meeting the same requirements were also to be permitted by the High Authority. Authorization of agreements could be both conditional and for a limited period, and could be amended or subsequently revoked if circumstances changed sufficiently.

The sanctions to be applied if agreements were made which did not obtain clearance under Article 65(2) were twofold. The first (contained in Article 65(4)) was that the relevant agreement, decision, or practice was 'automatically void' and could not be enforced in any Court or Tribunal in the Member States. The second sanction, contained in paragraph (5), was that a fine or penalty could be imposed by the High Authority on an undertaking which entered into or tried to enforce such an agreement. Such fines or penalties could be of a substantial nature rising to 10 per

[9] A point noted in the editorial preface to the first volume of Common Market Law Reports (1962), a rare example of such comment in the pages of a law report.

cent of annual turnover in respect of fines for past conduct or 20 per cent daily turnover in the case of penalties imposed to prevent the continuation of an offending agreement.

Article 65, therefore, dealing with agreements, decisions, and concerted practices anticipated in some respects the corresponding Article 85 of the Treaty of Rome. Article 66 by contrast dealt with the problem of future mergers or concentration of undertakings in the coal and steel markets; it is relevant to remember that problems of excessive concentration and oligopolistic structure were always a major issue in these markets which since the ending of the Second World War (and notwithstanding the efforts of the Western powers under their 'decartelization' programmes for Germany) had not enjoyed any period of genuine competition. Consent to such mergers could only be given by the High Authority if it found that the proposed transaction would not give power to the undertakings concerned to determine prices or to control or restrict production or distribution, or to hinder effective competition in a substantial part of the market, or otherwise to evade the rules of competition found in the Treaty so as to give an 'artificially privileged position involving a substantial advantage in access to supplies or markets' to any such undertakings. The High Authority was given authority by paragraph (3) of Article 66 to provide what in effect would be a form of group exemption from the need for prior consent for small undertakings wishing to merge and was also given sweeping powers to impose fines or order divestiture if concentration took place without its consent.

Finally, paragraph (7) dealt with the problem of an undertaking, private or public, which either *de facto* or *de jure* held a dominant position shielding it against effective competition in a substantial part of the Common Market and abused that position for purposes 'contrary to the objectives of the Treaty'. In such a situation, the High Authority could make recommendations to such undertakings to prevent such abuse of its position and in default could 'by decisions taken in consultation with the Government concerned' determine the prices and conditions of sale that that undertaking must apply and draw up mandatory delivery or production programmes with which it must comply. If it failed to comply, substantial fines could be imposed. In this last paragraph (7), we find for the first time the concept of 'abuse of a dominant position' which would appear again in the equivalent Article 6 of the Treaty of Rome.

Article 67 was headed generally 'Interference with conditions of competition' being intended to cover the problem of assistance provided by Member States to undertakings within the coal or steel industry without the approval of the High Authority. The High Authority was itself permitted to allow States to grant aid to individual undertakings but only on terms and for a period which it had laid down. This provision foreshadowed similar clauses to be found in Articles 92 to 94 of the Treaty of Rome.

To complete the array of powers granted to the High Authority, Article 60 prohibited pricing practices which either involved price reductions of a predatory nature which assisted the acquisition by an undertaking of a monopoly position or, alternatively, discriminatory pricing policies under which sellers applied different conditions to comparable transactions, especially on the grounds of the nationality of the buyer. To enable it to cope with any such practices, the High Authority could order that price lists and conditions of sale be made public.

This brief review of the powers conferred on the High Authority of the ECSC shows how the original simple conception contained in the Schuman Plan for unifying the coal, iron, and steel production and sales of the Six was converted through the Treaty of Paris into a set of working rules binding under international law on the Six; it embodied for the first time in an international treaty a set of rules of competition tailored to the special circumstances of particular industrial sectors in a group of sovereign states.[10] The executive authority for implementation of the rules lay firmly with the High Authority, but subject to the judicial control of an independent Court of Justice as well as the political control of the Council of Ministers, with whom the High Authority worked in close co-operation. This pattern of institutional structure, as well as some of the substantive provisions of Articles 60, 65, and 66 would, as we shall see, prove an important model a few years later when negotiations began which were destined to lead to the extension of the principles of the ECSC into a much wider field.

Ratification of the Treaty of Paris by the Governments of the Six was completed by the end of June 1952, and the Common Markets in coal and steel opened respectively on 10 February and 1 May in 1953.

2. The Treaty of Rome and the Formation of the European Economic Community

The first real step towards Western European integration in the economic area had now been taken. The success of the Six in agreeing upon a basis for co-operation within such crucial industrial sectors as coal and steel led naturally to further consideration of the possibility of the application of

[10] The definition of 'steel' in Annexe 1 to the Treaty does not, however, cover all steel products. For an explanation of those products not covered see the 1990 MMC Report on *British Steel Corporation/C. Walker & Sons* (Cm 1028), ch. 7, paras. 11–17.

these principles of integration on a broader front. The progress made in establishing the Coal and Steel Community had, however, to some extent been overshadowed by the subsequent failure of the Six to agree to the terms for a European Defence Community, the French National Assembly having rejected this project at the end of August 1954. Notwithstanding the temporary depression which this setback cast over those who favoured closer industrial and commercial arrangements linking the Six, once again Jean Monnet was ready with proposals for the development of further co-operation, on a wider economic and political level. To his individual contribution further impetus was added by the Benelux Memorandum[11] which was to have long-term significance in the establishment and development of a Common Market on a wider basis even than the Coal and Steel Community. A key extract from this Memorandum reads:

> All partial integration tends to solve the difficulties in one sector by measures which harm other sectors or the consumers' interests, and it tends to exclude foreign competition. That is not the way to increase European productivity. Furthermore, sector integration does not help to strengthen the feeling of Europe's solidarity in unity in the same degree as general economic integration. To strengthen this feeling, it is essential that the notion of the European States' joint responsibility for the common good be incorporated in an organisation adapted to the pursuit of the general interest and whose executive body is responsible not to Governments but to a supranational Parliament. . . .
>
> That is why it seems opportune for the three Benelux Governments to take a well prepared initiative which might usefully be announced at the ECSC Foreign Ministers' meeting. Such an initiative would aim at establishing a supranational community whose task it would be to achieve the economic integration of Europe in the general sense, proceeding by means of a customs union to the establishment of an economic union. . . .

A positive response was obtained to this document from the German Government, supporting the need for economic integration, liberalization of trade in goods and services, the elimination of tariffs, and the free movement of persons. The response also significantly emphasized the need for competition rules in order to establish normal market conditions within the integrated market, to protect competition from State intervention and also from private monopoly power.[12]

The result of this activity was the meeting held at Messina in Sicily on 29 and 30 May 1955 of the Foreign Ministers of the Six, principally to

[11] Submitted by Spaak on behalf of the Benelux nations to the Foreign Ministers of the Six on 6 May 1955 with a covering note to Monnet 'herewith your child' (Mayne, *Recovery of Europe*, p. 232).

[12] J. Kallaugher, 'The Influence of German Competition Law on the Development of the Competition Law of the European Communities', (unpublished article). The permission of the author to refer to it is gratefully acknowledged.

discuss the possibility of extending the principles of the ECSC on a wider front. The resulting Messina Declaration, which was substantially influenced by the Benelux Memorandum, set in motion a series of studies relating to the establishment of a European Common Market to be free from all customs duties and quantitative restrictions between its members. Preparation of treaties to shape these new ideas into detailed political and institutional form was entrusted to a group of senior officials under the presidency of Spaak. The importance of the preservation and promotion of a competitive structure of industry within such a Market was already well understood, and the final resolution of the Messina Conference contained *inter alia* the following passage: 'Its [i.e. the Treaty's] application necessitates a study of the following questions . . . *the elaboration of rules to ensure undistorted competition within the Community*, which will in particular exclude any national discrimination'.

The Spaak Report[13] is a seminal document of great importance, which comprises the most important of the various preparatory works (*travaux préparatoires*) upon which the subsequent Treaty of Rome is based. The necessity for freedom of competition within the new Market referred to by the Messina Declaration does not seem to have been in doubt, although the detailed drafting of the rules relating to competition to be contained in the Treaty was to be the subject of keen argument and negotiation between the different Member States involved, in particular between the French and German representatives.

The first paragraph of the Spaak Report sets the context within which competition policy will have to operate.

The object of the European Common Market should be to create a vast zone of common economic policy, constituting a powerful unit of production permitting continuous expansion and increased stability and accelerated raising of the standard of living and the development of harmonious relations between its Member States. . . . These advantages of a Common Market, however, cannot be obtained unless adequate time is allowed and means are collectively made available to enable the necessary adjustment to be effected, unless practices whereby competition between producers is distorted are put to an end, unless co-operation between states is established, to establish monetary stability, economic expansion, and social progress. . . .

At paragraph 55 of the Report, we find reference to the detailed requirements for competition policy in such a market.

In the final period, the elimination of trade barriers will lead to the disappearance of the opportunities for discrimination by competing enterprises. The problem only

[13] Kallaugher records in the same article that although the working group responsible for the report was chaired by Spaak, the chief draftsmen were a German, von der Groeben, and a Frenchman, Uri.

remains because there are enterprises which, owing to their size or specialisation or the agreements they have concluded, enjoy a monopoly position. The action against discrimination, therefore, links up with the action that will be necessary to counteract the formation of monopolies within the Common Market. The Treaty will have to lay down basic rules on these points. . . . More generally, the Treaty will have to provide means of ensuring that monopoly situations or practices do not stand in the way of the fundamental objectives of the Common Market. To this end, it will be necessary to prevent—

— A division of markets by agreement between enterprises, since this would be tantamount to re-establishing the compartmentalisation of the market.
— Agreements to limit production or curb technical progress because they would run counter to progress and productivity.
— The absorption or domination of the market for a product by a single enterprise since this would eliminate one of the essential advantages of a vast market, namely that it reconciles the use of mass production techniques with the maintenance of competition.

Already, we can sense here an anticipation of the final wording of Articles 85 and 86, indicating that the new Treaty will go well beyond the more limited objectives of Articles 65 and 66 of the Coal and Steel Community Treaty. The Spaak Report was indeed a document that combined compelling logic with a considerable degree of imagination and was accepted with little amendment by the Foreign Ministers of the Six at the Venice Conference on 29 May 1956. The officials responsible for its preparation resumed work some weeks later at Val Duchesse (just outside Brussels) for the preparation of the Treaty that was to embody the greater part of the content of the Spaak Report, subject to those slight modifications agreed to at the Venice Conference. Preparation of the detailed draft of its Articles then took place continuously over the next ten months prior to the signature of the Treaty of Rome on 27 March 1957.

The minutes and text of the relevant meetings during this period illustrate the ebb and flow of negotiations over the parts of the Treaty which deal with competition and related issues. The section of the Treaty dealing with competition rules is contained in Part III 'Policy of the Community: Title I—Common Rules: Chapter 1—Rules on Competition'. Chapter 1 covers the Articles which subsequently were numbered 85 to 94. Articles 85 and 86 are the main articles covering the rules of competition applicable to undertakings. The negotiators had, of course, the recent model of Articles 65 and 66 of the Treaty of Paris to follow but the provisions adopted in that Treaty to apply to the mainly oligopolistic market situation for both coal and steel would clearly not in their entirety be appropriate to the far wider range of products, processes, and markets that the new Treaty would have to cover.

One of the earliest points discussed and accepted was that the Treaty should contain a specific prohibition, applicable across the range of industrial activity, against discrimination on national grounds. This principle clearly had importance far beyond the rules of competition alone; and was ultimately not included in the later section concerned mainly with rules on competition but earlier in Article 7, as a 'principle', to emphasize its importance and wide application to all aspects of Community activity rather than just as one of many rules relating to competition.

The more detailed rules contained in the section of the Treaty dealing with competition were in fact drafted before the final form of the agreement was reached on the 'principles' set out in Articles 1 to 7 and in particular before the adoption in its final form of Article 3(*f*). Articles 1 and 2 declared the establishment of the European Economic Community and proclaimed its tasks of establishing a Common Market to bring about through the progressive approximation of economic policies, a harmonious development of economic activities, a continuous and balanced expansion, an increase in stability, an accelerated raising of the standard of living, and closer relations between States. Article 3 (equivalent to Article 3 in the Treaty of Paris) listed the policies of the Community necessary in order that the purposes set out in Article 2 could be achieved but in accordance with the wider objectives of the Treaty of Rome. They are considerably more comprehensive than in the earlier Treaty and in particular we find (as we do not in the Treaty of Paris) the following Article relating to competition: '(*f*). The institution of a system ensuring that competition in the Common Market is not distorted'.

In the following Article 4, were listed the institutions required to carry out the Community's tasks, namely the Assembly (subsequently known as the European Parliament), Council, Commission, and the Court of Justice (and in an advisory capacity only, the Economic and Social Committee); Article 5 laid down a further principle relevant to the maintenance of competition. In this, Member States are required

> to take all appropriate measures, whether general or particular, to ensure fulfilment of the obligations arising out of this Treaty or resulting from actions taken by the institutions of the Community. They shall facilitate the achievement of the Community's tasks. They shall abstain from any measure which could jeopardise the attainment of the objectives of this Treaty.

Under the terms of this Article, it is clear that Member States are required to take measures to ensure that competition in the Common Market is not distorted, this being one of the major objectives of Article 3, and in both a positive and negative sense to contribute to the implementation of competition policy.

We come now to the Articles in the Treaty, numbers 85 to 94, which deal with 'Rules on Competition'. Of these, Articles 85 and 86 contained the main substantive principles of law applicable to undertakings. Article 87 provided for the making of subordinate regulations and directives, and

Articles 88 and 89 for the responsibilities of both the Member States and the Commission during the transitional period before such subordinate regulations had been enacted.[14]

Since it would be unfair if Member States could at the same time effectively reduce the degree of competitive activity within their own economies by giving privileges to monopolies such as nationalized State bodies or authorities, Articles 90 to 94 inclusive provided the relevant rules governing the activities of Member States in maintaining public undertakings (Article 90), and in providing State aids in any form to undertakings (Articles 92 to 94): Article 91 dealt with the problem of the 'dumping' of goods within the Common Market.

The text of Articles 85 and 86 was finally adopted in the following form:

Article 85

1. The following shall be prohibited as incompatible with the common market: all agreements between undertakings, decisions by associations of undertakings and concerted practices which may affect trade between Member States and which have as their object or effect the prevention, restriction or distortion of competition within the common market, and in particular those which:

(*a*) directly or indirectly fix purchase or selling prices or any other trading conditions;
(*b*) limit or control production, markets, technical development, or investment;
(*c*) share markets or sources of supply;
(*d*) apply dissimilar conditions to equivalent transactions with other trading parties, thereby placing them at a competitive disadvantage;
(*e*) make the conclusion of contracts subject to acceptance by the other parties of supplementary obligations which, by their nature or according to commercial usage, have no connection with the subject of such contracts.

2. Any agreements or decisions prohibited pursuant to this Article shall be automatically void.

3. The provisions of paragraph 1 may, however, be declared inapplicable in the case of:

— any agreement or category of agreements between undertakings;
— any decision or category of decisions by associations of undertakings;
— any concerted practice or category of concerted practices;

which contributes to improving the production or distribution of goods or to promoting technical or economic progress, while allowing consumers a fair share of the resulting benefit, and which does not:

[14] The majority of original commentators on the Treaty assumed that Articles 88 and 89 had purely a transitional function, but we shall see in Chs. 6 and 23 that their life was prolonged as the result of the failure of the Council to adopt appropriate regulations in the areas of sea and air transport until 1986 and 1987 respectively. See the *Nouvelles Frontières* Cases 209–213/84 [1986] ECR 1425: 3 CMLR 173.

(*a*) impose on the undertakings concerned restrictions which are not indispensable to the attainment of these objectives;
(*b*) afford such undertakings the possibility of eliminating competition in respect of a substantial part of the products in question.

Article 86
Any abuse by one or more undertakings of a dominant position within the common market or in a substantial part of it shall be prohibited as incompatible with the common market in so far as it may affect trade between Member States.
Such abuse may, in particular, consist in:

(*a*) directly or indirectly imposing unfair purchase or selling prices or other unfair trading conditions;
(*b*) limiting production, markets or technical development to the prejudice of consumers;
(*c*) applying dissimilar conditions to equivalent transactions with other trading parties, thereby placing them at a competitive disadvantage;
(*d*) making the conclusion of contracts subject to acceptance by the other parties of supplementary obligations which, by their nature or according to commercial usage have no connection with the subject of such contracts.

The examples in Article 85(1) are certainly not intended to be exhaustive. They have some common features with, but cover a wider range than, those examples given in Article 65 of the ECSC Treaty, which are limited to agreements which fix prices, restrain production, technical development, or investment, or involve the sharing of markets, products, customers, and sources of supply. Of these provisions in Article 85(1), items (*d*) and (*e*) are unique to it, items (*a*) and (*b*) are more widely drawn than their equivalent in the ECSC Treaty, and (*c*) is rather more narrowly drawn. No great importance should, however, be read into these minor differences. Neither the Commission nor the Court has subsequently placed significance apparently on the fact that any particular type of agreement has been omitted from this list. Whilst inclusion in the list certainly may be of some minor relevance in that it indicates the type of restrictive agreement which the framers of the Treaty may have had in mind, the wide scope remaining for a declaration of inapplicability under paragraph (3) minimizes the significance of even that factor.

It is important to appreciate that these Articles do not provide all the rules under which the general principles of competition within the Treaty can be applied, and indeed offered only a basic framework leaving several important questions either obscurely answered or not yet dealt with at all. Any law relating to the control of competition has to provide answers to at least six questions, but the Treaty itself gives only limited answers to them.

These questions can be summarized as follows:

(i) What official authority will decide upon the application of the law?
Answer: From Article 155 of the Treaty, it is clear that the Commission has such a responsibility subject to the right of review by the European Court under Article 173.

(ii) By what administrative means will this official authority receive information, e.g. about the existence and terms of relevant agreements, to enable it to carry out its duties?
Answer: The Treaty does not tell us.

(iii) If the answer to (ii) is by some form of notification or registration procedure, is the notification or registration itself an essential element in obtaining a declaration of non-applicability (i.e. the German approach) or may undertakings themselves decide whether to notify or register (i.e. the French approach), subject to intervention by the official authority if it believes the undertakings have decided wrongly not to notify or register?
Answer: The Treaty does not tell us.

(iv) Is the substantive law (viz. Article 85) a law of 'prohibition' (i.e. forbidding automatically all agreements within its jurisdiction unless an exemption has been granted) or 'an abuse law' (i.e. where agreements are only prohibited if the responsible authority can show that they are in breach of the substantive law)?
Answer: Although the Treaty appears to indicate that Article 85 is a law of 'prohibition', there still remained an element of doubt: on the other hand, the Treaty gives a clear answer in cases arising under Article 86.

(v) Are the sanctions for breach of Articles 85 and 86 simply of unenforceability, or involving the payment of fines and penalties to the official authority, or damages and costs to undertakings injured by the agreement?
Answer: The Treaty indicates in Articles 85(2) and 87(2) (*a*) respectively that the first two sanctions will apply, but remains silent on the third.

(vi) Are the sanctions in any case retrospective or simply prospective?
Answer: The Treaty does not tell us.

It was not, of course, that those who negotiated this part of the Treaty overlooked the importance of these issues. The negotiators were well aware of the differences in national competition law, particularly between French and German law, and were concerned primarily to establish basic principles and to allow the Commission thereafter to deal with the remaining issues. These basic principles corresponding to the three sep-arate paragraphs of Article 85 were:

(1) that a defined category of agreement/concerted practice/decision of an association with anti-competitive objects or effects and the potential to affect trade between Member States would be prohibited; and
(2) that if prohibited, the agreement/concerted practice/decision would be void; and
(3) that the prohibition could nevertheless be declared inapplicable, i.e, an exemption be granted, if the agreement/concerted practice/decision satisfied four conditions.

The remaining answers to the six questions listed above would not be clearly given until after the *Bosch*[15] decision in which the European Court ruled in the case of an export ban imposed by Bosch on German companies which had bought refrigerators from it, that agreements prohibited under Article 85 which had not been notified to the Commission were automatically void from the entry into force of Regulation 17; and the Regulation, of course, itself provided answers to the majority of the remaining issues, other than the question of whether damages and costs could be awarded to an undertaking claiming to be injured by the making of an agreement found illegal under Article 85, an issue which would remain uncertain for a considerable period and is dealt with in Chapter 22.

A major difference between the new Treaty and the Treaty of Paris was clearly to be that the implementation of rules relating to competition would involve the co-operation of the relevant authorities in Member States, since the Commission would not be required nor indeed able (unlike the High Authority under the Treaty of Paris) to deal with the complete range of practices and agreements that would potentially fall under its jurisdiction. The content of Articles 87(2)(*e*), and of Articles 88 and 89, soon was to make clear that in enforcing competition policy a basis for co-operation was required between the Commission and Member States. At this time competition law in the Six was in a primitive state. Neither Italy nor Belgium nor Luxembourg had any such legislation on the statute book. The Netherlands had only an Economic Competition Act which required registration of restrictive agreements but granted exemptions on a relatively liberal basis. French competition law was almost entirely concerned with vertical agreements and in particular with preventing refusals to deal. Germany alone had a new but relatively comprehensive competition law and administrative procedures and machinery capable of enforcing it.

[15] *de Geus* v. *Robert Bosch* Case 13/61 [1962] ECR 45: CMLR 1. In this case, Article 1 of Reg. 17 also was interpreted so as to give provisional validity to agreements in existence at the date of the Treaty coming into effect, thus following the principle already established for agreements covered by Article 65 of the Treaty of Paris by the European Court in the *Stork* v. *High Authority* Case 1/58 [1959] ECR 17. See also *Brasserie de Haecht* v. *Wilkin-Janssens* (no. 2) referred to in Ch. 4, n. 3.

It is the more remarkable, therefore, in these circumstances, whereas the High Authority had exclusive jurisdiction over competition issues under Articles 65 and 66, the Treaty of Rome contained provisions for sharing jurisdiction over the equivalent area of law between the Commission and the Member States. Clearly much work still had to be done after signature of the Treaty in working out how this relationship would be implemented. Article 87(2)(*e*) stated that the relationship between national and Community laws would have to be dealt with in regulations or directives to be adopted by the Council, acting unanimously on proposals from the Commission and after consultation with the Assembly. Article 88 retained the right for authorities in Member States, until such regulations or directives had been adopted, to rule themselves on the status of agreements, decisions, and concerted practices and on abuses of dominant positions within the Common Market in accordance with national law applying, however, the substantive terms of 85(3) and 86. The Commission itself, however, was not allowed to be inactive during the interim period prior to the adoption of such regulations; Article 89 gave it authority to take steps to investigate cases of suspected infringement of the Treaty principles (see n. 14).

The Treaty of Rome came into effect on 1 January 1958. In Part II we shall be analysing the substantive content of Articles 85 to 90, but it is already clear from this brief account of their origins that the legal provisions of these Articles, whilst influenced by Articles 65 and 66 of the Treaty of Paris, went considerably beyond them in both width of application and in the specification of those grounds upon which exemption could alone be granted. On the other hand, the differences in the relative content of Article 66 of the Treaty of Paris and the new Article 86 appeared to have the result that under the latter Article the Commission lacked any control over concentrations or mergers as such, merely having jurisdiction over abuse of a dominant position by individual undertakings. It seems that those who negotiated Article 86 did not intend that it should provide the Commission with a control mechanism over concentration; had they intended to do so, it might have been expected that they would have made use of wording adapted from that already available in Article 66, paragraphs 1 to 5, rather than the more limited wording using only the terminology of paragraph 7 actually adopted.[16]

Nevertheless, in spite of this limitation, it was widely believed that these Articles in the Treaty, and in particular Articles 85 and 86, contained far-reaching and important rules affecting a wide range of agreements and practices that could have effect on trade between the Member States of

[16] But see the European Court's judgment in *Europemballage Corporation and Continental Can Co. Inc.* v. *Commission* Case 6/72 [1973] ECR 215; CMLR 199.

the Six. The primary task of that Directorate of the Commission responsible for enforcing competition policy (referred to throughout the rest of the book as 'DG IV') would be to establish the detailed content of these rules and exemptions, as well as procedural machinery adequate for enforcing them throughout the Community. The competition rules of the Coal and Steel Community would also continue to be applied, but in view of their specialized nature are not considered further in this work. It should nevertheless be noted that, when the Treaty of Paris comes to an end early in the next century, the products to which it applies may then be made subject to the competition rules of the Treaty of Rome.

4

Early Years of DG IV

1. The Need for a Procedural Framework: The Enactment of Regulation 17 of 1962

It is a sobering thought that the detailed application of important international treaties and basic constitutional documents, however lofty their inspiration or impressively ceremonial their inauguration, is determined largely by the subsequent thoughts and debates of anxious men and women sitting around a table, blank sheets of paper before them. In other words, the conversion of relatively abstract treaty principles into enforceable procedural and substantive rules adequate to implement the ideals which underlie the constitutional framework can be achieved only through the hard and unromantic task of drafting detailed working documents, and then obtaining sufficient acceptance of their provisions from all interested parties. In this process, the skill involved is to pay enough attention to those parties' special or national interests to obtain their consent without at the same time risking undue compromise over the basic underlying principles of the original Treaty.

The implementation of the Articles of the Treaty of Rome relating to competition itself involved this time-consuming but inevitable and necessary process. The Treaty had established the Commission as the basic executive body responsible under Article 155 for carrying out the implementation of the Treaty; this Article lays down a fourfold task for the Commission. Of these, two were directly relevant to DG IV with its particular responsibility for implementing Articles 85 to 94 inclusive. The first of these was to ensure that the provisions of the Treaty, as well as measures taken by the Community institutions under its authority, were applied; the second that it should exercise powers, to be conferred on it at a later date by the Council, for the implementation of subordinate legislation which the Council would prescribe. These general obligations were supplemented by the more specific references contained in Article 87. Paragraph (1) provided that the Council was to adopt appropriate regulations and directives to give effect to the principles set out in Articles 85 and 86 'within three years of the entry into force of this Treaty'. If that time period elapsed without the regulations being adopted, then the requirement for unanimity ceased, and a qualified majority of the Council could authorize the adoption of regulations; in both cases, however, regulations must be based on a proposal from the Commission, and the Parliament had the right to be consulted.

Such regulations were envisaged by Article 87(2) to deal with five items. These ranged from the general to the highly specific. The general requirements referred to the need to define the outer range of the jurisdiction of the provisions of Articles 85 and 86 so that the extent of their application to transport, agriculture, and other specialized sectors could be clarified. A further general requirement was the need to determine the relationship between national laws and the provisions of the Treaty, and the third general aim was to ensure that the respective functions of the Court of Justice and the Commission were clearly defined.

Two specific objectives were also defined under headings (*a*) and (*b*) of Article 87(2). Compliance with the requirements of these two Articles 85 and 86 could clearly only be enforced if sanctions could be applied by the Commission to undertakings failing to comply with the requirements of the Articles or with decisions made by the Commission in implementing its jurisdiction. It was, therefore, necessary that fines and periodic penalty payments be provided as sanctions and also that the application of the procedural rules for exemption of agreements contained in Article 85(3) was spelled out. This would require the Commission itself to regulate the proper implementation of these rules, balancing on the one hand the importance of its effective supervision against the need to keep administrative procedures as simple as possible, in compliance with the requirements of the Treaty.

For this reason, therefore, whilst Articles 88 and 89 provided for the temporary enforcement of the rules on competition pending the adoption of the relevant regulations by both Member States and the Commission itself, the preparation of a comprehensive procedural regulation was felt from the earliest days by the officials of DG IV to be an essential foundation for the exercise of its powers. The initial resources of DG IV were very small, growing slowly from a handful of senior officials appointed during the first year of its operations, gradually rising at the end of the second year to approximately twenty and then increasing steadily but slowly to a figure of seventy-eight by April 1964, just over one hundred by ten years later, and even now to no more than 210 'A' grade officials.[1] DG IV had received in its earliest years other responsibilities apart from the enforcement of the rules of competition, notably responsibility for harmonization of laws under Article 100 of the Treaty and certain fiscal matters, though these 'fringe' responsibilities were later removed in 1967 when the unification of the executives of the European Economic Community, the

[1] The total staff of DG IV at the present time is approximately 420.

European Coal and Steel Community and EURATOM meant that additional responsibilities relating to both these latter institutions became vested also in DG IV.

The first four years, therefore, of its existence were mainly committed to the preparation of this procedural regulation, later to become familiar as 'Regulation 17' which would need to provide answers to those questions listed in the previous chapter, for which the Treaty itself provided either no answer at all or merely an incomplete one. It had first to conduct a review of the existing national laws of Member States which, as we have seen, varied from the non-existent to the comprehensive but untried. Consultation had to take place on a wide basis with Member States and with numerous experts and organizations: meetings and conferences were held to discuss both the existing systems of law with Member States and the best method of adopting features from them for inclusion in the Regulation. The influence of German lawyers was particularly strong at this time, partly because they had recently seen the introduction of the German competition statute after long and fierce internal debate.[2] The German approach was to favour a system under which specified agreements in restraint of competition were void unless and until exemption was conferred upon them by the constitutive act of the Commission, which alone could have the power of granting exemption from the basic prohibition of the law.

In direct contrast was the French preference for specific exemptions which would be automatically applied if the necessary conditions were satisfied, without the need for any intervention or imprimatur by the Commission. This conflict was eventually resolved by the Council on a basis necessarily involving a compromise between the two approaches. In view of the wording of Article 85, the French argument could not be accepted, since it was impossible to reconcile the wording of Article 85 with a requirement that no prior decision of the Commission was needed before a restrictive agreement could be declared void. On the other hand, the terms of the Regulation as finally adopted left open the possibility that agreements in existence at the date of the adoption of the Regulation could be declared invalid following investigation by the Commission, not necessarily with retrospective but simply with prospective effect, and that exemptions could be granted on a retrospective basis rather than merely prospectively. This situation was in due course clarified by the European Court of Justice in the *Brasserie de Haecht* (no. 2)[3] case when the doctrine

[2] See J. Kallaugher, 'The Influence of German Competition Law on the Development of the Competition Law of the European Communities' (unpublished article).

[3] *Brasserie de Haecht* v. *Wilkin-Janssens* (no. 2) Case 48/72 [1973] ECR 77: CMLR 287. This decision of the Court of Justice followed the same general lines as earlier cases decided by it on cases arising under the ECSC rules referred to in Ch. 3 n. 15. The German view can

of provisional validity for agreements in existence at the implementation date of the Regulation was upheld, and the full rigours of the German preference for total illegality and unenforceability of all agreements pending the granting of the exemption not accepted.

Once the draft of the Regulation had been agreed on internally within DG IV, following this protracted period of consultation, it had to be approved by the full Commission, and then submitted in addition to the Economic and Social Committee and to the European Parliament. At the instance of the Parliament, whose committee reviewing the Regulation had as rapporteur Professor Arved Deringer, a German lawyer and expert on competition law, some important additional proposals were made. The early paragraphs of the Deringer Report contain a careful analysis of the three different possible approaches which the Commission could adopt in its approach to the implementation of Articles 85 and 86 in providing the answers to the essential questions which the Articles themselves for the most part had not dealt with. The first was that adopted in Dutch national legislation, under which all restrictive agreements have to be notified, but by notification obtain a presumption of validity until the competition authority is able to prove that any particular agreement constitutes an abuse, and then with prospective effect only. The second was the French approach, under which undertakings would themselves decide if their agreements were in accordance with the law, but where the competition authorities could themselves elect to challenge them and, if successful, declare them unenforceable with retrospective effect. The third was the German, and ECSC approach, of preventative control, under which exemption from the general law prohibiting agreements restraining competition was only effective if notification had been effected and an individual or group exemption pronounced by the competition authorities.

After discussing the various advantages and disadvantages of these alternative systems, the Parliamentary Committee accepted the third alternative as being that most consistent with the actual terms of the Treaty, and also as being the system most likely to give the Commission the widest knowledge of the types of agreement being made by undertakings within the Community. The Committee suggested that Article 86 should also be brought within the procedural framework of the new Regulation, although it had originally been intended to apply only to Article 85, and this was accepted in the final form of the Regulation. Notwithstanding the lesser probability that undertakings would wish to notify to DG IV the existence of dominant positions held by them in individual markets and possible abuses of which they were guilty, the inclusion of Article 86 within the scope of the Regulation enabled the powers given to the

be said, however, to have prevailed to the extent that the Court's decision excluded the possibility of provisional validity for new agreements.

Commission in respect of investigations and the imposition of fines and penalties to be applied likewise to abuses of a dominant position as well as in the control of agreements, decisions, and concerted practices.

It has proved certainly far more convenient for the Commission that both Articles are dealt with by the same procedural Regulation. Numerous other suggestions were also made by the Parliament's report including the introduction of a negative clearance procedure, this suggestion being supported by the Economic and Social Committee who drew attention to the value of a similar provision included in the Treaty of Paris.[4] Such a facility did not exist under German competition law but was accepted as part of the procedural machinery of DG IV, where it has proved its practical usefulness. It enabled the Commission to certify that the relevant Articles of the Treaty did not apply in an individual case because the jurisdictional requirements of Articles 85 or 86 were not satisfied. This might be because of insufficient effect on trade between Member States or because the sector concerned was not covered by Article 85 or because the agreement was insufficicntly restrictive to have any effect detrimental to competition in the Common Market, perhaps having an effect purely within one Member State. A negative clearance, however, would be valid only so long as the facts upon the basis of which it had been granted remained unchanged. If these altered, then the original negative clearance would automatically, and without further decision of the Commission, cease to have effect.

The concept of the 'negative clearance', therefore, had to be contrasted with the effect of the granting of a formal exemption under the provisions of 85(3). Once the Commission had taken a notified agreement through this procedure and satisfied itself that all the relevant four conditions for exemption had been satisfied, the parties were enabled, as with a negative clearance, to continue to operate the agreement without fear of fines, penalties, or legal unenforceability in the Courts. Unlike the position with a negative clearance, however, DG IV retained control over the agreement since it would have been exempted only for a specific period of time (not normally exceeding ten years) with a continuing right for the Commission to obtain information from time to time relating to its operation, and subject to renewal at a future date when the renewal conditions might be varied by DG IV in the light of any change in circumstances since the date of the original exemption.[5] If there were a major change in any

[4] Article 65(4).

[5] Thus in the well-known *Transocean Marine Paint Association* case the conditions imposed on renewal of the exemption were considerably more stringent than on the original decision. The two decisions of the Commission can be found at [1967] CMLR D9 and [1974] 1 CMLR D11 respectively, part of the latter being the subject of appeal by the Association to the Court as Case 17/74 [1974] ECR 1063: 2 CMLR 459. The exemption has been renewed again in 1980 and 1988.

material circumstances affecting the exemption, or the recipient could be shown to have provided false information to obtain it, then it could be withdrawn at once, and if necessary retroactively.

The negotiations over the terms of Regulation 17 were long drawn out, encountering a full measure both of legal and political difficulties. The Commission's attitude to the draft Regulation was understandably ambivalent. On the one hand it feared the sheer size and scope of the jurisdiction which the Treaty provided in this area, so that it welcomed the ability to share it to some extent with Member States and their national authorities. On the other hand, it was concerned not to allow Member States or their national courts to operate the exemption provisions nor to operate a system based on the 'abuse' rather than the 'prohibition' principle, for fear that loss of control over the system would weaken both its effectiveness and its consistency.

The final text of the Regulation was finally approved by the Council on 6 February 1962 and published in the Official Journal on 21 February, with the result that it became effective on the following 13 March. It has remained continuously in force from that date without major amendment[6] and is felt by the officials of the Commission to give them nearly all the procedural powers necessary for the implementation of competition policy, subject to only a few minor qualifications. It gives to DG IV indeed powers probably greater than would be accorded to it were the Regulation being now debated *ab initio*, when the delegation of such *de facto* powers to a single DG in the Commission might be made more subject to control and intervention by Member States as well as by other institutions of the Community.[7]

2. The Detailed Provisions of Regulation 17

Because of their continuing importance for the whole of the jurisdiction exercised by DG IV in enforcing Articles 85 and 86, and notwithstanding that this book is concerned primarily with substantive law rather than procedural issues, it is essential at this stage to refer in detail to the main powers conferred by the Regulation.

It opens with lengthy recitals setting out the background to the adoption of the Regulation by the Council. Under civil law as well as under the practice of the European Court, it is open to Courts interpreting such a Regulation to look to the terms of any recitals for assistance in interpreting points of difficulty arising from the construction of individual clauses.

[6] It has been slightly amended in Articles 4(2) and (3) by Reg. 2822/71.

[7] The history of the long struggle to adopt a Merger Regulation illustrates this point. See Ch. 20.

The Commission and Council have both adopted the practice, in regulations governing competition law, of setting out long (some would say, overlong) recitals giving the background to the circumstances in which they were adopted and their detailed objectives, in order apparently to assist the European Court of Justice and national courts in their interpretation. Some of the recitals in Regulations are purely precautionary, referring back to relevant Articles of the Treaty of Rome (notably 85, 86, 87, and 172) which establish the powers (*vires*) of the Council or Commission to act and set the framework within which the Regulation has to operate. Lengthy references are also made, however, to practical and policy issues which concern the implementation by the Commission through DG IV of competition policy. Thus in Regulation 17, there is reference to the fact that the number of agreements, decisions, and concerted practices to be investigated will probably be too numerous to be examined at the same time, whilst others may have special features rendering them less prejudicial to the development of a Common Market so that there is a need to provide more flexible arrangements for the treatment of certain categories of agreement. There is also reference to the need for liaison with the competent authorities of the Member States both in acquiring information and in carrying out investigations required to bring to light any agreements prohibited under Article 85 or any abuse of dominant position in breach of Article 86. Reference is also made to the need for procedural requirements to ensure that third parties, whose interests may be affected by a decision, are given the opportunity of submitting their comments before decisions are formally taken. Mention is made of the right of the European Court under Article 172 to exercise its unlimited jurisdiction in respect of any decisions where the Commission imposes fines or periodic penalties.

After this introduction, Article 1 of Regulation 17 launches immediately into a declaration that agreements, decisions, and concerted practices covered by Article 85(1) and abuses of dominant position covered by Article 86 'shall be prohibited, no prior decision to that effect being required'. The drafting of Article 1 in this way was probably required as a direct consequence of the terms of Articles 85 and 86 in the Treaty. Nevertheless, at least one national court (in Holland) had in 1961 declared that notwithstanding the terms of Article 85(2) agreements covered by Article 85 were not invalid until a decision to that effect had been taken by the Commission,[8] so that this Article was required to make clear to Member States and their Courts that no Commission decision was required as a

[8] See H. Smit and P. Herzog, *The Law of the EEC: A Commentary on the Treaty*, 6 vols. (New York: Matthew Bender, 1976), vol. ii, Pt. 3, para. 67. The original decision in this case had been given by the District Court of Zutphen in 1958, the decision being upheld by the Arnhem Court of Appeal in 1961.

necessary condition of invalidity. The introduction of the reference to Article 86 in this Article (as well as in the following Articles 2 and 3) owed its presence directly to the suggestion of the European Parliament already mentioned.

It is clearly in the interests of undertakings that they should be able to obtain confirmation where appropriate that their agreements or practices are not covered by either Article 85 or 86. Article 2 of the Regulation gives authority to the Commission to give such a negative clearance. In practice, undertakings will almost always apply in the alternative for either negative clearance or an exemption under Article 85(3).

Article 3 covers the important subject of complaints; paragraph (2) states that either Member States or natural or legal persons claiming a legitimate interest in the subject matter of the complaint may make application to the Commission which is then bound to investigate, though of course the volume of complaints and notifications to DG IV has over its lifetime gradually reached a level where the Commission cannot give priority to all applications and some have to be deferred for a considerable period. If, after investigation, the Commission wishes to require the undertakings concerned to bring legal infringements to an end, the Article permits them to do so by way of decision,[9] but paragraph (3) entitles the Commission to attempt to dispose of the complaint by way of informal negotiation and recommendation. In practice, the Commission does dispose of the majority of complaints in this way and indeed given its limited resources could not operate in any other manner.[10]

Articles 4 and 5 set the framework for the notification both of existing agreements and newly formed agreements coming into existence after the operative date of the Regulation. New agreements under Article 4(1) must be notified to the Commission and cannot, until after notification, receive the benefit of any exemption under Article 6. The clear intention of the Article is that new agreements should be notified when or immediately after they are made. The only exception relates to three groups of agreement which by their nature are considered less likely to affect competition or trade between Member States, namely:

[9] For an analysis of the Commission procedures in considering complaints including a definition of when a 'decision' is given, see *Automec* v. *Commission* Case T 64/89 [1991] 4 CMLR 177. See also the Twentieth Annual Report, pp. 117–18.

[10] DG IV has gradually developed the practice of negotiating settlements and undertakings from the parties to agreements and practices in order to avoid having to issue formal decisions in every case and also to avoid having to make extensive use of its powers to grant interim measures of relief prior to formal decisions. These practices, however, have not been without their critics. See, e.g. I. Van Bael, 'The Antitrust Settlement Practice of the EEC Commission', 23 CML *Rev* 61(1986), also Ch. 8, pp. 129–30.

(1) agreements where the only parties are undertakings from one Member State and the agreements do not relate either to imports or exports between Member States;
(2) where only two undertakings are party to the agreement which itself either
  (*a*) merely restricts the freedom of one party to the contract in determining prices or conditions of business upon which goods purchased from the other party are to be resold; or
  (*b*) imposes restrictions on industrial property licensees or persons obtaining know-how under contract or licence relating to the use and applications of industrial processes; or
(3) where the agreements have as their sole object standardization of joint research and development or specialization agreements and where the products the subject of the specialization do not in a substantial part of the Common Market represent more than 15 per cent of the volume of business done in that market so long as the total annual turnover of the participants does not exceed 200 million units of account.[11]

In all these cases, as notification is voluntary, grant of an exemption at a later stage can be retrospective to the original date when the agreement was made rather than, as with all other agreements, simply with effect from the date of notification.

One essential element of this Regulation was to establish a timetable for notification of the many existing agreements in force at the date when the Regulation became effective. The Commission elected to receive the forms of agreement required to be notified by two instalments. Multilateral agreements had to be notified to the Commission by 1 November 1962 whilst bilateral agreements (which were to comprise the vast majority of notifications) had to be lodged by the slightly later date of 1 February 1963, other than those, of course, for which exemption from notification had been given under Article 4.

These rather short time limits within which agreements had to be notified were the subject of criticism from both United States and European industrial and commercial concerns. They claimed that the volume of agreements to be notified would make it impossible in practice for the Commission to deal with them adequately, given its limited resources. This criticism turned out to be well founded. Originally, the draft of Article 5 dealing with the timetable and supported by the Deringer Report

[11] Amendments to the terms of this clause under Reg. 2822/71 included an increase in the maximum annual turnover permitted to benefit from the Regulation. The Commission's interpretation of the wording of this Article 4 tends to be strict. Of course many agreements that might benefit from it come also within the categories covered by group exemption and are freed from the obligation to notify for that reason.

provided also that all agreements notified within the time limit which were not specifically opposed by the Commission within a period of six months from notification should remain provisionally lawful and, therefore, enforceable in the Courts of Member States. Opposition procedure is a concept known to German law, under which the German Cartel Authority (the Bundeskartellamt) has to object to agreements within a fixed period from receipt of notification if it finds any objectionable features. The Commission had originally suggested that agreements of lesser importance (as listed in Article 4) should be subject to a similar procedure, the time limit being six months and with the consequence that if opposition were not raised by DG IV the agreements were provisionally deemed (not finally) to be outside the scope of Article 85(1)'s prohibition.

While this proposal certainly had its attractive features (particularly on the score of administrative convenience), it was felt that such a proposal would go too far in moving away from the principle embodied in Article 85 that all restrictive agreements (except those covered by the specified exemptions) were prohibited. For the effect of the opposition procedure being introduced could have been that agreements were allowed to be enforced by national courts, even in cases where no exemption was ultimately given. Another reason why the Commission ultimately declined at this stage to accept the opposition procedure as a basis for reducing its workload was that it is difficult for any administrative body to operate such a procedure without adequate case-law experience of the range of restrictive clauses and agreements which it may encounter. At this time, of course, the Commission had had almost no experience of the kind of agreements which would be notified to it for assessment. Whilst opposition procedures have since the start of the 1980s now become more common in group exemptions adopted by the Commission,[12] in all these cases there will have been adequate case by case experience of the type of agreements and restrictive clauses to which opposition may have to be notified. It is a characteristic of a mature competition authority, that it has such experience upon the basis of which decisions can be issued more speedily than is possible in its early days. Whilst the recitals had referred confidently to the fact that certain agreements 'have special features which may make them less prejudicial to development of the Common Market', hard evidence for assessing these special features was not easily obtained nor were the criteria for judging them agreed upon universally even within DG IV itself.

Articles 6 and 8 deal with the important subject of exemption under 85(3). It is clear from the combined effect of these Articles that any

[12] See e.g. the 'opposition' procedures now to be found in group exemptions covering patent licences (Reg. 2349/84), research and development agreements (Reg. 418/85), and know-how licences (Reg. 556/89).

exemption granted must operate from a specific date, which cannot be earlier than the date of notification (except for the 'less harmful' agreements referred to in Article 4), will be for a specified period, and quite possibly upon detailed conditions. No maximum time limit has been attached, though in practice it has rarely exceeded ten years.[13] Conditions and obligations imposed by DG IV must clearly also be those necessary to ensure compliance with the conditions set out by 85(3) and to make possible supervision of the conditions and general workings of the agreement by DG IV. Thus, information may be requested as to the names of parties involved in joint buying or selling arrangements or participating in distribution systems exempted on such a basis. Once these Articles entered in force, jurisdiction over the application of Article 85(3) was removed from national authorities of Member States and left entirely in the hands of the Commission (Article 9),[14] although under Article 9(3) until the Commission actually initiated a procedure against an undertaking, the application of Article 85(1) or Article 86 by a Member State could continue.[15]

The combined effect of Articles 1 to 9 is, therefore, to establish a procedural machinery under which both old and new agreements are to be notified and thereby brought within the knowledge of the Commission; they establish both a timetable for notification and a procedure for declaring either that the agreement infringes the Treaty or is entitled to exemption either with or without obligations attached. Breach of such obligations entitles DG IV to bring action to cancel the exemption.

Articles 10 to 14 contain powers required by the Commission in order to carry out its duties effectively. Article 14 is particularly important since it gives authority to the Commission to undertake all necessary investigations, empowering its officials to (*a*) examine books and other business records, (*b*) take copies of books or business records, (*c*) ask for oral explanations on the spot from undertakings investigated, and (*d*) to enter any premises or vehicle or land belonging to undertakings. These powers can either be exercised voluntarily with the consent of the undertaking or compulsorily if a decision of the Commission to such effect is granted.[16]

[13] An exception has been made in some recent joint venture cases where a fifteen-year exemption has been permitted, e.g. *Optical Fibres* OJ [1986] L236/30 and *Brown Boveri/NGK Insulators* [1989] 4 CMLR 610.

[14] This would apply of course only to those sectors covered by Reg. 17 and not to any which were expressly removed by Regulation from its scope, e.g. air and sea transport, where under Article 88 Member States' existing national authorities would retain jurisdiction to grant exemption until applicable regulations had been made. See the *Nouvelles Frontières* Cases 209–213/84 [1986] ECR 1425: 3 CMLR 173.

[15] The inability of national courts to grant individual exemption to agreements under Art. 85(3) was confirmed by the European Court in *Delimitis* v. *Henninger Brau*, Case C 234/89, a decision of 28 Feb. 1991.

[16] The voluntary and compulsory methods of proceeding are found in Articles 14(2) and

Liaison is required with the competent authority of the Member State in the territory of an undertaking if an 'on the spot' search is to be made:[17] a power absent from Article 14 is one to require employees or directors of undertakings to attend a hearing (in Brussels or elsewhere) and to bring relevant documents for that purpose.[18] This power would have been of considerable use in subsequent cases to the Commission, but it was no doubt felt that Member States would regard its exercise as too great an infringement on the rights and liberties of their nationals.

Considerable emphasis is placed on the obligation of the Governments in Member States and their competent authorities[19] to co-operate with the Commission. This obligation is found both in Article 11 (dealing with the right of the Commission to request information from Governments and competent authorities of Member States and undertakings in those Member States), and also in the right to require investigation by authorities of Member States under Article 13. Article 10 requires that Member States are sent copies of all relevant documents in each case involving an allegation of infringements of Article 85 or 86 or where notification has been made for purposes of obtaining negative clearance or exemption. Article 10(2) provides that all these procedures of DG IV must be carried out 'in close and constant liaison' with competent authorities, and that 'such authorities shall have the right to express their views upon that procedure'.

Article 10(3) was a clause which caused considerable difficulty during negotiations. It was the wish of France that the Member States would collectively have a veto upon the original decision of DG IV, before it was sent up for approval by the Commission, by requiring that any decision

(3) respectively, and the Commission has complete discretion which to adopt in any particular case. *National Panasonic (UK)* v. *Commission* Case 136/79 [1980] ECR 2033: 3 CMLR 169.

[17] The European Court has, in three leading cases decided in 1989, laid down important principles controlling the way in which these rights of the Commission are to be carried out. In particular Commission officials have no right to use force to obtain entry to business premises in order to examine documents, but must obtain appropriate authority from national courts: *Hoechst* v. *Commission* (Cases 46/87 and 227/88) [1989] ECR 2859: [1991] 4 CMLR 410. *Dow Benelux* v. *Commission* (Case 85/87) and *Dow Chemical Iberica* v. *Commission* (Case 97–99/87) [1989] ECR 3137, 3165 [1991] 4 CMLR 410. *Orkem and Solvay* v. *Commission* (Cases 374/87 and 27/88 respectively) [1989] ECR 3283, 3355 [1991] 4 CMLR 502. See also Ch. 22, p. 429.

[18] This would be a power equivalent to that possessed by English Courts to require witnesses to attend with relevant documents (know as a 'subpoena duces tecum').

[19] The definition of such authorities is discussed in Ch. 22, (p. 439 n.25). Both they and the Commission are limited in the use of information acquired under Articles 11 to 14 for the purposes of the relevant request or investigation. Nevertheless in the *Zwarteveld* case (Case 2/88) (1990) 3 CMLR 457 the European Court held that normally the Commission must supply national courts with confidential information in its possession unless the Court itself directs that the provision of such information would be detrimental to the functioning and independence of the Community.

should require the support of the majority of the Advisory Committee. This proposal was strenuously resisted by the officials of the Commission and also by the German negotiators because it was felt that it would both prevent development of competition law criteria applicable uniformly to all Member States and would also tend to make it far more difficult for the Commission to issue either decisions or exemptions in cases that were politically sensitive. The final solution adopted was to make the relevant committee, consisting of representatives of Member States, advisory only and without the power of veto, the committee having the right only to express its views rather than to pronounce any formal decision. In general the Commissioners at their meetings called to arrive at final decisions in competition cases will attach weight to the opinions expressed by the members of this committee. Nevertheless, apparently some decisions of the Commission have been issued notwithstanding an adverse vote (even on occasions, apparently, of a unanimous nature) in the Advisory Committee, but this is rare. Proceedings of the Advisory Committee are, however, confidential, and it is only, therefore, possible to speculate how often this has occurred.[20]

The remaining Articles of the Regulation contain a number of clauses of value to the Commission in its administrative task. Articles 15 and 16 deal respectively with the ability of the Commission to impose substantial fines and penalties. A fine is imposed for past conduct where undertakings have participated in agreements or practices forbidden under Article 85 or 86, or where incorrect or misleading information has been supplied in the course of investigation. In respect of the former type of fine, these may range up to a very substantial amount not exceeding 10 per cent of the turnover in the preceding business year of each of the undertakings found so to have participated. In fixing the fine, regard is to be had to the duration of the infringement and its seriousness.[21] Fines are not imposed in respect of any period after notification has been made unless a preliminary warning has been given under Article 15(6) that in the view of the Commission 'after preliminary examination' application of the exemption provisions in 85(3) cannot be justified. This procedure, though rarely utilized so far, is of value in the course of a very detailed investigation where the Commission is satisfied that ultimately no exemption will be available, but where the preparation of the necessary decision will require a substantial amount of work, and it is hoped in this way to put pressure upon

[20] In contrast the opinions of the Advisory Committee on the draft decisions under the Merger Regulation can be published if the Committee so request and this has occurred in several cases already.

[21] The size of fines imposed has increased substantially in recent years. Tables of fines are set out in *EEC Competition Law Handbook* (London: Sweet & Maxwell, 1991), and in practitioners' textbooks.

the undertakings concerned to terminate the agreement or practices without waiting for the issue of the formal decision. Penalties under Article 16 deal not with the past, but with the future, and are payable in respect of each day upon which an undertaking continues to refuse to put an end to an infringement condemned by a decision of the Commission, or to supply information required by a decision of the Commission made in the course of its investigation. Proposed fines and penalties are also considered by the Advisory Committee under Article 10, and the Commission's decision is in turn subject to the unlimited jurisdiction of the Court under Article 172, quite apart from the right of an undertaking to seek review of a decision under Article 173. The Court has powers to cancel, reduce, or increase the fine imposed.

Article 19 deals with the important issue of natural justice. Undertakings are entitled to be heard on matters where the Commission proposes to raise a statement of objections, and rights are also given to other natural or legal persons with a sufficient interest. This concept has been generously interpreted to allow competitors and customers standing to object to decisions by the Commission which are claimed to have prejudiced them.[22] Article 19(3) requires the Commission to publish any proposed decisions relating either to negative clearance or exemption in the Official Journal, allowing all interested parties to submit their observations within a time limit of not less than a month. All final decisions, whether for negative clearance exemption or prohibiting the infringement with or without fine, have under Article 21 to be published in detail in the Official Journal, subject only to the requirements of business secrecy.

3. Implementation of Regulation 17: The Early Experience of DG IV

The entry into force of this new Regulation was a major psychological boost for DG IV. At last it had a firm procedural foundation for its inquiries and investigations. The Directorate had previously felt itself inhibited in making inquiries and in carrying out investigations because of its lack of specific powers, notwithstanding the provisions of Article 89. Up to March 1962, therefore, the Commission had investigated only 33 cases and had issued no formal decisions, although in some of its cases it had taken steps to try and persuade the parties involved to end particular practices. In one case a boycott of a trader by a producer holding a dominant position was terminated as the result of such pressure, as was also the reservation of one national market to local producers and some preferred

[22] See, e.g., *Metro* v. *Saba* (no. 1) Case 26/76 [1977] ECR 1875: [1978] 2 CMLR 1.

foreign producers but to the exclusion of other producers within the Community. A magazine taking advertisements from various manufacturers refused to accept advertisements from competing manufacturers in other States and this practice was also brought to an end after intervention by DG IV.

Whilst therefore in early 1962 officials in DG IV had become in general pleased with the scope of their new powers, they had no doubt that the future pattern of administration would become a great deal more demanding than the comparatively tranquil period that had prevailed since its formation four years previously. The Regulation indeed could be (and was) criticized for bringing within coverage for notification too many routine and relatively innocuous agreements, without enabling DG IV to focus its attention on the kind of horizontal market-sharing and other cartels which are normally considered to have most effect on the health of the competitive process. The mesh of the net for catching notifications was so fine that it was anticipated that the number of agreements likely to be registered, particularly of a vertical nature, would be extremely high, notwithstanding the provisions of Article 4 with its scheme for exemption from notification of agreements of apparently less immediate threat to competition. The view of the critics was that the Commission, in its natural desire to assert a wide jurisdiction over undertakings within the Common Market, had really attempted 'to bite off more than it could chew'; they claimed that it rather should have proceeded more gradually in the selection of criteria applicable to notification requirements.

It is interesting at this stage in the history of the Commission to note the attitude to be found in the Action Programme for the second stage of the Community published in November 1962, on the eve of the implementation of the notification requirements of Regulation 17. The objectives of the Commission in this area are stated to be as follows:

> The Commission's task will now be to wield as efficiently as possible, through the administrative practice it will develop, this new instrument of European law. The Commission will base its interpretation on the two-fold aim laid down in the Treaty: to protect competition and the economic freedom of all who operate on the market. The need for certainty as to the law will be catered for by adopting precise clearly viable criteria, notably with regard to exempting agreements under Article 85(3) of the Treaty.
>
> The Commission will give particular attention to the question as to how certain vertical agreements and licensing contracts, frequently met with in business, come under the cartel regulations and how they can be effectively supervised with as little administrative control as possible. The European system of law against restraint of competition must be applied so as to eliminate first and foremost those restrictions which have a particular adverse effect on the development of the Common Market.

It is perhaps ironic that at this very moment the inherent consequences of the approach adopted by Regulation 17 was taking DG IV in the opposite direction of having to devote far too much of the energies of its limited staff in dealing with vertical agreements relating to distribution of goods and the licensing of intellectual property rights, leaving too little over for handling horizontal agreements and cartels that might themselves have been said to have a 'particularly adverse effect' on the development of the Common Market referred to. The Commission's own response at the time to such criticism would certainly have been that the integration of the Common Market itself positively required an emphasis on vertical agreements relating to distribution and licensing, the very numbers of restrictive agreements of the kind compelling the priority given to them.

It would be wrong, however, to give the impression that the entire energies of the Commission from the start of 1958 up to the end of 1962 were devoted to the negotiation of and implementation of Regulation 17. Substantial work was also done in other areas and the fruits of this work were contained in two Notices published on 24 December 1962. The first was relatively non-controversial and dealt with the subject of commercial agencies.[23] It took the form essentially of an unofficial 'group negative clearance' for contracts made by principals with commercial agents operating within the Community when these agents simply negotiated transactions on behalf of the principal either in their own name or in the name of the principal. For the clearance to apply, however, it was essential that the agent was indeed by the nature of his functions no more than that, a person who neither undertook nor engaged in activities 'proper to an independent trader in the course of commercial operations'. The decisive factor in determining the boundary between the commercial agent and the independent trader would be the degree of risk which the agent accepted in the course of his activities. A commercial agent to benefit from the clearance should not assume any risk arising from the transaction. Examples of the unacceptable forms of 'risktaking' for this purpose were given, including when the agent kept a considerable stock of products as his own property or was required to organize at his own expense substantial free services to customers or where he himself determined prices or terms of business.

The explanation given for this Notice was that obligations on manufacturers to have exclusive dealing contracts with their commercial agents had neither the object nor the effect of preventing competition within the Common Market. The agent was said to perform only an auxiliary function in the market, since he simply acted on the instructions of and in the interests of his principal. The obligations of exclusivity assumed by the

[23] JO [1962] 139/2921.

agent in return for the privilege granted to him by his principal necessarily involved some reduction of opportunity for other agents and other principals. Nevertheless, the Commission stated that it regarded these restrictions as the necessary result of the special relationship that existed between the agent and his principal for the mutual protection of their interests. The agent was really being treated as part of the undertaking of the principal in view of his lack of freedom to make independent pricing or other commercial decisions; and indeed such limited functions could almost as well be described as those of the principal rather than those of an undertaking independent from him. Though appearing straightforward, the Notice in fact has caused substantial difficulties of interpretation which we shall discuss in Chapter 12.

More controversial, and subsequently to prove an unfortunate venture (since the views of DG IV were later to change) was the December 1962 Notice on patent licensing agreements, also known as 'the Christmas Message'. It was believed that many of these agreements would be lodged with the Commission at the start of 1963 as the result of the notification deadlines of Regulation 17. To try to provide criteria by which a substantial proportion of these could be dealt with speedily, a group negative clearance was issued in the form of a Notice stating that a certain number of common clauses in patent licensing agreements were not caught by the prohibitions of Article 85(1). These included the following restrictions, imposed for a period no longer than the life of the relevant patent:

- obligations on the licensee which restricted the field of use and range of technical applications or were restrictions of the quantities allowed to be produced, or of the period during which the licence could be exercised, or of the territory in which it could be exercised;
- obligations with regard to marking of products produced with details of the relevant patent;
- quality standards imposed indispensable for the 'technically perfect exploitation' of the patent;
- undertakings relating to improvements provided these were not exclusive and were mutual upon licensor and licensee;
- undertakings on the part of a licensor not to authorize anyone else to exploit the invention nor to exploit it himself.

This negative clearance was not, however, to apply to agreements where there was joint ownership of patents or reciprocal or multiple parallel licensing or to any clauses other than those specifically permitted by the Notice. The objective of the publication was to indicate the considerations by which the Commission would be guided when interpreting Article 85(1). Provided that such licence agreements did not contain restrictions apart from those listed, it was considered that they were outside the scope of 85(1) and need, therefore, not be notified. Detailed explanation for the reason for each type of clause being given such clearance was set out. The Notice. however, contained the usual caveat, customary for all such Notices from the Commission, that it did not bind the courts or competent authorities of Member States.

The result of the time limits contained in the Regulation was that a total of 920 *multilateral* registrations were filed by the closing date of 1 November 1962, but the much larger and daunting total of 34,500 *bilateral* agreements by the slightly later time limit of 1 February 1963. The 'mass problem' of which the Commission had been warned had indeed arrived: and for some months all the officials in DG IV had to concentrate simply on giving a preliminary review to as many as possible of the vast number of documents which had arrived and were now piled throughout their offices. The notification form itself (known as form 'A/B') had been prescribed by a Regulation (no. 27/62) made by the Commission under the authority of Council Regulation 17, and a further Regulation (no. 99 of 1963) was shortly afterwards adopted setting out procedures for investigations and oral hearings.

Under Article 2 of this latter Regulation, the Commission was required, if it is considering reaching a decision against an undertaking, to send a statement of objections raised against it; the undertaking is then entitled to reply, setting out all matters relevant to its defence and attaching any relevant documents. When this reply had been received and considered by the Commission, an oral hearing is then called if the undertaking desires it and at the hearing both the parties themselves and other parties with a sufficient interest (in technical terms) in the case may attend and address the Commission. Representatives of the Advisory Committee are also present at this hearing and may themselves ask questions of the parties. The procedural framework, therefore, for carrying out investigations was now at last in place[24] and gradually over the following years decisions of the Commission begin to be issued, though initially at a very slow rate.

The first decisions came out during 1964 and, perhaps as a matter of deliberate policy, they concerned the granting of negative clearances and the interpretation of the scope of Article 85(1). The first case was *Grosfillex Fillistorf*[25] published in the Official Journal of 9 April 1964. This agreement related to an exclusive distribution agreement between a French manufacturer and a Swiss distributor, which contained a prohibition on the Swiss distributor from reselling the goods into the Common Market. If the Swiss distributor had wished to re-export the goods back into France or any other country within the Common Market, it would

[24] The leading textbook on the procedural aspects of the work of DG IV is C. S. Kerse, *EEC Antitrust Procedure*, 2nd edn. (London: European Law Centre Ltd., 1987).

[25] JO [1964] 58/915: [1964] CMLR 237.

have had to pay additional customs duties on such re-exports which would have rendered them unattractively expensive by comparison with competing goods in those markets. There was, therefore, no real likelihood that the distributor could hope to resell goods covered by the distribution agreement within the Common Market. In these circumstances, the Commission found that since the agreement could not, therefore, have an effect on trade between Member States, it fell outside Article 85(1). An equally straightforward negative clearance decision was issued on 10 June in the same year, in the case of *Mertens & Straet*[26] where a distribution agreement granted to Mertens & Straet by the Bendix Corporation for Belgium was on a non-exclusive basis and contained no clauses which gave any territorial protection to the distributor. Again the finding was on the basis that there were no provisions in the agreement likely to distort competition within the Common Market or affect trade between Member States.

The first case involving considerations of real substance was published in the Official Journal on 26 August 1964.[27] Here the restrictive clauses considered arose out of the winding up of a French company, Vitapointe, whose EC business in hairdressing products, including relevant trade mark, know-how, and patent licences, was sold to Nicholas (a French company) whilst its equivalent non-EC business, including the United Kingdom, went to Vitapro (an English company). Vitapro had previously acted as exclusive distributor in the UK for Vitapointe. Both Nicholas and Vitapro had agreed that for five years they would not sell nor manufacture products covered by the arrangement in the territory taken over by the other, with certain limited exceptions. Vitapro also agreed not to permit trade marks transferred to it to be used outside the non-EC territories to designate those products which it sold there.

At first sight, this kind of restriction would appear to have had an effect on trade between Member States, namely to prevent Vitapro from competing in the Common Market for hairdressing products. Nevertheless, a negative clearance was granted on three grounds; the first reason was that the restrictions only prevented Vitapro from competing in the Common Market for a short period of time and did not cover products that Vitapro bought from third parties or manufactured under licence, merely those which it manufactured itself or acquired directly from Vitapointe. Moreover, the effect of the restriction was to prevent the sale within the Common Market of products manufactured by two different companies

[26] JO [1964] 92/1426: [1964] CMLR 416. In the absence of any territorial restrictions on Mertens & Straet it is difficult to understand why the agreement had been notified at all, apart from *ex abundanti cautela* at a time when the likely attitude of the Commission was still unpredictable.

[27] *Nicholas–Vitapro* JO [1964] 136/2287: [1964] CMLR 505.

but bearing the same trade mark, which could have caused confusion to customers. Finally, the position of Nicholas within the Common Market as a seller of hairdressing products was relatively weak, with numerous other competitive products on the market.

4. The *Grundig* Case and its Consequences

Without question, however, the most important decision to be issued by the Commission in 1964, its first full year of active operation as a decision-making body, was the *Grundig* case. The outcome of this case,[28] following the appeal from the Commission's decision to the European Court of Justice,[29] was to have a major influence on the development of the competition policy of the Commission. The facts of the case, therefore, require to be set out in some detail.

The well-known German manufacturer, Grundig, had appointed a firm called Consten as its sole distributor in France, the Saar, and Corsica in respect of a range of products including wireless and television sets, tape recorders, and dictating machines. Consten accepted an obligation to place minimum orders, to set up a repair workshop with an adequate stock of spare parts, and to provide an effective guarantee and after-sales service. It agreed not to sell goods competitive with the Grundig range and also agreed that it would not make deliveries into other territories, having received an assurance from Grundig that similar restrictions had been placed on Grundig distributors in other countries. Consten was then enabled by Grundig to register under its own name in France the trade mark GINT which was carried on all Grundig appliances including those sold on the German domestic market. If Consten, however, ceased to be the sole distributor for France, this trade mark had to be re-assigned to Grundig.

A French competitor of Consten was UNEF, which found itself able to buy Grundig appliances in Germany, for resale to French retailers, at more favourable prices than those available to them in France from Consten. Consten therefore brought an action in the French Courts against UNEF both for unfair competition and for infringement of the trade mark GINT. As the result of a complaint by UNEF to the Commission, an investigation was carried out and a statement of objec-

[28] *Grundig–Consten* JO [1964] 161/2545: [1964] CMLR 489. It is interesting that on 12 Jan. 1962 the Dutch Supreme Court had already declared an export ban imposed by Grundig on its Dutch distributor to be unenforceable because it violated Article 85 and had struck down an injunction imposed by a lower court on parallel importers of Grundig products. The Dutch Court had refused to refer the case to the European Court under Article 177. For a contemporary assessment of the case, see C. Fulda, 65 *Columbia Law Review* 625 (1965).

[29] *Consten and Grundig* v. *Commission* Cases 56 and 58/64 [1966] ECR 299: CMLR 418.

tions filed against Grundig in accordance with the requirements of Regulation 17. The resulting decision of the Commission, issued on 23 September 1964, was that the agreement between Consten and Grundig violated Article 85(1). The Commission ruled that the rights of Consten both under the distribution agreement and under the linked trade mark licence must be limited so that they did not enable Grundig and Consten to prevent parallel imports of these products into France from other Member States. The Commission, therefore, declared the whole agreement between Consten and Grundig to be void and unenforceable.

An appeal was brought against this decision to the European Court. The Court heard the opinion of the Advocate General appointed to the case before considering its own decision. The opinion of Advocate General Roemer was that the numerous complaints of Consten and Grundig against the procedure of DG IV in investigating the case were justified, as it in his view had not been sufficiently thorough. He believed that it should be reopened to enable the parties to deal more fully with the allegations made. The Court, however, disagreed and upheld the Commission's decisions other than on two important points. The points upon which the Court overruled the Commission concerned the Commission's finding that exclusivity alone in a distribution system was sufficient to raise artificial barriers between Member States of the Common Market. It concluded first that such a finding was too sweeping and could not be justified. Nor in its view should the Commission have ruled that the clauses of the agreement which did not relate to restrictions on the parallel imports of goods be annulled. Notwithstanding the reference in the wording of Article 85 striking down 'all *agreements* between undertakings . . .', and the like wording in paragraph (2) ('any *agreements* . . . prohibited . . . shall be automatically void') the annulment should apply solely to restrictive clauses and the remainder of the agreement remain fully in force. The remaining findings of the Court, however, wholly supported the approach of the Commission and would have a major influence on future case-law and competition policy; they could be summarized as follows.

The Court's first ruling was that the expression 'affecting trade' in Article 85(1) meant that the Commission should consider whether the agreement was capable of endangering, directly or indirectly, in fact or potentially, freedom of trade between Member States in a direction which could harm the attainment of the object of a single integrated Market. Such an effect could be found even if the result of the agreement was to favour a large increase in Grundig trade through its appointed distributor, Consten, if at the same time other undertakings (such as UNEF) were prevented from importing Grundig products into France or if Consten itself was prevented from re-exporting. If interbrand and competition between Grundig dealers were weakened by the agreements, then it was not an answer for Grundig simply to claim that interbrand competition between manufacturers would thereby be increased.

Secondly, the Commission was commended for taking into account in its decision the whole 'system' set up in France for absolute territorial protection including the assignment of the GINT trade mark to Consten. Trade mark rights should not be used in such a way as to defeat the effectiveness of EC law in this area, nor should the domestic law of Member States relating to unfair competition or intellectual property rights be used for such a purpose. Other Articles of the EC Treaty had been cited in favour of Grundig and Consten to support their use of trade marks to protect the exclusivity of the distribution arrangement; in particular, Article 36 had been referred to in the course of the case because of its reference to the protection of 'industrial and commercial property', and Article 222 because of its statement, as a general principle, that the Treaty was not in any way to 'prejudice the rules of Member States governing the system of property ownership'. Neither Article, however, was adjudged sufficiently far-reaching in its scope to affect the Court's ruling.

The Commission as a whole had realized that the issues posed by the Grundig case were of central importance and needed the benefit of a clear ruling of the Court to provide a sound basis for its future policy, placing an emphasis on the elimination of export bans that would have the effect of segregating national markets, particularly if combined with ancillary restrictions on the use of intellectual property rights such as trade marks. This is but the first example of a number of situations in which a clear and authoritative ruling from the Court at a timely moment has enabled the Commission to go on and develop its own policy strengthened by the Court's support on a key issue. We shall see in our subsequent review in Part II of various areas of the development of competition policy how important the role played by the Court has been in assisting the development of the Commission's own confidence in the application of its policies. In particular, the wide interpretation given by the Court to the expression 'which may affect trade between Member States' has enabled the reach of the Commission to be extended more broadly than would have been possible if the Court's interpretation of these words had been more restrictive. It enabled the Commission to exercise jurisdiction in a number of areas where previously it had been thought the degree of effect on trade between Member States would be insufficient, even when the markets affected were mainly national with comparatively minor effect on imports or exports. In later chapters we shall see that subsequent case-law has further developed the original principles in this area of the law laid down in the *Grundig* case.[30]

[30] See also Ch. 15, p. 291 for a discussion of its influence on the distinction between the 'existence' and 'exercise' of rights.

The two points on which the Commission's judgment had been overturned were not damaging to it. The decision that only the restrictive clauses would be struck down and void under 85(2) was not itself limiting to the exercise of the Commission's jurisdiction, although it did appear to run contrary to the apparently categorical wording of Article 85(2). Nor did the refusal of the Court to accept that exclusivity arrangements alone would automatically be in breach of Article 85(1) cause major problems. For DG IV had already started to develop principles for distinguishing those situations where an element of exclusivity in distribution systems could in fact assist the process of interbrand competition and would eventually enable a large number of the bipartite distribution arrangements notified to it at the beginning of 1963 to receive exemption. To arrange for this to happen, however, would require the introduction of new administrative machinery to permit exemption by category rather than simply on an individual basis.

5

The Advent of the Group Exemption

1. Early Cases on Distribution Systems

While the *Grundig* case had been on appeal to the Court, DG IV began to issue a number of decisions which illustrated its developing approach to interpretation of the conditions for exemption set out in Article 85(3). The first two cases were relatively straightforward and dealt with bilateral distribution arrangements. The first, a decision issued on 22 September 1965, related to an exclusive distribution agreement covering Belgium granted by a German company, Hummel, to a Belgian enterprise, Isbecque, for the sale of small tractors, power driven cultivators, and accessories.[1] Negative clearance was not possible because there was a restriction on Hummel from selling to anyone else in Belgium apart from Isbecque, and Belgian undertakings other than Isbecque were unable to buy the products direct from Hummel. The importing of such products from Germany to Belgium was clearly, therefore, affected by the terms of this agreement.

Nevertheless, the Commission decided that the conditions set out in 85(3) were satisfied. The manufacturer found that it had only to deal with a single purchaser for a given territory, Belgium, and this enabled it to deal more easily with legal, linguistic, and other commercial difficulties in that country. It also gave the distributor an incentive to expend time and money on adapting goods to his local conditions and to provide an adequate after-sales service and stock of spare parts. The requirement found in 85(3), therefore, for an improvement in the distribution of goods was satisfied; and the Commission also found on this occasion that consumers had benefited from the agreement because they were more easily able to obtain agricultural machinery adapted to the Belgian market. Isbecque had no opportunity to increase prices as a result of the exclusive distribution arrangements because of the continuing possibility of the entry of goods from Germany by way of parallel imports. Having thus dealt with the two positive requirements of 85(3), little difficulty was encountered by DG IV with the two negative conditions; it held that no more restraints had been imposed on Isbecque than were necessary for the economic benefit of exclusive distribution to be obtained, and in par-

[1] JO 2581/65: [1965] CMLR 242.

ticular the parties had not sought to grant absolute territorial protection making parallel imports impossible. Similarly, interbrand competition remained because of the possibility of parallel imports, and indeed there was substantial competition on the particular market in Belgium from other manufacturers. Exemption was granted in this case for a period of six years. A similar decision, based on identical reasoning, was given in the case of *DRU/Blondel*[2] where the Dutch manufacturer had provided an exclusive distribution arrangement for Blondel, a French company, relating to the sale of enamelled iron oven dishes in France; here an eight-year period of exemption was given.

Even before the Court's judgment in the *Grundig* case (which came in the year following these 1965 decisions) it was already apparent how critical to the whole structure and operation of Article 85 was the protection of the right for undertakings to import goods outside the manufacturer's official distribution system. Neither the likelihood that consumers would obtain a fair share of the benefit of an exclusive distribution system nor the maintenance of competition alongside the official distribution system would be possible without the protection of the legal rights of dealers in one country to sell products outside these official channels freely into other Member States of the Common Market.

Early in January of the following year (1966) the development of the law was taken a stage further in the *Maison Jallatte–Voss*[3] case. Here again, an exclusive distribution arrangement was established, for safety shoes. The manufacturer, Maison Jallatte, was a French company which appointed for Germany as distributor Hans Voss KG and for Belgium a distributor known as SA Éts Vandeputte. Once again, the arrangement clearly fell within the jurisdiction of Article 85 (1) because Jallatte was unable under the arrangements to sell through any other distributors in Germany or Belgium. Moreover, under the terms of the arrangement, SA Éts Vandeputte (though not Hans Voss KG) had consented to an exclusive purchase clause so that they were unable to purchase safety shoes from the competitors of Maison Jallatte. The additional element of exclusive purchase had, of course, not been present in the earlier decisions issued by the Commission.

Nevertheless, 85 (3) was again held to apply because of the improvement which the exclusive arrangement gave to the distribution process in both countries: and in the case of Belgium it was decided that the exclusive purchase obligation on SA Éts Vandeputte forced it to exercise its commercial representation of Maison Jallatte on a more vigorous basis, for its profits had to come solely from the products of that company rather than from a range of competing or complementary products. The benefit to

[2] JO 2194/65: [1965] CMLR 180.

[3] JO 37/66: [1966] CMLR D1.

consumers was retained because the range of safety shoes available was considerably wider than would otherwise have been the case, that is if Maison Jallatte had itself been distributing direct to Belgium or Germany. The continuing possibility of parallel imports was said to make it impossible for prices to be raised unreasonably by either of the distributors. It was also suggested that the increased demand for safety shoes due to the improvement in distribution would soon allow both manufacturer and distributors to reduce their costs, and since the market was competitive, consumers would benefit from these arrangements. Here again an eight-year exemption was granted.

History does not relate the reason for the choice by DG IV of early decisions for these particular cases out of the many thousands of distribution agreements notified under Regulation 17. None the less, it became evident that DG IV would soon have to move on from such relatively simple cases involving bilateral distribution arrangements to more complex multilateral agreements. The only one of these with which DG IV had already dealt was the *Belgian Tile Cartel*[4] case. This had involved arrangements between two Belgian trade associations representing tile layers and tile dealers, as well as 29 manufacturers of ceramic tiles throughout the Community. Under the agreements notified, all these manufacturers were required to sell in Belgium only to dealers and tile layers accorded the status under the arrangements of 'recognized customers'. These dealers and layers in turn had a reciprocal obligation to buy their tiles only from manufacturers who were parties to the relevant agreements. General contractors and their agents were barred from membership so that manufacturers could not sell direct to them, nor to other large potential buyers of tiles, such as Government departments, hospitals, or factories. The admission of any new manufacturer to the agreements required the unanimous consent of all other manufacturers and the admission of any new trade associations of tile layers or dealers required the consent of the two existing associations.

The effect of this cartel was to split the Belgian market for ceramic tiles into two quite separate sections, covering members and non-members respectively. It substantially reduced both export opportunities from other Member States into Belgium and opportunities likewise for purchasers in Belgium to seek imports from the rest of the Common Market. The results of the arrangements were to place restrictions on every participant in the cartel, thus:

(1) Non-member manufacturers in the Common Market could now sell only to non-member customers in Belgium.

[4] JO 1167/64: the case is also referred to as the *Convention Faience* case.

(2) Member manufacturers could only sell to tile layers and dealers who were within the agreement.
(3) Non-recognized tile dealers or layers were cut off from the large number of sources of supply of tiles represented by the manufacturing members.
(4) Consumers, even those who employed a recognized tile layer, could not choose products for installation made by non-member manufacturers.

This type of cartel, as we shall see from later cases decided by the Commission and the Court, was common in the Common Market at the time of the signature of the Treaty of Rome and particularly in Belgium and Holland. The justification claimed for the arrangements, and the basis for requests for exemption under Article 85 (3), was that it brought about improvements in the proper use and distribution of ceramic tiles because of the higher standards that it was supposed to set for tile layers and dealers. The weakness in this argument was that the rules of the cartel actually contained no reference to minimum technical standards nor requirements for technical education, so that a manufacturer was actually free to sell tiles to any tiler or layer without any control over their end use. The cartel could have achieved its proclaimed objectives by only prescribing minimum technical standards for its distributors and placing other requirements on them for the benefit of their customers, such as the continual maintenance of a reasonable stock of tiles.

Clearly, therefore, it was impossible for the Commission to find that this kind of cartel satisfied the four conditions of 85 (3). Not only was there no apparent improvement to the distribution of goods, there was certainly no benefit reserved for consumers, and the restrictions imposed on both manufacturers and recognized customers went far beyond the level required for the improvement of technical standards. The conclusive argument against exemption, however, was that with such arrangements in force, the possibility of eliminating competition in the Belgian tile market was not only the aim but also the almost certain achievement of the cartel.

Unofficial publication of recommendations by the Commission to the parties for amendment of their arrangements became known about during the summer of 1963, but the first official reference to the case comes in the Official Journal of 13 May 1964 which carries a communication inviting third parties to comment on the request for negative clearance or exemption relating to the agreement.[4] Subsequently, no official decision was rendered by the Commission but the recommendations of DG IV were implemented by the association. Under these, whilst the cartel continued its existence and manufacturers were still entitled and required to supply only qualified customers, the qualifications for being a customer were laid down, with minimum objective standards to be evidenced by a certificate from a vocational or trade school. Moreover, such customers (qualified dealers or layers) were not debarred from purchasing from manufacturers who were not members of the cartel, and any person meeting the required technical standards had to be admitted to membership without being subject to any veto from existing members. General contractors were also now entitled to be included on the list of approved customers provided they could show that they had obtained the necessary standard of competence in handling such tiles. The 'closed circuit' market which had previously existed was thus opened up to a substantial extent, in a way that would have beneficial effects on the trading of the exports of tiles from other Member States into Belgium, with an equivalent effect on the freedom of Belgian tile layers and dealers to import them from a variety of sources.

This is an interesting and important early decision which shows that the Commission realized that the established European preference (in certain markets at least) for confining the operation of a national market to a group of approved undertakings (manufacturers and middlemen) which themselves retained a veto on the admission of new members at their respective levels ran completely contrary to the provisions of Article 85. It was a particularly clear case, since the parties to the arrangement included manufacturers not only from Belgium itself but from four other Member States. Later cases indeed would extend the principle of this case even to situations where all the members of the relevant association were merely trading within one Member State. The recommendation that entry to the categories of recognized customers should be open to all qualified entrants, assessed on an objective basis, would itself also have long-term effects on the jurisprudence of the Commission, notably in the area of selective distribution. The case also illustrated the point that horizontal agreements between competitors, especially when combined with restrictions on vertical relationships covering distribution of a product, could equally affect the integration of national markets within the Community as could vertical agreements of the *Grundig* type.[5]

Moreover, the way in which the Belgian tile cartel was disposed of by recommendation rather than formal decision illustrates that quite early in its operations DG IV recognized that the volume of cases to be decided made the availability of informal methods of disposal essential. A decision required a number of formal steps including the preparation of a statement of objections to the parties, probably an oral hearing of the issues, and a presentation of the draft decision to the Advisory Committee, leading up to the Commission's formal decision. All this took considerable

[5] See in particular cases referred to in ch. 21 on Trade Associations.

time. On the other hand, an informal negotiated settlement could be dealt with far more speedily; it was particularly suitable where an agreement contained only certain clauses regarded as contrary to Article 85, and which the parties were often willing to remove if it led to a comparatively quick clearance or exemption for the remainder of the agreement.

The Bulletins of the Community contain numerous examples of these settlements during the early years of the Commission. In 1963 to 1964 settlements were reported in the construction materials industry, where collective contracts of exclusivity between producers and distributors (similar to those found in the *Belgian Tile Cartel* case) were discontinued following negotiation. In May 1966, a report appeared of a complaint about the patent practices of two chemical companies, one French and one German. Both these companies held patents covering a process for use relating to plastic material. Each company had granted the other a free licence including the right to grant sub-licences to the other but these sub-licences could only be issued on the basis that the sub-licensees had to purchase from one of the two companies (and from no other source) the unpatented product required for the application of the process. This was a form of 'tying' arrangement which prevented competitors from offering supplies of the unpatented material, but without any justification being put forward that the unpatented material had to be of so exact a specification to ensure the technical perfection of the process that no other manufacturer was capable of supplying it. After investigation, this 'tying' arrangement was, therefore, ended by the participants.

In 1967, there is a further substantial increase in reported settlements including a case where a manufacturer sought to impose an additional charge on a product used for feeding livestock which the buyer intended to export to another Member State. The basis of the claim (similar to that made in the later *Distillers*[6] case) was that the manufacturer incurred higher advertising costs in the proposed country to which the goods were to be exported and that it wanted all suppliers (both locally based and importers) in that other country to bear these additional costs. The Commission succeeded in having this additional charge withdrawn, even though it bore some relationship to the additional advertising costs incurred in that country; the basis for the Commission's intervention was that the practice would have a tendency to maintain an artificial partitioning of markets between the Member States, since it would discourage retailers in one country from exporting to the other. Thus, the export ban (whether by way of direct or indirect measure) was coming to be recognized clearly as the principal sign of an agreement which (in the light of

[6] *Distillers Co.* v. *Commission* Case 30/78 [1980] ECR 2229: 3 CMLR 121.

the Court's *Grundig*[7] decision) the Commission could not accept. The Commission had become fully aware that it is the removal of export bans which enables the parallel movement of goods through unofficial channels of distribution, whose availability keeps downward pressure on price levels that otherwise, under exclusive distribution arrangements, might tend to rise.

Neither the mere handful of official decisions in these early years, however, nor even the large numbers of settlements could make substantial inroads into the large number of notified agreements. This problem had weighed heavily, not only in the filing cabinets but on the minds, of the officials in DG IV from the beginning of 1964 onwards when for some months all other tasks had had to be laid aside whilst an attempt was made to list and assess on a preliminary basis the great numbers of notified agreements. A very substantial proportion of these notifications were exclusive distribution agreements, and the next largest category were patent licences.

Quite clearly, the way forward would have to be the laying down by the Commission of broad criteria under which group exemption[8] could be given to these forms of agreement (particularly in the area of exclusive distribution where the largest number of notifications had occurred) which were regarded as having many pro-competitive features, and restrictions which on the whole fell within the letter and spirit of 85 (3). To apply to the Council for a group exemption, however, would be fruitless without being at least able to show some experience of the particular form of agreement for which the exemption was sought. The experience of the Commission through 1963 and 1964 on the nature and content of exclusive distribution was not extensive, but the experience gained from the *Grundig* case and the other minor cases already referred to had at least begun to show the nature of the distinctions that could be drawn. On the one hand there were relatively straightforward distribution agreements containing mainly the basic clauses essential to any system of exclusive distribution and which would not damage the competitive process so long as parallel imports remained possible; on the other hand there were a variety of more complex and restrictive agreements governing distribution where individual scrutiny of the particular circumstances of the product and the parties' market share would remain necessary before exemption could be allowed. The great advantage of the group exemption, moreover, would be that undertakings would have an incentive to frame their new agreements in terms that complied with it. They would thus eliminate the need for notification at all and thereby reduce further the administrative burden on DG IV.

[7] Cases 56 and 58/64 [1966] ECR 299: CMLR 418.

[8] Specifically referred to in Article 85(3) as 'exemption by category'.

2. The Enactment of Council Regulation No. 19/65

Work, therefore, had begun as early as the end of 1963 on the preparation of the draft of a regulation that the Council could adopt under the terms of Article 87, giving the Commission power to grant exemptions on a group basis by reference to objective criteria. Under normal consultative arrangements within the Community, once a draft had been approved within DG IV and by the Commission itself, it had to be circulated by the Council to the Economic and Social Committee and also to the European Parliament. Once their comments had been considered, and any consequential amendments suggested by the Commission, the Council could then enact the final version.

Picking up the reference in Article 85 (3) to 'any agreement or *category of* agreements between undertakings' the draft Regulation prepared and circulated for approval referred to two particular types of agreement to which DG IV wished to be able to declare that Article 85 (1) should be inapplicable. These two categories were:

(*a*) where one party agreed with the other to supply only to that other goods for resale within a defined area of the Common Market; or where one party agreed with the other to purchase only from that other goods for resale; or where two undertakings had entered into reciprocal obligations with each other for exclusive supply and purchase of goods for resale; or

(*b*) where restrictions were imposed on the acquisition or use of industrial property rights (patents, utility models, designs, or trade marks) or on rights arising out of contracts for assignment of or the right to use methods of manufacture or knowledge relating to the use or the application of industrial processes.

The draft made clear that the group exemption to be given by the Commission would only apply to bilateral agreements; each group exemption covering a particular category was required to specify both the individual clauses not to be contained in such agreements, and also those minimum clauses which had to be contained in the exempted agreements.

With regard to (*a*), clearly the intention was to cover cases such as those of *Hummel/Isbecque* and *Maison Jallatte–Voss* of which the Commission had already gained some experience. On the other hand, to issue group exemption for defined categories of patent licences (obviously the main type of agreement covered by (*b*)), would be more contentious, and considerably more experience would be required before the terms of a group exemption could be decided upon.[9]

[9] The first draft of a group exemption for patent licences was not in fact published until 1979. See Ch. 15 for a detailed consideration of this topic.

The Council recognized the difficulties of DG IV with regard to the 'mass problem' and relatively quickly authorized the enactment of Regulation 19/65 giving the required authority to the Commission to grant group exemptions for both categories of agreement. The recitals of this Regulation are important. After formal recitals referring to the authority of the Treaty, the proposal made by the Commission and the necessary opinions received from the Parliament and Economic and Social Committee, reference is then made to the large number of notifications received under Regulation 17. It is considered desirable 'in order to facilitate the task of the Commission' that the Commission should be able to declare by way of Regulation that the provisions of Article 85 (1) do not apply to certain categories of agreement. Reference is then made to the importance that the Council should lay down the conditions under which the Commission can grant group exemption 'in close and constant liaison with the competent authorities of the Member States' and also that such group exemptions are exercised only 'after sufficient experience has been gained in the light of individual decisions and it becomes possible to define categories of agreements . . . in respect of which the conditions of Article 85(3) may be considered as being fulfilled'.

This particular point is often overlooked by subsequent commentators who have criticized the Commission for not obtaining authority from the Council at an earlier stage for a wide range of group exemptions covering not only vertical but horizontal agreements. This ignores the fact that it is neither possible, nor indeed sensible, for a body such as DG IV to issue broad criteria for exemption until they have gained sufficient familiarity with the types of agreement which are being entered into and the types of restriction which they contain. Only by dealing with a reasonable number of individual requests for negative clearance or exemption could the Commission acquire the necessary experience to publish and implement an effective group exemption, based on criteria that would both make economic and commercial sense and command a broad measure of acceptance and understanding from European commerce and industry.

The Regulation goes on to authorize the Commission to make its own Regulation setting out group exemptions in categories (*a*) and (*b*) only, for a specified period of time. Before it could be issued it would have to be considered by the Advisory Committee and published in the Official Journal so that any interested party could submit comments. The Commission was also given the power by Article 7 to withdraw the benefit of the group exemption subsequently from any agreement to which it had been applied if it found that the agreement so exempted had 'nevertheless certain effects . . . incompatible with the conditions laid down in Article 85(3) of the Treaty'.

The enactment of this Regulation was of course essential for the

Commission because it provided the necessary powers for the issue of group exemptions which could deal at a stroke with a large number of those bilateral agreements on its books. It was not, however, to come into existence without challenge. Not long after its enactment on 2 March 1965 Italy brought a case to the European Court of Justice claiming that the Regulation should be annulled.[10] Italy's challenge was based on the ground that the passing of the Regulation by the Council was itself a *détournement de pouvoir* (abuse of power) in that it purported to extend the scope of Article 85 of the Treaty. Amongst other arguments raised by Italy were that the Regulation raised a presumption that those agreements not expressly referred to in it as eligible for group exemption were illegal, and also that Article 85 in any case was intended to govern only horizontal relationships between actual competitors, rather than vertical relationships concerning distribution or licensing to which Article 86 alone would apply.

Neither the Advocate General, Roemer, nor the Court, however, had great difficulty in rejecting these arguments. The Advocate General began by pointing out that the case was primarily concerned with a solution of a question of principle, not the mere application to any particular case of the Community law of competition. The passing of Regulation 19/65 had been the direct result of the large number of notifications received under Regulation 17, and the consequential demand from both DG IV and European commercial and industrial interests for a speedier treatment of the requests for exemption. The Advocate General continued

> For anyone who knows the difficulties of the law relating to agreements, it soon becomes clear that this is an area which does not lend itself at all to using legislative means to find a complete and comprehensive solution for all the problems which may occur. Therefore, the institutions empowered to deal with these matters are acting rightly in proceeding by stages. In proceeding thus, they are also acting rightly in directing their attention first to cases, such as exclusive dealing agreements, where relatively harmless restrictions on competition are involved and which, because of their number, call for a set of rules whereby they can be dealt with speedily in the interests of simplifying administration. Although it is not possible to avoid some initial uncertainty as regards various agreements not covered by Regulation 19/65 until decisions about them are taken individually, it must be admitted that difficulties of this sort are inevitable in the very complex matter of the law relating to agreements, particularly at the beginning of its evolution.[11]

He pointed out also that the Treaty was based on a wide concept of competition covering not only horizontal but vertical agreements, of particular economic importance in that the costs of distribution represented an

[10] *Italy* v. *Council and Commission* Case 32/65 [1966] ECR 389: [1969] CMLR 39.
[11] [1966] ECR 416.

important element in the total costs of consumer goods. He found no distinction in Article 85 between horizontal and vertical agreements, nor any limitation of Article 86 to vertical agreements, and indeed pointed out that the Court in the *de Geus/Robert Bosch*[12] case had already decided that Article 85 did apply to vertical agreements. He concluded by pointing out the duty of the Commission to make a conscientious examination of all relevant economic factors before granting an exemption under Article 85(3) whether individual or group.[13]

The Court in its judgment emphasized that in adopting Regulation 19/65 the Council had not enlarged the field of application of Article 85. It had merely given the Commission procedural powers to grant exemption to specific categories of agreement and concerted practices, a step clearly contemplated by the drafting of Article 85 with its reference to 'categories'. The definition of a category, however, provided only a framework for the Commission and DG IV and did not mean that all agreements of that type which failed to comply with the specific obligations laid down for group exemption would be automatically prohibited or be ineligible for individual clearance. It would be wrong, moreover, said the Court to read the Regulation as creating any presumption of law as to the way in which Article 85(1) was to be interpreted or as limiting the freedom of the Commission to spell out those conditions which an agreement had to satisfy in order to gain a group exemption. Italy's application, therefore, for annulment of the Regulation was rejected.

This case shows the Court fully aware of the administrative difficulties under which DG IV had been placed as the result of what had perhaps been the over-ambitious scope of Regulation 17 and of the immense task which DG IV would have in continuing to deal with, and prescribe appropriate criteria for, the large number of notified agreements. For the Court to have rejected the validity of the Council's Regulation would have been to place an impossible weight on the Commission's shoulders, by forcing it to proceed solely by way of individual decision. It happened that the decision of the Court was issued on the very same day as its *Grundig* decision, and the combined effect of the two cases was of importance not only in terms of the substantive law but also in raising the morale of the officials in DG IV. Moreover, only some two weeks before,[14] the Court had delivered a preliminary ruling under Article 177 in the case of *Société Technique Minière* v. *Maschinenbau Ulm*[15] on appeal from the Court of Appeal in

[12] Case 13/61 [1962] ECR 45: CMLR 1.

[13] It is interesting that in support of his opinion, the Advocate General drew attention to the recent *White Motors* case in the US Supreme Court where the Supreme Court had laid down precisely the same rule to be applied by Federal district courts in the USA when examining vertical agreements: 372 US 253 (1963).

[14] On 30 June 1966.

[15] Case 56/65 [1966] ECR 234: CMLR 357.

Paris which would also prove of importance to it. Maschinen-bau Ulm had supplied some large earth levellers to a French company, STM, engaged in the distribution of equipment for public works. STM found itself in difficulties in finding purchasers for all the levellers, of which it had agreed to buy a substantial number over a two-year period. It had also agreed to organize a repair service and spare parts stock and to meet the whole of the demand for the product in France, promising not to sell any competing products without Maschinenbau Ulm's agreement. In return, Maschinenbau Ulm had granted STM exclusive rights to sell these machines exclusively both in France and certain overseas territories. When Maschinenbau did not receive payment, it sued STM in France, and STM defended the claim on the basis that at least some clauses in the agreement were void under Article 85. The Court of Appeal asked the European Court of Justice to decide which, if any, of such clauses were so affected.

Both the Advocate General and the Court took the view that agreements of this kind could normally expect to receive exemption under 85(3). The Advocate General pointed out that it would be impossible in many cases for undertakings of a modest size to take the risk of entering foreign markets if they could not concentrate their distributive efforts in one set of hands, especially if the product were technical, requiring assembly and repair services and supplies of spare parts available at short notice to keep the machines working. The offer of exclusivity was in practice, therefore, essential to the setting up of a satisfactory distribution system and would not normally threaten the competitive process because of the same factors that we have seen in the earlier Commission cases such as *Hummel/Isbecque* and *Maison Jallatte–Voss*, namely that it would be difficult for the manufacturers to ensure that only products circulated and distributed through official channels could reach the relevant market.

The Court's view was that one had to look at the severity of the restrictive element in the clauses granting the exclusive right, and whether these clauses could have the effect of partitioning the market between the Member States. The Court, however, went on to make an important statement as to the interpretation of the words 'agreements . . . which may affect trade between Member States' in Article 85(1). The test applied by the Court was that 'it must be possible to foresee with a sufficient degree of probability on the basis of a set of objective factors of law or of fact that the agreement in question may have an influence, direct or indirect, actual or potential, on the pattern of trade between Member States'.[16] On

[16] [1966] ECR 249: CMLR 375. As the case was a reference to the Court under Article 177, the Court was not, of course, required to decide the issues between the parties but merely to provide a reply to the questions raised by the Paris Court. For further consideration of the words 'may affect trade between Member States' see Ch. 7, pp. 106–16.

this basis, the potential jurisdiction of DG IV was clearly not to be limited to agreements between undertakings in different Member States but could apply equally to agreements operating simply within one Member State but having wider effects, particularly on imports into or exports from that State.

3. The First Ten Years of Directorate-General IV: A Review

The combined effect of these early Court decisions and of its own experience in handling individual cases had now brought DG IV to a point in 1967, where some ten years after the signing of the Treaty of Rome it could be seen to be applying general principles reached as a result of its own experience and to start to operate on a legislative basis as opposed to simply applying the Treaty's general rules to the merits of individual notifications. Clearly progress had not been as speedy as might have been hoped, and the achievements of these first ten years were limited, if assessed solely by the number of cases decided both formally and informally. Nevertheless, at the end of its first decade there could be no doubt that progress had been made in the establishment of relevant principles which would provide sound foundations for DG IV and the Commission as a whole in the years to come. Apart from a number of informal settlements and recommendations, it had reached formal decisions during this period on four negative clearances and four 85(3) exemptions. It had issued Notices relating to commercial agency and patent licences and carried out a sectoral investigation under Regulation 17, Article 12 into the margarine industry. In the *Grundig* case it had obtained not only outright support for the disallowing of export bans but also a clarification of the expression 'which have as their object or effect . . .'. The Court had accepted the Commission's argument that it was sufficient to establish that the *purpose* of the restrictive arrangement or agreement was to separate national markets for distribution purposes, and that it was only if no such purpose could be found that it is necessary to consider separately the *effects* of the restrictive clauses in the agreement. It was clear too that the Court would not prevent Community law from having a restrictive influence on the free exercise of industrial property rights under national law, and that rights flowing from such national laws could not be used for purposes contrary to the basic principles of competition law laid down in Articles 85 and 86. The competition which the Court was protecting in dealing with vertical agreements was generally not the competition between the parties to that agreement (whose functions would be normally complementary to each other) but the competition between those parties and other parties at both the manufacturing and distribution level who desire to be in competition for that particular product. Article 85(1) was, therefore, to be applicable both to horizontal and vertical agreements.

As to the question of the meaning of the phrase 'which may affect trade between Member States', the Court as we have seen in the *STM* case had applied a very broad test which had the effect of maximizing the jurisdiction of the Commission. It is clear from *Grundig* moreover that in making this assessment the Court would not operate a 'balancing test'; even if the volume of trade between particular countries had been increased as a result of a particular agreement, the restrictive clauses in that same agreement could nevertheless be held to have affected trade in a way that would be in breach of Article 85. What had still to be worked out was the extent to which exclusive distribution agreements would be permitted when they no longer contained an absolute ban on exports but where other restrictions were included which might more indirectly make parallel imports, and, therefore, intrabrand competition, more difficult. This would indeed be one of the problems that would face the Commission when considering the exact phrasing required for the group regulation, shortly to be introduced, on exclusive distribution.

Of course, many other problems in the interpretation and application of Article 85 still remained to be solved. In two major respects, however, the first decade could be said to have achieved important steps forward. The first of these related to the expectation of industrialists and businessmen operating within the original Common Market. In the early years after the Treaty, there had been some expectation, and indeed hope, amongst European businessmen unfamiliar with a totally (or even substantially) competitive environment that the administration of Articles 85 and 86 would be relatively lenient, and that the interpretation given to the application of the exemptions and the interpretation of the phrase 'agreements . . . which may affect trade between Member States and which have as their object or effect the prevention, restriction or distortion of competition within the Common Market . . .' would catch only clear and obvious anticompetitive cartels involving price-fixing or market-sharing between horizontal competitors. This view had been encouraged by the relative weakness of the national competition laws in the Six Member States at that time, with the possible exception of the new German competition law which was only itself gradually gathering effectiveness over its first decade of operation in parallel with the growth of the Commission's own jurisdiction. The European tradition had been one which respected close links of loyalty between suppliers and their customers, and such relationships between manufacturers and their dealer or distributor systems were generally thought of greater importance than the free play of competitive influences. Maintenance of associations of manufacturers or dealers with a firm commitment to dealing on a regular basis with each other and excluding undertakings outside the scope of the agreement was considered respectable and normal, rather than as an artificial and objectionable division of national markets into 'insider' and 'outsider' groups. Agreements not to compete or not to export particular categories of product for fear of reprisals by other competitors in neighbouring States had also been considered normal.

Against this traditional fear of an outright rivalry between undertakings, expectation at the outset had been that though change would be necessary, it would be gradual, even gentle.[17] Although it took some years before the effect of the change in atmosphere was perceived, it had become clear by the middle of the 1960s that Articles 85 and 86, and the Regulations to be made under them, were going to introduce a totally new situation. Undoubtedly a major influence was that the Commission realized that a system of centralized planning of industrial and commercial objectives (akin to the centralized planning of agriculture carried out by the Commission to implement the Common Agricultural Policy) was impossible, firstly because of the size of the bureaucracy needed to enforce such a scheme throughout the Community, and secondly because the ground rules for such common policies could never have been agreed among the Member States in view of their widely differing views on industrial structures and their differing interpretations of national interest.

The second major development over this period was the establishment by DG IV of working relationships with the other institutions of the Community. DG IV, of course, was one of the smaller Directorates in the Commission with fewer resources than many of the other Directorates. At first sight, having scarcely 100 'A' grade officials even by this time, it would appear to have been totally inadequate to provide effective administration of Articles 85 and 86, even for the original Community of only six Member States. It had, however, received major help from other institutions. First, in the framing of Regulation 17/62 and also in assisting the enactment of Regulation 19/65, both the Economic and Social Committee and the European Parliament (especially through the work and reports of its relevant specialist committee) had made substantial contributions to the content and applicability of the Regulations to an extent that might not necessarily have been anticipated. Moreover, the European Parliament, even in its early stages when it had very limited powers, and well before the principle of direct elections had been implemented, was able to continue to apply pressure on the Commission as a whole by means of

[17] For an account of the European attitude to competition in the early 1960s, see E. Minoli, 'Industry's View of Trade Regulations in the EEC', *Proceedings of Fordham Corporate Law Institute* (1962).

questions tabled for answer by the relevant Commissioner relating to development of competition policy. The questions recorded illustrate the continuing concern of members of the Parliament that DG IV should continue its forward movement in the development of its policies.'[18] The value of the continuing individual questioning of the Commission as a means of ascertaining information as to its progress and as a means of goading it on to further efforts should not be underestimated. In later years the Parliament has continued to keep a keen interest in the development of competition law and does so on a methodical basis, especially in view of its ability to comment each year on the Annual Report of DG IV. This Report, which came into existence only as the result of persistent requests from Parliament itself, was first published in 1972 and now regularly appears each autumn with a review of the previous year's activities and developments in competition policy.

Relationships too between the Commission and the Council had matured substantially over the period. Whilst the Council is, of course, essentially a political body consisting of Ministers for the individual Member States, it had concluded during the early 1960s that substantial powers under Articles 85 and 86 would need to be delegated to the Commission if progress was to be made in enforcing the Articles; it would also be inappropriate for the Council, given its other responsibilities and the technical nature of most of the work to be carried out by DG IV in implementing Articles 85 and 86, to involve itself in individual cases. Few other Directorates received such wide delegated powers at that stage in the life of the Community as did DG IV with Regulation 17/62. Their existence has, however, enabled DG IV to continue its progress in enforcing policy both through its administrative (both investigatory and quasi-judicial) and legislative functions, even at times when political deadlock made it difficult for the Council to reach decisions on many other matters, to the general frustration of those working in other areas of the Commission. To this extent DG IV has been fortunate in that at a critical stage in its development it did receive these far-reaching delegated powers, not only to investigate but also to make decisions which would be final, subject only to appeal to the Court at Luxembourg on the grounds set out in the Treaty.

It is, however, the developing relationship of DG IV with the European Court of Justice that is the most important single factor in this first ten years of its life. There is no doubt that the support given by the Court to the work of the Commission was of critical importance in enabling its main lines of policy to be established. The outcome of the *Grundig* case was to lay a firm legal foundation supporting the Directorate in its aim to

[18] The questions, and the answers provided by DG IV, are recorded in the Official Journal.

establish market integration as the foremost principle to be considered in applying Article 85(3), to the extent that an export ban found in an agreement imposed on a distributor would alone certainly make it impossible for exemption to be obtained, unless the export ban were simply to countries outside the Common Market itself without any possibility of subsequent re-export of goods into the Market because of the existence of substantial tariff barriers or transport costs. So familiar has this concept become with the passage of time that it is perhaps too easy to overlook the consequences, had the European Court not so firmly supported the Commission on this principal issue. The Court's assistance was rendered, of course, through decisions under Article 177 on requests for preliminary rulings from national courts as well as in cases on direct appeal from the Commission's own decisions under Article 173. In the next five years (1967–72), a period when political progress would be slow largely owing to the influence of General De Gaulle, the role of the Court would continue to grow in importance in the establishment of further substantive legal doctrine in this area.

The end of this first decade was marked by the passage of the first Regulation by the Commission itself, namely Regulation 67/67 setting out the conditions for group exemption of exclusive distribution agreements, to be considered in detail in Chapter 12. At this point the Commission should have given adequate resources to consideration of the smaller number of multilateral agreements now notified, many of which involved horizontal agreements having at least an element of market-sharing on a national basis, and, therefore, representing a greater danger to the competitive process than the mass of bilateral agreements, largely relating to exclusive distribution and patent licensing, whose early notification by 1 February 1963 imposed so heavy a burden on the workings of DG IV for many years afterwards. It is arguable, though by no means certain, that in the long run this could have been done without doing substantial damage to the principle of market integration which is usually put forward as the essential reason for the early concentration on the bilateral vertical agreement. In the event, it was not until some years later that the Commission would get to grips with some of the major horizontal cartels that had operated with impunity for some years within the Community. The development of the substantive law to be considered in Part II was, however, materially affected by this early policy decision.

DG IV had, however, developed its policies towards vertical agreements far more quickly than towards horizontal agreements and cartels. It is a possible criticism of the policy of the Commission during its first ten years that the terms of Regulation 17 inevitably forced this initial concentration on vertical rather than horizontal agreements because of the extent to which the mass notification of the vertical agreements in accordance with the Regulation's requirements limited DG IV's freedom of action.[19] With hindsight, it might have been better for the Commission to have phased in the notification of agreements over a longer period of time.

[19] That this is so is confirmed by the subsequent figures produced by the Commission itself a decade later. The 9th Annual Report (pp. 15–16) confirms that 40,000 notifications were received by the Commission during the first years of application of Articles 85 and 86 of which as many as 29,500 concerned exclusive dealing agreements, and of which more than 25,000 were eventually disposed of on the basis of Reg. 67/67.

PART II

The Substantive Law of the European Community

Each man's experience starts again from the beginning. Only institutions grow wiser: they accumulate collective experience, and owing to this experience and this wisdom, men subject to the same rules will not see their own nature changing, but their behaviour gradually transformed.

Jean Monnet, quoting the Swiss philosopher Henri-Frédéric Amiel (1821–81), in the course of a speech to the Common Assembly of the European Coal and Steel Community in 1955.

6

Article 85 (1): Analysis (1)

In Part I, a historical account has been given of the early implementation by DG IV of the policies relating to competition contained in the Treaty of Rome. In retrospect the first ten years can be seen as a slow moving, but essential, gestation period for the development of a distinct Community competition policy, but the tempo of development begins to increase sharply with the implementation of the group exemption contained in Regulation 67/67 at the end of this first stage. Because of the sheer volume and complexity of substantive law developments since that time, it is not possible to give an account of the substantive law or the development of policy simply by reviewing cases as they occur in successive years. It is instead necessary to consider the law under a number of different subject headings and to review under each of them relevant major cases and policy developments.

To all these subjects, however, the structure of Article 85 itself is of central importance. It is, therefore, logical to begin our examination of the substantive law by analysing its content. Article 85 has not, however, been interpreted by the European Court of Justice, nor indeed by the Commission, in the way that an English Court or administrative agency would interpret an English statute. It has been treated in a far more teleological way, that is in a manner which gives great weight to the particular part that it is considered to play within the general framework of the Treaty, thus allowing the content of other Articles to influence its interpretation within the general scheme of the Treaty.[1] Correspondingly, whilst keeping this always in mind, less importance is placed on exact analysis of individual words and phrases. It is probably helpful, nevertheless, to analyse the Article under five separate headings, namely:

(1) its general scope and width, with special reference to those sectors partially or totally excepted from its jurisdiction;
(2) the meaning of 'undertakings' including the concept of 'group economic unit';

[1] Out of many possible cases concerning the interpretation of Article 85, the case of *Leclerc and others* v. *Au Ble Vert SARL* Case 229/83 [1985] ECR1: 2 CMLR 286 provides an example, though of course the best known example of the teleological approach comes from an Article 86 case. *Europemballage Corporation Can Co. Inc.* v. *Commission* Case 6/72 [1973] ECR 215: CMLR 199.

(3) the meaning of 'all agreements, decisions by associations of undertakings and concerted practices';
(4) the meaning of 'which may affect trade between Member States';
(5) the meaning of 'which have as their object or effect the prevention, restriction or distortion of competition within the common market'.

The first three items are dealt with in this chapter, the last two are considered in the following chapter.

1. The Width of its Coverage and Exceptions

The width of the coverage of Article 85 means that it is easier to understand its application by listing those sectors of the economy of the Common Market which have been specifically excluded from its scope, rather than in seeking to enumerate everything which it covers. In some cases, exclusion from its coverage has meant that there is partial or complete exemption for a sector from competition rules, but in other cases the exclusion from coverage by Article 85 has been with a view to providing an alternative set of competition rules. Some of the exclusions result directly from other specific provisions in the Treaty, whilst others arise from the passage of Regulations made under the authority of the Treaty.

In the first category of direct exclusion, we should refer first to Article 232(1) which states that the provisions of the Treaty (including of course Articles 85 to 94 inclusive) shall not affect the ECSC Treaty and 'in particular as regards . . . the rules laid down by that Treaty for the functioning of the common market in coal and steel'. Articles 65 and 66 clearly form part of these rules and are, therefore, applied in parallel to Articles 85 and 86 (but quite separately from them), also by a division of DG IV. The position of the EURATOM Treaty, however, is quite different since the terms of Article 232(2) merely state that the Treaty of Rome provisions shall not derogate from those contained in the EURATOM Treaty. This leaves the additional provisions of the Treaty of Rome, namely Articles 85 to 94 which have no counterpart in the EURATOM Treaty to be applied also in that sector.[2]

The competition rules may also be excluded by Article 223(1) (*b*) from applying to any product included in the list of *defence* items prepared by the Council of Ministers. This Article gives Member States the right to take necessary measures to protect their essential security interests connected with the production of or trade in arms, munitions, and war

[2] In practice the competition rules are applied to this sector with a comparatively light hand. See, e.g. *United Reprocessors* [1976] 2 CMLR D1: *KEWA* [1976] 2 CMLR D15; and *Amersham International and Buchler* [1983] 1 CMLR 619.

materials; the only qualification imposed is that 'such measures shall not adversely affect the conditions of competition in the Common Market regarding products which were not intended for specifically military purposes'. This distinction is presumably drawn to ensure that while undertakings may be allowed to enter into agreements or practices otherwise within the scope of Article 85, but for the fact that they relate, for example, to the supply of rifles or tanks, similar arrangements made to cover, for example, the supply of boots or thick blankets (potentially suitable for army or civilian use) are not so excluded. The tendency of the Commission in recent years has been to narrow the scope of this exemption. The sensitivity of this subject is, however, underlined by the fact that it is the Council rather than the Commission that is responsible for the drawing up of the original list of excluded items, or making changes to it, following proposals from the Commission under paragraph (3).

The position of *agricultural* products is covered by Article 42 of the Treaty. This declares that the competition rules shall only apply to agriculture to the extent that the Council determines, within the framework of and subject to the requirements of the Common Agricultural Policy, which is set out in outline in Article 43. The detailed objectives of Article 39 are moreover to be taken into account in framing these Regulations. The objectives set out in Article 39 for the Common Agricultural Policy include:

(*a*) the increasing of agricultural productivity by the promoting of technical progress, and by ensuring the rational development of agricultural production and the optimum utilization of the factors of production, in particular labour;
(*b*) ensuring a fair standard of living for the agricultural community;
(*c*) stabilization of markets;
(*d*) assurance of the availability of supplies;
(*e*) ensuring that supplies reach consumers at reasonable prices.

In working out these policies account is to be taken of the particular nature of agricultural activity, including its social functions and the disparities between different regions, the need to effect the appropriate adjustments gradually, and its close links with the rest of the economy of Member States. Some of these objectives, especially (*c*) and (*d*), would appear inevitably to require substantial intervention in free markets, and indeed the terms of the Common Agricultural Policy as at present in force illustrate this. Accordingly, Regulation 26/1962 was adopted by the Council, which in substance permits the competition rules to apply (Article 1) subject to two substantial exceptions which owe their existence and content to the requirements of Article 39 and the need to give support (in both senses) to national marketing organizations responsible for carrying out the Common Agricultural Policy. Article 2 of the Regulation defines these exceptions as including:

> Agreements, decisions and practices of farmers, farmers' associations, or associations of such associations belonging to a single Member State which concern the production or sale of agricultural products or the use of joint facilities for the storage, treatment or processing of agricultural products, and under which there is no obligation to charge identical prices, unless the Commission finds that competition is thereby excluded or that the objectives of Article 39 of the Treaty are jeopardised.

In this case, it is for the Commission, which has the relevant power, to prepare a list of exemptions from the rules of competition after consultation with Member States. In practice, the competition rules have been applied to several forms of undertaking engaged in the distribution of agricultural products. An example of the attitude taken by the Commission and Court towards the interpretation of Regulation 26 is the *FRUBO*[3] case. This involved an agreement between a trade association of Dutch fruit importers and another association of Dutch fruit wholesalers. Under this agreement, wholesalers were not allowed to seek alternative sources of supply into the Common Market of fruit and vegetables originating from outside the Community, other than through the Rotterdam fruit auctions. As a result, of all the apples, pears, and citrus fruit imported into Holland normally at least 75 per cent came in through the Rotterdam auctions, making it extremely difficult for Dutch importers to seek long-term import agreements with other parties within the Common Market. A rather half-hearted argument was put forward by the associations that, under Dutch law, the practices of the associations did not constitute contractual relationships between their members, comprising only a 'gentlemen's agreement'; even if this argument had been successful, this kind of agreement is in any case caught by the provisions of Article 85. The contractual nature of these rules, however, was emphasized that the fact that they contained penalties for their breach with the possibility of exclusion for defaulters from participation in the Rotterdam fruit auction, a serious detriment for the importers.

Both the Commission and the European Court of Justice rejected the argument that these arrangements were an integral part of a national market organization or necessary for the attainment of the objectives of Article 39. The Court concluded that it was possible the agreements might improve the distribution of fruit in Holland, and even that consumers might receive the benefit of some price reduction as a result, though it

[3] *FRUBO* v. *Commission* Case 71/74 [1975] ECR 563: 2 CMLR 123, considered also in Ch. 21 at p. 422. The first Commission decision (*Breton Cauliflowers* [1978] 1 CMLR D66) dealing with agricultural associations involved collective reciprocal exclusive dealing and other restrictive practices between cauliflower growers and dealers in Brittany.

remained somewhat sceptical on this issue. Nevertheless, the exemption under Article 85(3) was unavailable, for the restrictions placed on the parties were not indispensable to the agreements' claimed purpose, the improvement of the distribution of fruit throughout Holland. The restrictions also, of course, contravened the 'integration' principle laid down in *Grundig*, under which the wholesalers should have been left free to search for alternative sources of supply outside their own country rather than being kept effectively 'segmented' from the rest of the Common Market. The effect of the restriction was substantial, because in practice it meant that a Dutch wholesaler could not act as an importer and that importers established in the other Member States could not, without joining the association and participating in the fruit auction, deliver fruit to Dutch wholesalers. It is clear that at least in the area of the marketing of agricultural products, Article 85 has considerable application, and has also been utilized, for example, in the sugar, vegetable, and dairy products sectors.[4]

We turn next to the *transport* sector; transport is, of course, an element vital to the economy of the Common Market and, as a measure of its importance, has been given special provisions within the Treaty, Articles 74 to 84 inclusive. Article 84(1) states that the provisions of this Title of the Treaty apply to *road, rail, and inland waterway transport*, and paragraph (2) states that the Council may decide whether appropriate provisions may be laid down for *sea and air transport* (by a qualified majority since the Single European Act).

Until the Council had implemented Article 74 by laying down common transport policy, it was clearly advisable to postpone the application of the competition rules of Article 85 to the whole of the sector; a temporary Regulation was, therefore, introduced in 1962 (no. 141) to exempt the forms of transport referred to in Article 84(1) from Article 85 and to place these three forms of transport on the same basis for the time being as air and sea transport. Rules for transport by road, rail, and inland waterway were put forward six years later in Regulation 1017/68. The recitals of this Regulation are so lengthy as to comprise virtually a complete introduction to the transport policy of the Community. Effectively, however, by its Article 2, Articles 85 and 86 were stated to apply to transport undertakings, but with the following qualifications:

(*a*) Article 5 provided for a declaration of non-applicability (equivalent to that contained in Article 85(3)) for agreements which furthered technical and economic progress or improved the quality of transport services, or promoted greater continuity and stability in satisfaction of

[4] See n. 22 for references to the *Sugar Cartel* case. The marketing of Dutch milk was the subject of the Commission's *Meldoc* decision [1989] 4 CMLR 853.

transport needs in those markets, where supply and demand fluctuated to a major extent.

(*b*) Article 3 provided an exception for agreements for standardization of equipment or other technical matters including the co-ordination of timetables, routes, and organization of combined transport operations.

(*c*) Article 4 provided special exemptions for small and medium-sized undertakings subject to some maximum carrying capacities (10,000 metric tons for road transport, 500,000 metric tons for inland waterways). The exemption would not apply, however, in any case if effects were found by the Commission to be incompatible with the requirements of Article 5 and constituting an abuse of the exemption provided.

(*d*) The Commission might also provide specific exemption if there were disturbances amounting to 'a state of crisis' in a transport market; and even here no special exemption could be given unless the restriction imposed was indispensable to reduce these disturbances and did not make it possible for competition to be eliminated for a substantial part of the market concerned.

As a result of this Regulation, Articles 85 and 86 apply only marginally less in the field of road, rail, and inland waterways than to industry in general. The Court of Justice has for example applied the Regulation to prevent the application in France of a levy on barge operators which effectively discriminated against non-French operators.[5]

The position of air and sea transport was very different, as Article 84(2) recognizes. Here, there were a number of international treaties and conventions affecting the basis upon which such services were to be carried out, going back in the case of air transport to the Chicago Convention entered into as long ago as 1944, which had led to the creation of the International Air Transport Authority (IATA) to play a leading role in the worldwide co-ordination and fixing both of services and fares. The need felt to support national airlines had naturally led Member States themselves to become substantially involved in fixing details relating to the provision of air services. The provision of sea freight services also involved many nations outside, as well as within, the EC. Politically, both sectors are highly sensitive. The Commission's pessimism about its ability to produce a Regulation covering these sectors that could command acceptance by the Council is reflected in the recitals to Regulation 141 of 1962 already referred to.[6]

[5] *ANTIB* v. *Commission* Case 272/85 [1987] ECR 2201: [1988] 4 CMLR 677.

[6] 'Whereas on the other hand as regards sea and air transport it is impossible to foresee whether and at what date the Council will adopt appropriate provisions: whereas accordingly a limit to the period during which Regulation 17 shall not apply can be set only for transport by rail, road and inland waterway.'

The effect of Regulation 141, however, was not to provide a permanent exemption for air and sea transport from the provisions of the Treaty as this, in the view of the Commission, would have been inconsistent with its terms. There were, however, Member States prepared to argue that the rules of competition did not apply at all to this sector until at least an appropriate regulation had been adopted by the Council. Originally, the only case that appeared to have relevance to this issue was that of the *French Merchant Seaman.*[7] This case, which came before the European Court of Justice, had considered whether the French Government was in breach of Article 48 prohibiting discrimination based on nationality between workers of individual Member States affecting their rights to employment and conditions of work. The French Code du Travail Maritime had laid down that French merchant ships should be manned on the basis of not less than three Frenchmen to each seaman of other nationality. The French Government claimed that instructions had been given to the naval authorities to treat other EC nationals as if they were French, but that such instruction was a matter for unilateral French discretion rather than Treaty compulsion.

The Court ruled to the contrary that Part II of the Treaty (Articles 9 to 84 inclusive) was in principle applicable to the whole range of economic activities; these Treaty rules could be rendered inapplicable only as a result of some express clause in the Treaty to that effect. The Court said moreover that the reference to the 'objectives' of the Treaty in Article 74 meant those set out in Articles 2 and 3. So far as air and sea transport was concerned, both remained subject to the general rules of the Treaty until the provision of a transport policy under Article 84(2); in the interim period, the other provisions of the Treaty continued to apply, unless an express exception had been granted, including the principle of freedom of movement for workers contained in Article 48. The Court, therefore, ruled that the French legislation was discriminatory against foreign seamen in breach of this Article.

Subsequently, therefore, numerous references can be found in the Annual Competition Reports to Commission proposals for the introduction of some regulations to govern both these sectors. None of these proved, however, capable of implementation successfully owing to the strong opposition encountered from some Member States, and the Commission (with the aid of some Member States, including the United Kingdom) sought alternative means of opening up both sectors to a greater degree of competition, possibly by bringing legal action. The legal position was clarified by the *Nouvelles Frontières* case[8] (European Court 1986) which to

[7] *Commission* v. *France* Case 167/73 [1974] ECR 359: 2 CMLR 216.

[8] *Ministère* Public v. *Asjes* (but often referred to as the *Nouvelles Frontières* case). Cases 209 to 213/84 [1986] ECR 1425: 3 CMLR 173.

some extent strengthens the position of the Commission, although it did not give it the full authority that it had hoped for to impose the complete range of competition rules on Member States and air and sea transport undertakings within them. The case arose out of a request by the Tribunal de Police in Paris[9] which asked the Court to give a ruling under Article 177 on whether it was contrary to the Treaty for Member States to require Courts to apply the criminal law to undertakings which had failed to comply with mandatory tariffs for air transport 'if it were established that those tariffs were the result of an agreement, decision or concerted practice between undertakings contrary to Article 85'.

The Court, noting that no regulation had been made for air and sea transport under Article 87, concluded nevertheless that Articles 88 and 89 remained applicable.[10] Article 88 required authorities in the Member States to apply Article 85 including paragraph (3) and Article 86 until relevant regulations had been adopted. The authorities, however, referred to in Article 88 did not include criminal courts (whose task was simply to punish infringement of the law) but merely administrative bodies charged with administering competition law and civil courts having jurisdiction over such authorities or to apply such laws. Neither the French authorities nor the Commission had in fact exercised any of their powers under Articles 88 and 89 with regard to the air transport tariff agreements which had allegedly been broken by the defendant travel agents in the criminal proceedings before the Tribunal de Police.

In such circumstances, the Court, following earlier cases,[11] held that the two Articles were 'not of such a nature as to ensure a complete and consistent application of Article 85' so as to permit the assumption that Article 85 had been fully effective from the date of entry into force of the Treaty. The principle of legal certainty, therefore, to which the Court had always attached great importance meant that these agreements remained provisionally valid unless and until the Member State, or the Commission, had taken action under those Articles (88 and 89 respectively) which gave transitional powers to them. Only when such a ruling had been given under Article 88 or 89 could the French courts be required to hold, as a matter of Treaty obligation, that no criminal liability could be imposed for the breach of the relevant requirement of the French Code of Civil Aviation which required sales of air tickets to be at official approved prices.

[9] A court equivalent to a local magistrates' court in England and Wales.

[10] This had been considered the probable effect of *Commission* v. *France* (n. 7 above) but was here stated expressly by the Court for the first time. Agreements or practices which are merely ancillary to air and sea transport, such as stevedoring or baggage handling, have never been regarded as exempt from Articles 85 or 86. *Olympic Airways* OJ [1985] L46/51: [1985] 1 CMLR 730.

[11] Notably *De Geus* v. *Robert Bosch* Case 13/61 [1962] ECR 45: CMLR 1.

Within a few months after this decision in 1986, the Council adopted a detailed Regulation applying to sea transport, Regulation 4056/86. It included both procedural rules under which the Commission could carry out its duties under Articles 85 and 86, with powers largely equivalent to those conferred upon it under Regulation 17, and also special substantive provisions applicable to liner conferences. These provisions included group exemptions for certain categories of agreement (notably those laying down technical standards and co-ordinating timetables, the frequency of sailing and carrying capacity) provided that safeguards of publicity, non-discrimination and Commission supervisory rights were observed.

As a result of this case, it became clear that, until the Council did adopt a regulation dealing with air transport, the Commission would itself be able to make progress by implementing Article 89 and in no other way. The Commission subsequently gave notice during the summer of 1986 that it would be implementing Article 89 against ten airline undertakings in Member States with regard to certain forms of agreement, although it would prefer the Council to adopt a group of relevant measures, under which the rules of competition would be introduced in the first instance to only a limited extent, so as first to apply to fare and capacity agreements between airlines and then gradually to the full range of agreements between them. The Commission subsequently prepared separate reasoned decisions against three leading European airlines under Article 89, requiring them to end agreements on the setting of fares and sharing of route capacity and revenues. This had the desired effect of persuading those airlines to modify their agreements to the extent required by the Commission. Finally in December 1987 the Council adopted a package of measures applying Articles 85 and 86, including two Regulations (3975 and 3976/87), a Directive, and a Decision. Their collective effect is to place some limits on the terms of route and capacity sharing agreements between airlines, to enable some new discount fares to be offered and to authorize the Commission to issue Group Exemptions for some categories of agreement whose effect on competition is only minor (see page 445).

These regulations, which are in force until the end of 1992, apply only to international air transport between Community airports. Article 85 applies, therefore, only to such routes, and not to domestic routes wholly within a Member State nor to international routes to or from non-Member States.[12] The Commission too has intervened in a number of proposed mergers involving such routes where it has perceived potentially anti-competitive effects.[13]

[12] By contrast Article 86 has direct effect in respect of all air routes without the necessity for an authorizing regulation. *Ahmed Saaed* Case 66/86 [1989] ECR 803: [1990] 4 CMLR 102.

[13] For example in the *Air France/UTA* merger in 1989; See the 20th Annual Report, pp. 90–1.

In its earlier years the Commission took relatively little interest in and devoted a comparatively small amount of its resources to *financial services, such as insurance and banking*. In parallel, however, with the introduction of banking directives by the Commission to facilitate the development of banking on an interstate basis throughout the Community, it has since the mid-1980s devoted considerably more attention to these sectors. A number of cases have moreover resulted starting with that of *Züchner*.[14] A wide range of restrictions between groups of national banks have been reviewed; whilst some arrangements relating to systems for clearing cheques and providing guidelines for the treatment of foreign currency transactions and payment transfers have received negative clearance or exemption, agreements fixing common interest rates or commission payments have invariably been prohibited, as also various common rates agreed for the collecting of cheques or bills or providing safe deposit facilities.[15]

A similar development has taken place also with regard to insurance services where several Directives have also been promulgated since 1973 to enable insurers to establish branches or agencies in other Member States for certain kinds of business. In a number of antitrust jurisdictions, for example in the USA and in the United Kingdom, partial or total exemption from competition law rules is given to insurance companies and the agreements which they make with each other. The Commission, however, has had no hesitation in applying Article 85 to this sector and in particular to agreements between companies fixing common or minimum rates for policyholders. On the other hand the exchange of information between insurance companies, provided that this is limited to information relating to claims and risk experience rather than done for the purpose of fixing administrative charges, has been given exemption. The Commission has always regarded as important in these cases that each individual insurance company has to make its final decision on the total premium that it will charge.[16] Further experience of the special circumstances of this sector has led to the adoption by the Council of a regulation[17] which will allow the Commission to adopt group exemptions. These will cover co-operation between insurance companies on the establishment of common risk premium tariffs based on collectively ascertained statistics and will also permit the establishment of certain common standard policy conditions, common coverage of certain types of risk, and mutual assistance on the settlement of claims.

Finally, some degree of exemption from the scope of Article 85 is to be

[14] See p. 100 for the facts of this case.

[15] See for details of such cases e.g. *Belgian Banks* [1989] 4 CMLR 141 and *Dutch Banks* [1990] 4 CMLR 768.

[16] See *Nuovo Cegam* [1984] 2 CMLR 484.

[17] Regulation 1534/91.

found in respect of *public undertakings* carrying out a variety of governmental or quasi-governmental responsibilities, and to which Member States have granted 'special or exclusive rights'. Here, whilst the creation and support of such bodies is not prohibited, there are two specific restrictions on Member States; Article 90(1) requires Member States not to enact any measures contrary to the rules contained in the Treaty, in particular the rules contained in Article 7 and Articles 85 to 94 which relate to such undertakings. Moreover, under Article 90(2), such undertakings 'entrusted with the operation of services of general economic interest or having the character of a revenue producing monopoly' remain subject to the rules of competition 'insofar as the application of such rules does not obstruct the performance in law or in fact of the particular tasks assigned to them'. This paragraph continues, however, with the further qualification that 'the development of trade must not be affected to such an extent as would be contrary to the interests of the Community'. We shall be considering the interpretation given by the Court of Justice to Article 90 in the course of Chapter 23.[18]

2. The Concept of an Undertaking and of Group Economic Units

Essential to an understanding of Article 85(1) is to know the practice of the Commission and of the Court with regard to the definition of the expression 'undertaking'. In the absence of a definition in the Article, Deringer[19] refers in this connection to a 1961 German Federal Cartel Office definition of the subjects of competition law as being 'any activity aimed at an exchange of economic values (goods and services) that is not limited to an individual fulfilment of personal needs . . . any independent activity of a person that is not purely personal and that is not outside the economic sphere, regardless of its legal form and the intention to make a profit'. For the purpose of ECSC, the European Court had defined the concept of an undertaking as 'a single organisation of personal, tangible and intangible elements, attached to an autonomous legal entity and pursuing a given long term economic aim'.[20] An undertaking, of course, must be a body capable of having legal rights and duties acting in co-operation with

[18] We shall also in Ch. 23 consider the extent to which the Treaty imposes an obligation on Member States to refrain from the implementation of legislation which has the effect of reducing competition in particular sectors of the national economy.

[19] A. Deringer, *The Competition Law of the European Economic Community: A Commentary on the EEC Rules of Competition* (CCH Editions Ltd., 1968), 5. The corresponding French word is 'entreprise', the German word is 'Unternehmen'.

[20] *Mannesman* v. *High Authority* Case 19/61 [1962] ECR 357, 371. The definition found in Article 80 of the ECSC Treaty is not intended to be exhaustive.

other parties. An examination of the substantial number of cases in which the term has actually been considered (or has been held to apply without detailed consideration) illustrates that such broad definitions have been largely accepted, with the exclusion on the one hand of activities carried on by a State in the exercise of its sovereign function (as opposed to merely commercial functions) and at the other extreme, of those that neither involve the conduct of a business of any kind at all nor the co-ordination of commercial activities to be carried on by others. It, therefore, applies to State bodies engaged in commerce, nationalized industries operating as separate legal entities, municipalities, federations, trade associations (regardless of specific function), publicly owned undertakings, and undertakings with special rights granted by the State; equally, it also includes private individuals engaged in any form of business, commerce or profession, partnerships, co-operatives, companies, and performing rights societies. The objective of making a profit is not essential. Identification, therefore, of an 'undertaking' for the purposes of Article 85 should not in the vast majority of cases present any difficulty.

Nevertheless, over the years, two main issues have emerged in this area. The first is whether it is possible for the Commission to bring a statement of objections under Regulation 17 not only against corporate bodies but also against individuals as the result simply of their connection with those bodies. Corporate bodies act through the agency of such individuals, who may hold authority as directors, employees, or agents, or alternatively may be simply exercising influence or conduct in their capacity as shareholders. The issue was raised by a German Court under Article 177, in a case where a German company, Hydrotherm, claimed that a manufacturing and distribution agreement was void under Article 85(1).[21] The other parties to the agreement were a firm known as Compact and a company called OSA, both owned by an Italian individual, Dr Andreoli. Hydrotherm had agreed to place a substantial order with Compact for the purchase of aluminium alloy radiators; it was claimed that Hydrotherm had not placed an order as large as that required by the contract and the Italian parties sued Hydrotherm for breach of contract before the German Court. An issue in the case was whether the agreement could have been covered by the group exemption contained in Regulation 67/67; the Court held that the term 'undertaking' designated an economic unit even if in law that economic unit consisted of several persons, natural or legal. The requirement of 67/67 is that there shall be only two undertakings as parties, but the Court held that this condition was satisfied, even if one of the parties has contracted through the medium of several undertakings having identical interests and controlled

[21] *Hydrotherm Gerätebau* v. *Andreoli* Case 170/83 [1984] ECR 2999: [1985] 3 CMLR 224.

by the same natural person. In this case Dr Andreoli controlled all the relevant businesses in his operations, and the Court, adopting the economic analysis suggested by Advocate General Lenz, found that the agreement had been made by only two undertakings, namely Hydrotherm on the one side and Andreoli and his businesses on the other.

The position of agents was dealt with early by the Commission in its 1962 Notice already discussed in Chapter 4. The Notice does not state that the commercial agent is not an 'undertaking' for the purposes of Article 85, merely that the true commercial agent who undertakes for a principal to conclude transactions does not fall within the scope of the Article so long as the function performed by the agent is only auxiliary. From the viewpoint of the economic significance of the transaction as a whole, the part played by the agent is only to act as an arm of the principal company itself, which legally if not commercially could equally well operate through a salaried employee. There are, of course, many varieties of agent; some, as the Notice contemplates, may have a degree of independence which may render them capable of entering into agreements which will have an effect under the terms of Article 85(1). This is particularly likely where such agents keep substantial stocks of the product at their own risk, organize services and guarantee facilities to customers, or have discretion over conditions of business or prices charged.

The issue of agency, however, has received consideration from the European Court in the *Sugar Cartel*[22] case, where some of the undertakings involved acted both as agents for each other in the sugar market, as well as principals on their own account, and were adjudged by the Court to be essentially independent traders to whom Article 85 applied. An element in this decision was clearly the ambiguous relationship between those particular undertakings and their alleged principals, and it does not seem to be the Commission's practice to refuse the status of agent to a business simply because it also trades on its own account or for other principals in respect of other products. The actual economic function of the so-called agent has in each case to be fully analysed.[23]

At the end of 1990 the Commission issued on an unofficial basis a preliminary draft Notice which sought to provide amended guidance on its attitude towards agency agreements under Article 85. This Notice is considered in more detail in Chapter 12 but is to be welcomed in so far as its approach seems to indicate an intention to havc a more flexible approach towards the application of Article 85 to these relationships. In particular it

[22] *Suiker Unie* v. *Commission* Case 40/73 [1975] ECR 1663: [1976] 1 CMLR 295.

[23] In Dec. 1986, the first Council Directive on self-employed commercial agents defined them as persons who have a continuing authority to negotiate sales or purchases on behalf of a principal. It laid down certain mandatory clauses applicable to the appointment of such agents relating (*inter alia*) to remuneration, termination, and restrictive covenants. OJ [1986] L382/17.

seems to be prepared to accept a broader definition of the 'risk' which the agent may incur without losing its status as an agent and in extending the restrictions that may be placed on 'integrated agents' having a particularly close relationship with their principals without jeopardizing their position under Article 85(1).

It is clear, however, that the mere description of an individual or company as an 'agent' is insufficient to prevent the application of Article 85. In the *Pittsburgh Corning Europe* case[24] (Commission 1972), certain distributors for Pittsburgh Corning which supplied cellular glass in Belgium and Holland were described in relevant documents as agents, although the evidence showed that they were actually distributors. The Commission found that the status of 'agent' referred to in the agreement between Pittsburgh Corning and its Belgian representative company was in fact given simply for tax reasons.

The second important issue arising on the definition is whether separate corporate bodies comprised within a group are themselves separate undertakings or whether the group itself constitutes the relevant undertaking. The normal view taken here by the Commission and the Court, which is of course closely linked with its analysis referred to above of the status of agents, is that each of the companies within a group is itself treated as an undertaking capable of entering into an agreement or concerted practice. On the other hand, the essential condition that both entities concerned are capable of independent economic policy making before a relationship between them can be classified as an agreement or concerted practice is not satisfied if the agreement or concerted practice reflects no more than the allocation of functions within a corporate group under the legal and actual control of one company. An early example here was the *Kodak* case[25] (Commission 1970). This involved a request for negative clearance by various wholly owned European subsidiaries of the United States company Eastman Kodak, following the introduction of new uniform conditions of sale which permitted export sales by them to other Member State countries at the same price as that charged there by the local Kodak subsidiary, whilst imposing restrictions on sale outside the Community. The Commission took the view that, since these subsidiaries could not act independently of their parent company, their individual agreements within the group were not between independent economic policy making units and, therefore, an effect on competition was neither likely nor possible.[26]

[24] *Pittsburgh Corning Europe* [1973] CMLR D2.

[25] *Kodak* [1970] CMLR D19.

[26] A similar approach was taken by the Commission in *Christiani and Nielsen* [1969] CMLR D36 where the agreements were found to be likewise the allocation of functions between parent and subsidiary, which could have been organized without the need for contractual arrangements or concerted practices, simply by unilateral direction from the Danish parent company.

This approach adopted by the Commission (and often referred to as the doctrine of the 'group economic unit') was confirmed as correct by the Court in its *Centrafarm* cases in 1974,[27] though the Court indicated that Article 85 might well be applicable if the relevant agreement or practices went beyond the limited aim of establishing an internal distribution of tasks within the group. Thus, if a company is accorded a measure of independence, either *de facto* or *de jure*, it may well be capable of falling within the scope of the Article, even if its degree of independence is restricted. Difficult issues can also arise in joint venture cases, where the degree of independence of the joint venture company from its parents can present analytical problems for the Commission, as for example in the *Peroxygen*[28] case.

In the *Martell-DMP* case[29] fines were imposed on both the French Cognac producer Martell and DMP, its 50 per cent owned subsidiary which carried out distribution of the product in France, for conspiracy to prevent parallel exports to Italy (where prices were substantially higher). The parties had acknowledged to the Commission that they regarded themselves as separate rather than as parts of the same economic unit. Moreover, the other 50 per cent owner of DMP, Piper-Heidsieck, had exactly the same shareholding and representation on the supervisory board of DMP as had Martell. DMP also distributed brands not belonging to its two parent companies and acted independently from them in determining the terms of sales to buying syndicates in France.

The adoption by the Court of the 'group economic unit' theory has a dual advantage for the Commission. It enables it to apply a test to the relationship between undertakings within a corporate group which gives a realistic degree of emphasis to the actual economic relationship between them, rather than relying on formal tests relating to legal indicia which might more easily be manipulated by the parties. It also enables the Commission (and Court) to justify the extension of its jurisdiction to parent companies apparently operating from outside the Community, so long as they actively control subsidiary companies resident or doing business within it, an issue which we shall need to discuss further in Chapter 24.

3. Agreements, Decisions by Associations of Undertakings, and Concerted Practices

Article 85 has been drafted so as to cover three related, though conceptually distinct, situations. Undertakings can be said to have reached an

[27] *Centrafarm BV* v. *Sterling Drug Inc. and Winthrop BV* Cases 15 and 16/74 [1974] ECR 1147, 1183: 2 CMLR 480.

[28] *Commission* v. *Solvay and Laporte* (often referred to as the *Peroxygen* case) Case 85/74 [1985] 1 CMLR 481.

[29] OJ L 185/23 (11 July 1991).

agreement when there can objectively be said to be a sufficient consensus between them as to the bargain to which they have mutually committed themselves. Identification of that consensus is ultimately a matter for courts of law, not merely for the undertakings themselves. This task is easier, of course, if the contract is in written form but an agreement may nevertheless exist and can be recognized by a court or administrative body even if it is purely verbal or partly written and partly verbal, or even if the parties claim that whatever its outward form it was not meant as legally binding but simply as a 'gentlemen's agreement'. The case of *Atka A/S* v. *BP Kemi A/S*[30] (Commission 1979) illustrates that even if the relevant document has been neither signed nor dated, it will still be treated as an agreement if it has been acted upon by both parties. In this case, De Danske Spritfabrikker (DDSF) was the only producer in Denmark of synthetic and agricultural ethanol, used in the production of several products including vinegar and cosmetics. About 45 per cent of the total synthetic ethanol production in the Common Market in 1971 was manufactured by BP, and it had begun to supply certain customers in Denmark through its Danish subsidiary (BP Kemi). DDSF, to protect its market position, then sought an exclusive purchasing and co-operation agreement with BP and its subsidiaries; a document was drawn up, though never signed nor dated, under which:

(i) DDSF would buy all its requirements for the product from BP Kemi up to an agreed maximum, provided that BP Kemi had the opportunity of supplying it also with excess quantities on similar terms, and
(ii) BP Kemi would be entitled to sell yearly in Denmark up to 25 per cent of the combined aggregate sales of the product; but if BP Kemi exceeded this limit, it would pay compensation to DDSF.

Subsidiary agreements related to the pricing of the product (which was to be kept identical by both companies) and agreed rules for the allocation of customers within Denmark. Atka, a competitor of DDSF, challenged these arrangements, and the Commission ruled that the application by the two parties of identical prices (which continued after the termination of the original agreement) would not by itself necessarily prove the continuing effect of the agreement, but would amount to strong evidence of it if pricing of the product would have been unlikely to follow the same pattern had normal market competition prevailed.

A *decision* of an association on the other hand is any provision of the rules of a trade association (either having members which are themselves trade associations or more normally separate undertakings) or any decision or recommendation made under those rules either for or by its mem-

[30] [1979] 3 CMLR 684.

bers, or arrived at informally within the framework which they provide. Again, it is normally quite easy to identify the existence of such decisions.[31] The concept of *concerted practices* however, is more fluid; it is unquestionably the widest, and the vaguest, of the three expressions, intended to cover co-operative activity of any kind between undertakings but which falls short of an actual agreement. For safety's sake, 'concerted practices' are often alleged by the Commission as an alternative even when technically there is probably some kind of 'agreement' also in existence, as for example in the sales cartel operated in the *Floral* case[32] (Commission 1979). In an early decision, *Brasserie de Haecht* v. *Wilkin-Janssens* (no. 1)[33] (European Court 1967), the Court indicated that the three concepts need not be separately distinguished in the Commission's analysis of the situation; the constituent elements of the three possible forms of relationship could be considered together as a whole if factually they were hard to disentangle.

In this area, Article 85 again has drawn for part of its content on Article 65 of the ECSC Treaty where all three expressions are likewise found. Of these, both 'agreements' and 'concerted practices' require the existence of at least two undertakings as parties to the agreement or practice; a 'decision', however, of a trade association could involve simply the association (which may be a single legal person) although usually in practice it will also involve the participation of a number of undertakings, almost certainly the relevant members of that association. Even if, however, two undertakings have to be identified as parties in a case involving agreements or concerted practices, restrictions need only be identified that affect one party, though the fact that only one party out of two accepts restrictions may permit exemption from the requirement of notification in the case of some agreements.[34]

Cases dealt with by Commission or Court have involved agreements both of a horizontal and vertical nature. Horizontal agreements are those made between undertakings that compete with each other (or at least are believed to do so) because each operates at the same level whether as manufacturer, wholesaler, distributor, or retailer; on the other hand, vertical agreements are those made between undertakings at different levels in the commercial chain whose relationship to each other is complementary. The vertical agreement may involve, therefore, manufacturer and distributor, or wholesaler and retailer, or in another context patentee and licensee. Both types of agreement are covered by Article 85.[35]

[31] A number of cases relating to 'decisions' taken by trade associations are considered in Ch. 21.

[32] *Floral Düngemittelverkaufs* [1980] 2 CMLR 285.

[33] *Brasserie de Haecht* v. *Wilkin-Janssens* (no. 1) Case 23/67 [1967] ECR 407; [1968] CMLR 26.

[34] See Reg. 17, Article 4(2) (*a*).

[35] As we have already seen in Ch. 5, p. 65, an early attempt by Italy to have vertical agreements read out of the Article failed.

Even documents which look less like agreements than a form of unilateral instruction may nevertheless be held to constitute an agreement, if the recipients of the instructions have been required to acknowledge its receipt, even if they are not required expressly to acknowledge its contents. In *WEA-Filipacchi Music SA*[36] (Commission 1972). WEA, a subsidiary of the United States corporation Warner Brothers, was engaged in the sale of pop records and sent out a circular to its various French distributors informing them that sales of such records outside France were forbidden. In sending this circular out, WEA was influenced by the fact that the French price level was considerably lower than the equivalent price in Germany, so that sales or distribution by way of 'parallel imports' into Germany could have proved extremely profitable for the French distributors. The distributors were merely asked to affix their official stamp to the circular by way of acknowledgement, without being asked specifically to acknowledge that they were bound by the terms communicated. The Second Annual Report of the Commission refers to the finding of an agreement in this case as evidence that the Commission will not be prevented from finding an agreement merely by the use of carefully worded clauses in circulars purporting to evidence the unilateral decision of an undertaking which then relies on apparently tacit acceptance from its distributors .[37]

In the case of *Sandoz*[38] invoices for the supply of pharmaceuticals used by this producer bearing the words 'export prohibited' which in practice had been accepted by its trade customers was held by the European Court (confirming the Commission decision) to represent not simply a unilateral request, but part of an agreement of which the words endorsed were the documentary evidence.[39]

In the case of *BMW Belgium SA* v. *Commission*[40] (European Court 1979), a circular to discourage exports of cars to Germany was distributed (without authority from the parent company in Germany) by the BMW Belgium subsidiary. The circular went out after consultation with, and under the signatures of, the BMW Belgium dealers trade association's executive committee, and a number of the Belgian BMW dealers acknow-

[36] [1973] CMLR D43.

[37] pp. 47–8.

[38] Cases 227/87 [1990] ECR 145: [1990] 4 CMLR 242.

[39] Nevertheless it should not be supposed that the Commission's findings are always confirmed. Thus in the *Italian Flat Glass* cases (Cases T-68/77, 78/89: Judgment of 10 March 1992) a number of findings against the appellants were rejected for lack of evidence. The original Commission decision is reported at [1990] 4 CMLR 535.

[40] Cases 32/78 and 36–82/78 [1979] ECR 2435: [1980] 1 CMLR 370. It is noteworthy that in this case the BMW subsidiary was fined, but that its German parent which had played no part in the arrangements was not even made a defendant, an illustration of the fact that the doctrine of the 'group economic unit' is not, and should not be, applied automatically without analysis of the relevant facts. The rulings of the Commission were upheld by the Court.

ledged in writing that they had received it. These were held by the Commission to have in this way acknowledged participation in the arrangements, so that fines were imposed for the attempt which the circular represented to prevent exports of BMW vehicles from Belgium. It is clear also that, once such an agreement has been made, it is difficult to satisfy the Commission that it has been completely terminated since even if the parties claim to have done so, its effect may be such as to continue to influence their conduct. It may in such circumstances, therefore, be held still to exist, or alternatively to have been transformed into a number of concerted practices.[41]

An interesting development of the jurisprudence of the Court on the question of the definition of an 'agreement' has come in cases where a manufacturer has apparently taken unilateral action, but in a way so closely connected to an agreement that it has been interpreted, together with that agreement, as forming part of the general complex of contractual arrangements between the manufacturer and the various distributors or dealers with whom these individual agreements (whether notified or exempted) have been made. The two leading cases in this area relate respectively to the distribution in Germany of high quality television, electrical and electronic goods, and the distribution of Ford cars. In the first case *AEG Telefunken* v. *Commission*[42] (European Court 1983), AEG had its selective distribution system for consumer electronic products in Germany and other countries notified to the Commission and exempted under Article 85(3). This distribution agreement contained no right for AEG to lay down resale prices for its dealers. Numerous complaints were, however, then received by the Commission from small traders claiming that AEG had sought to discriminate against some of them and in particular those who would not maintain resale prices at the level considered appropriate by AEG. The Commission imposed a substantial fine, which was upheld by the Court. The Court ruled that the activities of AEG Telefunken alone in ensuring that at least a substantial proportion of its dealers did not cut prices below the recommended level was not simply unilateral conduct but on the contrary formed an integral part of its contractual nexus with the distributors which, by remaining part of the distribution network, had themselves confirmed their adherence to the AEG policy.

In the *Ford* v. *Commission* case[43] (European Court 1985), Ford likewise had notified to the Commission its distribution system for Germany.

[41] Another example is found in *Binon* v. *Agence et Messageries de la Presse (AMP)* Case 243/83 [1985] ECR 2015: 3 CMLR 800, where the effect of previous arrangements for the distribution of newspapers and periodicals in Belgium was held likely to continue even after a new system had been introduced.

[42] Case 107/82 [1983] ECR 3151: [1984] 3 CMLR 325.

[43] Cases 25–26/84 [1985] ECR 2725: 3 CMLR 528.

Originally, the products covered by the agreement included both left-hand drive and right-hand drive cars but by a circular taking effect on 1 May 1982 right-hand drive cars were withdrawn from availability, so that they would not be available to be resold to British buyers for import into the United Kingdom at a lower overall price than that then prevailing in the UK domestic market. The Commission alleged that exemption should be refused for the system of distribution because its effect was to partition national markets contrary to the well-known *Grundig* principle. Ford claimed that its decision was simply unilateral, having no necessary connection with the main dealer agreement for Germany, and that its distribution system was entitled to exemption under Article 85(3).

The Court again held, upholding the Commission's decision on this substantive issue (although it had earlier granted Ford's appeal against the Commission's order for interim relief[44] which had required Ford to continue to supply right-hand drive cars in Germany pending the Court's final ruling) that the decision to change the product covered by the dealer agreement was not simply unilateral. It was rather part of the whole contractual framework between Ford and its dealers; admission to the dealer network in Germany implied acceptance, in the view of the Court, by all parties of the policy pursued by Ford with regard to the choice of models to be delivered to the German market. The Court said that the Commission was entitled, in coming to a decision on whether to grant exemption, to take all the relevant facts into account including the effect which Ford's refusal to supply right-hand drive cars to Germany could have on the partitioning of the Common Market.

In its decision, the Court also laid down an important general finding as to the correct approach to the interpretation of Article 85(1). The Commission was held by the Court not to be required to rule on each individual element of the agreement so long as the agreement considered as a whole fell within the coverage of Article 85(1). Ford had claimed that the Commission must first identify each provision of the agreement in breach of Article 85(1), and then consider that provision in isolation against all the relevant criteria including the four conditions for exemption contained in Article 85(3). The Court, however, took the view that a more 'broad brush' approach would be justified in such circumstances.[45]

It is the experience of all systems of law that seek to restrain anti-competitive behaviour that it is difficult to do so unless the prohibitions apply effectively not only to agreements, but also to the many ways which can be utilized to dispense with the need for actual agreements. Those who negotiated the terms of Article 85 would have been well aware of the like-

[44] For this aspect of the case, see Cases 228–229/84 [1984] ECR 1129: 1 CMLR 649.

[45] The Court adopted a similar approach in the *Windsurfing International* case, considered in detail in Ch. 15, Case 193/83 [1986] 3 CMLR 489.

lihood that in many sectors of European industry and commerce (particularly in those where there was a strong oligopolistic element) effective concertation of commercial policy could be arranged without the need for formal agreement, indeed often without the need for creation of any external evidence at all. Desired results could often be achieved by far simpler or subtler methods.

From the viewpoint of DG IV as an enforcement authority, its problem in all such cases has been to find evidence of the arrangements made which have the effect of enforcing uniformity of price or other undesirable consequences falling within the range of Article 85. The best evidence, of course, in all such cases would be documentation comprising, if not a formal agreement, at least correspondence, fax copies, minutes of meetings, or other commercial documents, in other words the kind of document which is the routine product of the ordinary business office. To enable the Commission through DG IV to have ability to reach these more informal arrangements, the concept of 'concerted practice' was introduced into the Treaty and has, thanks to the wide definition given to it by the European Court, proved vital to the implementation of the Commission's competition policy.

It is noteworthy that at the time of the enactment of the Treaty of Rome in 1957, none of the Member States, even if they had competition legislation in their domestic law, had a concept in their domestic law equivalent to 'concerted action' or 'concerted practice'. Its introduction can undoubtedly be attributed to the influence of United States antitrust law, where the concept of 'concerted action' was already familiar in both horizontal and vertical relationships. When parallel conduct by enterprises was found, not simply because of independent business decisions but as a result of interdependent decisions raising the inference that there was a tacit agreement between the parties, then United States Federal Courts were entitled, and inclined, to look for 'plus factors' to distinguish the latter situation from the former. Among the factors which they took into account were any actions taken by companies contrary to their own normal business interests. These might include the raising of prices at a time when there was a substantial supply available of the relevant product, an artificial limitation of the supply of products, the imposition of unusual conditions of sale, refusal to attend sales of goods sold at auction unless other leading competitors were present, or an agreement to price identical goods to be transported long distance only on identical fixed basing points. [46]

[46] The Supreme Court laid down clear guidelines to prevent the courts from inferring agreements merely from parallel conduct. The *Monsanto* v. *Sprayrite* case, 104 Supreme Court Reports 1551 (1984), has finally established that a high degree of collaboration is required between a manufacturer and its distributors if a distributor whose appointment had

The issue of the proof of concerted practices arose in the first three of the major horizontal agreements cases to reach the Court, all being appeals brought against decisions by the Commission to impose fines on groups of manufacturers from a number of separate Member States. The first of these cases was the *Dyestuffs* case[47] (European Court 1972). Between 1964 and 1967, there were three general and uniform increases in the price of dyestuffs sold by leading European producers. In January 1964, there was a 15 per cent increase for nearly all dyes based on aniline.[48] In January 1965, there was a 10 per cent increase in prices for all dyes and pigments not covered by the increase of the previous year. Because major Italian companies did not participate in this increase, the other European companies did not then maintain the price increase for the Italian market even though they had already been announced. In October 1967, the parties uniformly stated an 8 per cent increase for dyes sold in Germany, The Netherlands, and in Belgium and Luxembourg, and a 12 per cent increase in France.

Advocate General Mayras characterized the market for such products as oligopolistic, controlled by a small number of producers with national markets 'walled off' from competition. Customers tended to have direct contacts only with producers in their own country, and the only contact with foreign producers was through their subsidiary companies' representatives or agents. There was, therefore, no true transparency of price or awareness on a European-wide basis from the viewpoint of the individual customer. He felt that in a more competitive market, price increases would have followed on a more individual basis; whereas oligopolistic pressures might account for downwards movements of prices on a concerted basis, it would be less likely without some form of agreement, to have been responsible for such concerted moves in an upward direction. He, therefore, felt sufficiently suspicious about the contacts between the parties to reach a finding that the price moves had been co-ordinated.

The Court agreed that there had been concerted practices by the defendants. There was an element of interconnection between the three separate occasions for price increases operating in a virtually simultaneous way. In

been ended is to establish the necessary proof of an agreement. The mere fact that complaints have been made by other distributors against that distributor is insufficient, since it is the normal practice of distributors to discuss such matters with their manufacturer, and the mere making of a complaint is not necessarily decisive in the manufacturer's decision to terminate an appointment. Whilst it is not the practice of the European Court of Justice itself to refer in its decisions to Court decisions or any jurisdiction, let alone those outside the Community, there is no doubt that the US experience has had an influence in the significant development of the concept of concerted practice through the cases. Advocates General are not so restricted, and examples can be found of cases where their opinions have made reference to relevant US cases, e.g. in the *Dyestuffs* case in n. 47.

[47] *ICI and others* v. *Commission* Cases 48–57/69 [1972] ECR 619: CMLR 557.

[48] There was a slight delay before the application of the increase in Germany.

1964 the increase had proceeded as planned but on the two later occasions the advance announcements from major undertakings had allowed each of them to observe the reaction of the others and to eliminate thereby the risk that any one producer's increase might get out of line. Although every producer was legally free in changing its prices to take into account the present or foreseeable conduct of its competitors, it was a breach of Article 85 to co-operate with competitors in a way which enabled a co-ordinated course of action relating to price increases to be put into effect. It was even more of a breach to ensure the success of that course of action by deliberate prior elimination of all uncertainty as to each other's conduct over such matters as percentages of increase, date, place, and exact range of products covered. The term 'concerted practice' was defined by the Court as a 'form of co-ordination between enterprises, that had not yet reached the point of true contract relationship but which had in practice substituted co-operation for the risks of competition'.[49]

In 1975, a more detailed definition of concerted practices was given by the Court in the *Sugar Cartel* case as follows:

> The concept . . . refers to a form of co-ordination between undertakings which, without having been taken to the stage where an agreement properly so-called has been concluded, knowingly substitutes for the risks of competition practical co-operation between them which leads to conditions of competition which do not correspond to the normal conditions of the market having regard to the nature of the products, the importance and number of the undertakings as well as the size and nature of the said market. Such practical co-operation amounts to a concerted practice, particularly if it enables the persons concerned to consolidate established positions to the detriment of effective freedom of movement of the products in the Common Market and of the freedom of consumers to choose their suppliers.[50]

This case was immensely complex,[51] involving sugar manufacturers throughout the Community and required the examination of the market situation in each of the relevant markets. Since sugar came within the scope of the Common Agricultural Policy, there was considerable Member State involvement in the marketing of sugar. In the case of Italy the Court found that the involvement was so substantial as to justify a finding that the restrictive practices of the parties in attempting to preserve the principle of *chacun chez soi* could not of themselves have affected the competitive position, so little room for manœuvre having been left by the

[49] [1972] ECR 655: CMLR 622

[50] [1975] ECR 1916: [1976] 1 CMLR 405.

[51] The total length of the case report in ECR is 491 pages and in CMLR 195 pages. One suspects that this means the report is rarely read in its entirety. It is the kind of case for which the existence of the Court of First Instance would have been invaluable, since its detailed analysis of the complex facts (had it then been in operation) would have allowed the European Court (if an appeal had been brought) to concentrate on the legal issues involved rather than having to review individually so many factual issues.

stringent domestic legislation. In other jurisdictions, notably Germany and The Netherlands, the scope for individual agreements and practices was greater and after a painstaking examination a number of findings of concerted practices were made, although by no means identical with the findings of the Commission. Some of the fines imposed were therefore reduced.

An important element, however, in establishing the existence of concerted practices is some form of contact between the parties which must involve an intentional communication of some information between them either directly or through an intermediary. There must be some positive intention on the part of one party to direct the information to the other or at least to launch it into the relevant area where the other is likely to receive it.[52] It is also essential for the Commission to be able to show that the party receiving the information is aware of having done so not accidentally, but as a target. Such giving and receiving of information is clearly not the same as mere mutual awareness by competitors that the terms which they quote to their customers will subsequently become known to their competitors, who will naturally make use of this knowledge in facilitating their own pricing and other competitive strategies. If even the awareness of what one's competitor is doing constitutes a sufficient level of contact to justify the claim that concerted practices have commenced, there could never be accorded to independent undertakings the right merely to adapt intelligently to the conduct of other undertakings. This would be particularly serious in a market whose operations are normally relatively transparent and where there are a substantial number of competitors. The distinction is clearly drawn in the case of *Züchner* v. *Bayerische Vereinsbank*[53] (European Court 1984). The facts of this case were that Herr Züchner had an account with the Bayerische Vereinsbank at Rosenheim. He challenged the service charge made for the transfer of funds to Italy, on the basis that the debiting of what appeared to be a uniform charge by a number of banks constituted a concerted practice. The issue was referred under Article 177 to the Court, which confirmed that the banking services were indeed covered by Article 85 and that any such agreement, if proved, would have been in breach of it.

[52] A difficult issue raised in the *Woodpulp* case (see n. 60) is whether concerted practices can occur without communication between the parties, if their individual systems for quoting prices contain sufficient common features, e.g. regular quarterly adjustments of prices in a single currency.

[53] Case 172/80 [1981] ECR 2021: [1982] 1 CMLR 313. With this one may contrast the *Eurocheques* case, OJ [1985] L35/43, where uniform charges by banks operating this system were approved because of the substantial advantages that the new system was shown to provide to tourists and other travellers. The Twentieth Annual Report indicates (p. 82) that when renewal of the exemption was sought a number of difficulties arising out of the operation of the system were raised by the Commission. See also *Arjomari–Prioux* v. *Euro-cheque International* (1989) 4 CMLR 907.

The tests which the Court laid down for the determination of whether the banks had operated sufficiently independently in applying a uniform charge to customers making foreign transfers were:

(*a*) whether contacts between the various banks which had imposed such charges merely related to the charges which were made in the past or also to those to be made in the future;
(*b*) whether in a state of free competition different rates of charge would have applied;
(*c*) the number and importance of the participating banks in the market for monetary transfers between Member States; and
(*d*) the volume of transfers on which a uniform charge was imposed, as compared with the total number of transfers made.

In conclusion, the Court said that parallel conduct in debiting uniform charges on transfers by banks from one Member State to another could amount to a concerted practice if it was established both that the parallel conduct exhibited the features of co-ordination and co-operation characteristic of such a practice and that such conduct was capable of having a sufficient effect on competition.

In such cases, the Commission will normally allege that the facts established cannot be explained other than by concerted action. The parties will then have to provide a rational explanation, which the Commission will then have to accept or reject. This is well illustrated by the case of *Compagnie Royal Asturienne des Mines and Rheinzink* v. *Commission*[54] (European Court 1984). These two companies supplied rolled zinc to a Belgian company, Schlitz. Schlitz then resold the zinc to a German company, the market price in Germany being higher than that in Belgium. Schlitz, however, had obtained its supplies from CRAM and Rheinzink by a false claim that the product would be resold to Egypt; when CRAM found out the deceit and refused to make further supplies, Rheinzink came to a similar conclusion to cut off Schlitz. Evidence showed that there was regular contact between CRAM and Rheinzink.

The Commission found that there was sufficient evidence of concerted practices between the two companies to refuse to continue supplies. The suggested motive was the fact that they wished to continue their own profitable supplies to the German market and to prevent Schlitz from continuing to be engaged in it. Unfortunately for the Commission, the European Court found on the contrary that the reason for the refusal of supplies was in fact the poor credit record of Schlitz, accepting that there was sufficient evidence that this was the major ground for the decision to refuse supplies in future.

[54] Cases 29–30/83 [1984] ECR 1679: [1985] 1 CMLR 688.

The final case of interest as an example of concerted practices between parties in competition with each other is the *Peroxygen* case[55] (Commission 1984). This involved a limited number of producers in Europe of hydrogen peroxide and its derivatives which are used as industrial bleaches. There was evidence that for a number of years major European companies had participated in agreements which had the effect of reserving each national market for supply by its major producer,[56] these arrangements being enforced by regular meetings of the members where any marginal disputes were resolved, arrangements similar to those found in the *Sugar Cartel* case. Full exchanges were made of information about the production of these products so that each knew the others' general commercial policy. It was held that these arrangements constituted a concerted practice, for although the parties had not necessarily agreed upon a precise or detailed plan in advance, it was sufficient that by their mutual involvement they had departed from the basic requirement, that each must determine independently the policy which it intended to adopt on the market.

When the participants in alleged widespread industry arrangements for identical basic products (such as particular chemicals) comprise virtually all the European manufacturers in this industry, the Commission has, of course, a major problem in finding evidence to link all the participants sufficiently closely to the arrangements. In these cases such as the *Polyvinyl Chloride (PVC)* case (Commission 1988, reversed on procedural grounds by the Court of First Instance)[57] the attitude of the Commission is that it is sufficient for it to show that all the parties had reached a consensus on a plan which limited, or was likely to limit, their commercial freedom, by determining the general lines of the action (or indeed abstention from action) on the market. The fact that some participants were more assiduous attenders at the relevant meetings or took a leading role to an extent not shared by other companies (a factor possibly relevant in assessment of fines) would not absolve the less frequent attenders from being found to have participated in a common enterprise. Responsibility for the operation for the cartel as a whole would in the view of the Commission apply to both the 'central' and 'fringe' players alike; this approach seems to be confirmed by the judgment of the Court of First Instance in the *Polypropylene*[58] case in 1991.

We have so far concentrated on the establishment of concerted prac-

[55] *Commission* v. *Solvay and Laporte* Case 85/74 [1985] 1 CMLR 481.

[56] An arrangement similar to that in the *Sugar Cartel* case, though here without the added complications of a common Community policy to give Member States a legitimate opportunity of becoming involved in the market 'arrangements'.

[57] [1990] 4 CMLR 345. This issue was not dealt with by the Court of First Instance because the Commission's decisions were declared void on procedural grounds.

[58] Case T-7/89 [1992] 4 CMLR 84, 310–1313.

tices between horizontal competitors, but it is also important to define this concept for vertical arrangements. The leading authority is the well-known *Pioneer* case[59] (European Court 1983). Pioneer was a leading Japanese manufacturer of high fidelity equipment and had a number of European distributors in Germany, France, and the United Kingdom. Distribution agreements allowed the dealers to export to other countries subject to the usual requirement not actively to solicit business outside their territories. Price levels in the French market were, however, considerably above those in the German and UK markets so that there was an incentive for parallel importing into France from both the United Kingdom and Germany. A German distributor was approached for an order for delivery in France. Following pressure from the main distributor in Germany, Melchers, this contract was cancelled. Likewise, the United Kingdom distributor stopped various sub-distributors from exporting to France and Holland.

The parent company, through its Dutch co-ordinating company, Pioneer Electronic Europe NV became concerned at all this activity and called a meeting in Antwerp in January 1976. No records were kept of the meeting, which was attended by all the European distributors, but it was established that various matters were discussed relating to the distribution of Pioneer products, including parallel imports. The Commission obtained evidence that the main complainant was the French distributor, which used the meeting as an opportunity to pass on complaints about parallel importing into France by other distributors. Pioneer Electronic Europe NV then used its influence as co-ordinating Pioneer subsidiary for Europe to damp down the zeal for such sales. The Commission later investigated the arrangements and reached a decision that a reduction in such sales was directly attributable to concerted practices between Pioneer and its distributors. Relationships here were not simply horizontal since Pioneer Electronic Europe NV had a co-ordinating function to ensure that supplies were available to its various distributors and was not in direct competition with them, whereas in theory all the distributors, although allocated to separate territories, were in competition with each other.

The Court naturally referred to its earlier judgments in the cases both of the *Dyestuffs* and *Sugar Cartel*. In this case, concerted practices had been established by circumstantial as well as direct evidence. The presumption and inference from the facts set out above was that the co-ordinating Pioneer subsidiary had called the relevant meeting in Antwerp to deal with the problem of parallel imports as the result of which the practice had been substantially reduced, at least for a time. It is in the context of

[59] *Musique Diffusion Française* v. *Commission* Case 100/80 [1983] ECR 1825: 3 CMLR 221.

parallel imports that co-ordination between parties having a vertical relationship is most likely, but it does seem that on occasions the Commission is inclined to take a view in the vertical context which goes even beyond the wide scope allowed by Court decisions. Thus it has sought on occasions to argue that extra-contractual intervention by one economic partner, which induces or is likely to induce action to restrict competition, is itself a concerted practice. This could mean, for example, should a distributor choose to suggest to its manufacturer that disparity in prices in different territories of the Common Market was such as to increase the risk of parallel imports, and to request the manufacturer, therefore, to bring his prices for the individual territories more closely into line, that this mere request could constitute such an 'extra-contractual intervention', if the reaction of the manufacturer was then to adopt a policy itself in breach of Article 85. No Court decision has yet given any support to this argument, however, and it seems unlikely to do so.

In summary, therefore, there is substantial case-law to illustrate the great importance of findings of 'concerted practices' from the viewpoint of the Commission as an enforcement agency. Further light may be shed on the width of this definition by an important case now on appeal to the European Court, originally decided by the Commission in December 1984, *Woodpulp*.[60] Bleached sulphate pulp used in the manufacture of many different kinds of paper can be made from either hardwood or softwood. A very large number of companies were alleged to have concerted prices for this product either directly or indirectly. The defendants included eleven United States, six Canadian, eleven Swedish, eleven Finnish, one Norwegian, one Portuguese, and one Spanish company who collectively supplied about two-thirds of the EC market. The Commission found that all the companies involved were subject to EC competition law because their conduct had an effect on the EC market and that the deliberate transparency of prices charged on the European market by such a large number of competitors, although in theory making concertation more difficult, in practice might itself make it easier if carried out in ways which helped anticipation and knowledge of the pricing policy of competitors. Apparently prices were charged on a quarterly basis, the vast majority being quoted in dollars rather than in the local currency of the company giving the quotation.

The Commission's decision was based on what the Commission characterized as the 'deliberate transparency of prices'. It argued that in an open competitive market one would find a variety of approaches to pricing

[60] [1985] 3 CMLR 474. Appeals have been lodged with the European Court by many of the defendants (Cases 89, 104, 114, 116, 117, 125–129/85). The Court's judgment on the jurisdictional issues was given first ([1988] ECR 5193: 4 CMLR 901) and its ruling on the substantive issues is expected in 1992.

strategy with prices being quoted in different currencies with substantial reductions for quantities supplied and long-term contracts. It said that there was a suspicious lack of variation between the quoted price levels and the transaction price and that the quoting of prices in dollars made it easier for any variation from the arrangements to be checked. Moreover, some of the European agents acted for a number of producers from different countries so as to enable the level and changes in pulp prices to be speedily known. The Commission accepted that the presumption of concerted practices could be displaced if evidence was produced of a genuine equilibrium[61] of price required as the result of independent action by the participants or as a result of the existence of a market leader upon whom other firms had aligned their prices, but said it was not satisfied that these explanations held good.

It is on the other hand the case for some of the parties charged that transparency is not necessarily evidence of a concerted practice, and that the choice of the quarterly price announcement was a traditional compromise between the buyer's desire to avoid uncertainty about future prices and the necessity for sellers to change their prices on a reasonably regular basis to reflect changes in costs, as well as the balance between supply and demand. The parties also pointed out that woodpulp is an item that can be stored and is not perishable, and this enables suppliers to store overproduction to meet changes in demand. Buyers are concerned not only about price but about the important issues of reliability of future supply and speed of delivery. It will be significant if the opportunity is taken by the Court to pronounce further on the interpretation of concerted practices, in the context of so large a number of sellers operating from so many different countries both within and outside the Community, in a factual situation possibly more complex than in any previous case.

[61] The equilibrium of such markets here is of what is called, in technical terms, the 'Cournet/Nash' rather than the 'Bertrand' version: in the former, no enterprise can increase its own profitability by deviating from the strategy of maintaining an equilibrium price whereas in the latter there is a continuous process by all of undercutting their competitors' prices until every enterprise is selling at approximately marginal cost.

7

Article 85 (1): Analysis (2)

1. 'Which may affect trade between Member States'

Neither agreements, decisions, nor concerted practices have any significance under Article 85 unless they may affect trade between Member States, quite apart from the second and further requirement that they should have as object or effect the prevention, restriction, or distortion of competition within the Common Market.

The interpretation of this phrase has been developing gradually over the last twenty-five years. Initially, it was believed that agreements and practices that had effect merely within a single Member State, and upon undertakings whose business was operating only there, could always claim to be outside the scope of the Article. On the other hand, if an agreement contained a transnational element, either because one or more of the undertakings involved was incorporated or resident in a different Member State of the Community or because the agreements concerned altered the actual or potential flow of goods or services between Member States, then the requirement of the Article was satisfied. This was clearly set out in one of the earliest preliminary rulings given under Article 177 by the European Court, in the *STM*[1] case of 1966 where the phrase was interpreted in the following way: 'For this requirement to be fulfilled, it must be possible to foresee with a sufficient degree of probability on the basis of a set of objective factors of law or of fact that the agreement in question may have an influence, direct or indirect, actual or potential, on the pattern of trade between Member States.'

Scarcely two weeks later the European Court supplemented this ruling by its findings in the *Grundig*[2] case: after pointing out that the purpose of this requirement is to establish the boundary between Community law and that of Member States in the context of the competition rules, it continued by stating that

> it is only to the extent to which agreement may affect trade between Member States that the deterioration in competition caused by the agreement falls under the prohibition of Community law contained in Article 85: otherwise it escapes

[1] *Société Technique Minière* v. *Maschinenbau Ulm* Case 56/65 [1966] ECR 234, 249: CMLR 357, 375.

[2] [1966] ECR 299, 341: CMLR 418, 472.

the prohibition. In this connection, what is particularly important is whether the agreement is capable of constituting a threat, either direct or indirect, actual or potential, to freedom of trade between Member States in a manner which might harm the attainment of the objectives of a single market between States. Thus the fact that an agreement encourages an increase, even a large one, in the volume of trade between States is not sufficient to exclude the possibility that the agreement may 'affect' such trade in the above-mentioned manner.

Little difficulty, of course, was found in that case in establishing the necessary effects on volume of trade because Consten was clearly restricted from exporting Grundig products to other countries of the Common Market, whilst purchasers of Grundig products in Germany were restrained from importing Grundig products into France.

At that time, the tests that seemed to be required as the result of these two cases appeared fairly simple to apply. The necessary approach involved imagining a complete overview of trade within the Common Market, as a series of flows and counter-flows crossing and recrossing national boundaries, under the effect of seasonal and other changes in supply and demand, costs, prices, and other external factors. It was necessary to compare this picture, as it existed in the absence of any restriction on competition, with the picture after the restriction had been introduced; in theory such a comparison should have produced a definite answer. We shall see, however, that this comparatively simple test was to be complicated later by further qualifications to the initial basic rule.

As we have learnt, the Commission was soon itself faced with a more immediate problem of how to exclude from the coverage of Article 85 a number of agreements considered of relatively minor importance, which it was especially anxious to remove from consideration in view of the volume of notified agreements received at the beginning of 1963 and whose continuing effect was felt throughout the following years. It received some assistance from the decision of the European Court, under an Article 177 reference from the Munich Oberlandsgericht in the case of *Frans Völk* v. *Vervaecke.*[3] Völk was a small German company that manufactured washing machines and entered into an exclusive distribution agreement for Belgium with Vervaecke, on normal terms including an obligation on the Belgian distributor to take minimum quantities. Subsequently Vervaecke was sued in Germany by Völk for alleged breach of contract. The question raised by the German Court under Article 177 was whether, in interpreting Article 85, it was entitled to pay attention to the very modest fraction of the relevant geographic market (Belgium) in which the product was sold.

The answer which the Court gave was that, in principle, insignificant

[3] Case 5/69 [1969] ECR 295: CMLR 273.

agreements could escape the prohibition of Article 85, since it was essential to be able to show a reasonably probable expectation that the agreement would exercise an influence direct or indirect, actual or potential, on trade trends between Member States to an extent that would harm the attainment of the objectives of a single market between States. The total market share that Völk had obtained in West Germany was about 0·2 and 0·05 per cent in 1963 and 1966 respectively, and the number of machines annually sold in Belgium and Luxembourg by Vervaecke was about 200 machines. The essence of the decision was, therefore, that a *de minimis* rule could be applied in such cases. This would mean at any rate that in assessing effect on trade flows between Member States, agreements affecting a market share of 1 per cent or below would be most unlikely to have the required effect. It should be noted, however, that in this case both parties were small undertakings, and the same generosity of treatment may not be accorded to major public companies even where their market share of the relevant product is less than 1 per cent.[4]

The Commission followed up this case with its 1970 Notice concerning 'Decisions and concerted practices of minor importance which do not fall under Article 85(1) of the Treaty', which stated that agreements whose effects on trade between Member States and on competition are negligible did not fall under the ban of Article 85(1), which only applies if the agreements 'have an appreciable impact on market conditions, that is if they appreciably alter the market position, in other words the sale and supply possibilities, of non-participating firms and of consumers'. This definition followed carefully the approach of the Court in the *STM* and *Grundig* cases. In this Notice the Commission sought to quantify this principle by referring to the conditions that (*a*) the product and its immediate substitutes should not constitute more than 5 per cent in a substantial part of the Common Market (which, of course, could comprise either a Member State or possibly even a well-defined region) and (*b*) that the aggregate annual turnover of the undertakings participating in the agreement did not exceed, originally, twenty million units of account for distributors or fifteen million units of account for manufacturers. An excess of up to 10 per cent was allowed during not more than two consecutive financial years; but the application of the Notice is reduced by the fact that in calculating the turnover it is not only the turnover in the goods which are the subject of the agreement which counts but the total turnover of goods and services of the parties. These figures were sub-

[4] *Distillers Co.* v. *Commission* Case 30/78 [1980] ECR 2229: 3 CMLR 121. Here the Court stated that Article 85 would apply and that no exemption would be granted of even a product like Pimm's which held only a fraction of 1 per cent of the relevant market, if it constituted merely one of the products of a large company which sold a wide range of products and whose total sales as a percentage of the combined markets exceeded the *de minimis* level.

sequently raised in 1977, and again by the most recent Notice of September 1986, under which the minimum aggregate turnover figure now stands at 200 million units of account.

These Notices are consistent with the concern expressed by the Commission generally (not simply DG IV) for the interests of small and medium-sized enterprises, affectionately known as SMEs, a concern which subsequently would be shared by the European Parliament and lead to a number of Commission initiatives in later years designed to provide help for such enterprises. Nevertheless, for several reasons such Notices have to be treated with great caution. They cannot, of course, bind the European Court or Courts in Member States. Moreover, the Commission itself reserves an 'escape route' by stating that the quantitative definition of 'appreciable' given in the Notice is not an absolute yardstick. This comment has two effects. Some agreements may exceed the quantitative figures stated and yet still not have the required effect on trade between Member States, but on the other hand the Commission retains the right to aim that undertakings with less than the total turnover mentioned in the Notice may still be unable to benefit from it.

The Notice, therefore, does not have a major effect upon the interpretation of 'may affect trade between Member States'. It is particularly difficult to apply in a vertical context where the agreement with an individual distributor may separately refer to quite small percentages or numbers of products, but may in aggregate reach a substantial total, if there are substantial networks of dealers or distributors in a particular Member State whose turnover must be taken into account.

In a case where there is a network of agreements, it is not permissible to treat them individually, ignoring the existence of the network, as the Court emphasized in *Brasserie de Haecht* (no. 1) (European Court 1967).[5] A brewery in Belgium had made loans to a married couple called Wilkin-Janssens, proprietors of a café in Esneux. In return for the loan, the café owners had agreed to take all their requirements for beer, lemonade, and other drinks exclusively from the brewery for the term of the loan plus a further two years. The brewery subsequently discovered that the agreement had been broken and sued for the return of the loan and damages. The Wilkin-Janssens then claimed that the agreement was void under Article 85(1), and the Belgian Court at Liège asked the Court for a ruling whether, in assessing whether 85(1) applied, it was possible to take into account the economic context of the agreement and the surrounding

[5] *Brasserie de Haecht* v. *Wilkin-Janssens* (no. 1) Case 23/67 [1967] ECR 407: [1968] CMLR 26. In a subsequent case the Court, however, emphasized that even if an agreement is part of a network of similar agreements the individual contribution of that particular agreement has also to be taken into account, *Delimitis* v. *Henninger Bräu* Case C 234/89, a decision of 28 Feb. 1991. (See Ch. 12, pp. 239).

circumstances affecting this particular type of agreement in the Belgian beer market. The Court, closely following in this case the opinion of its Advocate-General Roemer, ruled that when assessing whether an individual agreement was in breach of Article 85(1), it was necessary not to look simply at the individual agreement but to look at the context, both economic and legal, in which the agreement had been made. It placed importance on the degree to which the effect of the network would be to foreclose the brewer from finding sufficient outlets for its own products, whether by sales to existing pubs or cafés or through the acquisition of its own chain of outlets. If, therefore, the whole network of similar brewery 'tying' agreements made in Belgium was shown to have an effect on trade between Member States, e.g. if it made it more difficult for breweries in France or Germany to arrange for export of their beers into Belgium, then this fact could be taken into account both in determining whether the original agreement with the Wilkin-Janssens affected trade between Member States, and also with regard to its effect upon competition. The existence of this further qualification upon the rule of *Völk* v. *Vervaecke* likewise served to limit the scope of the application of the Notice on agreements of minor importance.

The subsequent history of the interpretation of this requirement for effect on trade between Member States shows a steady widening of its reach, and the pattern of interpretation accorded to it both by the Commission and the Court has close parallels (although in a totally different environment) with the development by the United States Federal Courts of a constantly expanding definition of 'interstate commerce', for purpose of interpretation of the United States Constitution.[6] In the great majority of cases, however, the effect of the restrictions concerned on trade between Member States is obvious and involves little difficulty for Commission or Court.

The definitions already adopted were further relied on in a number of cases involving widely-based trade associations within a single Member State which draw their members from those engaged at the different levels respectively in manufacturing, wholesaling, importing, and distribution of a specific product in that country, but where the effect of the restrictions was considerably to reduce the attractiveness of that market for imports from other Member States by increasing the difficulties of breaking into the market. This was so even though the relevant trade association rules or related agreements purported to be motivated by other objectives, e.g. protection of quality standards. We shall examine such cases in Chapter

[6] For a detailed and comparative analysis of the process of interpretation of the 'Interstate Commerce' clause in the USA, see *Courts and Free Markets: Perspectives from the United States and Europe*, ed. T. Sandalow and E. Stein, 2 vols. (Oxford: Clarendon, 1982), esp. chs. 1 and 2 of vol. 1.

21; in the great majority of cases the necessary effect on trade between Member States was easily established.

One exception, however, was the *Belgian Wallpapers*[7] case (European Court 1975). Whereas in the other similar cases the Commission's decision had either not been appealed against or had been upheld on appeal by the European Court, on this occasion the fine imposed by the Commission was quashed on the grounds that its analysis of the effects of the agreement was insufficiently detailed to justify a finding of an effect on trade between Member States. Indeed the Commission's reasoning consisted of simply the following words:

> The agreement and the decision based upon it may also affect trade between Member States, since wallpaper manufactured outside Belgium and sold in Belgium by the members of the Groupement is also included. Apart from this the aggregated rebate scheme hinders the importation of wallpaper into Belgium. The agreement and the decisions based upon it directly affect freedom of trade between Member States in a way which is detrimental to the attainment of a single market. By its very nature, an agreement extending over the whole area of a Member State has the effect of reinforcing the compartmentalization of markets at national level; it thereby prevents the economic interpenetration which the Treaty is designed to bring about and protects domestic production.[8]

Nevertheless, the reasoning both of the Court and of the Advocate-General gave little encouragement to those who would have restricted the ability of the Commission under this Article to investigate this kind of arrangement within a single Member State. The facts of the case indeed on the contrary illustrate just how unlikely it is that any attempt within a single Member State to regulate, at a number of different levels in the chain of distribution, the basis upon which a product is distributed can in practice avoid affecting trade between Member States.

The Belgian wallpaper manufacturers had detailed internal regulations governing the basis upon which wallpaper was distributed in that country, under the supervision of an independent trustee whose duty was to ensure that the rules were adhered to. Price scales were laid down for different qualities. Goods had to be sold on standard conditions of sale, and cash discounts prohibited. Certain kinds of customers could not be dealt with at all, and wholesalers were required to place minimum orders in each year; no deliveries could be made on Saturdays. Wholesalers to whom manufacturers sold were required to comply with mandatory resale prices, and a rebate allocated between them was based on a progressive scale related to the total purchases from all members of the manufacturers' group, thus encouraging the wholesalers to deal substantially or

[7] *Groupement des Fabricants de Papiers Peints de Belgique* v. *Commission* Case 73/74 [1975] ECR 1491: [1976] 1 CMLR 589.

[8] [1974] 2 CMLR D114.

exclusively with those manufacturers and not with other manufacturers. Manufacturers in the association supplied some 60 per cent of the domestic wallpaper product on the Belgian market, 50 per cent representing their own manufactures and 10 per cent their imports. Originally, the appeal was against the entire decision of the Commission, but was subsequently amended so as to relate solely to the fine imposed in respect of the collective boycott of a price cutting wholesaler by members of the association.

The Court pointed out that the proper interpretation of Article 85 with regard to requirements of effect upon trade between Member States was not simply one of preventing the partitioning of the market into separate national markets, but was now principally one of maintaining competition in a healthy state throughout the Common Market; this aim would be prejudiced by imposing an aggregate rebate system on wholesalers which would discourage them from importing foreign products which did not come through the members of the group and would, therefore, reduce the level of exports into Belgium.

Nevertheless, the decision imposing the fine for the collective boycott was overruled and in effect the Commission received a 'rap on the knuckles' for inadequate investigation of the effect on trade between Member States. There can be little doubt that this was a signal from the Court to the Commission that it should not be content merely to repeat standard phrases in those parts of its decisions dealing with the allegation that trade between Member States was affected; rather it should in each case carry out a sufficient economic analysis of the relevant market to show how imports of the relevant product had been affected. It had failed to do so in this case, but would be concerned to produce a less stereotyped explanation on future occasions as has indeed proved the case. As the Fifth Annual Report stated somewhat ruefully

> An important consequence of this judgment is that a more detailed explanation of the reasoning of the European Commission will be required when there is any broadening of the concept of effects on trade between Member States in cases involving national agreements. It will have to be shown from all the facts to what extent and in what manner the national market is protected by the national measures taken under such agreement.[9]

Yet further confirmation that even agreements of an apparently domestic nature within a Member State, and without the element of attempting to control the operation of a substantial part of an entire national market, can still affect trade between Member States is seen from *Salonia* v. *Poidomani and Baglieri*[10] (European Court 1981). This was a reference

[9] p. 30.

[10] Case 126/80 [1981] ECR 1563: [1982] 1 CMLR 64. Another good example of the

under Article 177 by a Court at Ragusa in Sicily. Salonia was the proprietor of a retail business in Ragusa which dealt in stationery, newspapers, and books. She wanted to be supplied with both newspapers and periodicals by Poidomani and Baglieri from their wholesale warehouse in Ragusa. They, however, refused on the grounds that she was not on the list of retailers approved by the Italian publishers' association. Mrs Salonia brought an action in Italy to force the defendants to supply her with newspapers and periodicals, claiming that the failure to do so constituted unfair competition.

The European Court was troubled by some uncertainties as to the factual situation in the case, and in particular whether the relevant agreement laid down by the publishers' association in Italy was still in existence, and whether the defendants adhered to it. Nevertheless, in dealing with the issues involved, it held that a distribution system operated in accordance with a national trade agreement which restricted the supply of relevant products to approved retailers would infringe Article 85, if it were shown that authorized retailers were not chosen on the basis of objective criteria, such as their abilities as retailers or their suitability with regard to staff, trading premises, and experience. Following its earlier rulings in selective distribution cases,[11] criteria were to be laid down uniformly for all potential retailers and not applied in a discriminatory fashion. Another relevant fact would be whether the product involved was simply 'local and national newspapers circulating within Italy', or whether foreign newspapers or periodicals were affected and were less easily distributed because of the operation of the scheme.

In the light of these developments, the early debates over the correct interpretation of the word corresponding to 'affect' in the English translation now seem relatively academic. At a time when there were only the four original authoritative languages for the Treaty of Rome, it was apparent that whilst the Dutch word required an unfavourable influence to be shown on trade between Member States, the equivalent expression in German and Italian was more ambiguous, and the French text 'affecter' could be used properly in both a neutral and a pejorative sense. The combined effect of the slightly different flavours of these four translations was to leave the impression that an effect upon trade which was of a purely neutral kind, i.e. having both good and bad consequences, would not necessarily be covered by Article 85(1). Deringer,[12] commenting on these expressions some ten years after signature of the Treaty, indicated that he

same principle is *Belasco* v. *Commission* Case 246/86 [1989] ECR 2117: [1991] 4 CMLR 96 involving a cartel of Belgian roofing felt manufacturers.

[11] See Ch. 13 for a review of cases on selective distribution.

[12] A. Deringer, *The Competition Law of the European Economic Community: A Commentary on the EEC Rules of Competition* (CCH Editions Ltd., 1968), 22–3.

found the wording alone of the Article insufficient to support a definite conclusion, and that the best means of interpretation was to consider the essential purpose of the expression. This was to promote and protect the free flow of trade within the Common Market unhampered by private restraints of competition. If the restraint changed the intensity or the direction of the flow of goods, artificially diverting it from its normal and natural course, then the Article would certainly come into play, and the question of whether the effect was of an adverse or of a neutral kind would not normally arise. The Court's approach in *Grundig* v. *Consten* would also be consistent with this view.

Finally, it should be mentioned at this stage that similar wording is found in the following Article, 86, prohibiting abuses of dominant positions within the Common Market. The two leading cases in which the European Court has confirmed the interpretation of these words in the context of Article 86 are *Commercial Solvents* v. *Commission*[13] (European Court 1974) and *Hugin* v. *Commission*[14] (European Court 1979). In the former case the Court stated that if an undertaking within the Common Market exploited its position abusively in such a way as was likely to eliminate a competitor, it did not matter whether the conduct complained of related specifically to the exports to be made by the competitor or its general trade within the Common Market, provided only that it was established that its elimination by the dominant company would have an effect on competitive structures within the Common Market. However, the conduct complained of in the case was a refusal to supply a particular chemical substance essential to the business of the smaller company and for which there was no other source of supply, so that ultimately complete elimination from the market would have followed for that smaller company.

The *Hugin* case by contrast was one of those rarities where the decision of the Court was that there was insufficient effect on trade between Member States to justify a finding of abuse or breach. The facts, however, were rather unusual since the product market was of spare parts for cash registers manufactured by Hugin, a large Swedish company, having at the relevant time some 13 per cent of the United Kingdom market for the registers. Lipton was a company acting in the servicing and maintenance of such machines; following disagreements between the parties further supplies of spare parts were refused, thus preventing Lipton from carrying on its independent business of servicing them.

Lipton claimed that the refusal by Hugin to supply spare parts violated Article 86 as, without them, Lipton was unable to carry on its business.

[13] *Istituto Chemioterapico Italiano and Commercial Solvents Corporation* v. *Commission* Cases 6-7/73 [1974] ECR 223: 1 CMLR 309.

[14] *Hugin* v. *Commission* Case 22/78 [1979] ECR 1869: 3 CMLR 345.

No objective reason had been offered for the refusal to supply the parts, now supplied only to Hugin's own subsidiaries. The Court agreed that there was no reason in theory why these facts should not justify a finding of abuse of a dominant position but found that there was insufficient effect on trade between Member States since Lipton only traded in its capacity as a service company within a 50-mile radius of London and made no exports of any kind. Moreover, the characteristics of the market were that Hugin produced many different models for the various countries in Europe which were affected by the different requirements of language and currency, so that the exporting of machines or spare parts from one Member State to another was uneconomic and unlikely. The Court's finding, therefore, was that an independent undertaking such as Lipton would not find it in its interests to buy spare parts in other Member States rather than obtaining them from the parent company in Sweden. There was, therefore, in the Court's view no normal pattern of trade in such spare parts between Member States that could be disrupted by the commercial decision made by Hugin not to supply.

Finally, it should be mentioned that 'trade' has received a wide interpretation; it includes not merely the normal industrial and commercial activity of manufacture and distribution but applies to commercial services of all kinds including banking, insurance, and financial services. It also applies to a variety of other activities of a commercial nature including the provision of exhibitions and trade fairs, the maintenance of performing rights societies, the provision of television programmes and other cultural facilities. Its application too is not limited solely to trade with the twelve Member States, but extends also to trade with all those EFTA countries still outside the Common Market (Austria, Finland, Iceland, Norway, Sweden, and Switzerland) which are likely to form part of the European Economic Area (EEA) and to Canada, Israel, Turkey, and certain other small associated territories having free trade agreements with the Community.

In the case of each of these countries, the absence of tariffs between it and the Common Market as the result of these agreements means that restraints relating to trade with such a country (and in particular any form of export ban on export from that country) will be likely to affect also the extent of trade with the Common Market. There may, of course, be other circumstances, such as transport costs, which would nevertheless justify a particular decision that trade with the Common Market would not be affected by such a restraint.

In summary, therefore, the effect of these words in the Article lay down a threshold qualification which, as a matter of jurisdiction, the Commission must in each case satisfy before proceeding further. It is possible to surmise that the Court might have originally interpreted these words more restrictively, but in the interests of the unity of the Common Market, especially in its early years, chose not to do so. In practice, the Commission will now assume that trade between Member States is affected by virtually any practice which brings about some noticeable effect on market conditions or structures and involves undertakings of a size above the level affected by the current Notice. The onus of proof will then effectively shift to the parties to prove the negative, in most cases a difficult task.

2. 'Which have as their Object or Effect the Prevention, Restriction or Distortion of Competition within the Common Market'

The difficulties inherent in defining competition have already been discussed (in Chapter 2). If the competition referred to in this Article were to be of that rarely encountered atomistic variety, requiring a very large number of market participants unable themselves by their own policies to influence prices or the level of demand, then to test whether any particular agreement or concerted practice between undertakings would have an effect on the competition within that market would be comparatively simple: its effect would be clearly and immediately perceptible, like the result of throwing a stone into a smooth pond. Since, however, virtually all the product markets encountered within the Community are imperfectly competitive, the task of assessing whether an agreement or practice does cause or have as its object the prevention, restriction, or distortion of competition becomes considerably more elusive.

Another problem arises, however, with the interpretation to be given to the words 'prevent, restrict or distort competition within the Common Market'. Too strict a definition might bring in almost every agreement for the sale of goods or services. Too liberal a definition would reduce the jurisdiction of the Commission, since it would exclude from its control a substantial number of agreements that may either individually or cumulatively have an effect on competitive structures or processes. In determining the correct approach to these words within Article 85(1) it is necessary to decide therefore whether the Article is a provision primarily designed to assert jurisdiction or alternatively is itself an assessment of whether individual agreements can be justified. Normally an Article that seeks to establish initial jurisdiction would be expected to have a more comprehensive interpretation than one that is concerned simply with the assessment or substantive justification of the agreements. After all, it is obvious that many more agreements will need scrutiny by the Commission than will ultimately be adjudged to have failed to pass whatever substantive criteria are applied to them. This will not necessarily imply, however, that a 'jurisdictional' Article should be read with total adherence to a strictly literal meaning, bringing within its grasp every kind of agreement or concerted practice with the smallest impact on competition. It is clearly sensible for some limit to be placed on the scope even of such a phrase read in a broad sense.

The European Court has taken the view that Article 85(1) is a jurisdictional clause but has nevertheless on occasions placed a limit on the width of interpretation to be placed upon it. This limit is sometimes described as 'the rule of reason' although it is misleading to use this phrase since it is not properly to be compared directly with the same phrase used in describing the interpretation by US Federal Courts of Section 1 of the Sherman Act.[15] Section 1 of the Sherman Act is, of course, a compression of both 'jurisdictional' and 'assessment' clauses and the relevant phrase used of 'restraint of trade' therefore necessarily requires a more liberal approach. Over its more than one hundred years of existence the section has seen a variety of different approaches from Courts, though the liberal approach to interpretation has been more in the ascendant in recent decades, particularly in the treatment of vertical agreements.

The assessment element contained in Article 85 is, of course, found primarily in Article 85(3) and to a much more limited extent in Article 85(1). To discuss therefore the alleged 'rule of reason' in the context of 85(1) and in particular in the context of the scope of 'prevention, restriction or distortion of competition' leads to confusion rather than clarity. What is important for a clear understanding of these words is knowledge of the way in which the Commission and, more importantly, the European Court and the Court of First Instance have interpreted them in a variety of different contexts. Initially this was done with strictness but in recent years with a greater degree of flexibility, but without ever giving an escape from the control of the Article to more than a narrow range of clauses primarily found in joint ventures and certain licensing agreements. Only, however, after the reader has considered the application of this phrase in all the different contexts which decided cases present will it be possible to determine whether there exists under Community Law a single principle to explain all the exceptions to Article 85(1) and to which a general title such as 'rule of reason' could be given, if misleadingly, or whether these examples will remain relatively limited exceptions to a normally rigorous general interpretation of these words in Article 85(1).

The meaning, however, of 'prevent, restrict or distort' itself becomes clearer if one considers the normal mental processes of any businessman

[15] See A. D. Neale and D. G. Goyder, *The Antitrust Laws of the United States*, 3rd edn. (Cambridge University Press, 1981), describing this issue, at pp. 23–30.

seeking to enter a specific geographic and product market.[16] If there are no agreements or concerted practices then he will have complete freedom of choice as to the territories in which he can sell, the availability of outlets through whom he can distribute his goods, the prices and conditions of sale at which he can distribute, not to mention a range of other choices in other matters such as marketing techniques. If, however, there are agreements between other undertakings which have the effect of limiting his freedom of choice so that he is, for example, foreclosed from using his first choice of distributors, or cannot (because of earlier exclusive appointments) be appointed a franchisee for particular goods, cannot be appointed an exclusive dealer himself because there is a quantitative limit on those nominated within a given territory, or if he cannot manufacture successfully in a particular Member State because sole or dominant suppliers have entered into agreements which restrict their ability to sell to him, then in all these cases the restraint of trade is directly responsible for the reduction in freedom imposed on his business by the agreements in existence between others already in the market. The object of Article 85 is to eliminate or reduce the number of such agreements and also to try and distinguish those agreements without sufficient redeeming virtues from those whose effect on competition, or whose purpose of reducing competition, is nevertheless accompanied by substantial advantages of the kind set out in Article 85(3).

In some cases, the restraint will affect all manufacturers or all wholesalers or retailers of a particular product in a particular Member State: in other cases, the range of undertakings affected will be smaller. An important preliminary point, however, is that fully free competition is not possible in all markets because of government intervention in their workings.

The best known example of this is the *Sugar Cartel*[17] case where the Italian Government had laid down complex rules to govern the marketing of sugar; the Italian public authority in charge of sugar prices, the CIP, adopted a series of regulations intended mainly to benefit Italian sugar-beet producers and exporters. These aids were financed by a levy applied both to Italian sugar manufacture and also upon imported sugar, but with reduction of the levy permitted in a few cases, to encourage a certain level of imports, so as to fill up the gap between Italian production and the target level of demand. Foreign producers were invited to tender to supply certain quantities of sugar for import, and complicated regulations determined the extent to which the levy was charged in full or only in part. The effect of the regulations together with the risks attaching to the

[16] For an assessment of the strategic business approach to this situation, the works of Professor Michael Porter of the Harvard Business School are worthy of careful study. See Ch. 2, n. 4.

[17] Case 40/73 [1975] ECR 1663: [1976] 1 CMLR 295.

invitations to tender had the result that foreign manufacturers would often agree to allocate between themselves the right to supply into Italy, because if they did not do so they individually risked being excluded from the market altogether. Certain imports were permitted outside the formal system of invitations to tender, but only in limited quantities and further regulations controlled the price that could be obtained by foreign concerns for imports into Italy. The regulations also had the result of concentrating demand in Italy in the hands of the large producers, since they alone were likely to be able to fulfil on a reliable basis the requirements of purchasers.

The real consequence of these regulations was to match supply closely to demand, which itself removed a vital element of normal competition, and had a considerable effect both on the buyer's freedom to choose his supplier (and vice versa) and the price at which the goods were supplied. The Court concluded that

> All these considerations show that Italian regulations and the way in which they have been implemented had a determinative effect on some of the most important aspects of the course of conduct of the undertakings concerned which the Commission criticises, so that it appears that, had it not been for these regulations and their implementation, the co-operation which is the subject matter of these proceedings, either would not have taken place, or would have assumed a form different from that found by the Commission to have existed.[18]

The Court's ruling in this instance was, therefore, that the sugar producers concerned, who themselves operated restrictive agreements in response to the Italian 'sugar regime', did not prevent, restrict, or distort competition because of the limited scope left for it by the result of the Italian State intervention.[19]

It was established in the early days of the Court that an agreement could satisfy the requirements of these words if it either had as its object a restriction on competition or if (whatever its object) the restriction had that effect; it was unnecessary to prove both object and effects in any individual case.[20] This wording is significantly different from the terms of Article 65 of the Treaty of Paris which refer to agreements 'tending directly or indirectly' to the prevention, restriction, or distortion of competition. The requirement for either an 'object' or 'effect' of such a restriction in the Treaty of Rome is more specific. The agreement must either be intended to have such a result on competition or, regardless of the parties' intentions, actually does so. This has over the years proved both a

[18] [1975] ECR 1923: [1976] 1 CMLR 410.
[19] See Ch. 23 for consideration of other leading cases where the degree of competition, in particular national markets, was reduced by Governmental measures, without, however, necessarily allowing undertakings affected by it to avoid the application of Article 85(1).
[20] See the *STM* Case (see n. 1).

comprehensive and workable definition. The Court stated that it was first necessary to look at the object or purpose of the agreement by looking at the provisions actually stated in it, considered in its economic context. Only if it did not appear that the purpose of the agreement was really to restrain competition was it then necessary to go on to consider its effects. If effects had to be considered, then the issue was whether the agreement restricted competition perceptibly, comparing the results of the agreement with the likely state of affairs which would exist in the absence of that restriction. Restrictions had always to be considered in the circumstances of their economic background.

The same issue came before the European Court in the *Lancôme* v. *Etos* case[21] (European Court 1980). Here a selective distribution system for perfumes in France had been notified to the Commission, but no formal grant of exemption had been given; the parties had received merely an administrative letter (comfort letter) confirming that the system was not regarded as infringing Article 85(1). Various retailers of perfume who had unsuccessfully tried to obtain supplies from Guerlain and other manufacturers brought action under French law on the grounds of refusal to supply. It was held that such agreements might be covered by Article 85(1) if the test for admission of retailers to the distribution system went beyond objective and qualitative criteria and were based simply on limiting the numbers of dealers, that is a quantitative basis for selection. To decide if there was an actual breach, however, it would be necessary to apply a test of whether in the actual economic circumstances of the markets concerned the nature of the competitive process would be effective absent the restrictive agreement or practices. To be taken into account were the nature and quality of the goods, the quantity available and whether this was limited either naturally or artificially, and the position and importance of the parties on the relevant market (in this case perfumes). Finally, the issue of whether the agreement was an isolated one or part of a large network would have to be considered.

Whatever the need in certain cases like this to examine the consequences of the agreement in detail to see if it has an effect on competition, there are some contractual restrictions which case-law has established are so prima facie likely to affect competition that this will be presumed without specific proof. Clearly, those categories of agreements which are listed under sections (*a*) to (*e*) inclusive of Article 85(1) would be likely to fall into this category. This is so even though these lists are only examples and are in no sense an exhaustive enumeration of the kinds of agreement which the Article prohibits. We will in later chapters

[21] Case 99/79 [1980] ECR 2511: [1981] 2 CMLR 164. There were a number of other European Court cases also concerned at this time with selective distribution of perfumes in France, but this is the leading case on this issue.

be considering some individual examples of agreements falling within these categories. The category which has come closest to being called *per se* illegal is the agreement which, like the restrictions in *Grundig–Consten*, bans parallel imports directly by imposing export bans on distributors, or which in arrangements between horizontal competitors bans the export of goods to other countries within the Common Market or into territories outside the Market from which they would be likely in practice to be re-imported into the Market again.

There are many cases illustrating this important feature of the jurisprudence of the Common Market, deriving from the original judgment of the European Court in *Grundig*, which has proved of such subsequent importance. Mention should at least be made, however, of two important cases where such export bans were presumed to restrain competition. In the *WEA-Filipacchi*[22] case, German manufacturers sought to impose a restraint on various French dealers in pop records from reselling the records outside France, particularly in Germany. One of the arguments raised by the manufacturers was that they would not in practice have been able to enforce these restrictions and that, therefore, they were without effect. The Commission held, however, that any restriction in an agreement which effectively operated as an export ban was so inherently the type of agreement that would have as its objective the prevention of free movement of goods across national boundaries by way of parallel imports that it was unnecessary to consider the further issue of whether the manufacturer concerned had the actual power to enforce such restrictions. A similar decision was reached by the Court itself in the later case of *Miller International* v. *Commission*[23] (European Court 1978); again a record company had imposed an export ban on its French distributors. On this occasion the Court disregarded the opinion of the Advocate General who had recommended a cancellation of the fine imposed by the Commission and remission of the case to the Commission on the grounds that the undertakings has been misled by advice given by its lawyer. The Court found that the export ban imposed was by its very nature a restriction on competition, whether adopted at the instigation of the supplier or of the customer, since the whole basis of the agreement between the parties was an endeavour to isolate a part of the market.

To affect competition within the Common Market, it is not necessary that all the parties to the agreement are themselves resident within it. This can be illustrated by reference to Commission decisions relating to trade between the Common Market and Japan. In the *Siemens/Fanuc* decision[24] (Commission 1985), a fine was imposed on both a German and a

[22] [1973] CMLR D43. [23] Case 19/77 [1978] ECR 131: 2 CMLR 334.
[24] [1988] 4 CMLR 945.

Japanese company which had entered into reciprocal exclusive dealing agreements covering numerical controls for machine tools for Asia and the EC respectively, and under which a substantial range of restrictions was imposed and extremely high prices charged to all customers, who could not obtain the controls from other sources. By comparison, in the *Europair–Durodyne*[25] case, the appointment of a Belgium company as the sole distributor for US manufactured air-conditioning units in the Community was upheld on the grounds that it was essential for effective coverage of the territory, although the exemption was doubtless favourably influenced by the fact that the US manufacturer remained able to compete for business in the distributor's territory.

Agreements, however, between parties in a horizontal relationship will nearly always be found to raise greater problems, when a similar market analysis is carried out. Thus, in the *Franco-Japanese Ballbearings*[26] case (Commission 1974) it was proved that meetings had taken place between the manufacturers of ballbearings both in France and Japan to discuss the sales and pricing policy of Japanese manufacturers with regard to imports into France. Considerable pressure was applied upon the Japanese manufacturers to bring export prices to France into line with French prices, and to inform the French manufacturers of the prices and discounts that the Japanese would grant to purchasers in France. Here the restraint on competition was clear in that customers for the ballbearings would find themselves as a result of the agreement paying a higher price to Japanese manufacturers than would otherwise have been the case. The flow of trade between Member States within the Common Market would also be affected, since Japanese ballbearings were supplied to other Common Market countries and the likely effect would be both a reduction in direct imports from Japan and also an increase in parallel imports to France from other Common Market countries, as well possibly as an overall reduction in the purchase made by France from Japanese sources. The approach of the Court of Justice in the *Woodpulp*[27] case has been to focus on the place where the agreement in question was implemented, regardless of where it was made or where the parties to it were situate.

In summary, therefore, the Commission has in nearly all cases to conduct a careful market analysis of the effect of particular restrictions or agreements on the patterns of trade. There will be occasions when it may be excused from doing this in detail either because the restriction is of the type, e.g. an export ban, so clearly established as having the required effect on competition as not to need substantial proof of its individual effect on the particular market concerned; in other cases the objective of

[25] [1975] 1 CMLR D62.

[26] [1975] 1 CMLR D8.

[27] Cases 89, 104, 114, 116–17, 125–29/85. [1988] ECR 5193: 4 CMLR 901. See Ch. 24 for a further account of this case.

the agreement may be so clearly anti-competitive that the Commission is excused the need to analyse the effects of that restriction.

The reader will also note when considering cases reviewed in the following chapters that the Court has found itself on occasions able to characterize a number of restrictions on competition as merely 'ancillary' to legitimate business purposes, and, therefore, outside the scope of the requirement that there be either substantial purpose or effect upon competition. Such a restriction is one which is regarded as necessary because, in its absence, the principal transaction proposed would not take place at all, because of the inability to prevent the other party to the agreement from acting later in such a way as in effect to deprive that transaction of its commercial benefit. These restrictions have included those enabling a manufacturer to control the methods of business adopted by its franchisees so as to protect its business secrets and to adopt qualitative standards as criteria for the appointment of distributors of certain specialized classes of goods (see Chapter 13), and those enabling purchasers of businesses to place 'no-competition' restrictions for a reasonable period on their vendors (see Chapter 9). In the field of intellectual property rights, the *Maize Seed* and *Coditel* cases considered in Chapters 15 and 16 offer licensors of intellectual property rights likewise a broad interpretation of this expression in cases where some form of 'open exclusivity' was essential to the launching of a new product protected by such rights.

8

Article 85 (3): Conditions for Exemption

The main task of the Commission in the area of competition policy is the application and enforcement of the two balanced paragraphs 85(1) and 85(3), the former setting out the prohibition, the latter limiting its application. They are linked by the important statement in paragraph (2) of the legal consequences of applicability of the Article to a specific agreement if the relief given by paragraph (3) is unavailable. The meaning of paragraph (2) will be considered at the end of this chapter. Before we reach it, however, it is necessary to understand both how the Commission goes about its task of applying Article 85(1) (and much of this description would be equally applicable to the enforcement of Article 86 relating to abuses of dominant position) and in particular how it operates the exempting provisions of Article 85(3).

The Commission itself exercises its powers through a number of Directorates-General,[1] which in turn are divided into directorates; each of the Directorates-General has, under Article 155, and subsequent regulations made by the Council, some policy-making functions. In some areas, e.g. those concerned with the harmonization of corporate and tax laws of Member States or in connection with measures to control environmental pollution within the Community, little can be achieved by the individual Directorate-General without active support on issues of policy from the Council of Ministers. This is so even if the body of Commissioners (17 in number since the admission of Spain and Portugal on 1 January 1986) are supportive of the proposals put forward. This may mean that the staff of a directorate within a Directorate-General may consult widely and work up proposals for long periods, yet find ultimately that all their carefully prepared suggestions (already the subject of detailed negotiations over many months with individual Member States and relevant organizations) are ignored or rejected or, at the least, postponed after proposal as a draft directive or regulation because of political difficulties arising at the relevant meeting of the Council.

It is notable that DG IV, as the Directorate-General responsible for competition matters, manages to avoid some, if not all, of this frustration and operates in many respects differently from other DGs. The reason is that in Council Regulation 17/62 it obtained a procedural statute which delegates

[1] Normally abbreviated to DG in both oral and written usage.

considerable powers from the Council to the Commission itself; the Commission in turn will normally act on the advice and recommendation of the Commissioner having special responsibility for competition. He in turn will normally be guided by the advice of his officials which will, of course, have been thoroughly discussed in advance with the Commissioner's 'cabinet', the group of individual officials operating as his private office, as well as having been the subject of prior consideration by the Advisory Committee, which comprises official representatives of Member States entitled to comment both on proposed individual decisions and on proposed Community legislation on competition issues.[2]

So far as competition issues are concerned, the Council will now only normally become involved if new legislation is needed, for which authority has not already been given to the Commission.[3] In general, therefore, the Commission is able to proceed with its regular decision-making function without political intervention. It is, however, important to be aware that all the decisions taken by the Commission are subject to a right of review for interested parties under Article 173 of the Treaty and in respect of alleged failure by the Commission to act under Article 175. Article 173 may be invoked by any person challenging a procedural or substantive error by the Commission, including a failure to provide economic analysis into the factual background to any decision and may lead to that decision being quashed on one of the following grounds:

(*a*) lack of competence (or authority);
(*b*) infringement of an essential procedural requirement;
(*c*) infringement of the terms of the Treaty (or any rule of law relating to its application); or
(*d*) misuse of powers.[4]

The functions of DG IV acting on behalf of the Commission in the application and enforcement of Articles 85 and 86 can conveniently be divided into administrative, legislative, and decision-making. The administrative function itself includes both the investigation process and also all areas of executive control, including the recording and review of notifications, and the correspondence, meetings, and negotiations necessarily involved with undertakings involved in notifications or complaints which ultimately may lead to the preparation of a draft decision. Since the latest reorganization of DG IV in 1986, officials operate in one of the three specialized directorates (B, C, and D) which carry out all necessary enquiries and deal with both formal and informal decisions, but advised

[2] Under the terms of Article 10(3) of Reg. 17.

[3] For example in finally adopting the Merger Regulation 4064/89 after many years of negotiation.

[4] This phrase represents the expression *détournement de pouvoir* used in French constitutional and administrative law.

and guided by directorate A which plays a co-ordinating role on policy issues.[5] There is in addition the Merger Task Force which, operating as a separate division, is responsible for the administration of the Merger Regulation.

1. Group Exemptions

Important legislative functions have been delegated to the Commission by the Council, and the powers (*vires*) which the Commission has received and which are exercised in the area of competition policy by DG IV come from two sources. Regulation 17, Article 24 gave power to the Commission to deal with all procedural matters arising in the course of its duties; Regulation 19/65 made by the Council authorized the Commission under Article 1, as we have already seen in Chapter 5, to issue its own Regulations providing group exemptions for categories of agreement covering distribution agreements and patent licences; subsequently Regulation 2821/71 (as amended by Regulation 2743/72) has given similar powers to the Commission relating to specialization agreements. So far, the Commission has under this authority issued several Regulations of its own (see Table 8.1) and has under the terms of Council Regulation 3976/87 been given power in addition to issue Group Exemption Regulations in the Air Transport sector. By contrast the Council's Regulation on Maritime Transport (4056/86) contains its own group exemption, whilst allowing the Commission to grant individual exemp-

Table 8.1 Principal Group Exemption Regulations

Reg. No.	Year	Subject Matter
67	1967	Exclusive dealing agreements (now replaced by 1983/83 and 1984/83 below).
3604	1982	Specialization agreements (now replaced by no. 417/85 below).
1983	1983	Exclusive distribution agreements.
1984	1983	Exclusive purchase agreements (dealing both with these agreements in general and also containing special provisions relating to such agreements for beer and petrol).
2349	1984	Patent licence agreements.
123	1985	Exclusive distribution of motor vehicles.
417	1985	Specialization agreements.
418	1985	Research and development agreements.
4087	1988	Franchise agreements.
556	1989	Know-how licence agreements.

[5] There is an additional directorate E, dealing with State aids.

tions. In the insurance sector the Council has adopted a Regulation (1534/91) authorizing the issue of certain group exemptions by the Commission.

These eight current group exemption Regulations together comprise the principal delegated legislation of the Commission on Articles 85 and 86, and together with the Treaty of Rome and the other Regulations issued by the Council itself form the body of statutory law which the Commission has to apply.

The process of finalizing the terms of any regulation is a lengthy one. It involves a great deal of consultation with interested parties including the authorities of all Member States (through the Advisory Committee) and relevant commercial and industrial interests. If the Regulation is to be made by the Council, then it is originally prepared in draft none the less by the staff of DG IV (with specialist advice also from the Legal Service of the Commission) who themselves consult widely with Member States and with all interested bodies before sending a draft forward to the Commissioner. When the Commissioner has himself given his approval, it then reaches the agenda of the full Commission. As we have seen in the discussion of the development of the draft of Regulation 17, this draft then has to go on from the Council to be considered on a consultative basis by the Economic and Social Committee and the European Parliament before being returned with their comments for reconsideration by the Commission. Though the Commission is not legally bound to accept comments or suggestions made by these other bodies, it will normally pay close attention to the suggestions made, and it will normally be in a revised form that the final draft is then sent forward to the Council for approval.[6]

If the Regulation is, on the other hand, to be made by the Commission itself under delegated powers, it does not go to the Council at all but has instead to go to the Advisory Committee of representatives of Member States as established under Article 10 of Regulation 17, usually both before and after its publication in the Official Journal. The Economic and Social Committee and the European Parliament will also be shown the draft on a consultative basis. It is of importance for DG IV to be able to obtain approval from the Advisory Committee for any draft regulations it puts forward, since this means that at a later stage when it comes before the Commission it will be possible to confirm that the national interests of the individual Member States have been taken into consideration.[7]

[6] It is also noteworthy that the Single European Act has increased the powers of the Commission in implementing Council Regulations (Article 10 amending Article 145 of the Treaty of Rome).

[7] The recitals to Reg. 19/65 refer specifically to the need for the Commission to prepare its group exemption regulations in close and constant liaison with the competent authorities of Member States.

A group exemption contains more than merely rules for the administrative convenience of DG IV; sufficient attention has to be paid to the preferences, susceptibilities, and suspicions of the individual Member States. The importance of such exemptions is great; their creation, however, is the long and laborious process of trying to establish and frame rules that both meet the policy needs of the Commission itself and also comply sufficiently with the requirements of the business and commercial community in Member States and the many other pressure groups, regional, national, and sectoral that have a voice in the process. However long drawn out the labour, it is essential for DG IV to persevere with trying to obtain the exemptions since this is the main method (and perhaps the only effective one) of reducing the burden of pending notifications which has remained in the range of three to four thousand for a number of years, even after the bulk of early notifications which was the consequence of the initial 'mass problem' in early 1963 has now largely been dealt with by a variety of means.

Group Exemptions now have a clearly recognizable form. After extensive recitals setting out the policy considerations which have led to the introduction of a particular measure there follows a statement of the basic scope of the exemption. There will then follow a list of restrictions permitted to be included in the exempted agreements (often popularly called 'the White List'[8]), matched immediately afterwards by a list of those restrictions whose inclusion would prevent the group exemption applying ('the Black List'). Some but not all group exemptions also contain a provision for 'opposition'. This means that the parties to the agreement may put forward clauses restraining competition which are found neither in the White nor Black List and which, therefore, might well be described as 'grey'. The Commission, after notification, has normally six months in which to oppose the inclusion in an individual agreement of such clauses, and has to do so if requested by Member States on substantive competition law grounds. If opposition is raised by the Commission for any reason then the agreement will be treated as if application had been made for an individual exemption.[9]

It is normal for all group exemptions, however, to be subject to certain general conditions which must continue to apply throughout the life of the group exemption. These conditions normally relate to the free movement of goods between Member States and the continuing existence of

[8] The number of clauses contained in the 'White Lists' has grown substantially in the more recent Group Exemptions, notably those for Franchising Agreements (4087/88) and Know-how Licensing Agreements (556/89).

[9] Opposition is applicable to the regulations relating to patent and know-how licensing and to franchising as well as to specialization and research and development agreements. It is not provided for in the exclusive distribution and exclusive purchasing regulations where the 'White List' is stated to be exhaustive.

competition to those goods protected by the group exemption. For example, in the case of exclusive distribution agreements the exemption can in principle be withdrawn if the exclusive distributor is, without objectively valid reasons, refusing to supply within his contract territory certain categories of purchaser who are unable to obtain the goods elsewhere. If any of these conditions are not satisfied, the Commission has the right to withdraw the benefit of the group exemption from individual agreements. To do so, of course, it will have to go through the normal procedure laid down by Regulation 17 for the making of any decision, and in practice the Commission has scarcely used this 'safety valve' power.

Finally it should be noted that group exemptions have to have a time limit. This would normally be for ten years or at most fifteen. When review occurs when the expiration date becomes close, changes to the terms of the regulation can be proposed and it is possible that the experience of its operation may lead in some cases to pressure for change or even, in extreme cases, for non-renewal of the regulation as a whole.

2. The Individual Application for Exemption under Article 85(3): The 'Four Conditions' for Exemption

The administrative function of DG IV is to deal with the notifications and review them, and at the same time to become aware from a variety of sources of any potential breaches of Article 85 or 86 within the Community. Information may arise from public sources, such as newspapers or trade periodicals, or often arises by reason of a complaint either from an individual consumer or a rival business concern. Sometimes the investigations arise as the result of information contained in a notification by another undertaking, and other investigations follow information provided by members of the European Parliament, many of whom take a lively interest in its work.

As we have seen in Chapter 4, Articles 11 to 14 inclusive of Regulation 17 give extensive power to DG IV to carry out not only general enquiries into particular sectors (these are comparatively rare) but also to obtain information direct from authorities of Member States, and the right to require such authorities to carry out investigations on behalf of the Commission. Most importantly they authorize DG IV itself to carry out investigation if necessary in the territory of Member States. In addition to this right of investigation 'out in the field', the administrative function covers many other responsibilities necessary for implementing the notification system set up under Articles 4 and 5 of Regulation 17 and applying Articles 85 and 86 to such notifications.

What is essential, however, for an understanding of the way in which

DG IV operates is to realize that the process which it sets in hand is in most cases a process of negotiated settlement, designed to reach a point at which formal or informal clearance can be given to the agreement or concerted practice after removal of those features which appear unacceptable under the terms of Article 85(1) and 85(3). As the European Court itself put it in the 1984 *Dutch Books*[10] case, 'The purpose of the preliminary administrative procedure is to prepare the way for the Commission's decision on the infringement of the rules on competition, but . . . also presents the opportunity for the undertakings concerned to adapt the practices at issue to the rules of the Treaty'. It will only be the exceptional agreement, not covered by a group exemption and individually notified, which is either so harmless that it can be approved in its entirety, or so restrictive and anti-competitive that it will have to be rejected by way of formal decision to that effect. The majority of agreements notified individually fall between those two extremes and are therefore potentially suitable to form the subject for negotiation.

The range of choice for DG IV, therefore, when it comes to consider an individual agreement is as follows. First, it may decide that the agreement falls outside Article 85(1) altogether. There may sometimes be insufficient evidence that there is an agreement or concerted practice between undertakings, or that the relevant trade association made a decision or recommendation; alternatively, it may on occasions not be possible to show sufficient effect on trade between Member States or on competition within the Common Market. The parties to the agreement may be small and possibly covered by the *de minimis* rule contained in the Notice on minor agreements; or alternatively covered by one of the other Commission Notices interpreting the application of Article 85(1) to particular types of agreement such as commercial agencies or subcontracting agreements. If, however, the agreement comes potentially within the scope of Article 85(1), then the next question is whether it is covered by an existing group exemption. If so, then the restrictions in it can be disregarded because Article 85(1) has by reason of that group exemption been determined as inapplicable; indeed there is no need for an agreement falling within that category to be notified at all.[11] Finally, however, if the decision is that the agreement is within Article 85(1), but not covered by an existing group exemption, then the question of the availability of individual exemption must be considered.

Regulation 17, Article 9(1) gives the necessary authority to DG IV to grant individual exemption, subject to the right of the European Court to

[10] *VBVB and VBBB* v. *Commission* Cases 43 and 63/82 [1984] ECR 19, 68: [1985] 1 CMLR 27, 95.

[11] Certain other bilateral agreements not necessarily covered by group exemptions may also be free from the requirement to notify under the terms of Reg. 17/62, para. 4(2).

review any such decision if an appeal is brought under Article 173. Once a procedure has officially been started by the Commission, the jurisdiction of a Member State to deal with such an agreement and to declare it either within or without Article 85(1) or 86 will cease. The preliminary question is whether the agreement is entitled to negative clearance because it falls outside the terms of Article 85(1) for one of the reasons already stated. If not, the function of DG IV then becomes the application of the four conditions in Article 85(3) which alone provide the criteria for the assessment of whether exemption can be applied either with, or without, amendments to the agreement.

The conditions for exemption briefly set out in Article 85(3) are probably best regarded as four separate conditions. In each case, four separate questions have to be asked and answered: exemption can be available only if the Commission is able to answer both 'Yes' to the first pair and 'No' to the second pair. Thus, *positive condition 1* will be 'Does the agreement, decision, or concerted practice improve the production or distribution of goods or promote technical economic progress?' *Positive condition 2* is 'If so, does it also allow consumers a fair share of the resulting benefit?' *Negative condition 1* is 'Does the agreement impose restrictions on the undertakings not indispensable to the attainment of the objectives already referred to?' Finally, *negative condition 2* runs 'Does the agreement afford the undertakings the possibility of eliminating competition in respect of a substantial part of the products?'

If Article 85 were a United Kingdom statute, we would at this point have to conduct a textual analysis of each of these conditions. This, however, would not be a useful or productive way of understanding how they are dealt with either by the Commission or by the European Court. The four conditions are treated more like provisions in a civil code, broad statements of principle to be read in the context of the remainder of Article 85 and the other Treaty provisions, as well as in the light of the many cases in which they have already been applied. On the other hand, there is no principle of *stare decisis* binding in its application on all subsequent cases. In theory (though not always in practice), the Commission is free to look individually at each case, though it will apply well-established principles of interpretation, based on the underlying objectives of Article 85, to the particular restrictive clauses encountered.

There is some ambivalence in the way in which DG IV addresses itself to the analysis of individual agreements. Its general approach, and that which would appear to be required by the terms of Article 85, is to take each condition separately and proceed logically through to see if an appropriate answer can be given to each of the four questions. Only if an appropriate answer can be given in each case will the question of non-applicability of 85(1) arise. On the other hand, onc frequently obtains the impression from the terms of Commission decisions that the positive question 1 and, to a lesser extent, question 2 demand priority of attention; if a satisfactory answer can be given to those questions from an analysis of the factual background, then lesser difficulties will be raised by negative questions 1 and 2. Sometimes again one feels from reading a decision that the way in which the Commission's answer to the first two questions is phrased has been deliberately chosen in order to make it easier to give an appropriate answer to the two following negative questions.

What happens in practice is that the Commission will first analyse the agreement to see if it is outside the terms of any one group exemption and therefore liable to individual notification, and if so whether it contains any of those restrictive clauses which have, as the result of analysis in earlier cases, become almost automatically deemed to lead to the splitting up or segmentation of geographical markets. This may be caused by a specific clause to that effect between parties in competition with each other, or alternatively as the result of export bans contained in vertical agreements, perhaps patent licences or distribution agreements, which prevent the movement of goods at a lower level in the distribution stage from moving freely from country to country within the Market by way of parallel imports. If any clause of this kind is found in the agreement, then it is unlikely that the requirements of even the first positive question can be satisfied. Only if no such presumptively illegal clause appears will DG IV then be able to go on to the next stage of the inquiry, to try to appreciate the main objectives and effects of the agreement, and to carry out the balancing test of whether the anti-competitive restraints are outweighed by the advantages claimed as likely to result from them. In carrying out this process, the four questions may then be approached in order. If, after a first review, the agreement does not seem tainted by any restrictions or restraints on competition found within it, an attempt will be made to see if it can be saved after removing its undesirable features as mentioned in the *Dutch Books* case. The process here being carried out is not a court hearing but a review by an administrative body seeking to apply the provisions of paragraph (3) in a flexible way, in the light of the objectives of the Treaty and the Regulations made under it, taking also into account the considerable case-law that now exists as the result of both Court and Commission decisions.

Positive Condition 1: 'Which contributes to improving the production or distribution of goods or to promoting technical or economic progress'

Of the four conditions, this is normally considered first and is probably the most important in the eyes of DG IV. It is essential that the economic benefit is clearly identified, to be balanced against the detriment the various restrictions will or may place on the competitive process. The Commission is required to examine the objective behind the agreements and to use its resources sensibly in the establishment of the relevant facts and circumstances.[12] Nevertheless, though the Commission must be constructive in its approach to the investigation, the burden of proof in respect of all these conditions is upon the parties to the agreement. The 'improvements' or 'progress' may either arise immediately at the moment of production by the undertakings, or at a later level following processes in which value has been added or distribution has taken place. In other words, the improvement or progress does not have to be at the same level, e.g. manufacturing, as that in which the parties to the agreement are involved. Improvement could include almost any kind of beneficial alteration to the operation of industry or commerce, including the elimination of barriers to entry, increasing output from a given number of inputs, better quality in production, greater speed or quality control in manufacturing output, a greater range of products produced from the same inputs, or the possibility of making a wider range of products on the same plant or range of machine tools. As we have already seen in early Commission cases, distribution systems likely to improve the service of local markets with their particular features will also qualify.

Whilst such improvements are all of a fairly tangible nature, the assessment of the alternative benefit, 'technical' or 'economic' progress, is a good deal more shadowy. It is an expression familiar in French law, requiring a reasonable conjecture to be made about the likely outcome of an agreement. It does not, however, involve a broad social balancing of economic advantages and disadvantages. This was illustrated by the *Fedetab*[13] case (European Court 1980). This case involved the majority of those engaged in the tobacco trade in Belgium including manufacturers, wholesalers, and retailers where Belgian legislation relating to excise tax and price controls virtually made it impossible for undertakings selling cigarettes and tobacco at the retail level to compete with each other on selling prices. Nevertheless, the Commission found, and the Court agreed, that the legislation did not prevent competition entirely, since this was still possible on terms of payment, profit margins for both wholesalers and retailers, and on rebates payable. The rules of Fedetab for classification of wholesalers and retailers into a number of categories, fixing discounts for each category, and laying down in great detail the terms of business between each level of distribution, brought rigidity even to those areas which had been left unaffected by the Government regulations. A

[12] This was laid down in the early case of *Grundig–Consten* discussed In Ch. 4. The Commission cannot therefore insist that it can adopt a passive or 'wait and see' attitude until the parties to the agreement have produced sufficient relevant evidence.

[13] Cases 209–215, 218/78 [1980] ECR 3125, 3278; [1981] 3 CMLR 134, 247.

justification put forward for the restrictive systems involved was that it enabled a very large number (approximately 80,000) of retail outlets to be preserved, and it was argued that this enabled the retailer to offer customers a much larger number of brands of cigarettes and cigars than would otherwise have been the case, as well as improving the freshness of products for the customer. The Commission took the view that if the services which specialized wholesalers and retailers offered were as valuable as the association maintained, customers and retailers would ensure by their buying practices that wholesalers and retailers of this specialized nature would survive, and that, therefore, it was unnecessary for the association itself to give them more favourable treatment to ensure their survival (especially if there was no guarantee that in return for this special treatment they would actually provide services of the desired standard).

It had been hoped that the European Court would, therefore, pronounce in the *Fedetab* case on whether 'social' arguments relating to the protection of small businessmen and shopkeepers in the tobacco trade were relevant considerations properly to be taken into account by the Commission in applying Article 85(3). The Commission was unsuccessful, however, in obtaining a clear answer because the Court avoided the question. It is possible that it found itself divided and unable to reach a clear-cut decision upon it, though such indications as it gave were that it would not allow such an argument to be adopted under 85(3). In this context, of some significance was the statement 'the number of intermediaries and brands is not necessarily an essential criterion for approving distribution within the meaning of Article 85(3) which has above all to be judged by its commercial flexibility and capacity to react to stimuli from both manufacturers and consumers'. The Court preferred to decide the case on the fact that restrictions gave the parties the possibility of eliminating competition almost entirely on the Belgian cigarette market, so there was no need in upholding the Commission's decision to reach a finding on whether any other of the four conditions were specifically satisfied.

The 'improvement' or 'progress' that is required must not simply be for the benefit of the parties themselves. This is clearly laid down in the case involving exchanges of information between the Dutch Paper Industry Federation (VNP) and its Belgian counterpart organization (Cobelpa).[14] Nearly all the important manufacturers of stationery were members, accounting for approximately 80 to 90 per cent of domestic output in both countries. The Commission found a clear breach of Article 85(1) to have been established and refused exemption under paragraph (3). The extent of the information exchanged between the association and its members was so detailed and

[14] *VNP/Cobelpa* [1977] 2 CMLR D28.

identifiable to the transactions of specific companies that it had become more than simply the provision of an information service for a trade association. It had become rather so detailed a source of information as to provide artificial market stability, which in turn eliminated a large number of the competitive risks involved for members of both associations. This was particularly so as the benefit of the information went solely to sellers of paper, not to buyers, so that transparency of prices and terms of trading were still absent from one side of the market.

Both associations claimed that their arrangements meant that progress and improvements were available to members of the association, who were able to plan output and investment on a more stable basis because of the information in their possession about the future trends in the market and prices being charged by their competitors. The Commission ruled that any such benefit could not be taken into account, since a benefit to be treated as either an improvement or as progress must have the potential at least of helping third parties; the benefit could be either a short-term benefit or one whose results would take longer to become apparent. The fact that the improvement of production or distribution or the promotion of economic or technical progress was only one of the effects of the agreement is not necessarily fatal to an exemption, however, provided that overall the balance of detriment and benefit is seen as sufficiently favourable to have some recognizable value to such third parties.

In conclusion, therefore, a broad interpretation has generally been given to this condition extending to a wide variety of situations and responsive to many different types of economic progress and technical advance.

Positive Condition 2: 'While allowing consumers a fair share of the resulting benefit'

There was at one time considerable debate as to whether it was strictly 'consumers' or 'users' who were entitled to benefit, in the light of the language used in the four original versions of the Treaty. It soon became clear, however, that the word was to be interpreted broadly and would not only be limited to final consumers or retail purchasers. The use in the French text of the expression *utilisateurs* was always a cogent indication that the class of 'consumers' was wide enough to cover those who acquired the goods at any stage of the distribution process, whether in order to add value or merely to make use of the product for any purpose. This has now been confirmed by the Commission.[15] Again, a broad definition of 'benefit' is taken not limited only to reductions in purchase

[15] In e.g. *Kabel und Metallwerke Neumeyer/Luchaire* [1975] 2 CMLR D40.

price, but to any other economic advantages which a consumer may enjoy such as improvements in numbers or quality of outlets from which the goods can be purchased, better guarantee and service facilities, quicker delivery, greater range of goods, or more responsive distribution systems. So long as the agreement is capable of producing some or all of these advantages in the course of normal trade, it is sufficient that the potential is there and the actual proof it has done so already need not be given, provided there is some connection between the agreement itself and the subsequent 'benefit' to users of the relevant goods or services.

Sometimes one feels that this condition is satisfied rather too easily, in that the Commission tends to assume that, if the first positive condition is satisfied, then the mere existence of at least a moderately competitive market will ensure that the benefits of progress will be passed on, or 'trickle down', to the 'consumers'. The suspicion remains that sometimes this part of the analysis is not carried out with the thoroughness accorded to the first condition. At one time there were two officials in DG IV with specific responsibility for looking after the interest of 'consumers' in the traditional sense, but this responsibility was passed some years ago to a division of another Directorate. The direct contact between DG IV and consumers' organizations in connection with the balancing process carried out by the Commission under 85(3) was thereby lost. It is doubtful whether this has actually altered the final outcome of any cases, but psychologically it has had the effect that consumers' associations and organizations have come to feel rather distant from the decision-making part of the Commission on such issues.

Negative Condition 1: 'Which does not impose on the undertakings concerned restrictions which are not indispensable to the attainment of these objectives'

The first negative condition deals with the effect of the restrictions on the undertakings themselves. The objectives which the condition mentions are those set out as justification for the exemption under the first positive condition. It is essential for the undertakings to be able to show that the individual restrictions, individually and collectively, are tailored strictly to the valid purposes of the agreement and that the damage to the competitive process caused by the restrictions does not spill over to wider effect. The concept of necessity is very closely related to the principle of proportionality, familiar as a general principle of law which plays an important part in the jurisprudence of the European Court of Justice.[16] In this more limited

[16] For an assessment of the wider significance of this concept in the jurisprudence of the Court, see T. C. Hartley, *The Foundations of European Community Law*, 2nd edn. (Oxford: Clarendon Press, 1988) 145–7.

context too, proportionality is highly relevant. Agreements are quite often able to satisfy the two positive conditions, but fail to clear the hurdle of this particular condition, sometimes possibly because the effect of the restrictions has not been carefully enough analysed by the parties to the agreement.

An example of the two positive conditions being satisfied, but where the agreement failed on this first negative condition was the *Rennet* case[17] (European Court 1981). In this case, a co-operative at Leeuwaden produced animal rennet and colouring agents for cheese. It accounted for 100 per cent of the Dutch national output of rennet and 90 per cent of colouring agents; of this production, 94 per cent of the rennet and 80 per cent of the colouring agents went to its members, who between them accounted for over 90 per cent of the total output of Dutch dairy products. The members were obliged to purchase all their requirements from the co-operative, and if in breach of this obligation, were fined substantially. A further large payment had to be made by members if they resigned from the co-operative, calculated on the annual average quantity of the rennet which they had purchased in their previous five years' membership. Resignation, therefore, was made extremely expensive, whether the member did so so as to obtain supplies from another source or because he preferred to produce his own rennet in competition with the co-operative. Although the Commission agreed that the first two conditions were satisfied by the restrictive clauses, since it meant that there was greater efficiency in the production of rennet within Holland and that there had been some benefit in terms of price to the purchasers, it considered that the two negative conditions (including the issue of indispensability) were not satisfied; the economic advantages could be obtained by less sweeping restrictions on the members. For example, an obligation to take a rather smaller percentage of its requirements, or an obligation to give reasonable notice of the intention to withdraw, would have been adequate to protect the interests of the co-operative without the penal infliction of a large fine. The Commission's decision was upheld by the Court.

Another and well-known example is that in the early *Grundig*[18] case where, although the distribution contract under which Consten was entitled to distribute Grundig products on an exclusive basis in France was found likely to lead to an improvement in distribution, the use of trade mark law so as to ensure absolute territorial protection provided unnecessarily broad restrictions of competition for the benefit of the distributor. It was felt that Consten could have derived sufficient strength simply from its exclusive appointment without the additional protection conferred and

[17] *Cöoperative Stremsel-en Kleurselfabriek* v. *Commission* Case 61/80 [1981] ECR 851: [1982] 1 CMLR 240.

[18] Cases 56 and 58/64 [1966] ECR 299: CMLR 418.

the attempted use of French unfair competition law to keep out all parallel imports from Germany and other Member States.

A more recent example is that of *Bayo-n-ox*[19] (Commission 1990). Bayer had manufactured a growth-promoting product for piglets, which was normally added to their feed. The relevant patents had expired and Bayer was concerned that parallel imports of the product were increasing. It therefore sought to sell it to feeding-stuff manufacturers subject to the express condition that they only used it in the preparation of their own mixes and would not sell it on separately. Bayer's claimed justification for this restrictive clause was that it was necessary to prevent the mixing of the products with other products of lower quality, with possible ill effects to the piglets. The Commission in refusing an individual exemption, however, pointed out that the restriction was actually so wide as to have prevented feeding-stuff manufacturers from supplying the product even to perfectly reputable undertakings in Germany and elsewhere in Europe, to whom Bayer would have had no reason to refuse direct supplies. The restriction was therefore too widely drawn to allow an individual exemption to be granted, even if Bayer's original objective of limiting parallel imports of the product had been considered valid (itself most unlikely).

Negative Condition 2: 'And which do not afford such undertakings the possibility of eliminating competition in respect of a substantial part of the product in question'

This final condition relates to the external effect of the agreement and clearly indicates that just as with the application of Article 85(1) there is a need for market analysis by the Commission on both the product range and geographic range of that market. Analysis of markets, in spite of the volume of economic and legal writings on the subject, is far from an exact science, and the Court has made it clear that it will only interfere if the Commission has abused the wide measure of discretion given to it for the purposes of paragraph (3). It is probably primarily the result of this condition (as well as under the first positive condition) that specific market-sharing arrangements between competitors, as well as export bans in vertical arrangements, will fail to obtain exemption because of the priority given rightly by DG IV to retaining the possibility of parallel imports as a controlling influence upon price levels. Similar considerations will apply in respect of dual pricing under which goods for export are priced at a level different from those for home consumption since DG IV normally regards these also as likely to discourage exports.[20]

[19] [1990] 4 CMLR 905.

[20] The well-known *Distillers* case, discussed in Ch. 12 at p. 223, falls into this category: see n. 31 for references.

The Commission will look at the nature of the market involved which could be the entire Common Market, or one or more Member States, or simply a substantial region of a Member State. The absolute size of the enterprises concerned, and their market shares in the relevant product, will then be considered; and the structural condition of the particular market will be taken into account; DG IV is always interested in knowing whether the trends in the market are pro-competitive or in the opposite direction. It will be necessary to ask whether there are to be found particular forms of competition, possibly in promotion or product innovation, which may be particularly damaged by the agreement. The Commission here is normally concerned with interbrand competition, and reductions in intrabrand competition, i.e. between rival distributors of goods of the same brand, will not normally weigh heavily so long as there is still reasonable competition between the different brands and the possibility of these brands moving across national boundaries.

Finally a question often asked is whether the implementation of Article 85(3) by the Commission amounts to the equivalent of the 'rule of reason', for many years applied by Federal Courts in the United States to a wide variety of agreements not covered by the *per se* rule. If the US rule of reason is to be compared at all it is (as already explained) more appropriate to do so with Article 85(3) rather than 85(1). A superficial answer might be that the broad effect of the decisions rendered over the years since DG IV was first established has been similar to that of the case-law jurisprudence of these United States Courts; the restrictions to which Article 85(1) has been applied have been analysed in the context of their particular markets, product and geographic, and in their function and consequences (trivial or weighty) within the framework of the relevant agreement. The Commission, like the United States Courts, has been aware in its application of Article 85(3) of the importance of distinguishing between the restriction that is merely ancillary to a legitimate commercial purpose and that which by contrast provides a measure of relief for the participants from the rigours of the competitive struggle.

Nevertheless, at a deeper level, the fundamental difference between the two jurisdictions remains pronounced. The United States Courts have a liberty to take into account all the positive and negative features of the restraint, as well as the context in which it is applied, remaining as free from statutory restriction as are the courts of common law in assessing the local validity of contractual restraints as between vendor and purchaser or employer and employee. The Commission, by comparison, must operate within a rigid conceptual framework which allows less freedom of manœuvre and instead requires the restriction to pass not a single balancing test, but rather a cumulative series of four separate tests. Credit obtained by a restraint for passing any particular condition by a considerable margin cannot be taken advantage of at a later stage (at least in theory) if the restraint fails to satisfy a subsequent condition for exemption. There cannot be an 'overall' balancing of debits and credits under the system laid down by Article 85(3). If the consumer does not receive a fair share of the improvement to production or distribution of goods, or the restrictions appear disproportionately great to the benefits obtained, even if shared fairly with consumers, then no 'rule of reason' can in principle prevail to exempt the restriction.[21]

3. The Application of Article 85(3) in Individual Cases: Some Examples

To understand the way in which the Commission interprets the four conditions in relation to facts of an individual case, it may be helpful to consider the application of Article 85(3) in five widely differing examples. The first case chosen is that of *Gerofabriek*.[22] This company was the largest manufacturer in Holland of spoons and forks specializing in stainless steel though it also sold silver plated cutlery. Ninety per cent of its turnover was sold in Holland and only six per cent in Belgium. Goods were sold both direct to large buyers such as hotels and hospitals and also to wholesalers for resale to retailers in the normal distribution pattern. The company had 60 per cent of the market in Holland in 1971, falling to 50 per cent in 1975 and approximately 15 per cent of the Belgian market at the later date.

The strict terms imposed upon retailers and wholesalers were that goods could be resold only at the prices laid down by Gerofabriek and all discounts or rebates of any kind whatsoever were forbidden. Retailers moreover were barred from selling on to other dealers, being restricted to sales to end-users; this prevented some dealers, who were not allocated the complete range of products, from receiving additional supplies from other dealers so as to be able to offer a full range. Moreover, dealers in Belgium could not sell to The Netherlands and vice versa, nor was either type of dealer allowed to make exports sales to other Member States. The restrictions not only affected the distribution of products in Holland and Belgium but more generally throughout the Common Market.

This relatively simple case did not give the Commission great difficulty. Gerofabriek argued that the arrangements assisted the better distribution of their products, but the Commission found that on the contrary the arrangement acted as a barrier preventing dealers obtaining goods from

[21] Deringer in *The Competition Law of the European Economic Community: A Commentary on the Rules of Competition* (CCH Editions Ltd., 1968), 134.
[22] [1977] 1 CMLR D35.

other dealers in the same Member State and also preventing dealers in Belgium getting hold of the goods from dealers at the slightly cheaper prices at which they were available in Holland. The speeding up of availability of supplies to consumers was also restricted by the various restrictions upon sales by retailers to other dealers. The Commission, therefore, refused an exemption on the basis that the first positive condition was not satisfied but stated that on the facts presented the agreement would not have satisfied any of the four conditions. The test which the Commission stated should apply in such cases was whether the market situation with the restraints was considered objectively an improvement of the situation that would have existed in their absence, here easily answered in the negative.

A recent example of the application of the four conditions was in connection with the exemption granted to the eleven-year exclusive purchase agreement between Grand Metropolitan (the parent company of Watney Mann and Truman Brewers) in the United Kingdom, and Carlsberg Brewery, a United Kingdom subsidiary of this Danish company.[23] The notified agreement provided for the exclusive entitlement for Grand Metropolitan to sell Carlsberg in the United Kingdom and to be allowed to brew Carlsberg beers of the lower gravities, whilst Carlsberg accepted an obligation to supply all remaining requirements of Grand Metropolitan for the product. The agreement did restrict the freedom of other breweries, since not only were they unable to supply lager in substantial quantities to Grand Metropolitan, but a substantial proportion of the beer of higher gravities brewed by Carlsberg in the UK was no longer available for sale on other markets within the Community.

In reviewing the agreement and its background, the Commission found some special factors. The first was a strong tendency towards concentration of brewery groups in the United Kingdom, the second the existence of the 'tied-house system', and the third the scarcity of independent wholesalers for the distribution of beer. The Commission found that the existence of the agreement enabled Carlsberg to establish itself more quickly over a wider area than it would have done on its own in the United Kingdom because of its ability to use the large network of tied outlets controlled by Grand Metropolitan. This meant that Carlsberg's Northampton brewery was fully utilized more quickly than could otherwise have been the case, and there was no reason why after 1991 (when the eleven-year agreement would finish) Carlsberg could not then become independent of any other large brewery and able to distribute its own production. In the interim period, however, the purchase and supply and forecasting obligations mutually shared by the two parties provided stability for the

[23] [1985] 1 CMLR 735. See also the 14th Annual Report, p. 75.

production and distribution of Carlsberg, which had at the relevant time only 4 per cent of the total UK beer market.

The Commission was satisfied in this case that consumers would receive a fair share of the benefits resulting from the agreement, since it made possible the brewing of Carlsberg in the United Kingdom, so that it could be purchased there in a more plentiful, fresher, and cheaper form.[24] Taking the objectives into account, the restrictions that were contained in the agreement were considered indispensable, after deleting some minor restrictions. Carlsberg and Grand Metropolitan set their own prices and conditions of sale, which were not the subject of any agreement between them; and all the other restrictions were necessary in the special circumstances of the UK beer market. Finally, the degree of competition in the United Kingdom beer market appeared unaffected by these arrangements. The total share of the beer market in the UK enjoyed by Watney Mann was 11·5 per cent, and it could not be said that there was not still substantial competition on the UK market with the other major brewers. A relevant fact was that the volume of Carlsberg lager which Grand Metropolitan was committed to taking represented only one-third of the total of its own annual lager sales.

By contrast, in the *Floral Sales* agreement case[25] (Commission 1980) exemption was refused to an agreement between the three leading French manufacturers of fertilizer who co-operated to form a German subsidiary company to carry out their exporting to Germany of compound fertilizer. Although these three French undertakings sold to their German subsidiary at varying prices, the product was resold in Germany at a uniform price and on uniform terms and conditions.

The parties argued that the formation of the joint sales subsidiary was necessary to improve the efficiency of the distribution system on export sales. Whereas in the *GM/Carlsberg Brewery* case the Commission had reached a conclusion that it would not have been possible for Carlsberg to have successfully established their Northampton brewery without the support of the long-term requirements contract with Grand Metropolitan, it was not by contrast felt that these large French companies had the same commercial need to have joined together for the purpose of export sales to Germany. They were indeed the largest producers of these fertilizers in France, with excess production capacity, and as the market in France was itself oligopolistic the effect of their joining together was an appreciable limitation of competition *inter se*. The joint selling arrangements had not only no compelling commercial justification, but also no beneficial effects on either production or distribution to offset those competitive disadvant-

[24] The cost of transport of beer from Denmark was substantial, beer being an expensive product to move about in large quantities.

[25] *Floral Düngemittelverkaufs* [1980] 2 CMLR 285.

ages. Use could have been made of the normal wholesale and retail distribution facilities provided for the sale of both the straight compound nitrogenous fertilizers which they sold, and no reduction in buying prices was obtained by their customers in Germany as the result of these arrangements which could have been regarded as a consequent benefit to 'consumers'.

Clearly, the market strength and share of these companies weighed heavily against them; and the case in this respect may be contrasted with the weak market position and small size of the companies involved in the *Transocean Marine Paint*[26] case, often cited as a good example of the flexible use of Article 85(3) in allowing joint sales arrangements to take place between small companies with a weak market position, even where these amount to a horizontal market-sharing agreement. It is, however, always a requirement that the restrictions imposed are the minimum essential for carrying out effective marketing, and only for products where there are technical reasons for allowing allocation of territories, as in the case of *Transocean Marine Paint Assoc.* when customers would be travelling from port to port and would need a range of standard quality paints available at a number of ports of call. The Commission in that case accepted that such an arrangement would not be possible for small companies without this kind of market-sharing agreement. On renewal of an individual exemption, however, restrictions exempted on an earlier occasion may be considered to have become unnecessary for the achievement of the original objectives of the agreement. See Ch. 4 n. 5 relating to *Transocean Marine Paint Association.*

The approach of the Commission to the application of these four conditions can be additionally illustrated by a unique and complex example, the 1983 decision relating to the Energy Programme of the International Energy Authority (IEA).[27] The OECD had established a separate agency known as the International Energy Authority which carried out this International Energy Programme.

The objectives of the International Energy Programme were to take common measures effective to meet any oil supply emergency by providing a plan for self-sufficiency that would restrain demand and allocate available oil on an equitable basis between the members. It was agreed that since each country's supply position was slightly different, any crisis leading to shortages would affect them differently. Some would have a large refining capacity and would normally import mainly crude oil,

[26] *Transocean Marine Paint Association* v. *Commission* Case 17/74 [1974] ECR 1063: 2 CMLR 459.

[27] [1984] 2 CMLR 186. The members of the IEA were the 21 members of OECD including all the Member States of the Community other than France. The Community itself had observer status at the IEA.

whilst others would normally import only refined products. Some countries would have oil provided from only one source, others from many, and others again might be largely self-sufficient having their own production. Supply rights were established by IEA for each country and then in the light of any particular disruption in any part of the world the rights of each country would be compared with the overall supplies available in the world. If it was determined that the country had more than its equitable share in such circumstances, it would be deemed to have an 'allocation obligation' but if less then an 'allocation right'. Those with allocation obligations would then yield up excess supplies to those with allocation rights.

To make this system function, an International Advisory Board (IAB) had to be established from the oil industry to assist the IEA in providing effective emergency measures. This International Advisory Board had represented on it 16 separate oil companies and two oil company trade associations, and altogether there were 46 oil companies reporting to the IAB giving regular information relating to imports, exports, levels of production, refining, etc. to both the IEA and to national governments during any crisis period. Other non-reporting companies were asked likewise to submit data to their national governments, who would then pass the aggregated information on to the IEA.

A group of expert oil company employees co-ordinated these arrangements on a monthly cycle in the event of a crisis. Transactions were placed in three categories. Type 1 was a voluntary rearrangement of supply between a reporting and non-reporting company. A type 2 transaction was the rearrangement of supplies by reporting companies in response to a specific request from the IEA. It could be effected by either special arrangements between two companies or by the company receiving the request making an open offer to provide supplies capable of acceptance by any company having an allocation right. Type 3, the rarest transaction, was one required to supplement the voluntary system, if the situation was sufficiently serious that participating countries had to meet and discuss further allocations, if and only if type 1 and type 2 arrangements could not meet the deficit.

These arrangements came into force only once the overall shortfall to all countries within the plan was at least 7 per cent and once this level of supply had again been reached the allocation process ceased. Nevertheless, members of the oil industry, because of their knowledge of the market, were involved in the scheme before it was activated and continued to be so after it had been laid to rest. It was clear indeed that the International Energy Programme could not operate satisfactorily without the assistance of the oil companies in carrying out the emergency oil allocation system. The consent of the oil companies to participating in the system was at least a concerted practice, if not an agreement *stricto sensu*, even though the rules had been laid down by the IEA on an inter-governmental basis. The purpose of the involvement of the major oil companies was to arrive at a specific objective, namely the redistribution of oil products that would enable the target of allocation rights and obligations to be met and ensure that no company and no country was unfairly treated. To achieve this result, extensive information would be needed as to origins of oil, both in general, and as to particular cargos, with supporting detail relating to its quality, the location of tankers, refinery capacity, and storage capacity. All this exchange of information would inevitably provide company representatives with considerable data from their rivals that normally would have remained confidential.

Clearly, such concerted practices could have had the effect of distorting competition. Oil under arrangements type 2 and 3 would be sent to destinations it would not have gone to in a free market, and the existence of the information in the hands of many more companies than would normally know it might cause oil companies to behave otherwise than would have been the case if the scheme had not existed. If fully implemented, an International Energy Programme would probably involve redistribution of about 10 to 15 million tonnes every month out of a total monthly oil supply to IEA members of approximately 120 million tonnes.

Not surprisingly, many of the oil companies, notably those from the USA, were unwilling to participate in such arrangements unless a specific clearance under Article 85(3) could be obtained. The oil industry is one with a long history of cartel arrangements, and the size and power of many of the participants in the industry was substantial by any standard. Nevertheless, after complex negotiations, an exemption was forthcoming from DG IV. It was tailored as carefully as possible to the precise needs of the International Energy Programme. Some of the conditions imposed to ensure that the agreement did not contain any unnecessary broad restraints were that:

(*a*) The programme did not apply to any exchange of price information except in so far as it was absolutely necessary for the negotiation of a specific bilateral transaction, e.g. the price of a single cargo in a tanker available at a particular port on a particular date. Price information on a more general basis could not be exchanged.

(*b*) These arrangements could only be put into effect at the time when the emergency oil allocation system was operating, except that consultation was allowed to take place in preparation for the commencement of an actual allocation period. Test runs to ensure the efficiency of the system could also be dealt with on this basis.

(*c*) To ensure that discussions on oil prices and other sensitive commercial information did not stray beyond the permitted limits, the Commission was itself entitled to attend meetings of the International Advisory Board and be shown copies of relevant documentation.

With these restrictions a decision was ultimately issued that the requirements of Article 85(3) were satisfied. The concerted practices were stated to contribute to improving the distribution of oil throughout Western Europe and thereby promoted economic progress and substantially reduced the inconvenience (possibly even major economic distress) that would be caused if supplies to any one participating country were substantially interfered with. Consumers would benefit because they knew that their own country would receive proportionately equal supplies in any such crisis, and the agreement would minimize impact of shortages on the general economy of each of the countries involved. Great care had been taken to ensure that the restrictions imposed were limited to those indispensable to the scheme, and that competition in respect of 90 per cent of oil supplies would still remain on the same basis as previously. In fact, with the majority of type 1 and type 2 transactions the companies themselves would still establish the commercial terms for supply of the product. The negotiation of this agreement indeed represents a considerable achievement for the Commission and the participants in the agreement. To adjust an agreement of such importance to the members of OECD within the 'four conditions' without either compromising the integrity of those conditions or imposing impossible requirements upon the oil companies and governments participating was no small task.

It would, however, be wrong to assume from the *IEA* case that the Commission's interpretation of paragraph (3) is such that any programme of mutual assistance will always be upheld. For example, in the *Rheinzink* case[28] (European Court 1984) a reciprocal supply agreement between three large suppliers of rolled zinc was not given exemption. The arrangement here was that if the companies suffered disruption which caused a loss of supply of 20 tonnes per day, or a total of 200 tonnes, each of the other parties would make up the shortfall up to a total of 1,500 tonnes maximum so long as its own production was not disrupted. This agreement would be for a contract year and automatically renewed for another year.

The Court in assessing the arrangements concluded that it

deprives the parties of their independence of action, of their ability to adapt individually to circumstances and of a possibility of benefiting, by increasing direct sales to customers, from production stoppages or reductions in output sustained by

[28] *Compagnie Royal Austurienne des Mines and Rheinzink* v. *Commission* Cases 29 and 30/83 [1984] ECR 1679 [1985] 1 CMLR 688.

the other undertakings. The contract could moreover compel the parties to supply each other with considerable tonnages.

The Court concluded that a contract of such general scope and of such long duration 'institutionalises mutual aid in place of competition and was likely to prevent any change in the respective market position'.[29] It indicated, however, that it would not necessarily have disallowed the mutual supply agreement if applicable simply to cases of *force majeure*, that is a loss of supplies directly attributable to an external natural disaster or act outside the control of the parties. It was fearful, however, that the agreement in question could be implemented in a way different from the limited circumstances which the parties claimed to have envisaged.

It is in this perhaps that the agreement differs from the International Energy Programme case, even though the extent of the assistance to be provided in the latter case was also extensive. Nevertheless, some points of difference can be seen. First the involvement of national governments in the *IEA* case meant that it was unlikely that the agreement could be used as a cover for market-sharing arrangements wider than those specifically referred to in the actual agreement. Second, the threshold of a 7 per cent reduction in supplies before the crisis arrangement came about meant that the 'trigger' for the arrangement had to be circumstances at least moderately critical before the co-operation arrangements could come into force. Finally, the external factors causing the intervention were more clearly defined in the *IEA* case, as being an interruption of supplies in circumstances falling normally within the *force majeure* situation.

In agreements relating to services the Commission has to go through precisely the same analytical process as with those relating to goods. This is well illustrated by the *Concordato Italiano*[30] case. This involved an agreement between twenty-eight Italian insurance companies setting up a non-profit association in order to co-ordinate some aspects of their activities in the industrial fire insurance sector, which controlled about 50 per cent of the Italian industrial fire insurance market. Originally members who did not use a standard form of agreement had to notify the association but this obligation was removed at the request of the Commission. The association's responsibilities include the definition of various key terms relevant to the policy, the preparation of relevant statistics required for the calculation of premiums, the updating of the required premium rates in the light of the risks shown, and a study of preventive measures to reduce risks. Membership of the association was open to all insurance undertakings in Italy and was initially for a term of two years.

[29] para. 33 of the Court's decision.

[30] [1991] 4 CMLR 199. The facts of this case differ from *Verband des Sachversicheres* v. *Commission* Case 45/85 [1987] ECR 405: [1988] 4 CMLR 264 where recommendations on actual percentage increases were made to association members.

The Commission had no difficulty in reaching a conclusion that the agreement fell within Article 85(1); undoubtedly its effect would be to limit competition between members, which would tend to standardize premiums, and in general to use the recommended terms and conditions. There would be an effect on trade between Member States because of the existence of foreign insurers operating in these markets in Italy, as well as the fact that some of the property insured by members of the association would be owned by undertakings outside Italy. In reviewing the application of the four conditions the Commission stated that the acquisition of specialized knowledge by the association was a means of improving the services provided by it, and that the statistical and other assistance which the association could provide could make it easier for new entry to the market to quote sensible premiums for industrial fire cover. Consumers would, therefore, be able to have a choice of a wider variety of insurance cover. The existence of standard conditions would also make it easier for them to compare the policies offered to them by the different insurers both in terms of extent of cover and premium paid. No part of the agreement permitted the different insurance companies to agree the final premium to be quoted to potential customers.

The Commission took the view that the restrictions on competition contained in the agreement were indispensable to the attainment of the association's objectives. Only by establishing certain common practices with regard to the premiums and conditions would it be possible for the insurers and their clients to have a proper basis for evaluating the quality of management and of service offered by the individual companies. This should lead to higher standards within the fire insurance industry in Italy. The Commission did not feel that the agreement would afford the members the possibility of eliminating competition for a substantial part of the services. Members were free to fix their own premium rates and could vary the terms of the standard policies to suit particular cases. In addition there were a number of other powerful insurance companies outside the agreement which were in active competition with members of the association. Exemption for a period of ten years from the date of notification was therefore granted.

The range of cases dealt with above under Article 85(3) shows the inherent flexibility of the conditions. Whilst its application to individual cases can, of course, always be criticized, the structure of Article 85(3) itself provides a suitable framework for the assessment of exemption. One point that cannot be stressed too carefully is that exemption can only be given by the Commission not by any Member States or authority and that no exemption can be provided for an agreement that has not been duly notified.[31]

[31] Reg. 17, Article 9(1). See *Distillers Co.* v. *Commission* Case 30/78 [1980] ECR 2229: 3 CMLR 121. Many agreements may, of course, avoid the need for notification either because

Further examples of the operation of these conditions will be seen in following chapters. It is important to note, however, that a decision granting an exemption need not be absolute, but can be made conditional in a number of respects. The condition may relate to the period of exemption, to supervision by the Commission, or to reporting and even attendance conditions. The exemption is normally granted for a fixed period which will usually be for a minimum of three years and with a maximum extending as long as twelve or fifteen years. The supervision required by DG IV will vary according to the circumstances and may in some cases be limited simply to the provision of annual accounts by undertakings, lists of members in or of persons refused membership of an association, ranging up to the review in considerable detail by the Commission of individual contracts or the business dealings of the parties exempted. Reporting requirements may be imposed on the working of particular aspects of an agreement and, as in the IEA case, the attendance of representatives of the Commission may be made a specific term of approval. In at least one case, the *IBM*[32] case, it was a condition of a settlement that the operation of the relevant business transactions of IBM were the subject of an annual report to the Commission, and this has enabled close contact to be maintained between the company's senior executives and those supervising officials of the Commission.

4. The Effect of Article 85(2)

It is logical to conclude our examination of the application and enforcement by the Commission of Article 85 by briefly considering the important linking clause, paragraph (2), which simply states that any 'agreements or decisions prohibited pursuant to this Article shall be automatically void'. This broad statement of legal effect should, of course, be read in conjunction with the provisions in Regulation 17 about the ability of the Commission to grant retroactive exemption from Article 85(1) where appropriate notification of the agreement has been made. In all other cases, 85(2) means that agreements which infringe Article 85(1) cannot be treated as valid in any Court within the Community except in those three cases where after notification they benefit from provisional validity by reason of having been in existence at the date of accession of the relevant Member State to the Treaty. One would think from the bare words of the text that the quality of voidness applies to the entire agree-

they contain no restraints not permitted by the terms of a relevant group exemption, or because they fall under the provisions of Reg. 17, Article 4(2).

[32] For an account of the unusual terms of settlement in the *IBM* case, an Article 86 case, see Ch. 18 and the 14th Annual Report, pp. 77–9.

ment in which any prohibited clauses were to be found, and this was the original attitude adopted by the Commission in the *Grundig* case. The Court, however, adopted a different interpretation, finding that those other clauses of the agreement which did not contain restraints in breach of Article 85(1) were not rendered void by the terms of either 85(2) or Regulation 17. In all subsequent cases, therefore, the Commission has challenged the individual clauses in restraint of trade rather than the agreement as a whole.[33]

This sanction of voidness is necessarily of greater consequence to parties to agreements whose basic purpose is legitimate but which may contain certain clauses, for example as to royalty payments, exclusivity, or limits on methods of sale, which may be on the borderline of what can be individually permitted under Article 85(3). Such parties may find themselves, if they implement their agreement without obtaining individual exemption, in the unfortunate position of continuing to be bound by a contract now very different from that originally negotiated between the parties. This is especially likely to occur in the case of a patent or know-how licensing agreement or distribution agreement. Of course, a protective clause entitling either party to terminate the agreement in such circumstances is often included, though the consequences as a result of this may also be serious. If such a clause is not included, however, the parties may have to continue with a 'limping' agreement which in its cutdown form may be unduly favourable to one or other.

By contrast, the parties to a price-fixing or market-sharing arrangement of the kind dealt with in the next chapter are normally little concerned about the consequences of Article 85(2), since they will not have expected in any case to enforce it through Court proceedings, whatever other unofficial sanctions may be contemplated or threatened against parties who deviate from it. Moreover, there is little likelihood after substantive restrictions relating to price-fixing or market-sharing have been removed from the agreement that there will remain any enforceable continuing contractual provisions between the parties. For such agreements the only sanctions of any value or deterrent effect are the fines and penalties which the Commission is entitled to impose under Articles 15 and 16 of Regulation 17.

Normally, the validity of agreements containing restrictions in this category falls to be considered by national courts, and the application of Article 85 by such courts is dealt with in Chapter 22. The difficult question, therefore, of whether parts of an agreement containing prohibited

[33] On the other hand, in several cases the Court has pointed out to the Commission that it need not make in its decision a finding that specific restrictive clauses are each in breach of Article 85(1) so long as the agreement taken as a whole is in breach of it. See, e.g. Ch. 6 n. 45.

clauses can remain valid after these clauses have been deleted is a matter of national law, not of Community law. The normal approach adopted by United Kingdom Courts is to ask whether the relevant clause in the agreement can stand by itself without the restrictive words and, if so, then to allow deletion (often called severance) of those words from it without affecting the validity of what remains.[34] It would, however, be risky to assume that severance will always be possible, since in a case where the restraints were of the essence of the contract between the parties and the agreement once the restraints had been removed bore no recognizable relationship to the original, a UK Court might well declare the whole agreement void. Courts of other Member States would necessarily apply their own rules to such a situation, and there is no guarantee that the answer would be the same.

[34] *Chemidus Wavin* v. *TERI* [1978] 3 CMLR 514. Buckley LJ pointed out in this case that the test in this situation is not necessarily identical to that applied in other more traditionally familiar situations of 'illegality'.

9

Horizontal Agreements: Defensive Cartels

At first sight horizontal agreements are less likely to have redeeming virtues than vertical agreements, for by definition parties to a horizontal agreement are already in actual, or at least potential, competition with each other. Nevertheless, even though in some such agreements resulting benefits to the consumer are hard to find, there are some categories of horizontal agreements from which advantages arise which have to be balanced against the restraints on competition which they also involve.

Many agreements are primarily defensive in their objectives and usually involve the introduction of what, to most businessmen, is regarded as welcome certainty and stability into an uncertain world in order to replace some (even if not all) of the unpredictable elements of competition. In this category we can place price-fixing between manufacturers or by collective agreement in order to impose resale price maintenance on distributors or retailers, also allocation of markets or customers, collective refusals to supply, collective boycotts, and other collective exclusionary arrangements. Methods may vary considerably, from simple written agreement to concerted practice following secret meetings or by the open exchange of information about deals and customers. Nevertheless, in nearly every case the objective is similar, to protect an existing position or to prevent a feared deterioration in the relevant market.

This fear could be caused by competitive pressure, shortages, falling demand, changing technology, or lack of confidence in the ability of the undertaking to adjust to new circumstances. The latter part of the twentieth century has been marked by instability within many markets, especially those involving high technology: the products of last year may in those markets become obsolete before those of next year have yet been fully developed. Price cartels are often to be found here, sometimes as a response to overcapacity in a sector where demand has fallen substantially and is unable to revive. Often the enforcement of this kind of agreement is marked by machinery for tight control with a punitive element to ensure that all participants stay in line. Nevertheless, external and internal influences on individual members mean that often such agreements are broken, so that they either collapse completely or remain in existence but are only partially effective.

The other major category may combine some features of the first kind of horizontal agreement with outwardly more worthy purposes. This category would include co-operation agreements, specialization agreements, agreements for sharing research and development, and promotional joint ventures; one could describe them as all overtly creative in their objectives but nevertheless containing often ancillary restraints which may have the effect of weakening their appeal for exemption under Article 85(3). The difficult task for competition authorities such as DG IV is to ascertain if the anti-competitive restraints are no more widely drawn than is necessary to the main object of the agreement, whilst analysing the alleged benefits from the agreement and the extent to which they are passed on to consumers.

There are, of course, a large variety of horizontal agreements of different kinds that may contain restraints of trade, and to which Article 85(1) may apply. Sometimes a restraint of trade is merely a minor (indeed normal) element in a routine contract, for example, the sale of an existing business including its goodwill. The covenant restraining the subsequent trading activities of the vendor may be strictly ancillary to the need of the purchaser to protect the goodwill sold under the agreement. In the majority of cases coming before DG IV for individual exemption, however, the anti-competitive restraint plays a more significant part. The analysis required of such restraints has to be thorough and probing, sensitive to the nuances and special circumstances of particular industries. It is possible for the parties to seek to conceal a naked cartel in the fine clothes of what appears as a creative joint venture: the officials of DG IV have to be ready and able to distinguish substance from surface. In this chapter we are primarily concerned with examining some examples from the first category, that is agreements having primarily a defensive quality, and will consider some of the leading cases in order to learn how Article 85 has been applied to them by the Commission and Court.

The Article itself does give examples of the kind of agreement which will normally fall within its prohibition; they provide little assistance as to the interpretation of the basis upon which the benefits and burdens of the restraints concerned are to be assessed, referring simply to certain general types of agreement. Whilst it is sometimes possible to allocate notified agreements neatly within these headings, there is little value in trying to do so in every case since many agreements will fall within more than one category; and there are agreements falling within each category which will deserve and receive exemption under Article 85(3).

The first category (*a*) refers to agreements which 'directly or indirectly fix purchase or selling prices or any other trading condition'. We know from *Italy* v. *Council and Commission*[1] (European Court 1966) that both horizontal and vertical agreements are comprised within the prohibition.

[1] Case 32/65 [1966] ECR 389: [1969] CMLR 39.

Clearly, however, (*a*) deals primarily with agreements or concerted practices in which agreements are made between horizontally related competitors as to prices to be charged, markets to be allocated, or conditions of sale to be applied. It could also apply to such agreements between rival manufacturers to enforce resale price maintenance within a particular sector, or even to an agreement between a manufacturer and a distributor or dealer requiring resale price maintenance over the sales by the latter. One of the earliest horizontal cartels within category (*a*) dealt with by the Commission (whose decision was later upheld by the Court) was in the *Quinine Cartel*[2] case.

The participants in this cartel were companies in France, Germany, and Holland. The companies were engaged in the sale of quinine, which is used for making medicines to treat malaria, and also used in synthetic quinidine which is an ingredient for making other medicines. Both products are obtained from cinchoma bark, which grows in a number of third world countries. Co-operation between the companies in the European industry dated as far back as 1913 when agreements were made in order to try and stabilize market conditions for the supply of the raw material. After the Second World War, new plantations of cinchoma were established in Congo and Indonesia, and as a result the supplies of the bark rose steadily, with the result that prices by 1958 had reached a lower level than had prevailed twenty years before. At the same time the US Government elected to dispose of surplus held by the General Services Administration (GSA) at auction, having accumulated very large stocks over the previous year. In the face of this unstable market situation, the defendant companies decided in 1958 to put together a defensive cartel, involving collaboration on the prices to be paid for the bark, the allocation of purchases of the GSA stockpile, and the protection of individual national markets for their respective national producers. It was also decided that particular export markets would be reserved for particular companies, and that quotas would be allocated in respect of those markets when they were permitted to 'compete' there.

The original agreement was for a period of five years and in order that compliance with the agreement could be policed, members had to notify each other of their quarterly sales and prices. If a company was unable to reach its quota for export sales, compensating sales would then have to be made to other members which would be supervised by the Dutch company (Nedchem), which took prime responsibility for administration of the cartel and for the enforcement of restrictions on the manufacture of quinidine by companies other than the German and Dutch participants.

[2] *Boehringer Mannheim* v. *Commission* Case 45/69 [1970] ECR 769. UK companies were mentioned in the report as participating in the agreement, but were not proceeded against by the Commission. The Commission's decision is reported at [1969] CMLR D41.

Some of these arrangements were contained in so-called 'export agreements' and others in so-called 'gentlemen's agreements'. Although the Common Market was excluded from the scope of the export agreement, in practice all companies set their Common Market prices on the basis of those charged for exports outside the Market so that pricing effects were felt, as a result of the arrangements, within Europe. Market prices within Europe at the time were generally higher than those found in world markets. Later on in 1961 a buying pool was set up to stabilize the cost of the cinchoma bark and quotas were allocated. The evidence in the case deals interestingly with the reaction of the participants to the entry into force of the Treaty of Rome. They had to decide whether to notify the agreements, to terminate them, or to keep them in existence but secret. They determined to amend the agreements slightly, but to keep them in force without notification to the Commission. Subsequently market conditions changed sharply as supplies of bark were reduced, and the prices charged by all companies rose substantially. A contributing factor was the requirements of the United States Defence Department for use in Vietnam during the early Sixties. Eventually, however, the arrangement came to an end in 1965.

The Commission had imposed substantial fines against the participants, and these were only slightly reduced on appeal by the European Court, on the basis of a finding as to the relevant period during which the principal agreement had affected the market price within the industry. The Court made it clear, however, that it would not accept any legalistic distinctions between the effect of the export agreement and the gentlemen's agreement and stated that it regarded both as a single set of concerted practices.

We have already seen in Chapter 6 that the definition of 'concerted practices' lay at the heart of the next important horizontal cartel case decided by the Court. This was the *Dyestuffs* case[3] (European Court 1972). Undoubtedly, the evidence was more circumstantial than in the *Quinine Cartel*[4] case, and in particular the proof of concerted practices between the dyestuffs manufacturers depended largely on inferences to be drawn from the similarities of price increases and subsequent retraction of individual increases as a response to withdrawals of the increases by other parties to the agreement. As with quinine, the function of the dyestuffs agreement was both to protect national markets from outside intervention and to maintain price levels generally. The number of companies engaged in the Common Market in the dyestuffs business was relatively small (approximately ten) but all were of a very substantial size, and the nature of the market distinctly oligopolistic. Whereas *Quinine* had dealt with a small

[3] Cases 48, 49, 51–57/69 [1972] ECR 619: CMLR 557 (*ICI and others* v. *Commission*).
[4] [1972] ECR 661: CMLR 627.

number of related medicinal products, dyestuffs manufacturers produced between them several thousand different dyes, with a large variety of different customers having slightly different needs. It was felt essential by the manufacturers to maintain control of their local markets by ensuring that local national producers maintained direct contact with all their large national customers, being attentive to their needs and having a natural advantage of location and nationality over foreign producers who were likely to be represented locally only by subsidiaries, representatives, or agents who could not give the same complete service as the local manufacturers. The price of dye is only a small element in the total price of clothing and other items to which it is applied; therefore the market was not very price elastic, quality of service counting for much in the competitive process. Each leading manufacturer within its own country tended to behave as a price leader, but was faced always with the risk that some undertakings from outside the country might decide to make a positive effort to penetrate the national market by selective price cutting. The purpose of the cartel was to remove this area of risk.

In its judgment, the Court considers the need for competition in such markets against the perceived needs and wishes of the producers for stability. With regard to the former, it says that

> the function of competition in relation to prices is to maintain prices at the lowest possible level, and to encourage the movement of products between member-States so as to permit an optimum sharing out of activities on the basis of productivity and the adaptive capacity of undertakings. The variation of price rates encourages the pursuit of one of the essential aims of the Treaty, namely the inter-penetration of national markets, and hence the direct access of consumers to the sources of production of the whole Community.

but the purpose of the cartel was deliberately designed to counter these aims

> in these circumstances having regard to the characteristics of the market in these products, the behaviour of the applicant, in conjunction with other undertakings . . . was designed to substitute for the risks of competition, and the hazards of their spontaneous reactions, a co-operation which amounts to a concerted practice prohibited by Article 85(1) of the Treaty.

The argument of the manufacturers was that uniform price changes over the relevant years 1964, 1965, and 1967 were the result of price leadership by different undertakings making their choice of pricing policies which in turn produced a response required to maintain the equilibrium between the major manufacturers. The weakness in this explanation of the conduct of the parties was pointed out by Advocate General Mayras, namely that while such equilibrium could perhaps have explained a downward adjustment of price to keep in line with the price leadership, it was less probable that substantial parallel increases would be explained in this way especially when there were so many different varieties of dyestuffs at such a range of prices.

Elimination of the risks of competition lies thus at the heart of many horizontal cartels. This element was also at the heart of the *Sugar Cartel*[5] case, though here the factual complexities were compounded by the decision of the Commission to proceed under both Articles 85 and 86, arising from the fact that some of the European sugar producers had engaged in unilateral action to protect or extend dominance within specific Member States (or important regions of Member States), quite apart from the concerted practices taken by the parties with a view to the protection of national markets and price levels. The European Court judgment considers each of these geographical markets in turn and (just as in the *Dyestuffs* case) found that the principle of *chacun chez soi* (to each his own) dominated the attitude of the participants. The lengthy judgment takes us through a variety of individual national or regional markets examining the evidence to an extent possibly unique for the European Court. As we have seen in Chapter 7, Italy proved a special case where the degree of government intervention meant that the residual element left for restraints of competition by arrangement between the sugar companies was so limited that it was not sufficient to affect competition so as to constitute a breach of Article 85.[6] In the remaining jurisdictions, the position was that whilst Common Market agricultural policy, including the intervention and target prices set for producers, provided the background to the arrangements, there was still considerable scope left for competition between them. There was indeed a considerable surplus of sugar production over the level of demand within the Common Market. France and Belgium in particular had surplus capacity and in a freer market would have been able to sell substantial quantities to major end-users and sugar brokers or wholesalers within the Community; there were also collusive arrangements for obtaining tenders for sales into export where export refunds were available from the Commission to cover the difference between world prices and Community prices. The purpose and the effect of the cartel was to restrain the degree to which sugar was bought and sold across national boundaries within the Common Market and to allocate the export sales between the participants. As the Fifth Annual Report of the Commission comments

When there are tendering procedures to determine export refunds to non-Member States and European Community firms act in concert for refunds and quantities, this will restrict Common Market competition and trade between Member States,

[5] *Suiker Unie* v. *Commission* Cases 40–48, 50, 54, 56, 111, 113, 114/73 [1975] ECR 1663; [1976] 1 CMLR 295.

[6] At pp. 18–19.

because some firms would have been awarded smaller lots than those actually awarded and would thus have been given an incentive to sell more sugar in other Member States.[7]

In giving strong support to the majority of the findings of the Commission in these important cases early in the history of the Community's competition policy, the Court made an important contribution to underpinning the confidence of the Commission in preparing and bringing substantial cartel cases. The wide definition which the Court was prepared to give to concerted practices, and the willingness of the Court to take a broad view of the exercise of the Commission's discretion in reaching conclusions from the evidence available to it, encouraged it to bring other cases of similar magnitude in following years. Such cases often, however, require several years investigation and preparation before decisions can be issued.

By way of parallel to the three early cases described, we can consider some more recent important decisions, issued by the Commission in 1984. In three of them no appeal was taken to the European Court, but in the fourth case an appeal has been made by substantially all the defendants, namely the *Woodpulp*[8] case which is undoubtedly the most complex of the four. The three cases not involving an appeal were hydrogen peroxide, aluminium, and zinc. In the *Peroxygen*[9] case, a number of European producers of hydrogen peroxide and its derivatives, used as industrial bleaches, were found to have participated for several years in an allocation of markets that (as in *Sugar and Dyestuffs*) gave preference to home producers on their own national markets. As a result prices for these products were kept at a higher level than would have prevailed under competitive conditions. There were arrangements for meetings between the companies to settle any disputes about the allocation of territories or about sales made over quota. Production and sales details moreover were freely exchanged between companies so that the operation of the cartel could be easily controlled. The Commission pointed out in its decision that if the parties were prepared to take such trouble to enter and maintain such arrangements, and to concert their policies and practices so as to produce such convenient results, it was to be presumed that they believed it gave them an actual trading benefit. The mere fact, therefore, that the end result in terms of prices might be much the same as could have been expected from an oligopolistic market without the exchange of information would not itself provide any defence to a claim under Article 85(1).

[7] 5th Annual Report, p. 28.

[8] *Wood Pulp* [1985] 3 CMLR 474. Appeals were lodged against this decision by all the 12 Finnish companies named as defendants (Case 89/85), by 9 US companies (Cases 104, 114, and 117/85), and by 6 Canadian companies (Cases 116 and 125–129/85).

[9] [1985] 1 CMLR 481.

The *Aluminium Cartel*[10] was a particularly long-lived arrangement between several European and North American producers between 1963 and 1976. The object here, however, was not to protect individual home markets for specific companies but to protect Western markets generally from the supplies of aluminium ingot produced in Eastern Europe, which could have undercut Western producers on a price basis, and forced world prices down if released without control on to the market. To prevent this, Western producers arranged through a broker, Brandeis Goldschmidt and Co. Ltd. and its Swiss subsidiary, that all the Eastern European production would be purchased by this broker, who would then allocate it to the participants in the cartel, in a ratio corresponding to the share that each had of sales in Western Europe. Each manufacturer within the cartel would allow both its own allocated quota and those of other members. To disguise the source of the metal, and in order to check whether the Eastern Bloc countries nevertheless exported any metal direct to the West, all participants agreed that the Eastern metal purchased through Brandeis Goldschmidt would not be resold outside the cartel until it had been melted down.[11]

Many aluminium producers are also engaged in the production of semi-manufactured items such as aluminium plates, sheets, strips, and foil (known generally as rolled products) as well as rods and sections, tubes, and wire (known as extruded or drawn products). Those remaining manufacturers of aluminium products, therefore, without access to the raw material of aluminium ingot, and who wished to produce semi-manufactured or manufactured items (rolled or extruded) in aluminium, were very dependent upon the integrated producers for their supplies; this control enabled the integrated producers within the cartel to increase substantially the prices they charged for semi-manufactured products. Moreover, the parties in the cartel did all they could to prevent aluminium being traded freely or quoted on the London Metal Exchange, since this again could have weakened the effect of the main agreement by introducing a greater element of volatility to prices. The agreement had a substantial impact on competition in Western Europe, affecting directly the price of between 13 and 20 per cent of the total consumption in that territory. No fines were imposed by the Commission, allegedly for technical reasons relating to the age of the agreement, and this no doubt influenced the decision of the participants not to appeal.

There were some similar features in the *Zinc Cartel*[12] case (*Rio Tinto Zinc Corporation and Others* v. *Commission* (1984), an arrangement lasting for

[10] [1987] 3 CMLR 813. See also the 14th Annual Report, pp. 58–9.

[11] There uere also technical arguments raised by the participants that metal from Eastern Bloc sources required mixing with other aluminium of a higher grade before commercial use.

[12] [1985] 2 CMLR 108.

approximately the same period as had the aluminium cartel. This involved a producers' cartel which decided that the price fluctuations of the metal on the London Metal Exchange (often as much the consequence of speculation as of changes in supply and demand) were too uncertain to enable long-term contracts to be entered into with security. They, therefore, decided to agree that members of the cartel would only buy zinc at an agreed 'producers' price' and to abandon use of the London Metal Exchange for fixing prices. The parties also agreed to limit production and sales, if necessary through export quotas.

The Commission ruled that a common price agreed to be paid for the production of zinc necessarily involved a breach of Article 85 since it restricted the freedom of an individual smelter to negotiate its purchase price from the producer. The fact that this common producer price was not invariably applied, or that discounts were sometimes available, made no difference since the underlying stability of the market price was undoubtedly still preserved by the arrangement. Members had also agreed that they would sell zinc only to bona fida users for consumption and would not sell zinc for resale. As the product was homogeneous and the number of participants limited, the argument was not surprisingly again raised that the effect of the agreement was no different from that produced automatically by the price equilibrium that can prevail in an oligopolistic market. Again, the Commission refused to accept this argument saying that whereas mere parallel pricing behaviour was insufficient evidence of a concerted practice, it was here combined with sufficient other indicators of regular contact and a wide exchange of information between the participants which provided sufficient evidence of such practices to justify a finding against the parties.

In the later 1980s there were several important cases involving the great majority of large chemical producers in Western Europe where it was alleged that price cartels had existed and that important production information had been exchanged by the manufacturers respectively of polypropylene, polyvinylchloride (PVC), and low density polyethylene (LDPE). In each of these cases[13] the Commission decision imposed substantial fines on participants; regular meetings of the leading producers were proved though clearly there were both 'leaders' and 'followers' in respect of each product, and not all producers were at all times as closely involved in the industry's continuing efforts to protect its price levels against the background of oversupply and excessive production capacity. Uniform price increases and closely matching list prices appeared to be the outcome of these arrangements. The individual cases have been taken on appeal to the Court of First Instance, which is of course a more suitable

13 *Polyvinylchloride (PVC) Cartel* [1990] 4 CMLR 345; *LDPE Cartel* [1990] 4 CMLR 382; *Polypropylene Cartel* [1988] 4 CMLR 347.

forum for the investigation of detailed evidence in cases of this kind than the European Court itself. In the *PVC* case the fines and decision of the Commission were quashed on procedural grounds.

Whilst the problem of successfully maintaining over a number of years a price or market-sharing cartel is obvious, an arrangement between only two or three companies may be less difficult to maintain and indeed may survive unchallenged for a very long period. This was illustrated by the *ICI/Solvay*[14] cases (Commission 1991) in which ultimately heavy fines totalling 47 million ECU were imposed on the two companies for making agreements covering the sale of soda-ash (an essential ingredient in glass-making). The original agreement, a secret document known as 'Page 1000', had been signed by the parties shortly after the end of the Second World War. Under this Solvay had agreed not to sell the product in the United Kingdom or Ireland provided that ICI did not actively compete in Continental Europe. The agreement had technically been brought to an end in 1972 but the Commission found that the parties had actually continued it in effect for a number of years after that date. Article 86 findings were also made in respect of fidelity rebates and fines were also imposed in respect of these findings. The case is also on appeal to the Court of First Instance.

A common feature in all these cases is the attempt to control the price of essential raw materials. It would, however, though understandable, be wrong to assume that such agreements are always inevitably struck down under Article 85(1) by the Commission. It is possible for exemption to be given under paragraph (3) if there are the required compensating advantages which relate to improvement of production or distribution and where consequent benefits arise not only for the participants but also for purchasers and consumers. In the *National Sulphuric Acid Association* case[15] (Commission 1980), we find such an example. Sulphur in solid form is heavy and expensive to transport from the United States, where almost all of it is produced. Manufacturers of sulphuric acid in the United Kingdom had formed a buying pool for sulphur which nearly all the producers within the United Kingdom had joined. The existence of the buying pool had made it much easier for substantial dock installations to be built in England enabling sulphur to be imported by tankers in liquid form, enabling the smaller undertakings within the pool to receive small loads at convenient ports; the agreement also enabled the price of sulphuric acid to be reduced and continuity of supplies to be protected. The Commission, however, successfully objected to restrictions in the original

[14] 1991 OJ 152/1, a decision dated 19 Dec. 1990.

[15] [1980] 3 CMLR 429. The UK 'buying pool' had been upheld by the UK Restrictive Practices Court under the Restrictive Practices Acts. See *National Sulphuric Acid Association's Agreement* [1963] 3 All ER 73: LR 4 RP 169 and [1966] LR 6 RP 210.

agreement imposed on resales to non-members and against use being made of the sulphur for other than acid making purposes. Moreover, the members of the pool were not obliged under the finally amended agreement to be required to take more than 25 per cent of their total needs from the pool, whereas this figure had originally been 100 per cent of their requirements. In practice some 30 per cent of the sulphuric acid produced by pool members was sold to third parties and competition between members of the pool was thus maintained. The terms of the final agreement, therefore, whilst maintaining the essential advantages of the pool, limited the restrictions contained in it to those the Commission felt were necessary for its objects to be achieved. A different result might have been applicable had the effect of the buying pool been to give the buyers disproportionate market power over the sellers.

The effect of many of the agreements referred to so far could have been to 'limit or control production' referred to in (*b*) of Article 85(1). Perhaps the clearest example of agreements that have this effect directly are so-called 'crisis' cartels where the participants agree that they will close down part of their productive capacity, in a situation where demand has fallen so far that normal competition is thought to be impossible. This arises because of the disparity between the total productive capacity of the industry and the actual demand for output of those factories. Without some form of industry-wide agreement the weaker units in the sector may close down altogether, leaving only a small number of active undertakings, and accompanied inevitably by a sharp increase in unemployment and social dislocation.

Under normal economic theory, the productive capacity of such an industry would have to decline to the point where it matched available demand, notwithstanding the painful nature of that transition period. The Treaty of Rome itself came into being at a time when economic expansion seemed set to continue indefinitely, and no qualifying clauses limit the operation of Article 85 in the event of such crises arising. The agreement entered into to mitigate the effect of such substantial falls in demand would be essentially 'defensive', rather than creative, and, therefore, open to suspicion. Moreover, it was not immediately apparent how the four conditions could be applied without considerable strain upon their language. In particular, it was difficult to see how a crisis cartel could be shown to 'improve the production or distribution of goods' or 'promote technical or economic progress' particularly if consumers had to receive a fair share of the resulting benefit. The furthest that the Commission went during the first years of its operation was to reduce or dispense with fines in cases where the background to proven agreements or concerted practices were crises in various industries affected by major recession in demand, such as iron and steel, lead, glass, and cement.

The continuation of the recession through the late 1970s led eventually, however, to a reconsideration of policy which surfaced officially in the Twelfth Annual Report[16] where the Commission gave guidance on how far the solutions for restructuring of individual industries were compatible with Article 85. A Commission definition of the situation in which it would be prepared to tolerate restructuring agreements was given, not, however, in the form of official notices of the kind issued in respect of commercial agencies and sub-contracting on earlier occasions, nor in the form of a group exemption as this would have required preparation of a draft whose content would have posed particular difficulties and would in any case have had to be the subject of the usual extended consultation on a wide basis. The announcement in the Twelfth Report was rather on a more informal basis, indicating that the thinking of the Commission was to treat such agreements in a more tolerant way, and providing guidelines to enable those preparing such agreements to be aware of the limitations the Commission would impose upon them.

The minimum requirement laid down by the Commission for permitting this kind of agreement was that there must be shown to exist a structural overcapacity which had led to all the undertakings concerned having experienced, over a prolonged period, a significant reduction in their rate of capacity utilization and leading to a drop in output accompanied by substantial operating losses without any expectation of lasting improvement in the medium term. The conditions upon which the Commission would insist before granting exemption for an agreement containing mutual reductions in capacity and output would include:

(*a*) that the reduction in overcapacity was permanent and irreversible and of an amount which would enable the existing participants in the industry to compete at this lower level of capacity;[17]
(*b*) that the cutback would facilitate moves to specialization by individual companies; and
(*c*) that the timing of the reduction in capacity would be carried out in such a way as to minimize the social dislocation caused by the inevitable loss of employment.

Since the first announcement of the change of policy, there have been a number of decisions by the Commission permitting agreements in this category. A leading example is the *Synthetic Fibres* case[18] (Commission 1985). The companies involved in this agreement manufactured synthetic fibres, principally polyester and polyamide. The factories owned by the

[16] 12th Annual Report, pp. 43–5, where the issues that such cases place before the Commission are well stated.
[17] So that it would not simply become an allocation of existing production levels between them.
[18] [1985] 1 CMLR 787.

undertakings had been operating for the period 1981 to 1983 with between 54.7 and 77.6 per cent of the various markets for the different fibres and comprised 85 per cent of the installed products capacity for the EC. As a result of a serious imbalance between this capacity and existing demand, they had tried in 1978 to implement a 13 per cent reduction in capacity but at that time clearance had been refused by the Commission because of the degree to which production and delivery quotas had been an integral part of the scheme. Nevertheless the scheme, although refused clearance, had been provisionally implemented. The subject of the new and later Commission decision was a supplementary agreement signed in 1982 which had also already been partly implemented. Under this each company would commit itself to achieving its own reduction in capacity by a given date and would agree not to increase this until the end of 1985. All the figures would be checked by independent experts and the overall target over the period would be a reduction of 18 per cent of the capacity originally scheduled to remain in existence at the end of the operation of the 1978 agreement.

In its decision approving the exemption, the Commission referred to the definition given in the Twelfth Annual Report of continuing over-capacity and the fact that market forces had tried but failed to achieve the reduction necessary to re-establish and maintain in a longer term an effective competitive structure. In these circumstances, the need to organize on a collective basis an adjustment to the new levels of demand could itself be argued, taking a longer view, to be in the benefit of consumers; for the industry that would emerge from the restructuring would in theory at least once again have potential for fresh competition, the capacity remaining could be more intensively operated, and trends to specialization encouraged once profitability had returned, since the least profitable plants were those which were likely to be scheduled for closure under the scheme. There was little possibility that the parties would increase prices as the result of the approval of the plan because they operated in what was a very competitive market under pressure from available substitute products. Unlike the earlier 1978 agreement, there were no restrictions on the parties as to their level of production from the reduced capacity or the total deliveries which they could make.

The history of the alteration of the policy of the Commission towards crisis cartels does show the flexibility inherent in the use of Article 85(3) and in particular that its application may vary at different times against differing economic backgrounds. These cases, however, do pose major difiiculties in DG IV and the Commission generally because of the substantial political pressures which inevitably accompany such cases. Such pressures may come not only from national and industrial sources, but also from other DGs within the Commission, whose differing responsibilities may place them in a position where they may wish to urge policy arguments that in their view should take precedence over normal application of Article 85(1).

Rather less demanding on the Commission's decision-making process have been a substantial number of cases where in individual Member States close control of the operation of particular national markets, without the background of recession or fall in demand, has been maintained simply on a basis of protectionism. From a large number of cases, mainly involving Dutch and Belgian undertakings and associations, some only can be mentioned by way of example. The 1972 *VCH* case[19] (European Court 1972) concerned the rules imposed by the association of cement dealers in Holland. This association laid down extremely tight rules on its members, including mandatory pricing for sales of less than 100 tonnes (later abandoned shortly before the European Court decision) and target pricing for quantities over 100 tonnes. Members of this group marketed some two-thirds of the cement sold in Holland. The association laid down strict qualification for membership, listed the approved manufacturers from whom they could purchase and the different conditions applicable to sales to different categories of customer. For example, members were not allowed to deliver to construction contractors an amount believed to be in excess of the amount required for the particular site, so as to prevent the contractor building up its stocks. If any changes occurred in their business, its objectives, or its management, these had to be notified to the association; fines were imposed for any breach of the regulations. The Court described the arrangements as a coherent and strictly organized system the object of which was to restrain competition between members of the association and in effect made it much more difficult for outsiders to penetrate the Dutch cement market even though the members had only control of some two-thirds. The Court ruled that even though the agreement only extended over the territory of a single Member State, it would have by its nature the effect of compartmentalizing national markets for cement because the likelihood of other manufacturers being able to pen-etrate the market was so much reduced by the degree of regulation which the association had imposed on the industry.[20]

Numerous horizontal agreements have been struck down by the Commission which have attempted to provide either an exclusive purchasing arrangement or have provided for exclusive selling agreements between manufacturers in competition with each other. By contrast, bilateral arrangements made for the distribution of goods in particular territories have generally been regarded in a more favourable light by the

[19] *Vereeniging van Cementhandelaren* v. *Commission* Case 8/72 [1972] ECR 977: [1973] CMLR 7.

[20] A similar association imposing close knit regulations on a national market is of course found in the *Belgian Wallpapers* case discussed in Ch. 7, pp. 111–12.

Commission. An example of a case of exclusive purchasing unfavourably received by the Commission is that of *Atka A/S* v. *BP Kemi A/S and Danske Spritfabrikker*.[21] Here, the element in the agreement to which the Commission objected was the fact that DDSF was required to purchase all its requirements of synthetic ethanol from BP Kemi up to a maximum amount, and BP Kemi moreover would have the opportunity of supplying excess quantities on similar terms. This would give BP Kemi a substantial advantage over all other competitors in the market, especially as in return for its assurance of its supplies BP Kemi was guaranteed by DDSF at least one-quarter of the combined sales of ethanol in Denmark; if these exceeded certain percentages, compensating payments would have to be made to DDSF in return. In effect this was a market-sharing agreement which assured to DDSF supplies for a period of six years and would in practice relieve it from having to seek supplies from any other foreign source.

The cement industry seems to be one where restrictive agreements of a horizontal nature affecting sales are commonly found. In the *Netherlands Cement* case[22] (Commission 1972) a sales agency was formed by a number of German cement manufacturers to handle the sale of their cement in that country. These manufacturers bound themselves to sell only through this company, at the price fixed by the trade association of Dutch cement traders. Quotas were granted to the individual German suppliers and detailed provisions governing the conditions of sale, quality guarantees, freight charges, and many other contractual details were imposed, with penalties to be paid if a member deviated from the rules. The manufacturers retained no freedom as to the prices to be charged and had to follow the decision of the Dutch cement traders. The application for exemption under Article 85(3) was refused because the obligation of the manufacturers in Germany to sell only on these conditions restrained their freedom substantially, whereas each was perfectly capable of selling direct into Holland. The buyers, rather than having a number of suppliers, would have to deal with only a single supplier, and the volume of business was again controlled not by the individual manufacturers responding to demand but by a single agency working in close association with the traders' association. Even if the arrangements did make the distribution process more efficient, there was little evidence that it would benefit buyers.

By contrast, the case of *Société de Vente de Ciments et Bétons de l'est (SVCB)* v. *Kerpen and Kerpen*[23] (European Court 1985) involved a bilateral agreement. A French company (which later went into liquidation) had

[21] A case already discussed, in Ch. 6, p. 92.

[22] *Cementregeling voor Nederland* [1973] CMLR D149. See also the case of *Cimbel* [1973] CMLR D167 and *Nederlandse Cement-Handelsmaatshappij* [1973] CMLR D257.

agreed to supply 40,000 tonnes of cement to Kerpens in Germany annually for the next five years. The German company had agreed as a condition of the contract that it would not resell the cement in Saarland and would also pay attention to the interests of the supplier's partly owned works at Karlsruhe if asked to make deliveries in that area. When the liquidator of SVCB claimed payment for the cement, the defendant argued that the contract was void under Article 85. In a preliminary ruling under Article 177, the Court held that the agreement was indeed void since it had as its object the reduction of competition in a part of the Common Market; moreover as the total quantity on an annual basis amounted to between 10 and 15 per cent of the total exports of cement by France to Germany this was adjudged sufficient to have an appreciable effect on trade between Member States.

Not all such arrangements, however, necessarily fall within the prohibition of the Article. In the *Kali und Salz* cases[24] (European Court 1975), the Commission had ruled that Article 85(1) applied to an agreement under which Kali-Chemi sold its surplus potash production, which it did not require for the manufacture of compound fertilizer, to a much larger company, Kali und Salz, which was the only other German potash manufacturer, having approximately 90 per cent of the total production, Kali-Chemi having the other 10 per cent. There had been a gradual decline in the surplus quantities that Kali-Chemi had made available to Kali und Salz, and in defending the agreement Kali-Chemi argued that it could not have disposed of the potash in any other way because of the expense involved in setting up a separate sales organization solely for the sale of potash. The Court ruled that the Commission's decision here had failed to give sufficient weight to the fact that the surplus supplies which Kali-Chemi had agreed that Kali und Salz should dispose of were gradually reducing, nor had it given sufficient weight to the expense involved for Kali-Chemi in establishing a separate sales organization. For this reason the decision was annulled.

In other sectors, however, the benefit to purchasers may nevertheless be held sufficient to justify the grant of exemption to joint selling arrangements, notwithstanding the Commission's normal long-standing objection to such arrangements. In *United International Pictures* (Commission 1989)[25] three major US film producers (Paramount, MGM, and MCA) with a total market share of 25 per cent in the Common Market had formed a joint sales company for distributing their films. Each shareholder/producer had to give the sales company a right of first refusal for the distribution of its films and moreover could require it to distribute particular films if it chose. The Commission found that the agreement had

[23] Case 319/82 [1983] ECR 4173: [1985] 1 CMLR 511.
[24] Cases 19–20/74 [1975] ECR 499: 2 CMLR 154.

not eliminated all competition between the three producers since each retained considerable control over distribution in certain Member States; it did, however, enable more effective distribution through the Community as a whole so that consumers in each Member State could enjoy a wider range of films.

Relevant also was the volatile nature of the film market which saw substantial reductions in both admissions and revenue over the 1970s and 1980s as the result of competition from television and videos. Exemption was therefore granted for the comparatively short period of five years after certain changes had been made to the agreement to ensure that the shareholder/producers were not able to share profits from the distribution of each other's films but only from their own. As films are not a homogeneous product but are normally unique, their price cannot be fixed under the joint selling arrangement, unlike a normal bulk product such as sugar. This meant that notwithstanding the joint selling company each shareholder/producer's return would depend, therefore, on the success of its own films. To that extent the arrangement did not substantially impair competition between the three of them.

The width of Article 85(1) is such, however, that it extends to cover some agreements where one would not have anticipated problems arising, the restraints on competition being found where they are ancillary to the main object of the agreement. This category includes cases such as *Reuter* v. *BASF*[26] (Commission 1976). Dr Reuter had sold a business to the German chemical company, BASF, which involved the manufacture of polyurethanes, synthetic products used in the manufacture of plastics and varnishes. As part of the agreements, Dr Reuter accepted a number of covenants restricting his own future activities, and which he subsequently complained were contrary to Article 85. The principal covenants restricted him for eight years from engaging directly or indirectly either in Germany or elsewhere in research, development, manufacture, use, or distribution of any chemical product required in the production of polyurethanes or any related products. He was also put under a secrecy obligation for eight years relating to any confidential technical matters known to him before the sale. The Commission ruled that any restrictive covenant on the sale of a business or know-how has to be limited in length and width to an extent proportionate to the goodwill and business sold. In assessing reasonable duration and extent for such a covenant, the factors to be considered included the nature of the know-how, opportunities for its use, and the transferee's own technical knowledge. Any clause should normally be restricted to the markets in which the vendor was active before the sale or on which he could be regarded as a potential

[25] [1990] 4 CMLR 749. [26] [1976] 2 CMLR D44.

competitor; any clause that extended to cover new developments could not be of greater length than those properly inserted to protect existing markets. If clauses were limited in this way they might well not even be caught by the provisions of 85(1), and would not, therefore, have to be subjected to the individual exemption process.

Applying these principles to Dr Reuter's agreement, however, the Commission held that the contractual clauses imposed were invalid because they covered non-commercial as well as commercial research and development, and exceeded the limits of what was required to safeguard the interests of the purchaser in preserving the value of the goodwill and assets acquired. Moreover, it was felt that the obligation on Dr Reuter to keep information confidential for as long as eight years might be used to prevent him later from competing again with the purchaser through his further development of basic know-how not covered by the sale agreement. The Commission also ruled that a competition-restraining agreement between a vendor and purchaser, even those situated in the same Member State, would be likely to affect trade between Member States if it affected goods or services which if put on sale by the vendor of the business could be the subject of trade between Member States. The order of the Commission was ultimately that the agreement should stand for a period of only five years rather than the original eight-year period imposed.

A similar intervention occurred in the case of *Remia and Nutricia*[27] (European Court 1984) involving the sale of two businesses in Holland for the production of sauce and pickles. Here the Court upheld a Commission decision that covenants for ten years restraining the vendors from competitive activity were reduced to four years in one case and from five years to two years in another, on the basis that even such apparently normal clauses may impose restraints upon trade between Member States which the Commission is required to consider. The Court, however, emphasized that clauses of this kind which do not extend in length, scope, or geographical application more than is necessary to permit the agreement in which they occur at all are not caught by 85(1), in view of the fact that they are simply ancillary to a necessary and legitimate business transaction.

The wide range of cases considered in this chapter shows the extensive effect of Article 85 and the influence which it has on the validity of horizontal agreements. This in turn requires very careful attention to be given to its effect by all businessmen and lawyers when drawing up commercial agreements having even an indirect effect on trade within the Community. The extensive case-law now available on the interpretation of

[27] *Remia and Nutricia* v. *Commission* Case 42/84 [1985] ECR 2566: [1987] 1 CMLR 1. The Court pointed out that the effect of such sales or businesses is to increase the number of potential competitors in the market.

the Article[28] means that the major guidelines are now clearly understood, but nevertheless borderline cases will continue to arise, and that the Commission will be required to continue to exercise its judgment in these in a way which builds on the case-law that already exists. A major problem is, however, that the type of horizontal agreement considered in this chapter is just that for which the delays inherent in the granting of individual exemption are most likely to be encountered. The parties will not until a formal decision has been given be absolutely certain of the validity of their arrangements, and this may take several years. There are sometimes possibilities of an informal decision or administrative letter from DG IV, confirming that no adverse reaction is likely if some specified amendments are made; but this solution still leaves the status of the agreement in doubt before national courts.[29]

Nevertheless, the support given by the Court to the Commission in clarifying the outlines and details of the application of Article 85 to what we can generally describe as the 'defensive' form of horizontal agreement has been essential. In the next two Chapters, 10 and 11, we shall consider the rather less clear-cut situation that exists with regard to co-operation, specialization, and joint venture agreements.

[28] This approach of DG IV to such ancillary restrictions relating to the sale and purchase of businesses should be compared with the similar commentary contained in the Notice published by the Commission in August 1990 on 'restrictions ancillary to concentrations'. OJ C/203, 14 Aug. 1990

[29] On this point the *Perfume* cases are considered in Ch. 13, pp. 249–50.

10

Horizontal Agreements: Specialization and R & D

1. Co-operation Agreements: A General Introduction

Undertakings which have the ability to manufacture a particular product or range of products or to provide a service will normally try to do so independently, making use of their own resources of personnel, technical know-how, and finance supplemented by buying-in specialist services or products. In the real world, however, those directing such undertakings must necessarily recognize that often the development of existing products, as well as the research and development required to produce new products, demands investment of time and money for which their own resources are inadequate. The undertaking may for example have sufficient well-trained engineers and scientists but insufficient finance; alternatively, while finance may be no problem, its personnel may lack technical expertise in particular areas. This is particularly likely in areas of quickly developing high technology such as electronics and communications, in which an increasing proportion of both manufacturing and service industries are now involved.

In these circumstances it is likely that the directors of such an undertaking will give careful scrutiny to the known abilities and resources of those undertakings which are, or might be expected to be, their competitors in the search for such new products. This scrutiny may lead on some occasions to a proposal for a co-operative project which will necessarily involve some reduction in the degree of competition between them. There are many different varieties of integration and linkage in such joint ventures, which may intentionally fall well short of a full merger. In the comparatively rare cases where the degree of integration becomes total, the undertakings concerned normally have to agree to give up their individual right to engage in the market to be allocated to the joint company. Whilst Article 86 may occasionally be applicable in these situations, Article 85 is relevant in practically all such cases, of which there have been many over the last thirty years. From the beginning Article 85(1) has been widely interpreted by the Commission so that it applies to a very wide range of co-operative agreements of all kinds, with the consequence that having taken jurisdiction it can consider the advantages and disadvantages of the individual agreements at the next stage, that of applying Article 85(3). Nevertheless, the Commission also understands that too broad an interpretation of Article 85(1) will mean some agreements involving only minor elements of co-operation will be caught so as then to occupy the time of the Commission, when it should be applying its limited resources to agreements with greater anti-competitive potential. An early Notice[1] was, therefore, issued by the Commission indicating the way in which it proposed to apply Article 85(1) to such agreements. The Commission, however, was not prepared to transpose into Article 85(1) the detailed consideration of benefits and consequences of individual agreements that has to be carried out when applying Article 85(3).

These categories mentioned in the Notice therefore only cover a wide variety of the more limited kind of co-operation where the sole purpose of the agreement consists of one of the following:

(1) simple exchange of information (statistics, market research, or comparative studies) provided that it (*a*) does not involve any restriction on the actions of the undertakings involved and (*b*) does not in any way involve recommendations (explicit or implicit) that might induce parties to behave in an identical way;
(2) agreements for the joint handling of accounting matters, debt collecting, credit guarantees, or other services having no effect on supply and demand;
(3) agreements for the joint implementation of research and development projects or the placing of such contracts among participating undertakings;
(4) agreements for the joint use of production facilities, storage, and transport;
(5) agreements for the setting up of consortia to carry out orders, provided the undertakings do not compete with each other or cannot do so;
(6) agreements for joint sales or after-sales and repair services, again provided that the undertakings participating are not competitors;
(7) joint advertising agreements;
(8) joint use of common labels designating a specific level of quality and available to all competitors on the same conditions.

In issuing such a Notice, the Commission found itself not for the first nor last time in a dilemma. It had been pressed to give guidance as to how far Article 85(1) was intended to be taken literally, in its apparent application to almost any possible kind of agreement in which two competitors or potential competitors enter into some degree of co-operation; on the other hand, if such a Notice were to be issued, it would have to be carefully qualified, so as to prevent a number of comparatively 'undesir-

[1] JO [1968] C75/3: CMLR D5. 'Notice concerning Agreements, Decisions and Concerted Practices in the Field of Co-operation between Enterprises'.

able' agreements from slipping through the net under cover of the large mass of unobjectionable proposals. Each of the eight headings, therefore, is subject to careful qualification and has been strictly construed by the Commission, particularly that relating to research and development projects. An example of the careful approach adopted by the Commission at this time even towards the more limited type of co-operation is found in the case of the *Steel Tubes Association*[2] (Commission 1970) which featured an association of which companies involved in the manufacture of electrically soldered steel tube were members. The rules of the association provided the possibility of using a quality label sponsored by the association; membership of the association was open to all producers who met objective quality standards. Nevertheless, at the same time, members of the association were quite free, if they chose, to market their products solely on the basis of their individual label. Negative clearance was granted because the rules of the association contained no restrictions relating to production, marketing, or pricing by the members.

The extent to which companies may properly exchange commercial information and statistics with each other has been considered in a number of cases; the majority of these involve trade associations and are considered in Chapter 21. Nevertheless, it is important to realize that such agreements may exist between undertakings without there being any form of common membership in any such organization, whilst the organization itself may have only a tenuous existence when it is dominated by a few large producers in an oligopolistic setting. The *Fatty Acids*[3] decision shows that the Commission is capable of a sophisticated analysis of this type of arrangement and will be able to distinguish the 'pure exchange' of information from that which is linked to anti-competitive involvement in the details of a competitor's trading arrangements, including market shares, pricing, and order book. In that case, the three leading European manufacturers of oleine and stearine were fined for information exchanges which had allowed them to adjust their production and marketing policies with knowledge of each other's plans and thereby avoid 'destructive' competition.

2. Specialization Agreements

The essence of this variety of co-operation is that there is an allocation of the products to be produced by the parties, so that each can specialize in manufacturing part of a range of goods. Thus, for example, in the case of two motor manufacturers, one might agree to limit its activities to cars of

[2] *Association pour la Promotion du Tube d'Acier Soudé Electriquement* [1970] CMLR D31.
[3] [1989] 4 CMLR 445.

2,000 cc and less and the other to cars of an engine capacity above that level. This allocation is normally combined with a close degree of technical co-operation and a mutual obligation to supply each other with the specialized products produced, so that each can sell the full range in those territories allocated for distribution, which may often be on an exclusive basis. This kind of agreement may lead to a reduction in the number of producers of a particular model or type of product, but overall may enable both companies to realize the benefit of longer production runs and better use of their manufacturing capacity (leading in turn to a lowering of the fixed costs for each product and, therefore, the final price to be paid by the consumer). It may well also have the advantage of improving product quality, as the result of the combination of technical co-operation and specialization.

There are, of course, a wide range of detailed provisions to be tailored to meet the facts of each particular commercial situation. An example, however, of the kind of agreement approved by the Commission in its early years which may serve as a useful model for the kind of co-operation which we are describing is the *Jaz/Peter* case[4] (Commission 1969). Jaz was a French company specializing in electric clocks and alarm clocks, whilst Peter was a German company specializing in large mechanical alarm clocks. Under the arrangements, each agreed to supply the other with its special range of products and spare parts and not to supply any other customers. Each agreed that the other would be entitled to sell the full range of combined products in its own territory, Jaz in France, Peter in Germany. Each agreed not to buy from third parties any clocks or watches of the same type as those covered by the agreement. The effect of these arrangements was to allow each of the two companies to manufacture in much larger quantities than before the range of clocks in which they specialized, whilst still permitted in its own territory to sell the whole range of clocks and watches. The increase in their production led to a reduction in costs but the agreement itself did not involve itself with pricing restrictions, the choice of prices being left to the individual companies.

Clearly there was a restraint on competition involved by the agreement. A purchaser in France would not now choose between goods supplied by Jaz or Peter, since in France Jaz was the only distributor for the entire range of clocks, likewise Peter the only distributor for this range in Germany. On the other hand, the fact that both companies could produce a greater volume of clocks meant that prices could be and were reduced during the period of the agreement. The exchange of technical information between the parties also proved of mutual benefit; the Commission found moreover that none of the restrictions went beyond the level that

[4] [1970] CMLR 129.

was required to support the specialization agreement. When the parties applied for a renewal of the exemption in 1977, they were able to show that Jaz had quadrupled its production of pendulum clocks and electric alarm clocks, and that Peter had increased its production of large mechanical alarm clocks by two and a half times. Not surprisingly a further exemption was granted. An important element of course in the Commission's decision was that at all times customers had an adequate range of such clocks to choose from made by competing manufacturers; as even such an apparently beneficial element of co-operation could have been dangerous had there not remained substantial competition in the relevant markets, as indeed required by the second negative condition in Article 85(3).

A slightly different form of specialization is that where the parties combine not to split products in a particular range between them, but where they allocate the responsibility for the production and development of individual components for a particular product they wish to produce jointly. *ACEC/Berliet*[5] is one of the earliest examples (Commission 1968). Here, ACEC, a Belgian company specializing in electric transmissions, had developed a low weight and high yield model particularly useful in commercial vehicles and buses. They entered into a collaborative arrangement with Berliet, a French company, which had experience on the manufacture and selling of buses. Responsibilities were divided on the basis that they would co-operate for ten years in joint development of a bus which would have the ACEC transmission, but the basic structure of the Berliet bus. When a prototype had been made, ACEC would deliver its transmission equipment in France only to Berliet, and Berliet would restrict its purchase of electric transmissions to those made by ACEC, though remaining free to sell its products incorporating such transmissions anywhere in the world. ACEC would deal with Berliet on a 'most favoured client' basis and would guarantee performance by the transmission and continued availability of spare parts.

Clearly some restrictions on competition were inherent in the arrangement as third parties engaged in the manufacture of electric transmission were no longer able to sell them to Berliet, nor were some bus manufacturers who might have wanted to buy the ACEC transmission now able to do so. The restrictions, however, which ACEC and Berliet had accepted were in the view of the Commission reasonable, given the commitment which each entered into with the other.

If successful, the specialization would permit the manufacture of buses in longer production runs, giving a reasonable chance of producing a new model that simplified the mechanical construction and would give better

[5] [1968] CMLR D35.

performance and comfort to its users. Though the outcome of the technical research could not be predicted with certainty, the fact that it was being carried out on a collaborative basis by two specialist companies was such as to increase the chances of a useful outcome; there was a sufficient degree of probability that the results would be successful to justify the collaboration and the restraints ancillary to it. ACEC remained free throughout to make contracts with other bus manufacturers outside France and any such buses would be able to compete with those produced by the joint venture. Formal exemption, therefore, was again granted by the Commission.

As the Commission acquired greater experience of such agreements during the 1960s, it became clear that it was advisable to seek the issue of a group exemption to cover the relatively straightforward forms of specialization agreements engaged in by smaller undertakings, rather than have to subject each of them to individual scrutiny. As the Commission was itself without legal authority to issue such a regulation, it was necessary for the Council to provide it with necessary powers under Article 87, by adopting Regulation 2821/71[6] which authorized the Commission to issue its own group exemption covering, *inter alia*, the application of standards or types and specialization agreements. The Commission, thus provided with the necessary powers to introduce its own group exemption, did so a year later in Regulation 2779/72.[7] The definition of specialization agreement was, however, rather narrower than might have been expected, namely

> agreements whereby with the object of specialization, undertakings mutually bind themselves for the duration of the agreement not to manufacture certain products or cause them to be manufactured by other undertakings and to leave it to the other contracting parties to manufacture such products or cause them to be manufactured by other undertakings.

This would cover the kind of collaboration referred to in the *Jaz/Peter* case, but not the shared research and development on a single new product contemplated by *ACEC/Berliet*. In Article 2, the sole restrictions that could be imposed were listed, and it was made clear that the group exemption could not apply if any others were included. The permitted restrictions were:

(*a*) not (except with consent) to make any specialization agreements covering the same products with any other undertaking;
(*b*) to be required to supply the relevant products at a minimum quality standard;

[6] Made 10 Dec. 1971. JO [1971] L285/46.
[7] Made 21 Dec. 1972. JO [1972] L292/25.

(*c*) to purchase products the subject of specialization solely from the other undertakings (except where more favourable terms of purchase were available elsewhere);

(*d*) an obligation to grant to the other party the exclusive rights of distribution so long as they do not limit, by the exercise of industrial property rights or otherwise, intermediaries or consumers from purchasing the products from other sources within the Common Market;

(*e*) an obligation to maintain minimum stocks and to provide after-sales and guarantee services.

The exemption, however, was not to be available to large or even medium-sized companies, it being intended for smaller undertakings which had an aggregate annual turnover (covering all the participants in the arrangements) of 150 million units of account, and also with no more than 10 per cent of the relevant volume of business for the products covered by the specialization (or considered by consumers as similar by reason of characteristics, price, or use) in any one Member State. The group exemption came into force on 1 January 1973 and gradually achieved its objective in reducing the flow of relatively small specialization agreements since each specialization agreement tended to have individual characteristics that did not easily fit within the fairly narrow framework of the group exemption, a number of individual arrangements still came to be considered.

In the *Prym/Beka*[8] case, Prym was a German company having over 4,000 employees which decided nevertheless that its capacity for production of sewing needles was inadequate; it made an agreement with a much smaller company in Belgium, Beka, employing only 350 people that Prym would itself no longer make sewing needles and would transfer the relevant parts of its plant and equipment to Beka, provided that Beka guaranteed to supply all Prym's requirements. Prym would take a 25 per cent interest in the share capital of Beka and would agree not to purchase its needles from elsewhere, unless Beka proved unable to supply its requirements. Originally the agreement also contained market-sharing restrictions, both by definition of end-users and on a geographical basis, but the Commission insisted that the exemption was only available upon the removal of these restrictions. The advantages from the specialization agreements were that the Belgian plant could be more intensively used, by increasing by at least 50 per cent its production of needles, whose unit cost was substantially reduced. Though this was not a true specialization agreement (since Prym itself was not allocated a specific product in whose manufacture it would itself specialize) the effect of the new arrangements for centralizing production at the factory of Beka had many of the same

[8] [1973] CMLR D250.

advantages and was, therefore, accepted within 85(3), subject to the deletion of the clauses dealing with the allocation of markets.

Rather more unusual was a case involving allocation of markets in the nuclear industry, the *United Reprocessors*[9] cases. These involved an agreement between three undertakings, in the UK (British Nuclear Fuels), France (the Commissariat à l'Energie Atomique), and Germany (KEWA). The agreement notified involved the setting up of a joint subsidiary company to cover the market of the reprocessing of nuclear fuels. It was agreed that they would not invest in any other business for this purpose and would allocate all their reprocessing work between the three companies. Other restrictions were also accepted including an obligation on KEWA not to build a new reprocessing plant until the throughput of the French and United Kingdom plants reached certain levels. In granting an exemption, the Commission held that the co-ordination of the capital investment between the three companies would ensure that uneconomic plants would not be set up, and would enable the companies to wait until market conditions were more favourable as well as reducing costs through increasing the production of the existing reprocessing plant. Dealing moreover with a product as delicate and dangerous as nuclear waste, it was felt essential by the Commission that the reprocessing service should be firmly established. The necessary finding of benefits to consumers was based on the fact that the proposal should enable the cost of electricity to be reduced because of the improved stability of the reprocessing service. The Commission, however, itself insisted that it should be closely involved in periodically reviewing the arrangements in order to ensure that consumers continued to receive their proportionate share of resulting savings. It also insisted on participation in the monitoring of the assessment of the throughput of the reprocessing work of the French and United Kingdom plants.

One senses from the terms of the decision that the Commission had major doubts about the effect on competition of such an arrangement, and that it was only the combination of the stringent conditions imposed and the fact that this was a particularly sensitive industry in which to insist on completely free competition which enabled an exemption to be granted. The Commission indeed acknowledged in its decision that for a period competition would actually be reduced.

In recent years, the number of specialization agreements being granted individual exemptions has been reduced, although there has been no shortage of notifications involving other kinds of joint venture proposals. The specialization agreements that still come before the Commission seem

[9] [1976] 2 CMLR D1: cf. *KEWA* [1976] 2 CMLR D15, a similar decision granting exemption to a joint subsidiary being set up to reprocess nuclear fuels in Germany by four companies and which would actually be one of the shareholders in United Reprocessors.

to fall into two categories. One category deals with industries of particularly advanced technology, where the investment costs are particularly high, and where there is a substantial element of research and development involved. Of this kind, *Bayer/Gist-Brocades*[10] (Commission 1975) may serve as an example. Both Bayer, a German company, and Gist-Brocades, a Dutch company, were large drug manufacturers; each produced raw penicillin and also intermediate pencillin products for processing into ampicillins and other semi-synthetic products. In order to increase overall production, a specialization agreement was entered into under which Gist-Brocades would specialize in the production of raw penicillin, whilst Bayer would concentrate on intermediate products. The majority of the raw penicillin to be produced by Gist-Brocades would then be forwarded to Bayer for processing into intermediate products, although Gist-Brocades would retain some of the raw penicillin for traditional penicillin preparations. Both companies kept the freedom to carry out their own research and development, subject to an obligation to exchange information; various non-exclusive licences for the use of specific chemical processes relating to the production of penicillin were also granted. Although the companies were both large, with substantial financial resources and knowledge of the market, the Commission agreed that production could be carried out more economically as the result of the specialization.

Exemption was, therefore. granted but with one important amendment to the original proposals. Originally, the individual factories were to have been transferred to joint subsidiary companies in which both parties would have a 50 per cent interest and an equal number of directors. Inevitably this would have led not only to joint research, but to joint control over the production and total investment in the individual plants. The Commission felt that this would have restricted competition between the companies more than was essential to the specialization proposals. The proposal to form joint subsidiaries was then abandoned, each company administering the arrangements through their existing corporate structure. Having thus eliminated the restrictions that appeared too wide to be indispensable to the arrangement, the Commission found that on other grounds the exemption could be granted. The specialization agreement would increase availability on world markets of both raw penicillin and intermediate products. The Commission fixed an eight-year period for the exemption but attached conditions, so that it would remain able to check the results of co-operation on the competitive process in the relevant markets.

[10] [1976]1 CMLR D98. A subsequent extension of the exemption was later granted until 1995.

The second variety of specialization agreement now more commonly found was one brought about by the economic recession of the 1970s and early 1980s. The more generous treatment afforded to crisis cartels at that time illustrates the concern felt by the Commission in such situations, that it should not be seen to take too rigid a line and should be accommodating where there was a genuine long-term situation involving a substantial fall in demand and consequent excess production capacity. An alternative to a crisis cartel is an agreement whereby 'swaps' are made between major companies enabling them to eliminate production capacity they no longer require and to specialize in those fields where they remain strong. The process is well illustrated by *Imperial Chemical Industries/ British Petroleum Chemicals*[11] (Commission 1984). Both ICI and BPC had manufacturing plants producing PVC (polyvinylchloride) and low density polyethylene (LDPE). Arrangements were made under which ICI disposed of its modern LDPE production plant in the UK and all related technical information and patents to BPC whilst in return BPC sold its most modern production plant for PVC in the UK and related technical information to ICI. The actual notified agreement related, therefore, simply to a swap of individual plants enabling ICI in the future to specialize in PVC and BPC likewise in LDPE, neither continuing in the production process of the other product.

Although not specifically referred to in the agreement, consequential closures were then carried out by both ICI and BPC of the remaining plants in the UK producing LDPE and PVC respectively. The final outcome was that for the United Kingdom ICI had completely left the production of LDPE and BPC had completely left the production of PVC. It is difficult to imagine that at an earlier stage in the Commission's history an exemption would have been available for such an arrangement. The object and effect of the agreement was the restriction of competition eliminating an important producer in the United Kingdom for each of two main chemical products, PVC and LDPE. Nevertheless, following the precedents of the crisis cartels, exemption was granted under Article 85(3), for the undoubted structural overcapacity in both sectors (PVC and LDPE) made it essential that the older plants should be closed, and this was only possible if each company specialized in one of the two products.

As a result of the agreement, BPC and ICI were each able to increase the capacity and the usage of their more modern plant thereby reducing unit costs, always an essential feature in specialization agreements. In the long run, a more healthy industrial structure was likely to be promoted, and although the agreement in the short run would not improve competition, the Commission was satisfied that in the longer term effective com-

[11] [1985] 2 CMLR 330.

petition would be maintained and consumers not deprived of a range of choice.[12]

BP subsequently followed up these arrangements by making a further agreement with Bayer, under which BP utilized its special knowledge of LDPE to become the Community distributor in place of Bayer for their joint subsidiary which produced both LDPE and other naphtha derivatives. BP was required to accept an annual minimum quantity of LDPE as distributor and to provide technical information which would enable the eventual building of the new plant in Germany for LLDPE (linear low density polyethylene), which would gradually replace LDPE. The Commission granted the exemption because it enabled the restructuring of the chemical industry in this sector in a way that would lead to the production of a superior product in a way not possible for the individual companies from their own resources, thereby promoting both technical and economic progress.

As a result of the considerable experience gained by DG IV in analysing the kind of agreements referred to above, further group exemptions were made possible now on a rather more generous basis. Regulation 3604/82[13] replaced the previous Regulation 2779/72 and extended both the categories of agreements covered and the financial limits involved. The exemption was extended to agreements where the undertakings agree only to manufacture or have manufactured certain products on a joint basis, so that cases such as *ACEC/Berliet* would also now come within its scope. The financial limits were raised from 150 million to 300 million ECU, and the relevant percentage share of trade in a market within the Common Market (either a Member State or substantial part of a Member State) was raised from 10 per cent to 15 per cent.

The group exemption took effect from 1 January 1983 and was originally intended to remain in force for fifteen years, a rather longer period than for any earlier group exemption. Whilst in the Twelfth Annual Report the Commission had expressed its opposition to expanding the criteria for exemption beyond the figure of 300 million ECU, it clearly changed its mind soon afterwards. A further version of the group exemption, namely Regulation 417/85[14] came into effect on 1 March 1985, replacing Regulation 3604/82 after less than three years. Influenced by a desire to increase the ability of small and medium-sized companies within the Community to enter into such arrangements, the maximum figures

[12] A similar decision was rendered in the case of an agreement betueen Shell and another Dutch company, AKZO, under which Shell disposed of its PVC plants in Holland in favour of AKZO which in return disposed or its vinylchloride monomer plant to Shell. See the 14th Annual Report, pp. 71–2.

[13] Made 23 Dec. 1982. OJ [1983] L376/33.

[14] Made 19 Dec. 1984. OJ [1985] L53/5.

have again been substantially raised to 500 million ECU and the market share percentage from 15 per cent to 20 per cent. Moreover, the turnover threshold of 500 million ECU is rendered less rigid than previously, as agreements between undertakings whose aggregate turnover exceeds that figure can now still be exempted, if the agreement is notified and not objected to by the Commission within six months. This is through the now familiar opposition procedure, whose aim is stated to be to provide legal security for agreements which whilst breaching the turnover limit do not pose major problems from the viewpoint of competition policy. No such provision, however, applies to the 20 per cent market share threshold. It is important to remember that this percentage applies not just to a share of an individual Member State but could apply to a lesser area, such as a region within a larger Member State. In this respect, the test is more stringent than that contained in the original Article 2779/72 where the percentage was calculated simply on the basis of market share in individual Member States.

3. Research and Development Agreements

Many of the agreements which we discuss in this chapter do not fall neatly into separate categories. Nevertheless, research and development agreements form a distinct group which can be recognized without difficulty; the main purposes of such agreements are to arrange for the carrying out of a number of functions, all of which are essential steps in the lengthy process which elapses between the first creative step of the inventor and the ultimate delivery of a finished product to customers at a competitive price. They are both more common, and more important in their economic and commercial effects, than specialization agreements.

The development of a new product actually begins with the basic research carried out in a laboratory or workshop, often taking years rather than months to complete. If the research has, however, proceeded successfully, the next stage is the application of its findings to the development of a product that could be saleable, either in competition with existing products serving the same function or possibly as a novel product without immediate competitors. Once development has been completed (this again especially in high technology areas may take several years) the remaining stages follow a pattern. Production facilities are first set up and manufacturing commenced, coupled with the setting up of distribution systems and suitable promotion, especially if it is a consumer product. Once the product is on the market, it is necessary to have backup arrangements to provide service and support for guarantee claims, this applying whether the product is a consumer or capital item.

It is, of course, possible to have a joint venture concerned solely with research and development which finishes when this has been completed; often, however, the co-operation extends into one or more areas of subsequent functions. Indeed, the terms of the recent group exemption for such agreements (Regulation 418/85) apply both to:

(i) joint research and development of the products or processes without any subsequent joint exploitation of the results;
(ii) joint research and development of products or processes with joint exploitation of the results; and
(iii) joint exploitation of the results of research and development carried out as the result of earlier research and development agreements between the same undertakings.

In this context, exploitation includes the manufacturing of the product or even its licensing to third parties but does not cover the provision of promotional and distribution facilities. Only a minority of R and D agreements can, however, meet the stringent requirements of the group exemption, especially if joint exploitation of the results is required, and there are, therefore, a majority which nevertheless still require individual exemption.

In its First Annual Report,[15] the Commission indicated that it would adopt a sympathetic attitude to agreements for joint research and development and later utilization of the results. It stressed that agreements particularly between small and medium-sized enterprises relating only to joint research and development did not generally present any danger to competition. It was rather at the later stage, that of the exploitation of the results of the research and development, that problems could arise. One would have expected to find research and development agreements given favourable treatment under Regulation 17, and this is so to a limited extent. Article 4(2) (iii) (*b*) provided in its original form that joint research and development agreements could be freed from notification, provided that they were concerned specifically with improvements to techniques and provided that the results of the research were accessible to all parties and usable by them. It was later found that this definition of agreements not required to be notified was too narrow, and under Regulation 2822/71 (which dealt both with specialization and research and development agreements), the wording of Article 4(2) (iii) (*b*) was simplified to eliminate the qualifications and to provide that all joint research and development agreements would be free from notification requirements provided that the objects did not extend beyond joint research and development.

[15] 1st Annual Report, p. 45.

The first reported case in which the Commission addressed itself to the possible effect on competition of research and development agreements was the *Eurogypsum*[16] case in 1968. Here, negative clearance was given to the rules of an association which had its seat in Geneva but its administrative headquarters in France. It had 31 members from 16 countries including five of the then Member States. The majority of undertakings within the Common Market which manufactured plaster or gypsum or their constructional derivatives belonged to this association, either directly or through a national trade association. Its objectives were the promotion of the development of the plaster and gypsum industry by carrying out research on scientific and technical problems, by giving lectures and demonstrations, and also by issuing publications and providing advice to individual members. All the members of the association were free to use the results of the work, and indeed it was often made public. The open nature of the constitution of the association, and the fact that no restrictions on the use of the results of the research were placed upon its members, made the decision that the agreement fell outside Article 85(1) inevitable. The members had not yielded up any of their own freedom to carry out their own research and development, and any benefits which they derived from membership of the association were in addition to their own efforts, not by way of substitution.

This decision preceded by some six months the Notice on co-operation between enterprises, and was initially thought to provide a justification for confidence that many research and development agreements would fall outside Article 85 altogether. The Notice was, however, to prove a disappointment, limiting its scope, however, to assist only a narrow group of R and D agreements. It accepted that the kind of agreement that is clearly outside Article 85 is where there is simply an exchange of experience and results on the basis of information (as in the *Eurogypsum* case) without any restriction on the undertakings from carrying out their own research and development. Agreements of such a pure and limited nature are, however, rare. If there is no right to mutual access to the results, or if the practical exploitation of its results is covered by the agreement, then a restraint on competition may occur. Whilst it is possible to have an arrangement under which there are major and minor participants in the research project, who may receive access to the results in proportion to the degree of their participation in the research work, again a restraint of competition arises if some of the participating enterprises are unable to exploit the results either totally or to an extent that is not proportionate with their participation. Moreover, if there is any restriction on the granting of licences to third parties, competition is again likely to be affected,

[16] [1968] CMLR D1.

though it is permissible to provide that a majority decision of the participants is required for the grant of any individual licence.

From this Notice alone, one might have assumed that an agreement which imposed no restrictions on the parties from continuing their own research and development activities would not normally be in breach of Article 85(1), but *Henkel–Colgate*[17] (Commission 1971) showed even such an assumption to be fallacious. The two companies involved were both of importance in the detergent market, Colgate holding some 10 per cent of the EC market and Henkel nearly three times that amount. In three of the Member States their combined market share was over 50 per cent, but they had major competitors, notably Unilever and Proctor & Gamble. The proposed jointly owned company was to be limited to research and development, and not involved with distribution and marketing. It was apparent that in this particular industry the 'cutting edge' of competition lay principally in the ability of firms to improve the technical quality of the product, which, of course, had the effect of placing great importance on their capabilities for research and development. The effect of the agreement was in practice to eliminate the likelihood that the firms concerned would carry out their own research, since each party was bound almost certainly to encourage the joint subsidiary to exploit any successful progress which it might make on its own. Any successful results from the collaboration would improve the competitive position of the two parties jointly, whilst eliminating competition in this area between themselves.

In view of the size and strength of both Henkel and Colgate, an unfavourable reaction from the Commission was not unexpected, but an exemption was given, perhaps surprisingly, to the agreement limited to research and development on the basis that it might enable technical and economic progress and contained no restrictions on either of the parties indispensable to the arrangements proposed. The proceeds of the collaboration would be available to both on payment of a royalty not to exceed 2 per cent of the net price of the product and on a non-exclusive basis. Licences to use the information could only be granted if both parties agreed, but this limitation was felt to be an inherent consequence of the setting up of the joint research organization for the benefit of the two participants. In order, however, to prevent the joint research from leading to further integrated activities between such substantial companies (such as market-sharing or arrangements over production and sales policies), the Commission imposed stringent conditions on the exemption granted. These included an obligation to pass full information to DG IV on their policy relating to the licensing of patents and know-how resulting from the joint research, and the provision of information about changes in the

[17] JO [1972] L14/14.

shareholding in the joint venture or in other links between the two companies.

It is an interesting tailpiece to this early case that seven years later the parties sought a further exemption for the exchange of research information, having not proceeded with the original scheme for a joint subsidiary.[18] On this occasion, however, the Commission refused to grant the exemption because the condition that mutual agreement was required for the granting of licences to cover patents or know-how resulting from the association had too great an effect on competition between them. It is doubtful that application by companies in the same circumstances as those of the original case would today be likely to receive exemption because of the fear that close association on the research and development activities would spill over to reduce rivalry in production and marketing activities.[19]

Continuing experience with such R and D agreements gradually illustrated to the Commission that, although in nearly all cases it was finally possible to grant exemption to the agreement, it was often necessary to impose detailed conditions on their operation. This required the Commission to become involved in monitoring the agreements between the parties. In many other cases under Article 85(3), however, the Commission once having rendered a favourable decision would not be concerned with the details of its operation, unless it received complaints that the agreement was not being carried out in accordance with its notified terms.

During the 1970s, there was a substantial increase of cases in the high technology area. An interesting pair of cases were the *Vacuum Interrupters*[20] and the *GEC/Weir*[21] cases from the United Kingdom. The *Vacuum Interrupters* case involved the development of a key component for heavy electrical equipment, the 'circuit breaker', which in the event of a fault, cuts off in a fraction of a second the flow of electric current. AEI Ltd. and Reyrolle Parsons Ltd., who both made heavy switchgear, formed a joint company to continue research and development in this area which had previously moved only slowly. The research work was expensive and neither company would have considered it worthwhile to continue it, if unable to pool its resources with the other. The arrangements put forward were that Vacuum Interrupters Ltd. would only manufacture interrupters of a vacuum type and not the switchgear into which they would be

[18] For the sequel see the 8th Annual Report, pp. 77–8.

[19] Nevertheless, in some cases, the importance of the research being carried out, e.g. in the development of new drugs for the treatment of disease, might be adjudged to outweigh any risk of this kind; this was an important element in, e.g. *Beecham/Parke Davis*, n. 24 below.

[20] [1977] 1 CMLR D67.

[21] [1978] 1 CMLR D42.

installed. Switchgear is ordered on an individual basis, and the specification would be given by either AEI or Reyrolle Parsons to Vacuum Interrupters Ltd. who would then design and construct the interrupters for installation in the electrical switchgear specified.

In this case, the Commission had no difficulty in finding that Article 85(1) would apply to such an arrangement because even though AEI and Reyrolle did not actually compete in the market for vacuum interrupters in the Common Market (and indeed there were no manufacturers of that product in that market at the time), they were clearly potential competitors in the market. If they both had entered the market separately, they would have been in direct competition with each other and with other manufacturers of heavy electrical equipment; and the pattern of exports and imports between the United Kingdom and the rest of the Common Market would have been affected. The possibility of Common Market undertakings entering the UK market for vacuum interrupters would also have been considerably reduced.

Nevertheless, an exemption was given because the results anticipated from the joint venture would mean that a more durable and efficient low power interrupter could be manufactured at a reasonable cost.[22] The technical progress likely to arise from the joint efforts of the participants would, therefore, provide a fair share of benefit to the ultimate consumer of electricity. Vacuum Interrupters Ltd. remained able to manufacture and sell such equipment for any switchgear manufacturer and would handle all orders (whether from its parent company or elsewhere) on an independent and confidential basis. The research and development would not have proceeded at all if the joint venture had not been formed; and the only restraint on competition was that the parents had agreed not themselves to develop, design, manufacture, or sell such interrupters and to use reasonable efforts to obtain all their requirements of such a product from the joint venture. An exemption was given for ten years. Some years later the parties were on the point of abandoning the project when a large switchgear manufacturer decided to take an equity interest in the project and make available its own special expertise. A further exemption was granted for the revived joint venture.[23]

The *GEC/Weir* case had similar features; the product was a sodium circulator which is required to maintain circulation of the coolant for the high-powered density cores of fast nuclear reactors. These have to be designed to a very high degree of accuracy and reliability. Here it was not a joint company to be formed, but a joint committee in which GEC and

[22] The circuit breaker has to eliminate arcs which are caused by the sudden switching off of an electric current, and an interrupter involving a vacuum has technical advantages, e.g. a reduced risk of fire.

[23] [1981] 2 CMLR 217.

Weir Group were to hold an equal interest, the committee operating under a non-exclusive licence from both parties covering all necessary technical information for development. This was shared between the factories and the facilities of both parties; on termination of the joint venture both companies would have unrestricted rights to use the information obtained from the venture with the right to grant sub-licences. Royalties would, however, have to be paid on the use of this information. During the agreement neither GEC nor Weir would itself compete in the design or manufacture of the circulator. Exemption was again granted, subject to conditions that information about developments in the UK nuclear power industry which affected the terms of the collaboration of parties would be provided to DG IV. The Commission accepted that the non-competition clause was indispensable to the arrangements and was ancillary simply to the joint venture which could not have proceeded without such a basic limitation on the parties' freedom to compete with it.

The need for joint research into long-term problems too difficult for individual companies was illustrated also in the *Beecham/Parke Davis* agreement[24] (Commission 1979). Here the parties proposed to carry out joint research into drugs to prevent the impairment of blood circulation, none being available at the time. The time-scale for the work was considered to be at least a decade. Testing alone would take an extremely long period, and neither company thought it would be worthwhile to undertake the required investment alone. A joint research programme was set up to be followed if successful by a development programme. The development programme itself was expected to last five years before any marketing would be possible. The arrangements involved cross-licensing arrangements for patents on a non-exclusive royalty-free basis, and neither party was prevented from carrying out its own individual research. The parties agreed that they would exchange information relating to any improvements in manufacturing processes for a term of ten years after the product was marketed. Marketing would be carried out by each party individually.

The exemption was granted for a period of ten years, and the Commission indicated that a conclusive element in this decision was the fact that the new product was to be 'pharmacologically and therapeutically different from all known medicine' so that it was not an issue of improving production of existing products but of creating a new product. Even for companies as large as Beecham and Parke Davis such research and development would stand a better chance of proceeding if carried out on a joint basis. Moreover, both parties remained free to use the results of the research and development independently and could grant licences to third

[24] [1979] 2 CMLR 157.

parties without obtaining the prior consent of the other party. Deleted from the arrangements was an obligation to pay royalties for cross-licences since the burden of royalty payments proposed would have been a considerable disincentive for the parties to compete with each other.

It has, however, also always been a major concern of the Commission to prevent the use of an apparently harmless research and development agreement from changing into what in effect becomes a market-sharing arrangement. This concern arose in the case of *Carbon Gas Technology*[25] (Commission 1983), which involved a joint venture between three companies, Deutsche BP, Deutsche Babcock, and a German company, PCV, for developing coal gasification. The possible conversion of coal into gas had major industrial and economic significance for Europe in view of its dependence on imported oil. This venture limited to research and development appeared to have important strategic consequences for the European Community and was approved, but only subject to strict conditions. The commercial logic of the venture was that PCV specialized in basic process technology and the manufacture of the equipment required for conversion; Deutsche Babcock were the specialists in the constructing of large-scale plant in which such equipment should be placed; whilst BP's expertise was in refining oil, with a technology similar to that of gasification. The co-operation between these three large companies would effect considerable savings both of time and money, and it was likely that any technical advances made in adapting and improving the process would ensure that consumers would receive a share of the benefit, given the active competition with other fuels.

The parties accepted restrictions on their activity by agreeing to refrain from competition with a joint venture, which meant that in practice they would not compete with each other. If any of them withdrew from the joint venture for a period of five years they agreed not to disclose any know-how that had belonged to the joint venture whether or not the know-how originated from the parent company or from the joint venture itself. The exemption was granted provided all licensing agreements between the joint venture and its parents were submitted to DG IV for approval, in order to prevent them from being used for market-sharing arrangements for sale of the equipment or plant for coal gasification within the Common Market. This was considered a real possibility given the individual strength and contacts of the three companies in their specialized fields.

It had been noteworthy that although the enactment by the Council of Regulation 2821/71 had given the Commission the *vires* to issue a group exemption for research and development agreements, legal and political

[25] [1984] 2 CMLR 275.

difficulties prevented this, until finally publication of a draft occurred in early 1984, which subsequently after amendment was adopted as Regulation 418/85 effective from 1 March 1985. The content of the group exemption is highly detailed, much more so than in the equivalent group exemption for specialization agreements already mentioned which came into force on the same day. An early draft of the R and D group exemption had appeared at the beginning of 1984, and it is striking how great are the differences between the original draft and the final version adopted approximately one year later. The consultation process produced on this occasion both a marked improvement and a lengthening of the original draft.

The original draft provided that the exemption would only be available if the aggregate annual turnover of the undertakings did not exceed 500 million ECU, and if not more than one of the three actually or potentially leading undertakings in the sector to which the research and development programme relates at the conclusion of the agreement were a party to the agreement (although an opposition procedure was provided if either of these conditions were not complied with). The Commission was required to oppose such an exemption if it received a reasonable request to do so from any Member State within four months from the transmission of the notification of the agreement to that Member State. In the final version, all reference to aggregate turnover of the participants had been removed, although a market share limit of 20 per cent was included. In the original draft moreover there was no time limit for the exemption of the agreement, whereas in the final version a five year limitation applied once the contract products were first marketed within the Common Market. This marketing could not, of course, take place immediately at the end of any research and development programme. If the parties were not competing manufacturers, however, then the exemption period continued after the five years from first marketing so long as the parties' joint market share still did not exceed 20 per cent.

Although the permitted restrictive clauses were considerably extended in the final version beyond those contained in the original draft, the Regulation does contain a number of preconditions applicable to the framework of the joint venture, which did not appear in the original draft of Article 2. These preconditions are:

(*a*) that the joint research and development work is carried out within the clear framework of a programme that defines the object of the work and the fields in which it is to be carried out;
(*b*) that all parties have access to results of the work;
(*c*) in agreements providing simply for R and D, that each party is free to exploit the results of the work;

(*d*) that the joint exploitation of the research and development relates only to results which are protected by intellectual property rights or which constitute know-how which substantially contributes to technical or economic progress 'and the results are decisive for the manufacturer of the contract products or the application of the contract processes';

(e) that no joint undertaking or third party responsible for manufacture of the contract products is required to supply them only to the joint venture parties;

(*f*) that any undertakings required to manufacture by way of specialization in production are required to fulfil orders for supplies from all parties in the venture.

Another change between the two versions is the extensive definition of research and development agreements contained in the Regulation which was also not found in the original draft. The definition contained in the Regulation covers, as we have already seen, three categories, namely joint research and development alone, joint research and development coupled with exploitation but excluding distribution and sale, and exploitation of the product as the result of earlier research and development agreements between the same parties. Research and development itself is carefully defined to cover 'the acquisition of technical knowledge and the carrying out of theoretical analysis, systematic study of experimentation, including experimental production, technical testing of products or processes, the establishment of the necessary facilities and the obtaining of intellectual property rights for the results'.

Articles 4 and 5 of the Regulation contain the clauses which are permitted (the 'white list'). Restrictions which are permitted are:

(*a*) not to carry out on an independent basis research and development in the same field or a closely related field during the joint venture programme;

(*b*) not to enter into agreements with third parties of research and development in such fields during the joint venture programme;

(*c*) an obligation to procure contract products exclusively from specified sources;

(*d*) an obligation not to manufacture contract products or apply the processes in territory reserved for other parties (a form of territorial protection);

(*e*) an obligation to restrict the manufacture or application of the products or processes to specific fields of use, unless the parties in the joint venture are already competitors at the time it is signed;

(*f*) an obligation not for five years at the time when the contract products are first marketed within the Common Market to follow an active policy of marketing in territories reserved for the other parties so long as users of the products can obtain them from other sources and the parties do not themselves obstruct the entry of parallel imports or make it otherwise difficult for supplies to be obtained (a restriction similar to one permitted in the Patent Licence Group Exemption 2349/84);[26]

(*g*) a mutual obligation to communicate experience or improvements to each other and to grant non-exclusive licences;

(*h*) clearance is likewise accorded by Article 5 to obligations relating to confidentiality in the use of know-how and intellectual property rights, obligations to pay royalties in cases where the participants contribute unequally to the joint research and development or obtain unequal benefits from exploitation, and an obligation to supply minimum quantities of contract products and to observe minimum quality standards.

Article 6 of the Regulation sets out the 'black list' of clauses which may not appear in the agreement if the exemption is to apply. Some of these will already be familiar, being similar or identical to those required by the Commission to be deleted in approving earlier individual requests for exemption. Thus, the following clauses are prohibited:

(*a*) where the founders of the joint venture are restricted in their own freedom to carry out research and development either independently or with third parties in separate fields of activity during the joint venture programme or after completion of the research programme in the same field;

(*b*) where after completion of the research and development programme the parties are restricted from challenging the validity of intellectual property rights which protect the results of that research and development within the Common Market;

(*c*) where there are restrictions as to the quantity of the contract products the participants may sell or manufacture;

(*d*) where the parties are restricted in their free determination of prices or discounts within the contract products;

(*e*) when they are restricted as to the customers they serve, except so far as it is required by field of use restriction;

(*f*) when the parties are unable to pursue an active sales policy in territories within the Common Market reserved for other parties after the end of the initial five-year period;

(*g*) where they are prohibited from allowing third parties to manufacture contract products or apply contract processes when joint manufacture is outside the scope of the agreement; and

(*h*) when they are required to refuse 'without any objectively justified reason' to meet orders from sources within their respective territories

[26] See Ch. 15, pp. 304–6.

that wish to market the products in other territories within the Common Market, or are required to make it difficult for customers (users or dealers) to obtain the contract products from other dealers and to obstruct the free movement of products within the Common Market which have been lawfully marketed by another party.

An opposition procedure is included for restrictions in the agreement not within the 'black list', i.e. those which are not expressly exempted or prohibited by the Regulation. These can be notified to the Commission and will be deemed covered by the group exemption unless within six months from notification the Commission opposes exemption and notifies the parties that any exemption would have to be on an individual basis. It is also still possible for a Member State to request the Commission to enter opposition to proposed restrictions but such a request must now be justified on the basis of considerations relating to the competition rules of the Treaty rather than on an open-ended basis. Even if a Member State has requested opposition, the Commission may still withdraw it after consultation with the Advisory Committee.

The recitals to the Regulation are also considerably more lengthy and explicit than those to the original draft. Of particular interest is Recital 14 which states that agreements covered by the Regulation may also take advantage of provisions contained in other group exemption Regulations, notably those on specialization agreements, exclusive distribution agreements, exclusive purchasing agreements, and patent licensing agreements so that it would be possible to obtain a cumulative group exemption for agreements falling within more than one category. On the other hand, if any of the clauses contained on the black list of the research and development group exemption are found in the agreement, then these cannot be saved even by the specific provision of any other group exemption. Application of this 'overlap' rule is still in its early days and poses many practical difficulties to undertakings. For this reason, it would clearly be wise policy to frame agreements in such a way that they fall squarely within the terms of one or other of the individual group exemptions. Nevertheless, as the Commission indicated in its Fourteenth Report,[27] it regards the maintenance of research and development on an international basis as of considerable importance, to ensure that the improvement of competition as the result of new or improved products and services can continue. The Commission concedes that although innovative efforts should be regarded 'as a normal part of the entrepreneurial spirit' of individual undertakings, it cannot be denied that in many cases

the synergy arising out of co-operation is necessary because it enables the partners to share the financial risks involved and in particular to bring together a wider

[27] pp. 37–8.

range of intellectual and mental resources and experience, thus promoting the transfer of technology. In the absence of such co-operation, the innovation may not take place at all, or otherwse not as successfully or as efficiently. . . . The present situation in the Community demands a more rapid and effective transformation of new ideas into marketable products and processes, which may be facilitated by joint efforts by several undertakings.

The existence of the group exemption will not, however, mean that the Commission will be completely spared from the review of such new joint ventures. The nature of the collaboration between the parties may well take them outside the detailed requirements of this Regulation, and this is particularly likely in cases where the parties wish to carry their exploitation through into distribution of the products or services representing the end result of their collaboration. The *BP International/Kellogg* joint venture[28] (Commission 1985) underlines the continuing importance of this type of collaboration on an international basis.

British Petroleum had done considerable research into catalysts for producing ammonia. They wanted to exploit this work in plants to be operated by themselves or to sell the benefit of it to other producers of ammonia. They had, however, no experience in the design, construction, or commercial exploitation of large-scale plant whereas Kellogg was a large United States company specializing in the design of such plants. The two companies, therefore, decided that they would collaborate in the development of designs for the construction of ammonia plant using the BP catalyst. Each party would remain the owner of the know-how in its possession and would own jointly any know-how arising out of their work. Each would have an irrevocable world-wide non-exclusive right to use and sub-license their joint patents. Ancillary restrictions were imposed on both parties.

BP were restricted in that they could not disclose information about the catalysts to third parties and had also agreed not to supply catalysts during the life of the agreement to any other person without Kellogg's consent. Moreover, they had to use their best endeavours to ensure that Kellogg would become the design contractor for the first commercial plant using the new process to be built for BP.

Kellogg for their part had to agree not to carry out independently other research and development on competitive catalysts and not to spend more on developing other ammonia processes (except those using a traditional iron-based catalyst) without informing BP nor to commercialize any new ammonia process (using even iron-based catalysts) which was likely to be more commercially attractive than the process using BP's catalyst, unless BP was informed in advance. The purpose of these restrictions was to

[28] [1986] 2 CMLR 619.

ensure that BP was kept fully informed as to the commitment that Kellogg was putting into the development work. Kellogg was also required not to bid for the construction of ammonia plants other than by way of offering a jointly developed process, again unless BP had been informed.

If the development was successful, Kellogg would license the process to the plant operators and pay BP a percentage of the licence fee. The Commission noted in approving the joint venture that it represented a considerable technical advance, and success would make possible energy savings and lower ammonia prices. It was felt that BP and Kellogg were unlikely, had the joint venture not occurred, to have become competitors in the design of process plants using this new form of catalyst. The restrictions imposed were the minimum necessary to protect BP's cost in research effort over a number of years and to ensure that Kellogg devoted sufficient attention to the project to allow it a chance of success. This particular venture fell outside the group exemption since it included a number of restrictions not provided for in it, principally the restrictions on the exploitation of patents and processes that would not jointly be developed by both parties. The period of exemption granted was seven years.

In conclusion, the Commission is concerned that Article 85 is not applied to research and development agreements in a way that will discourage those that are genuinely playing a part in the Community's general policy to increase European competitiveness in world markets, especially in telecommunications and other areas of high technology. This applies particularly to those agreements arising out of the ESPRIT, BRITE, and RACE programmes which were implemented for this purpose in 1984 and 1985.[29]

[29] *ESPRIT*, adopted by the Council on 9 Mar. 1984 (OJ L67/54), is concerned with projects involving 'pre-competitive research and development' carried out by at least two undertakings not established in the same Member State and covers advanced microelectronics, software technology, information processing, integrated computer manufacture, and office systems. (Such projects would in any case appear to fall outside Article 85(1).) *BRITE*, adopted on 25 Mar. 1985 (OJ L83/8), involves research in the application of new technologies and their development in a variety of industrial techniques including lasers, metal shaping and forming, and testing of new materials. *RACE*, adopted on 7 Aug. 1985 (OJ L210/24), involves research and development in the area of telecommunication technology.

11

Horizontal Agreements: Joint Ventures

There are also a number of other agreements which could be described as 'joint ventures' but which fall outside the categories already considered, even if they share some common features with them. By reason of the great variety of these agreements, no group exemptions have yet been provided. As a result DG IV has had to develop criteria for assessing on a case-by-case basis whether they fall within the jurisdiction of Article 85(1) and, if so, whether the restraints on competition which they impose outweigh the advantages which they may provide in terms of the four conditions for exemption set out in Article 85(3). Similar criteria will, of course, have equally to be applied to those specialization and research and development agreements which for one reason or another cannot benefit from the group exemption.[1]

The Commission's handling of joint ventures entered a new phase with the implementation of the Merger Regulation in 1990, since it then became essential to provide detailed guidance to the business community on the definition of those joint ventures (concentrative) which would fall within the scope of the Regulation as compared with those which would fall outside, namely 'co-operative' joint ventures which would remain subject to Article 85 or even possibly Article 86.

It is necessary in order to understand the way in which the Commission assesses joint ventures by reviewing some of the leading cases in which the relevant criteria have been applied and relevant published material from the Commission and senior officials. Some of the main issues of the two previous chapters will, of course, reappear, though in a slightly different field of application. Many horizontal agreements could in a general sense be described as co-operation agreements, where the parties concerned normally agree to work for some limited common objective, but using their own resources and without normally establishing any separate company or even pooling resources in any way to achieve the purpose of their co-operation. On the other hand, a joint venture normally

[1] Failure to obtain the benefit of the existing group exemptions (417 and 418/85 for specialization and R and D Agreements respectively) arises for a number of different reasons: possibly the area of joint exploitation falls outside the restrictions covered by the group exemption involving, e.g. joint distribution or selling activities. Often, of course, the difficulty is simply that the aggregate market share or turnover of the parties is too substantial.

involves a greater degree of commitment, involving at least the following factors:

— the creation of either a separate legal undertaking or at least some recognizable joint committee or association clearly identifiable as separate from its founders;
— the transfer by the founders of personnel and assets (often including intellectual property rights) to the new undertaking; and
— the allocation to the new undertaking of responsibility for carrying out a particular function or functions decided upon by its founders.

Prior to the Merger Regulation Article 85 was only inapplicable in circumstances where the founders had irreversibly given up any thought of carrying on the business committed to the joint venture and had also deprived themselves effectively of the means of doing so. This required that the integration of the founders' interests in the particular manufacturing product or process or service had been irretrievably committed to the new venture. In that case, Article 86 began to operate in place of Article 85 since a merger was deemed to have occurred, provided at least that in other markets the parent companies remained independent of each other and able to compete. This was known as the 'partial merger' test. The continued use of Article 86, however, in such situations is unlikely following the adoption of the Merger Regulation.

The first major Commission decision in this area was the *SHV/Chevron Oil Europe Inc.*[2] case, where the Commission issued a negative clearance covering agreements under which Chevron (a subsidiary of a US oil company) set up a joint holding company with a Dutch group, SHV, which had interests in coal, chain stores, and transport, but had previously been unconnected with the production or refining of oil.[3] Under the agreement, a joint holding company was set up under the name of Calpam on a fifty-fifty basis. This company and its subsidiaries were to sell petroleum products in Belgium, The Netherlands, Luxembourg, Germany, and Denmark where previously each had had independent distribution networks. Both companies transferred into Calpam subsidiaries, for at least 50 years, all their distribution networks and the plant and equipment that comprised them. The joint venture, therefore, was limited to distribution: SHV ceased to do business as an independent wholesale buyer of petrol and related products, and neither SHV nor Chevron retailed these products separately. No restriction on competition was imposed in respect of other products not being distributed by the Calpam companies.[4]

[2] [1975] 1 CMLR D68.
[3] It had held a limited interest in the distribution of petroleum products.
[4] A clause to this effect covering certain products in the Benelux countries and Denmark was deleted at the request of the Commission.

The development of the thinking of DG IV in this area was later illustrated by its informal approval of the joint venture, known as Hinmont, set up for the production of polypropylene by two large chemical companies, Montedison of Italy and Hercules of the USA.[5] The original proposal had been a fifty-fifty jointly owned subsidiary with both parents retaining their right to remain as producers in some product areas in which Hinmont would also engage but subject still to detailed supply, purchasing, and distribution obligations with its parents. This caused difficulties to DG IV since the continuing close involvement of the three companies appeared likely to have the effect of reducing the degree of competition between the two parents and Hinmont itself. Under revised proposals, however, approved by the Commission, Hinmont was rendered a more separate and independent entity, 20 per cent of its equity being sold to the public, and the distribution and manufacturing arrangements altered so as to reduce mutual dependence. In this way, the proposal was adjudged to be outside Article 85(1), and under the less stringent tests contained in Article 86 was accepted by the Commission. The 'partial merger' exception then operative to the application of 85(1) to joint ventures was clearly thus of great practical importance.

Assuming, however, that the joint venture in issue does not now come under the Merger Regulation, the first question to resolve is whether it is indeed within the scope of Article 85(1). This depends on whether the undertakings are already in actual competition with each other or are potential competitors in that field of activity where the joint venture is to operate. If the answer is in the negative, then prima facie the joint venture is outside the jurisdiction of 85(1).

The Commission had from the outset taken a strict approach to the application of Article 85(1) in this context. Nevertheless there has been an indication in recent decisions that its approach is becoming more flexible. In particular the *Odin* decision in 1990 is likely to be particularly significant.[6] The creators of the joint venture were a Norwegian company, Elopak, and a United Kingdom company, Metal Box. The joint venture, known as Odin Developments Limited, had been formed to develop a new product. This was a container with a carton base and a separate metal laminated cap, that could be used for foodstuffs with a long shelf life. The container would be in competition with existing products such as metal cans and glass jars.

The Commission based its decision that the joint venture itself fell outside Article 85(1) principally on the lack of any actual or potential competition between the parent companies, and also the fact that neither had

[5] Announced in a press release by the Commission dated 26 Mar. 1987.
[6] [1991] 4 CMLR 832.

sufficient technical knowledge to develop the product on their own. Their existing production and selling activities were quite separate. The various restrictions in the joint venture agreement did not go beyond those reasonably required for the creation of a joint venture and could properly be described as ancillary to it. By the same process of reasoning, these restrictions themselves were also adjudged to fall outside Article 85(1) rather than simply being exempted (as in most earlier cases) for a period of time under Article 85(3).[7]

On the other hand, the nature of the relationships to be set up between the creators of their joint venture *inter se* and with the joint venture itself (possibly including patent and know-how licences with extensive reciprocal clauses relating to improvements), may itself suggest that the barriers thereby being raised to entry, either to the founders or to third parties, will themselves justify a finding that the agreement falls within 85(1). It will then need detailed consideration under 85(3) before clearance can be given. An example of this last situation was the *Mitchell Cotts/Sofiltra*[8] case. Here the Commission ruled that the proposed joint manufacturing operation of the subsidiary to be formed by English and French parents was outside 85(1) because those parents were neither actual nor potential competitors in the manufacture of high-efficiency air filters made of submicronic glass fibres, in which it was to be engaged. Nevertheless, 85(1) did apply to the distribution and sales of such products so that exclusive distribution rights given to the joint venture restricting Sofiltra from making active sales of such filters in nine countries had to be considered under the terms of 85(3).[9]

Another comparable example is the *Optical Fibres* case[10] where, though neither Corning nor its national partners in a series of joint ventures throughout the Common Market were actual or potential competitors, nevertheless the network of parallel joint ventures created by Corning in these different Member States gave it a definite potential for reducing competition between them. This feature was adjudged sufficient to bring the joint venture within Article 85(1), and provided the Commission with the

[7] Prof. Korah in her *Introductory Guide to EEC Competition Law and Practice*, 4th edn. (Oxford: ESC Publishing, 1990) at p. 209 indicates her belief that the Commission have for long taken too rigid a view of the likelihood of potential competition between the parents of joint venture companies and has welcomed the *Odin* case in particular as an illustration of a more relaxed approach to the interpretation of Article 85(1) in this context.

[8] [1988] 4 CMLR 111. See also *IGR Stereo Television* in 11th Annual Report, p. 63 and 14th Annual Report, p. 76. Similar problems can also arise upon the dissolution of an existing joint venture when the parties may wish to impose some degree of restriction on each other's activities for a transition period after the dissolution takes effect; as in *Roquette/National Starch*, an informal decision of the Commission reported in the 14th Annual Report, pp. 73–4.

[9] The Commission ruled that on the facts exemption could be granted for a period of ten years.

[10] OJ [1986] L236/30.

opportunity in applying Article 85(3) of requiring amendments to the detailed individual joint ventures to reduce this risk.

In the Sixth Annual Report,[11] the Commission pointed out that the application of competition rules in this area could not depend on the legal form chosen by the undertakings but on the economic and other realities of their situation, which may differ in many respects, including:

(*a*) *The relationship of the parties*. They may be direct competitors who in other products or geographical markets compete fiercely with each other. The issue is whether they will continue to do so when required to work in harmony in at least an area of the joint venture. The parties may have no prior relationship at all, each operating in a totally different business area and environment but having been brought together perhaps by no more significant a link than the accident of a common director or financial adviser. They may believe that synergy will somehow work successfully for them. Alternatively, they may have operated in fields that are separate but related by common links, e.g. telecommunications on the one hand, computers and office machinery on the other, or indeed have been engaged in a vertical relationship within the same industry.

(*b*) *The coverage of the joint venture*. The coverage of the joint venture may have a horizontal relationship to the markets in which the parents both compete. It may serve to fill a gap in their respective product ranges, of which both are conscious. On the other hand, a joint venture may relate to a product or process totally unrelated to the normal field of operation of its founders. Alternatively, the area chosen for joint venture may be for a derivative or intermediate product, related to the basic products in which both parents are specialists. The arrangement might represent scarcely more than an allocation of existing products and geographical markets (so that it is only a glorified specialization agreement) or be an entry into new product markets (product extension) or new geographical markets (market extension). In recent years, the tendency for joint ventures to be entered into in order to spread the time span and considerable expense of research into advanced technology has become prominent, as Chapter 10 has illustrated.

(*c*) *The extent and aims of the joint venture*. In each case the Commission has to consider which functions are to be covered by the joint venture. At its simplest it will provide simply a single limited aim, most easily illustrated as say the establishment of a jointly owned laboratory to provide information as the results of research for its founders, which both are free to use as they will without further commitment to each other and for which each pays on a fifty-fifty basis. The degree of involvement may, however, be substantially greater, culminating finally in the possibility of

[11] pp. 38–9.

a complete merger. The intermediate stages before this point is reached include joint development of research data leading to the development of products and processes available for commercial use or disposal, joint manufacture, joint promotion and advertising, joint sales, joint distribution, and joint service and aftersales facilities. If the collaboration extends as far as this, the parties will normally be sharing profits and losses, not merely expenses, though the allocation of profits themselves may be carried out in a variety of ways. This may include the payment by way of royalty, licence fees, or commissions to reflect the relative contributions and responsibilities of the parties. The founders can choose whether some or all of these functions are to be included in the venture and the inclusion of one or more functions does not necessarily require that other functions necessarily also be included. Even a joint sales agency could thus be described as a joint venture even if all the other functions referred to above were kept strictly separate. DG IV is aware from experience of the risk that collaboration at one or more stages has a tendency to induce informal collaboration (or at least an absence of true competition) in other areas not formally covered by agreement between the parties.

(*d*) *The freedom of action left to the founders.* At one extreme the participants may elect to retain complete freedom of action to engage in all the functions which the joint venture carries out, from basic research to full exploitation of its results and distribution function in addition. At the other extreme, the founders may accept complete restrictions on any activity at all in the relevant market, either by way of legal obligation contained in what will inevitably be highly detailed written agreements or, at the other extreme, as the calculated unilateral choice of an undertaking unwilling to compete with its own offspring, even if it comes in time to regard the maturing joint venture company as an undertaking completely at arm's length.

An experienced Commission official has emphasized the wide variety of restrictions that can be encountered.[12] Amongst these, he lists as particularly important the following:

— clauses prohibiting the joint venture from carrying on any other than the specified activities without consent of both parents or limiting it to a specific geographic market;
— a clause prohibiting the parents from competing with the joint venture (and therefore with each other) in a specified market;
— a clause obliging parents to buy from the joint venture all their requirements of its products or services;

[12] J. Temple Lang, 'Joint Ventures under the EEC Treaty Rules on Competition', 12 (new series) *Irish Jurist* 15 (1977).

— a clause obliging the parents to supply to the joint venture all its requirements of raw materials, components, industrial property rights, and know-how;
— a clause limiting the rights of the parent company to acquire more than a specified proportion of the production of the joint venture;
— a variety of restrictions relating to the terms on which exclusive or sole licences can be given by the joint venture to a third party or its parents.

Wherever the restrictions fall along this range, the Commission will seek to ensure that they are limited to those essential to the successful implementation of the joint venture.

(*e*) *The method of integration*. This, as we have already seen, is largely irrelevant in the eyes of the Commission since whether the joint venture operates by way of a jointly owned company or companies (with shares held either fifty-fifty or in other proportions) or by way of less formal association (possibly merely through a management committee containing representatives from both sides) seems comparatively unimportant. Nonetheless, the existence of a separate company has occasionally been considered by the Commission as itself a restriction which is unnecessary, for example, in the *Bayer/Gist-Brocades*[13] case. The advantages of limited liability will normally, however, lead the founders to seek to create a separate legal entity as the vehicle for their joint activities.

(*f*) Finally, to be considered is *the degree to which all these factors can or will restrain competition in the relevant markets*. Competition can, of course, be affected without there being any express restriction in the agreement between the parties or, at the other extreme, by a range of restrictions both explicit and implicit. Competition may be affected both simply by alteration in the conduct of the parties as a result of the creation of the joint venture or by structural changes following upon changes in their respective market shares. The way in which the degree of effect upon potential competition can be measured was considered in the Thirteenth Annual Report[14] as follows:

Input of the Joint Venture: Does the investment expenditure involved substantially exceed the financing capacity of each partner? Does the partner have the necessary technical know-how and sources of supply of input products?

Production of the Joint Venture: Is each partner familiar with the process technology? Does each partner itself produce inputs for or products derived from the joint venture product, and does it have access to the necessary production facilities?

[13] [1976] 1 CMLR D98. [14] pp. 50–2.

Sales by the Joint Venture: Is the actual or potential demand such that it would be feasible for each of the partners to manufacture the product on its own? Does each have access to the necessary distribution channels for the joint venture's product?

Risk Factor: Could each partner bear the financial risks associated with the production operation of the joint venture alone?

We should now consider the application of these principles to some cases. In the *De Laval/Stork* case[15] (Commission 1977), De Laval was a Delaware corporation which sold centrifugal compressors throughout the world and Koninklijke Machinefabriek Stork, a Dutch public engineering company. They formed a business association under the name of De Laval/Stork on a fifty-fifty basis. De Laval managed this joint venture under contract, but the consent of both parents was needed for any major policy decision.

It was agreed that the joint venture would produce steam turbines, centrifugal compressors, and pumps for nuclear power stations and provide essential aftersales service. De Laval had to provide know-how and industrial property rights to the joint venture under royalty-free licences covering the Common Market and other territories, so that the joint venture could respond promptly to manufacturing orders received, all products having to be produced on an individual basis. The agreement ran for five years and, if not renewed, it had been proposed that non-exclusive three-year licences should then be given to the party which carried on the business formerly carried on by the joint venture in return for reasonable royalties. Both founders continued to sell in the Common Market other products not covered by the joint ventures, but already, since the joint venture had started, it had expanded its sales to approximately 10 to 15 per cent of the market within the Community.

Exemption was given under Article 85(3) on the basis that the joint venture allowed De Laval to penetrate European markets more quickly than if it had acted alone, whilst likewise enabling KMS to expand its own business more quickly. The joint venture would be able to work at greater capacity than either of its parents, so reducing fixed costs per unit of output. Sharing of research and development would also produce joint savings; the result from the customer's viewpoint would be a better range of products and better aftersales service. There was no lack of substantial competition to the joint venture in the Member States. Trade between Member States was certainly affected as both parent companies were likely to refrain from selling there in direct competition with their joint venture; both were also potential competitors in Europe and would revert to being so in all likelihood after the joint venture had ended. Nevertheless, the agreement allowed each party to deal direct with a customer if

[15] [1977] 2 CMLR D69.

the joint venture could not supply his needs. Moreover, the Commission insisted that the contract should leave both parties free to compete after the termination of the joint venture and to have the benefit of industrial property rights, not limited (as in the original proposals) to three years only, but fully available to whatever extent was required for sale of the product that previously had been the responsibility of the joint venture. The royalties in such cases were not to exceed the lowest rate which had been charged to any third party by the joint venture.

Some similar features to this case could be found in the *Amersham International/Buchler* case[16] (Commission 1983). Here, the product was radioactive material for which Amersham International had appointed Buchler as its distributor in the German market since 1960. Prior to the formation of the joint venture, Buchler had competed to a limited extent with Amersham, but had a narrower range of products. Distribution of radioactive material required a very high level of know-how, since the product was perishable and very detailed safety regulations had to be complied with for its carriage and disposal. The normal distributor would not be able or expected to provide all these services, and, therefore, the formation of the joint venture with Buchler enabled it to have greater market penetration and to market a greater variety of products in Germany. The joint venture had already achieved sales of approximately 18 per cent of the German market at the time of the decision being rendered by DG IV in favour of an exemption; the only restriction contained in the agreement was that the founders would not compete with their joint subsidiary, and this was held to be indispensable to the arrangements. Each party remained, however, free to extend its activities into product lines not covered by the joint venture.

It may reasonably be supposed at this stage, in view of the cases so far cited, that no joint venture is ever totally rejected by the Commission, and that joint ventures considered by it are always permitted subject to the deletion of certain objectionable clauses. There have, however, been a few cases where the decision rendered has been to reject the joint venture in its entirety. The best known example is the *ICI/Black Powder* case[17] (Commission 1978). This was a production joint venture under which ICI entered into an arrangement with a German company, Wasag Chemie, covering this product which is a low-power explosive mainly used in the manufacture of fireworks. ICI no longer had production facilities in the United Kingdom but had a near monopoly of sales there, whilst Wasag Chemie continued to produce black powder in Germany and held around

[16] [1983] 1 CMLR 619. Amersham International had originally constituted the commercial department of the UK Atomic Energy Authority but had later been privatized. It was the majority (60%) shareholder in the joint venture.

[17] [1979] 1 CMLR 403.

50 per cent of the German market. The result of the joint venture was that two companies which were actual competitors in the same market no longer made independent decisions with regard to the manufacture and sale of the product. Through their control of the joint venture they would moreover have created a situation where it would be easy to align prices in related products, such as safety fuses. The effect of the agreement was that the UK black powder market would at least for a considerable time have been blocked to other producers, particularly those in France and Italy, who had excess capacity. ICI would have purchased black powder for substantially the entire United Kingdom demand only from the joint venture and would not have been free to purchase it from other producers on better terms.

Although some element of pooling of technology and resources was included in the arrangement, the Commission did not accept that this was sufficient to bring about product development, it being an industry where manufacturing methods in the twentieth century remained much as they had been in the nineteenth. Any increase in security of supplies of black powder to the United Kingdom was an insufficient benefit to justify an exemption against the background of considerable excess capacity in the industry as a whole in the Common Market. The main effect of the agreement would have been to eliminate competition for the supply of black powder in the United Kingdom. In view of the Commission's objections to the agreement, the parties decided to abandon it; in view of the importance of the decision and the reasoning underlying it, the Commission decided for its part nevertheless to publish its decision on a formal basis.

More recently a joint venture between Sky Television and a consortium of some seventeen national television authorities (members of the European Broadcasting Union (EBU)) for operating the Eurosport TV channel was refused an individual exemption.[18] This television channel is a transnational satellite service operated by Sky and received in twenty-two European countries in three languages. Under the joint venture agreement the consortium members would have been required to make available to the Eurosport channel all those sports programmes which its members acquired and to offer the satellite service priority for 48 hours after the end of any sporting event so covered. A company, Screensport, which provided a competitive sports programme on satellite, claimed that the joint venture would mean that Eurosport's preferred access to all these programmes would make its own competitive activities considerably more difficult and would make it almost impossible to compete actively with Sky. Sky's own principal argument was that the new arrangements would enable the owners of satellite dishes to receive more and better sports

[18] OJ L 63/32, 9 Mar. 1991.

programmes, some of which had previously only been available for national audiences.

The main element in the Commission's refusal to give the exemption was the effect it would have on competition not only between Screensport and Sky but also between Sky and the members of the consortium itself who were also actual or potential competitors in broadcasting sports events. It dismissed peremptorily the claims of the joint venture participants that they were under Article 90(2) engaged in providing a public service of general economic interest and therefore entitled to an exemption from Article 85(1).

The size of market shares had played an important part in the *ICI/Black Powder* decision. For this reason the approval of the joint venture in *Rockwell International Corporation/Iveco Industrial Vehicles Corporation*[19] came as a surprise. Rockwell, the United States corporation, and Iveco, a Dutch subsidiary of Fiat, entered into a complex joint venture agreement to produce single-reduction rear drive axles for trucks, apparently considerably more efficient that the previously standard double-reduction axles. Iveco had an existing factory in Italy which could be used for this purpose; and the joint venture was formed on the basis of a 60 per cent shareholding to Rockwell and 40 per cent to Iveco.

Both parents agreed to some limitation on their own competitive activities. Exports to Western Europe, Africa, and the Middle East were to be made through the joint venture only. If the joint venture were to invest in other manufacturing capacity apart from the original plant in Italy, the consent of both parents was required. Rockwell would transfer substantial technology to the joint venture which the joint venture could sub-licence and the Iveco subsidiary, Vispa, was also entitled and required to transfer its technology relating to rear axles to the joint venture company. Both parties agreed to place limits on licences for third parties; for eight years companies in the Iveco group must take all their rear axle requirements from the joint company. Rockwell agreed for its part that it would grant no further manufacturing licences in Italy or distribution licences within the Common Market. Iveco granted the joint venture an exclusive manufacturing licence for Italy in respect of Vispa axles, and all Iveco subsidiaries must get their rear axles from the joint venture only.

The parties claimed that their aim was to penetrate the hitherto undeveloped market for independently manufactured axles within the EC, and that this was only possible if truck manufacturers could be convinced that the joint venture was so well established and committed to manufacturing on such a long-term basis that the truck manufacturers could rely on con-

[19] [1983] 3 CMLR 709. Another joint venture involving Iveco with Ford for the manufacture of a range of trucks in the United Kingdom was also exempted. [1989] 4 CMLR 40.

tinuity of supply and a high technical level. The restrictions imposed on the parent companies and on the joint ventures were claimed as being indispensable to this long-term stability, and the exclusivity of the technical rights conferred on the joint venture was needed to justify the high level of capital investment which it would be required to make if the venture was to succeed. Iveco had a market share for trucks in the EC territories of 17 per cent, second to Daimler Benz which had 26 per cent. The free market for rear drive axles comprised only 5 per cent of the total production of axles (since the majority of manufacturers of trucks made their own) whereas in the United States the corresponding figures had been approximately 70 per cent. Although the new joint venture would have a considerable market share of the independent supply of rear axles (amounting to approximately 75 per cent), the Commission allowed an exemption.

The basis for the decision was that the only way that the truck rear axle market could be entered was by a very substantial investment committing both partners to a long-term relationship; mere sub-contracting would be insufficient to provide the confidence required, likewise assignments of patents or know-how. Truck manufacturers would only place orders with the joint venture once extensive testing had been completed and close technical co-operation established. The Commission was satisfied that the considerable advance in technology which the joint venture company would be able to achieve would enable cost savings to be made and would lead to more efficient production processes. Although the proportion of the free axle market held by the two companies was itself substantial, this share was itself very small in comparison with the 'in-house' proportion of the market controlled by the truck manufacturers.

DG IV, in applying Article 85(3), placed special significance on two of the four conditions, namely the first positive condition relating to the improvement of production or distribution and promotion of technical or economic progress on the one hand and the indispensability of restrictive conditions on the other. Interesting light has been shed on the way in which the Commission interprets both these criteria in a recent article by another official of the Commission.[20] This article, published in 1984, contains a list of some of the factors which have been found to give rise to relevant improvements in production or distribution or promotion of technical or economic progress within the meaning of the first condition. These include:

(*a*) production or distribution of products;
(*b*) exploitation of complementary technology, know-how, and expertise;

[20] J. Faull, 'Joint Ventures under the EEC Competition Rules', 5 *ECLR* 358, 364 (1984). Numerous additional articles on this topic in both US and European legal literature are cited.

(*c*) means of overcoming technical difficulties, including previous failures by one party;
(*d*) introduction of a new competitor into the Common Market or into individual Member States;
(*e*) ability to respond to major customer's requirements;
(*f*) sharing of costs in cases where research and development is long and expensive;
(*g*) provision of aftersales service for capital equipment essential to continuous production;
(*h*) ability of joint venture to obtain more advantageous fixed costs of production runs;
(*i*) enhancement of safety standards;
(*j*) security of supplies to be provided in markets where there is a shortage;
(*k*) elimination or avoidance of excess capacity.

Many of these issues have been considered in the cases already discussed. Although not all weigh equally, if several advantages can be combined and shown as the probable result of the joint venture then a strong case is likely to be made out for granting an exemption.

The second important condition, that of indispensability, has also arisen often in the cases referred to. The emphasis in recent cases has been upon whether the objective of the joint venture could be achieved in any other way by a less complete form of integration. This might enable there to be a lesser restriction on competition between the parties but with the risk that mutual commitment might not be so readily given, for example, if the parties remained at arm's length by way of licence agreements, or in circumstances where the parties remained free to compete with their joint venture. The answer may be in most cases that the apparently anti-competitive restriction is justified for a limited period in the interests of the joint venture although not on an indefinite basis.

Thus, in the *Carbon Gas Technology*[21] case the Commission's finding was that the non-competition provision binding all three founders was to be accepted because it would ensure that during the period of exemption the parents would allow the joint venture to specialize in the development of the relevant process without the parents individually trying to obtain business direct. In other cases the answer has been likewise affirmative provided that careful supervision of the arrangements remained in the hands of the Commission, for example in the *GEC/Weir*[22] and *Vacuum Interrupters*[23] cases. Alternatively, as in *Bayer/Gist-Brocades*, one might find that some alteration in the structure of the arrangements was essen-

[21] [1984] 2 CMLR 275.
[22] [1978] 1 CMLR D42.
[23] [1977] 1 CMLR D67.

tial if the indispensability requirement was to be met. At the end of the day, however, as Faull points out, the parents must then be able to compete freely with each other and to use the know-how and industrial property rights which the joint venture has acquired upon reasonable commercial terms.[24] It is, of course, unusual for joint ventures to continue indefinitely, although the Commission has been generous in granting exemptions for them for periods of between five and fifteen years.

Faull also mentions two additional elements which he feels play a major role in the analysis made by the Commission. He refers to the balancing of the advantages for competition against the loss of competition as the result of the restriction; the second element referred to by him is the importance of 'EEC criteria', the need for the Community to acknowledge the importance of developing independent sources of energy, such as was the subject of the *Carbon Gas Technology* case, and to make progress in important areas of high technology.

Thus in *Alcatel/ANT*[25] exemption was given on an individual basis to a co-operative joint venture dealing with all aspects of research and development, production, and marketing of electronic components for satellites. Apparently in this highly technical field a research and development joint venture without also a marketing function would be unacceptable to Government departments and other satellite manufacturers who were amongst likely purchasers. A satellite, once in orbit, can only be repaired with great difficulty and thus purchasers insist on a very detailed technical specification of any components which they buy.

For a number of years the Commission has been pressed to provide a group exemption for joint ventures but the complexities of the issues involved (as can be understood even from this comparatively brief account) proved too great to allow a comprehensive document to be produced in the form either of a notice or group exemption. A new situation has, however, now arisen with the publication of the Notice on Joint Ventures, which has been published along with the Merger Regulation in 1990. By placing *concentrative* joint ventures into the category of concentrations covered by the regulation, the Commission has in effect rendered approval of such arrangements by the Commission not only far more speedy (in view of the time limits under the regulation) but also less likely to be blocked or substantially amended. This arises because of the fact that the criteria which the Merger Regulation applies are less stringent than those set out in Article 85(3) and, as we shall see in Chapter 20, far more concerned with the risk that the joint venture will create or

[24] Unless, of course, there are substantial reasons why one or both of the founders are unable to enter the market.

[25] [1991] 4 CMLR 208.

strengthen a dominant position rather than with the familiar tests of Article 85(3).

The majority of joint ventures, however, which, under the terms of the Notice[26] are regarded as co-operative, will continue to fall within the jurisdiction of Article 85(1). The Commission's awareness of the comparative difficulty in obtaining individual exemption for them under Article 85(3) within a reasonable time was shown by the issue in January 1992 of a draft Notice and discussion paper summarizing the legal principles so far applied.[27] Although no dramatic change in policy was announced a sympathetic approach was indicated to those joint ventures creating substantial new production capacity by undertakings with relatively low market shares. A self-imposed time limit of five months for dealing with straightforward cases was also put forward, provided that only a comfort letter and not a formal decision was required.

[26] Normally referred to as the 'Interface Notice'.
[27] The text can be found at [1992] 4 CMLR 504.

12

Distribution: Exclusive Distribution and Purchasing

1. General Introduction: Distribution through an Agent

The development of any product for sale involves a substantial investment of both time and money. This effort and expenditure may, however, be wasted if an appropriate system is not then made available for the distribution of the product to the point of sale. This is particularly true of consumer goods, e.g. lawnmowers, clothing, or cutlery sold in large numbers primarily through wholesalers and then on to a variety of retail establishments including specialist shops, department stores, cash and carry outlets, supermarkets, or discount houses. The same applies also to goods for industrial use whether sold for use or consumption by industrial concerns or for incorporation by purchasers in their own manufacturing operation, e.g. nuts, screws, and bolts. In every case, commercial success depends to a substantial extent on the choice of an appropriate system of distribution.

Any manufacturer considering how to distribute products in the Common Market or any of its Member States for such purposes has a wide range of choice. The main decision to be taken relates to the extent to which its distribution system is to be integrated with the remainder of its organization. The greatest degree of control is generally given by making use of its own employees (whether based at home or abroad) to be responsible for ensuring distribution of its goods to the chosen form of outlet in the relevant geographical markets within the Community. Reliance on one's own employees for sales can, of course, be combined with other means of promotion such as exhibiting at trade fairs, maintenance of a local showroom or warehouse, and the mailing of catalogues and other sales information to prospective customers.

In many cases, however, direct representation may not be the best method of ensuring effective distribution in a foreign country.[1] The advantages of organizing distribution through persons familiar with local markets, languages, and conditions may well mean that local representation is chosen, either an agent or a distributor. The relationship of a

[1] For an assessment of the commercial advantages of different forms of sales organizations and descriptions of different varieties of sales agents, see A. Branch, *Elements of Export Marketing and Management*, 2nd edn. (London: Chapman and Hall, 1990).

manufacturer with an agent is quite different from that with a distributor; the agent is a person or company utilized to bring its principal (the manufacturer) into contractual relations with customers and normally paid for its efforts by a percentage commission on those sales which it effects. By contrast the distributor stands at arm's length from the manufacturer, purchasing goods from it on agreed terms normally at its own risk and reselling either direct to the public or through its own selected wholesalers and dealers.

Although competition policy in the EC has been far more concerned with the relationship between manufacturers and their distributors, particularly exclusive distributors, any review of the law relating to distribution of goods must, however, begin with a brief consideration of the legal position of the commercial agent. There are, of course, many varieties of commercial agent. These include the simple commission agent who does not hold stocks of the product but merely passes orders received back to the manufacturer after which goods are delivered direct to the customer. Another variety of agent is the 'stocking agent' so called because in addition to his selling function he provides warehousing facilities for a certain amount of the manufacturer's stock, for which he receives a fixed payment, but without, however, acquiring ownership of the goods. An agent may also help the manufacturer in other ways, including the provision of service facilities for which he will often have to carry a stock of spare parts purchased from the manufacturer. The agent may also take on a *del credere* responsibility under which he will accept liability for payment of the accounts of those customers from whom he has introduced business, thereby relieving the manufacturer of the worry that payment will not be made, but for which again the agent will expect an additional commission percentage.

The mere appointment of an agent, however, does not itself involve Article 85 unless the appointment involves some form of express or implied restriction upon either the agent or his principal. If an agent, therefore, is appointed simply as an agent for a product in a particular territory, Article 85 has no application since the essential element of a clause containing a restriction upon competition is lacking. Whilst manufacturers sometimes retain the right to appoint a number of agents in parallel in a territory, all competing with each other and where the agents themselves do not accept any restriction on the supply of competitive products to customers, this is a comparatively rare situation. It is far more usual in the Common Market to find that either the manufacturer or the agent, or both, accept restrictions on their commercial freedom. The restriction on the manufacturer may be that having appointed an agent for a given territory, say France, it agrees to appoint no other agent for that territory; for its part, the agent may agree that within France it will represent only this manufacturer and no other manufacturer selling identical or competitive products.

It is this latter kind of agency agreement with which the 1962 Notice is concerned.[2] This states that the decisive criterion which in the view of the Commission distinguishes an agent from the independent trader is the degree to which the agent has any responsibility for the financial risks connected with performance of his obligations. If the agent undertakes any element of risk on its own account, then its function becomes more similar to that of an independent trader and, for purposes or competition rules, must, therefore, be treated as an independent undertaking. As has already been noted, examples given in the Notice of where a sufficient element of risk is taken would be: when the agent either keeps a minimum stock of products covered by the contract at its own risk, or is required to maintain at its own expense services for customers free of charge, or where it determines or has the right to determine prices or conditions of business. If on the other hand the agent accepts none of these functions but simply performs the basic role of seeking customers for its principal, the agent's restriction to the goods of its principal is acceptable as the inherent consequence of the agency relationship.

The timing of the 1962 announcement was significant since it appeared only five weeks before the deadline for notification of bilateral agreements under the Regulation (1 February 1963) and was the first of a number of Notices issued by the Commission to provide guidance to enable undertakings to decide their legal position under Article 85(1) and to discourage them from filing unnecessary and purely precautionary notifications. The legal basis for the Notice was challenged by the Italian Government in the course of the proceedings at the European Court in the *Consten–Grundig*[3] case in 1966. The Court, in rejecting the suggestion that Article 85 applied only to arrangements between parties who were direct competitors with each other, ruled that this Article on the contrary referred in a general way to all agreements which distorted competition within the Common Market. The Article made no distinction at all between agreements between undertakings competing at the same stages or between non-competing undertakings at different levels. The Court went on to say that a distortion of competition could occur within the meaning of Article 85(1) not only by agreement which limited competition between the actual parties but by agreements which prevented competition which might occur between one of them only and third parties.

Nevertheless, the Commission will always look at the economic reality of a commercial relationship rather than the legal description applied by the parties to their agreements. This principle is highlighted by the

[2] For an account of the 1962 Notice relating to agents, see Ch. 4, pp. 48–9.

[3] Cases 56 and 58/64 [1966] ECR 299: CMLR 418.

decision in the *Pittsburg Corning Europe*[4] case. In this case, a Belgian subsidiary of a United States corporation manufactured cellular glass and distributed it in Belgium, Holland and Germany. The prices which it was able to obtain for this product in Germany were some 40 per cent higher than those which could be obtained in Belgium and Holland. It, therefore, tried to insist that orders from its Belgian and Dutch distributors always indicated the ultimate country of destination to which the order would be supplied, normally for incorporation in a new building, so as to obstruct any inclination of the distributors to send the goods by way of parallel imports into Germany at a price undercutting the prices being obtained by its German distributors. Because of the fact that parallel imports into Germany had occurred, the system of control was later tightened up still further by increasing prices for glass sold into Belgium and Holland substantially, subject to a reduction if it could be shown at a later date that the glass had actually been incorporated in a work site in Belgium or Holland rather than Germany.

The agreement, if made between independent undertakings, clearly represented concerted practices under which a single territory where a high price level operated for this product was kept separated from other territories with a lower price level. The argument raised, however, by Pittsburg Corning Europe was that the Belgium distributor, Formica Belgium, was merely an agent; the actual written agreement described Formica Belgium in this way. After examination of the facts, the Commission concluded that the status of 'agent' accorded by this agreement was accorded merely for tax reasons and did not genuinely reflect the actual relationship. For this reason, the agency concession contained in the Notice was inapplicable, and a fine was imposed on Pittsburg Corning Europe for breach of the provisions of Article 85 by its arrangements for differential pricing between the separate national markets.

The question has also arisen as to whether the benefit of the Notice can apply also to agents who act for more than one principal. The argument has been made that a principal cannot truthfully claim to have integrated an agent into its own sales network (as the Notice appears to contemplate) if that agent devotes only part of its time and effort to promotion of the products of that principal, reserving a substantial commitment also to competing manufacturers. The approach taken by the Court in the complex *Sugar Cartel*[5] case suggests that so long as the agent only undertakes

[4] [1973] CMLR D2. This case has been briefly referred to already in Ch. 6, p. 90.

[5] Case 40/73 [1975] ECR 1663: [1976] 1 CMLR 295. See also Ch. 6, p. 89. Similar issues arose in *Binon* v. *AMP* Case 243/83 [1985] ECR 2015: 3 CMLR 800, but the Court ruled here that, in an Article 177 reference, it was for the national court to make the legal assessment of whether a so-called agent was in economic reality playing the role of an independent distributor.

a limited function, it does not matter whether it takes this function solely for a single principal (with whom it normally becomes closely associated in the course of dealing) or whether it spreads its efforts over a number of principals for each of which it will remain only partially involved, provided only that in none of the cases does it accept an element of risk. If, however, the appointment is accepted by the agent not as its sole occupation, but at a time when it is also engaged as a broker or manufacturer, then this split status will tend to weaken the likelihood that the Notice can benefit it.[6] It is also unlikely that the Notice will apply if the company claiming to act as agent is so large and powerful that in practice it will have sufficient independence not simply to obey the requirements of the principal but to have substantial influence on the distribution policy of the principal. It is also essential if the Notice is to apply that the appointment of an agent does not involve the appointment of an actual or potential competitor, since clearly the required integration of function will not be able to occur if the economic interests of the two undertakings are opposed to each other.

In late 1990 the Commission issued a new unofficial draft Notice for consultation purposes. Clearly the original Notice no longer reflected the subsequent Court decisions[7] already mentioned nor the changes that had taken place in the Commission's own approach. The two new principal elements in the draft Notice which apply now both to exclusive and non-exclusive appointments and to goods as well as services are:

(i) a wider definition of the risk that commercial agents may accept without falling within the scope of Article 85(1) and
(ii) a recognition that agreements between 'integrated' agents and their principals may contain a wider range of restrictions than previously allowed, also without falling within the scope of Article 85(1).

So far as (i) is concerned, instead of having to act essentially without incurring the risk at all in order to qualify for clearance, the agent is stated now to be able to accept risks provided that it does not accept primary responsibility for performance of the principal's contracts. Any independent trading activities of the agent will, however, prevent him from benefiting from the Notice unless these activities are 'ancillary' to his main agency function. If these conditions apply, any contractual terms

[6] See also the Commission's decision in *ARG/Unipart* [1988] 4 CMLR 513 for an example of a case where the agent's role varied according to the category of product (various kinds of spare parts for cars) in which it dealt. Article 85(1) was held to apply to all the relationships except that where the agent accepted no risk at all. An individual exemption for seven years was, however, granted to all the arrangements.

[7] In particular the *Flemish Travel Agents* case (Case 311/85) [1987] ECR 3801: [1989] 4 CMLR 213. The MMC report on Foreign Package Holidays (1986) Cmnd. 9879 contains an analysis of the relationship between principal and agent in a service industry.

obliging the agent to negotiate or conclude transactions for the principal prices, terms, and conditions which the principal has laid down are not considered to restrict competition.

The definition of the 'integrated' agent under (ii) requires that the agent either be totally committed to activities for his principal or to have other activities that amount to no more than two-thirds of his commercial activities, and do not involve the sale of competing products (nor indeed an independent sales function in respect of the principal's product). Given such relationship, the principal is entitled to impose a number of additional restrictions upon the agent without bringing the relationship within Article 85(1). These include non-competition clauses (effective for up to two years after the termination of the agreement) and a restriction on the active promotion of sales outside the allotted territory. Also allowed are clauses to prevent the principal from appointing another agent for the territory or dealing direct with certain customers without paying compensation to the agent.

2. Exclusive Distribution: The Period 1967 to 1978

In this chapter, before reaching the present state of the law relating to distribution, it is important to examine its development during the period starting with the introduction of Regulation 67/67 and closing with its replacement by Regulations 1983/83 and 1984/83. The present state of the law on distribution agreements is more easily understood once the development of the subject over the 1960s and 1970s has been explained.

If for any reason the choice of an agent is inappropriate for the requirements of the manufacturer, the selection of a distributor, and particularly of an exclusive distributor, may be the best alternative. It has been with distribution agreements, especially those having an exclusive element, rather than with agencies, that DG IV has been primarily concerned. The advantages of exclusive distribution from the viewpoint of both manufacturer and distributor were quite early recognized by DG IV and indeed were set forth by the Commission in the recitals to the original group exemption which it subsequently issued in 1967. These refer to some advantages of such a relationship:

> The entrepreneur is able to consolidate his sales activities . . . he is not obliged to maintain numerous business contracts with a large number of dealers, and the . . . fact of maintaining contacts with only one dealer makes it easier to overcome sales difficulties resulting from linguistic, legal and other differences; . . . exclusive dealing agreements facilitate the promotion of the sale of a product and make it possible to carry out more intensive marketing and to ensure continuity of supplies, while at the same time rationalising distribution; moreover the appointment

of an exclusive distributor . . . who will take over, in place of the manufacturer, sales promotion, after-sales service and carrying of stocks, is often the sole means whereby small and medium-sized undertakings can compete in the market . . .[8]

The fact that the number of bilateral distribution agreements filed after the coming into force of Regulation 17 was so large meant that the priority for the Commission was to prepare a group exemption that would eliminate the need to review individually too many of the exclusive distribution agreements which amounted to nearly three-quarters of all those bilateral agreements notified. The Commission, of course, by this time had already issued a certain number of decisions on exclusive distribution agreements. The primary influence on the content of the Regulation came from *Consten–Grundig* which provided a clear model for future patterns of distribution by exclusive distributors. The principles laid down by the Commission in its decision in this case, as approved with only minor adjustments by the Court, formed the foundation for the original group Regulation itself.

This original Regulation is no longer in force, having been replaced after some sixteen years as from 1 July 1983 by Regulations 1983/83 and 1984/83. Nevertheless, the content of the original group Regulation remains important both because of the influence which it had in the development of competition policy on distribution agreements and also because subsequent changes in the 1983 Regulations were of minor rather than major significance, without altering the basic pattern for exclusive distribution established by the 1967 Regulation.

Before examining the detailed provisions of the 1967 group Regulation, it should be emphasized that not all distribution arrangements or even all exclusive distribution arrangements fall within Article 85(1). In the 1966 European Court case of *STM* (*Société Technique Minière*) v. *Maschinenbau Ulm*,[9] the Court had ruled that exclusive distribution agreements did not necessarily contain the elements to bring them within Article 85(1) and that in determining this issue the following considerations should be taken into account:

(1) the nature of the products covered, and whether the supply was of limited or unlimited amount;
(2) the importance of both the supplier and the distributor with respect to the relevant markets;
(3) whether the agreement was of an isolated nature or took its place among a network of agreements covering at least a substantial area or region of a Member State;

[8] Recital 10 of Reg. 67/67. The Regulation came into force on 1 May 1967.
[9] Case 56/65 [1966] ECR 234: CMLR 357.

(4) the degree of territorial protection afforded by the agreement, in particular whether the agreement aimed to give absolute territorial protection, and whether parallel imports could be brought into the territory by third parties.

The effect of this case was reinforced by a subsequent decision of the Court in the later case of *Béguelin Imports Co. v. GL Import & Export SA*,[10] when in another Article 177 reference the same criteria were applied to an exclusive distribution arrangement made between the Japanese producer of gas pocket lighters (Oshawa), which had appointed companies named Béguelin as exclusive distributors for France and Belgium. GL Import & Export of Nice had obtained lighters manufactured by Oshawa in Germany and tried to sell them in France; the Béguelin company trading in France then sought an injunction against it under the French unfair competition law. The Court held that a distributor was only entitled to enforce an unfair competition law to exclude parallel imports if the alleged unfairness of the behaviour of competitors (such as GL Import & Export) consisted of factors other than simply introducing such imports into a Member State to compete with the goods imported through official channels.

If, after applying these tests, it appears that there is either no effect on trade between Member States or no object or effect of imposing any restriction on competition within the Common Market, then the Commission's interest will in any case cease. If, however, as is often the case, there is some effect in either or both these respects, then the criteria contained in the group exemption had to be considered, as follows:

1. A supplier (S) and distributor (D) had to be the only parties to the agreement.

With regard to the first condition, the group exemption was limited to bilateral agreements in order to stop horizontal arrangements or collusion either between manufacturers or dealers *inter se*. Therefore, when a manufacturer wished to appoint several distributors each for separate territories, these appointments should normally be contained in separate agreements so that the limit of two parties is maintained for each agreement. It was perfectly possible for a whole network of bilateral agreements to be covered by the group exemption provided that each separate agreement was between only two parties. Parent and subsidiary companies were treated as a single entity for this purpose, provided the subsidiary did not enjoy *de facto* economic autonomy from its parent.

2. S agreed to supply only D with the relevant defined goods for resale within a specific territory forming part or all of the Common Market (T).

[10] Case 22/71 [1971] ECR 949: [1972] CMLR 81.

The second condition was important; it was clearly established that, however small, the territory must be an exclusive one even if in practice the distributor divided up the territory of a Member State between a number of undertakings. On the other hand, the territory could be either a single Member State or several Member States, so long as it did not comprise the entire Common Market. This followed from the wording of Article 1(1)(*a*) which refers to agreements which cover 'a defined area of the Common Market'. In the 1974 case of *Europair International–Durodyne Corporation*[11] the Commission confirmed that an exclusion distribution agreement for the entire Common Market covering heating and air-conditioning installations could only receive an individual exemption, and not a group exemption, even if all other conditions set out in the group exemption were satisfied. The goods moreover had to be supplied for resale and not for treatment or reprocessing or even for the use of the distributor. If the distributor merely repacked or relabelled or reassembled products which have been broken into component parts to make transport easier, then the group exemption continued to apply; but if the goods purchased were inserted as components in other goods or combined with them to form a separate product, then the application of the exemption was more doubtful. Difficult borderline cases could, of course, arise; the less identifiable the original products supplied to the distributor, the less likely the group exemption would have been to apply.

3. D agreed to purchase such goods for resale in T only from S (and not to sell competing goods during the term of the agreement or for one year thereafter), i.e. a restriction affecting both intrabrand and interbrand competition.

The third condition was the essential exclusivity restriction, that D accepted the obligation to resell the contract goods within T only as supplied by S, and that it would not sell either these goods or any competing products either during the period of the distribution agreement or for one year afterwards. In determining what were competing products, Article 6 could be called in aid, referring to the relevant 'properties', 'price', and 'intended use' of the goods.

4. S placed no restriction on D as to prices or customers but could require D to carry out certain obligations, namely to purchase complete ranges of goods or minimum quantities, to sell goods under S's trade marks and packaging, and to take promotional measures including provision of an adequate sales network, stock, guarantee services, and adequate trained staff.

The fourth condition was also of central importance. It stated the very limited extent to which the conduct of D's business could be controlled.

[11] [1975] 1 CMLR D62.

These were essentially positive obligations rather than negative restrictions, laying down what was regarded by the Commission as a reasonable set of basic obligations that any distributor could reasonably be asked to undertake. The distributor may well not have been able to function effectively if not required to make minimum purchases nor hold a minimum quantity of stock, nor would customers find a sufficient range of goods available for satisfactory selection. The distributor would not, therefore, be able to make the impact on the competitive situation in the territory that was reasonably required by the manufacturer. By the same token, the distributor was obliged if it was to be effective to take active promotional steps, not only advertising locally but investing sufficient resources in every aspect of its business.

What S could not prevent D from doing (save by obtaining an individual exemption) was also selling non-competitive goods alongside the goods covered by the distribution agreement even though S might reasonably feel that D's promotional efforts were being diluted by the existence of these other interests; clearly a case will have to be made out if an individual exemption to the Commission is sought that distribution cannot be effective other than on a totally exclusive basis, a difficult view to prove.

5. S remained free to supply other dealers or users of contract goods outside T (even if it knows that they intend to use or resell them in T).

6. D remained free to sell the contract goods outside T so long as it is simply a passive recipient of orders, i.e. it does no advertising or promotion and has no branches or depots outside T.

It will be noted that the fifth and sixth provisions of the model left both S and D with some degree of freedom with regard to their ability to supply the contract goods. For its part, S remained free to supply other dealers who were to receive the goods outside the contract territory even if they would be selling them in competition with D inside it, and S also remained free to supply users of the goods (as opposed to dealers) even within the territory. For its part, D remained free to sell the goods outside T so long as it does this as a passive recipient of orders rather than as an active seller. This freedom was, however, specifically stated to exclude any promotion or the opening or maintenance of branches or depots outside T. These conditions were regarded by DG IV as important to ensure that exclusive distribution arrangements did not contain unnecessarily restrictive clauses. It was important that neither S nor D refused orders that fell within the permitted category, since otherwise it could have been argued that the agreement was being operated in such a way as not to comply with these conditions. S, however, could within the framework of the group exemption still contract to impose on other exclusive distributors in other territories the same restrictions against 'active' selling outside their territories as were imposed upon D in its territory.

7. The group exemption applied only so long as S or D were not manufacturers of competing goods giving each other exclusive distribution rights on a reciprocal basis.

8. The group exemption also applied only so long as S and D do not exercise industrial property rights or otherwise act so as to make it difficult for dealers or consumers to obtain goods from elsewhere (possibly from obtaining parallel imports).

The two final conditions were also essential to the Commission's aim of allowing exclusive distribution to occur whilst preventing it from causing too much damage to the competitive position in any particular product and geographic market. The group exemption could not apply if the parties are or become manufacturers of competing goods giving each other reciprocal rights, because this would then tend to become a market-sharing agreement rather than a distribution agreement; the group exemption was intended rather for situations where the businesses being carried on were complementary rather than competing. The final and eighth condition was very much that which flows from the *Grundig* condition, under which S and D must not take any action either through use of their industrial property or otherwise to impede the free movement of parallel imports.

These eight elements comprised the model for permitted exclusive distribution; the more detailed comments show how they were interpreted both by the Commission and by Court decisions[12] during the subsequent period between the first implementation of the Regulations in 1967 and later replacement in 1983.

After 1967 a large proportion of the notified exclusive distribution agreements were brought under the terms of the group exemption and thereby enabled DG IV to turn to the large number of individual notifications already made. Nevertheless, many other distribution agreements that could not benefit from the exemption remained for consideration. There were some kinds of clauses which could make individual exemption difficult or impossible. The central importance of the integration of the national markets into the Common Market meant that the existence within such an agreement of any form of ban on exports, direct or indirect, was fatal to the possibility of an exemption on an individual basis. Normally, the Commission drew the attention of the notifying parties to this or any other restriction considered undesirable, and offered exemption only if such restrictions were removed or appropriately

[12] For example, the Commission had originally assumed that exclusive distribution agreements affecting only one Member State, and made between parties solely from that State, would fall outside Article 85 because of the lack of effect on trade between Member States; however, the European Court took a different view in *Fonderies Roubaix-Wattrelos* v. *Fonderies Roux* Case 63/75 [1976] ECR 111: 1 CMLR 538.

amended. The strong line adopted from the outset by the Commission in this area was fully supported by the Court. This, of course, runs directly contrary to the immediate interests of the distributor who would prefer to have absolute territorial protection for its own area into which neither the manufacturer nor any other distributor appointed for another Member State could make entry. This had been precisely the hope of Consten in the *Grundig* case and such protection had been sought deliberately both by use of the 'GINT' trade mark coupled with the French unfair competition law and also by the contractual restrictions placed on distributors in Germany not to export. The ban on exports if found in any exclusive distribution agreement is almost certain to require deletion if an exemption is to be given. This has been laid down both in a number of cases before the Commission and by several European Court rulings.

Miller International v. *Commission* (European Court 1978) illustrated the attitude of the Court.[13] A fine had been imposed by the Commission on Miller which was a German company producing records and tapes as a 100 per cent subsidiary of a United States company. It imposed a specific export ban on a French distributor in Strasbourg and on other French distributors. The business done by Miller was confined to cheap records and 'bargain offers', and they had just 2·5 per cent in value (and 5 per cent in volume) of the total German gramophone record market; Miller claimed that its business was so limited that it had no effect on trade between Member States and should fall within the *de minimis* category of *Völk* v. *Vervaecke*.[14] The fine was, however, upheld by the Court which said that there was evidence that, although the sales were not large, they made up a not inconsiderable proportion of the market and that in certain distinct categories of that market Miller could at least be described as important, even if not strong, certainly to the extent that it had the potential to affect trade between Member States. Moreover, the Court took the view that if its exports had been so unimportant, the export ban itself would have been unnecessary. Its existence showed that Miller was clearly afraid of reimportation of the products at lower prices than those prevalent in Germany. The mere fact that distributors for Miller preferred to limit their commercial operations to restricted regional markets could not justify the formal adoption by the parties of clauses prohibiting exports. It was unnecessary for the Commission, to justify its decision, to prove an appreciable effect of these individual clauses on trade between Member States.

Distribution could also be hindered indirectly rather than directly by clauses which provide a disincentive to export. If consumer goods were sold within one Member State, the Commission regarded it as essential

[13] [1978] ECR 131: 2 CMLR 334. See Ch. 7, p. 121. Advocate General Warner recommended that the Commission should review the amount of the fine.

[14] Case 5/69 [1969] ECR 295: CMLR 273.

that there be no restriction on their resale which discouraged them from being exported to other Member States. In the case of consumer goods, this resale was less easy if the purchaser in another Member State was unable to rely on the original guarantee. *Zanussi* was a large Italian manufacturer of electrical goods and originally the guarantee which it gave would be honoured only by the Zanussi subsidiary which had directly imported the appliance into its own territory. If the appliance had been used in a country other than that in which it had originally been imported, the guarantee would not apply. A negative clearance was granted by the Commission[15] to the new form of guarantee only on the basis that the parent company in Italy would guarantee all the products but that the relevant subsidiary in any country where the appliance was used would provide guarantee and aftersales service. Such service could only be refused if the appliance had been used in an abnormal manner or in a manner inconsistent with local technical or safety standards or after making improper adjustments. Guarantees cannot be restricted to goods supplied only through official channels and must be available for parallel imports.

There are other varieties of clauses which contain a disincentive to distributors to sell outside their own territory. It was long regarded, therefore, as most unlikely that the Commission would give approval for any pricing scheme which established different prices or conditions of sale for home and export sales respectively. This issue arose in the well-known *Distillers* case[16] (European Court 1980). Complaints were made to the Commission by Bulloch, a Glasgow whisky dealer, that Distillers had refused to sell whisky to it for export because it was alleged that Bulloch had been reselling 'Red Label' whisky in the EC after buying it in the UK at a price lower than that officially applicable for purchase for export. Distillers had notified their main conditions of sale to the Commission, but had not notified to it officially the supplementary terms of sale which included allowances and rebates which distinguished between goods to be resold in the UK and export sales.[17] The extra cost of 'export whisky' was alleged to represent the additional promotional cost which had to be paid by distributors in the EC territories against fierce competition from local brands of spirits. In the United Kingdom the whisky market was regarded as 'mature', and competition was almost entirely on price; the strength of

[15] [1979] 1 CMLR 81. See also *ETA Fabriques d'Ebauches* v. *DK Investment SA* [1985] ECR 3933 [1986] 2 CMLR 674, a Court Art. 177 case involving cheap Swiss watches. On the other hand, it is not a breach of Article 85 for a retailer to provide a special service for goods acquired through official distributors so long as the standard manufacturer's guarantee is available to all purchasers. *Hasselblad* v. *Commission* Case 86/82 [1984] ECR 883: 1 CMLR 559.

[16] Case 30/78 [1980] ECR 2229: 3 CMLR 121.

[17] An additional amount of £5.20 per case was payable on whisky to be exported.

the large brewing groups ensured that prices were kept at a relatively low level.

Since there was a clear disincentive in Distillers' supplementary terms of sale to the exporting of whisky from the United Kingdom to the remainder of the EC, one would have expected the Commission, as it did, to determine such terms unable to benefit from Article 85(3). Interestingly, the Advocate General on the substantive issues favoured Distillers, although on procedural grounds (failure to notify) he was unable to render an opinion in their favour. On the substantive issues, he felt that the investigation of the relevant markets in the EC was inadequate and that the price differential applied was indispensable to the promotion of whisky in the other Member States of the EC. In his view, the dual price system did not make parallel imports impossible nor would it be reasonable to expect Distillers themselves to carry out EC promotion. Equally, he felt that Distillers had little scope for raising its UK prices to match the export price without losing substantial market shares because of the strong buying position of the large UK brewery groups. The European Court, however, found against Distillers on the procedural ground alone without going into the substantive issues.

Problems would also arise with regard to the placing of bans or disincentives on the resale of goods outside the Common Market. If distributors or dealers outside the Common Market had as a result to pay higher prices than those within it, such clauses too might be held to affect trade between Member States if there was a possibility that the dealer outside the EC would have been able to import into the Common Market had it not had to pay the higher price. It is also clear that if an export ban is placed on a dealer operating in an EFTA country, whose Free Trade Agreement with the Community enables goods to pass without tariffs into Member States, it will be treated by the Commission as if it were an export ban on movement within the EC itself.

A similar form of obligation which may also raise problems for distributors is one which requires a payment to be made if they effect a sale outside their main territory. This is often known as a 'profit passover'. The normal justification for such a charge claimed by the manufacturer is that the distributor into whose territory the sale is made will incur costs in covering guarantee and service obligations which will not neccssarily be reclaimable from the manufacturer, so that it is only fair for the exporting distributor to make a payment of part of his profit on the sale to the other distributor into whose territory it is sold. It is clear that if the amount payable cannot be shown to represent a fair assessment of the extra cost to the latter dealer, then it will be treated likewise as an indirect disincentive to export sales and, therefore, unlikely to be exempted under 85(3). A similar payment was, however, permitted in the horizontal agreement between rival marine paint manufacturers in the 1967 *Transocean Marine Paint*[18] case where a member making a sale into the territory of another member paid a small commission on that sale, although no attempt appears to have been made there to justify the exact amount of the payment on a cost basis.

The type of clause that could be permitted on application for individual exemption would not differ greatly from those permitted under the group exemption. It could not, however, be assumed in the case of a non-exclusive distributorship that as wide a range of restrictions would be permitted to be imposed upon either the supplier or distributor, since it is in the nature of exclusive distribution arrangements that restrictions are accepted on both sides in order to ensure the commitment and investment required from the distributor to the interests of his supplier and vice versa. Thus, restrictions on the sale of competing goods by the distributor would be likely to have been allowed in individual requests for exemption under Article 85(3), also a ban on prospecting actively for sales outside the sales territory. An exception might arise, however, where a distributor had a very strong market position as a potential outlet within a particular Member State so that any manufacturer which could not make use of its facilities would be seriously disadvantaged in seeking to enter that particular market. This might particularly be the case if a product required special servicing or handling skills, for example, complex computers or advanced electronic equipment. If the services of this distributor were not available to a small or medium-sized undertaking trying to break into the market, competition in the whole of that market might be seriously affected.

It was doubtful whether in the case of individual applications under 85(3) DG IV would look sympathetically at any prohibitions on categories of customers to whom the distributor should sell particularly if the prohibition was broad enough to affect cross-border sales within the Common Market. Apart from the limited selectivity permissible for schemes of selective distribution which by reason of their criteria for selection fall outside Article 85(1),[19] such prohibitions were unlikely to be accepted save in the most unusual circumstances. Clauses falling into this category would include those banning distributors selling to other distributors, or those banning sales to certain retail outlets such as department stores, mail-order houses, and discount houses.

[18] [1967] CMLR D9 and [1974] 1 CMLR D11, on appeal Case 17/74 [1974] ECR 1063: 2 CMLR 459. The 'profit passover' payment, however, was not allowed by the Commission on the renewal of the original agreement in 1973. For individual exemption to be possible, consumers would have to receive a fair share of the resulting benefits, e.g. the provision of better servicing and maintenance facilities. Cf. recitals (4) and (25) of Regulation 123/85 relating to car servicing.

[19] See Ch. 13.

Restrictions on the quantities of goods to be sold to individual customers would also have been regarded with suspicion, as market-sharing devices between the manufacturer and its distributors, if the result of the limit imposed enabled the manufacturer in practice to reserve for itself a number of the larger and more important customers. Restrictions on use were also likely to cause great difficulty since although the manufacturer might have good commercial reasons for wishing to split the market between distributors of goods for separate end uses, it is felt that to allow such restrictions to be imposed on distributors would be interfering too far in their essentially independent business judgement.

During the ten years from the introduction of the group exemption in 1967 many notifications of distribution agreements were dealt with under the provisions of Article 85(3), and the experience gained in the course of assessing those cases gradually enabled the Commission to form a view on the extent to which the provisions of 67/67 were satisfactory. One conclusion reached by DG IV was that the territory permissible to be exclusively held by a distributor covered too wide an area and that it should be reduced below the 'substantial area of the Common Market' originally prescribed in 67/67,[20] which as had been confirmed in *Europair–Durodyne* did not include the whole territory of the Common Market. The Sixth Annual Report[21] pointed out that the aim of the Regulation had been to give the benefit of such a group exemption only to agreements that covered a part of the Common Market.

The Commission also felt strongly that the Regulation should not apply when the proposed agreements were between competing manufacturers even if the rights granted were not on a reciprocal basis, so that only one of the manufacturers became a distributor for the other in a particular territory. The *Fonderies Roubaix* case (European Court 1976) had held to the Commission's surprise that the group exemption was available for contracts which satisfied all the conditions laid down in Article 1 of the Regulation even if the two undertakings came from the same Member State and notwithstanding the apparent exclusion of such agreements from the coverage of the group exemption by the terms of Article 1(2). Whilst Commission consideration of the amendments to be introduced to this in the light of these considerations was still in progress, the *Concordia*[22] case was decided by the Court in 1977 on the interpretation of the Regulation, whose result also caused the Commission some alarm. This case involved the loan by the Concordia Brewery in Belgium to a

[20] Since it was so easy to grant the rights for substantially the entire territorial area of the Community by excluding simply a small area, e.g. Luxembourg, the practice was one that violated the spirit of the group exemption if not its letter.

[21] p. 20.

[22] *de Norre* v. *Brouwerij Concordia* Case 47/76 [1977] ECR 65: 1 CMLR 378.

married couple called Detant of 300,000 Belgian francs repayable over ten years at interest and containing a 'tie' under which the Detants and their successors agreed that their café would purchase only Concordia beverages. The Norre family, having taken over from the Detants, then sold beer and other drinks not produced by Concordia, and Concordia sued them for damages; the defence put forward by the Norres was that the contract itself was void under Article 85(1). Reference was made to the Court for a preliminary ruling under the Article 177 procedure.

The issue arose whether the agreement could be covered by the group exemption since it was not an agreement for exclusive distribution, but was an exclusive purchase arrangement under which the tenants of the café had to purchase all their requirements for beer and other drinks available from Concordia exclusively from it. The geographical location of the café meant in practice that the drinks were then resold in a fairly limited area, but the contract itself provided no specific territory. The Court took the view that the Regulation could apply even when there was no specific geographical area referred to in the contract within which resale of the product was permitted. Whereas in an exclusive distribution arrangement the territorial limitation has to be explicitly stated, in an exclusive purchase arrangement normally no territorial limitation is mentioned, it being determined in practice by the geographical scope of the purchaser's business. There was no reason, said the Court, that exclusive purchase agreements also could not fall within the scope of 67/67 so long as they complied with the requirements of Article 3 (the seventh and eighth of the basic elements of the Regulation as enumerated above).

3. Drafting and Implementation of Regulation 1983/83: The Group Exemption for Exclusive Distribution Agreements

Shortly afterwards, in February 1978, the first draft Regulation to amend 67/67 was published. In the new proposed draft the following changes were incorporated to reflect both judgments of the Court over the previous decade and the other changes felt necessary as the result of DG IV's own experience with this type of agreement:

(*a*) that exclusive purchase arrangements would be covered as well as those for exclusive distribution;
(*b*) that agreements simply between two national companies relating to distribution or purchase within a single Member State could likewise be covered; this would not simply be dependent upon the criterion whether they could be shown to affect trade between Member States;

(*c*) that in the future agreements between competing manufacturers would be excluded whether or not reciprocal (save in the case where the manufacturers entered into a non-reciprocal agreement where one had a turnover of less than 100 million ECUs);

(*d*) that greater emphasis would be placed in the recitals on the importance of encouraging interbrand competition between dealers;

(*e*) that the population of the territory to be covered by an exclusive dealing arrangement should not exceed 100 million, this being taken as a round figure to cover the territory of any of the larger Member States;

(*f*) that exclusive purchasing agreements could only be exempt if a producer who entered into such an agreement with one or more dealers did not in a substantial part of the Common Market account for more than 15 per cent of total sales or those goods and their close substitutes. This requirement was intended to prevent a large manufacturer from raising barriers to market entry by other companies through the creation of a network of exclusive purchasing agreements. This requirement was, however, controversial because of the difficulty in practice that companies would have in calculating the 15 per cent figure, given also that it was not applicable simply to a market for a single product but covered all close substitutes.

These proposals were subject to wide consultation and discussion with European industrial and commercial bodies and employers' associations; widespread opposition was encountered to the proposal not only for the 15 per cent market share limit under exclusive purchase agreements but also to the proposed limit of the exemption to territories with less than 100 million inhabitants. In its Ninth Annual Report,[23] the Commission admitted that its original aim of achieving urgent amendments to Regulation 67/67 as quickly as possible was no longer attainable: instead, a more fundamental recasting of the Regulation was put in hand to cover not only the changes required because of Court decisions but also to ensure that the group exemption covered additional points. The first was that the exemption must be confined to genuine vertical arrangements between manufacturers and distributors and that any agreements between competing producers should be omitted unless in the case of items of advanced technology only competing manufacturers could be found with the necessary know-how to carry out distribution. Moreover, territories must be kept from becoming too large so that intrabrand competition as well could be preserved. The Commission had always had doubts whether relying only on interbrand competition between manufacturers was sufficient, given their ability to reduce the degree of substitutability between similar goods by product differentiation, enhanced by massive promotional spending to

[23] p. 16.

create consumer preferences for strong individual brands, and given also the long tradition of economic nationalism in Western Europe.

It was not, however, until July 1982 that the second draft Regulation was produced. Following further lengthy consultations, DG IV confirmed that the proposals to limit territories to those containing 100 million inhabitants had been withdrawn, as also those confining exclusive purchasing agreements to those where the manufacturer supplies only up to 15 per cent of the requirements of the particular territory. Another key change included, which DG IV had decided sometime previously, was that the Commission would issue separate Regulations covering distribution and purchasing respectively. It would also issue specific provisions covering two sectors, beer and petrol, where there have traditionally been large networks of retail outlets in different Member States and which were difficult to deal with within the framework of general Regulations required to cover such a wide area of industrial and commercial activity. Finally, in 1983, both the group exemption Regulations appeared in final form and came into effect from 1 July.

The Thirteenth Annual Report[24] sets out the philosophy of the Commission in introducing the Regulations in their final form. The Commission felt it necessary to separate the treatment of exclusive distribution and exclusive purchasing because exclusive distribution agreements raised the problem that the assignment of such sales territories to dealers could lead to market-sharing at their level and thereby lessen competition between dealers in the same goods, i.e. intrabrand competition. On the other hand, exclusive purchasing agreements made access to the market more difficult for competing manufacturers and suppliers and thereby endangered mainly interbrand competition, i.e. between manufacturers. This danger was accentuated if the purchasing agreements formed a network which included a majority of the best sales outlets in a particular Member State.

Dealing first with 1983/83 covering exclusive distribution agreements, we should now consider in what respects it has departed from the eight basic elements of 67/67 and whether the Commission's aims referred to above were achieved. The eight main elements of the group exemption had undergone some detailed amendments but without losing their basic thrust. The major amendments can be summarized as follows:

(1) The territory can now include the entire Common Market; even though it remains necessary in the agreement to define the territorial area covered by the agreement, it is no longer necessary to select an artificial area that excludes some small part of the Market.

[24] pp. 35–8.

(2) The supplier can now be required not to sell to consumers in the territory of the distributor.
(3) It is only during the period of the agreement that the distributor can be obliged not to deal with the manufacturer of competitive goods (and not even for 12 months after as previously).
(4) In order to ensure that agreements are not entered into between substantial manufacturers even on a non-reciprocal basis, the only agreements that are permissible between manufacturers are where those are (*a*) non-reciprocal and (*b*) one of the manufacturers has a total annual turnover of less than 100 million ECU.
(5) Circumstances in which the exemption can be withdrawn have been widened, and the emphasis is placed not only on any restrictions placed upon competitive goods by the parties themselves but also upon restrictions which may occur as the result of other circumstances. In other words, the exemption may be withdrawn as the result of circumstances outside the control of the parties themselves, and the relevant Article[25] makes clear that the exemption does not apply where the effects of the exemption include any of the following:

 (*a*) where the contract goods are not subject, in the contract territory, to effective competition from identical goods or goods considered by users as equivalent in view of their characteristics, price, and intended use;
 (*b*) where access by other suppliers to the different stages of distribution within the contract territory is made difficult to a significant extent;
 (*c*) where for reasons other than those referred to in Article 3(*c*) and (*d*) (namely the seventh and eighth conditions contained in 67/67), it is not possible for intermediaries or users to obtain supplies of the contract goods from dealers outside the contract territory on the terms that are customary;[26] or
 (*d*) where the exclusive distributor either sells the contract goods at excessively high prices or without any objectively justified reason refuses to supply in the contract territory categories of purchasers who cannot obtain contract goods elsewhere on suitable terms or applies differing prices or conditions of sale.

For the first time, the Commission at the request of numerous industrial and commercial organizations and trade associations issued guidelines

[25] Article 6.
[26] It is, therefore, important in their own interests that both the manufacturer and distributor do not refuse orders placed with them from outside the exclusive territory, so that by making such sales, a real possibility remains that consumers within the territory can obtain the goods from a source other than the official channels.

giving its understanding of the interpretation of the individual clauses. These guidelines are extremely useful to any lawyer or businessman seeking to interpret the Regulations. Their length makes them difficult to summarize, but they contain a number of important features. First, they set out the Commission's understanding of the purpose of exclusive distribution agreements as already discussed. Although the views of the Commission cannot exclude either the jurisdiction of national courts to apply the Regulations, nor prevent the European Court of Justice from placing its own interpretation upon them, the guidelines will in the majority of cases have persuasive influence. An explanation is given of a number of the phrases which have caused problems such as the concepts of 'resale' and 'goods' and clarification given of the requirement that not more than two undertakings can be parties to the agreement. An analysis is also given of the obligations upon the distributor which are permitted and a full justification given of the reasons for reducing the scope of group exemption where the agreements are between competing manufacturers.

The guidelines also stress the importance of maintaining competition at all times within an exclusive territory, even though the exclusivity of the distributor within that territory has to be preserved. Thus, whilst a supplier can sell the contract goods to a reseller outside the territory, notwithstanding that the reseller is himself established within the territory, this sale would not be permitted if the supplier himself paid the cost of transporting the goods into the contract territory itself, for this would effectively amount to supplying another dealer within the exclusive territory of the first dealer (so that the agreement would no longer qualify as an exclusive agreement). By the same token, the distributor cannot be restricted so as to supply only certain categories of customers or stopped from selling to customers in other categories supplied by other resellers. This restriction could not be imposed even if the categories were defined by reference to separate end uses of the goods. The manufacturer himself can be restricted, as we have seen, from supplying the contract goods to consumers within the territory, but this obligation need not be absolute as long as the supply within the territory to these special customers is direct and is not a sale to an intermediary who will in turn resell within the territory. Sales for resale have always to be outside the territory, save of course for supplies made direct to the exclusive distributor.

4. Regulation 1984/83: Drafting and Implementation of the Group Exemption for Exclusive Purchasing Agreements

The development of competition policy relating to exclusive purchasing agreements has always been closely linked with that relating to exclusive distribution agreements. The growing importance, however, of exclusive purchasing agreements and the difficulty of providing a single set of group Regulations to cover them as well as exclusive distribution agreements was eventually recognized by the Commission. The new Regulations, therefore, draw a clear dividing line.

The close association between the two forms of agreement in the early years of the Commission is understandable. An exclusive distribution agreement will normally include a promise by the distributor to its manufacturer not to distribute or manufacture competing goods made by other manufacturers. Therefore, explicitly or implicitly, it is required to purchase all its needs for contract goods from the other party. In the exclusive distribution agreement, this exclusivity relates to a specific territorial area. In the true exclusive purchasing agreement, no territorial area is named, and the purchaser remains in competition with all other parties entering into similar agreements, though, of course, the manufacturer will normally take care to ensure that the number of distributors granted this concession is not unrealistically high, so that all can enjoy a reasonable level of turnover. Undoubtedly, such agreements potentially offer economic advantages to both parties. From the viewpoint of the manufacturer, it enables him to calculate the likely demand for his product over the period of the agreement. The distributor in return for committing himself for a reasonable length of time often receives special prices, preference in supply if shortages occur, and perhaps the provision of technical and financial assistance. In particular, the acceptance of an exclusive purchase arrangement is often the principal condition for the making of loans on favourable terms to individual businesses which cannot easily acquire these funds on comparable terms from other sources.

At an early stage in the development of competition policy, the European Court had taken a relatively relaxed attitude to such agreements. In the *Brasserie de Haecht* v. *Wilkin-Janssens* case (no. 1)[27] (European Court 1967) the Court of Justice stated that

> agreements whereby an undertaking agrees to obtain its supplies from one undertaking to the exclusion of all other do not by their very nature necessarily include all the elements constituting incompatibility with the Common Market as referred to in Article 85(1) of the Treaty. Such agreements may, however, exhibit such elements, where taken either in isolation or together with others, and in the

[27] Case 23/67 [1967] ECR 407: [1968] CMLR 26. Also discussed in Ch. 7, pp. 109–10.

economic and legal context in which they are made on the basis of a set of objective factors of law or of fact, they may affect trade between Member States and where they have either as their object or effect the prevention, restriction or distortion of competition.

In the case *Brasserie de Haecht*, a Belgian brewery had brought a claim against its tenants for breach of their agreement to sell only beers produced by the brewery; the tenants claimed that the agreement was null and void under Article 85 because it prevented them from selling other beers and thereby restricted trade between Member States, since it limited the outlets in Belgium for foreign breweries. Even if the effects would be trivial in the case of an agreement affecting a single café or inn, the Court of Justice held, on this reference for a preliminary ruling under Article 177, that the effects of such a tying agreement must not be considered in isolation but in the light of the fact that it formed part of the network of other similar agreements. On the other hand, if one or more parties to the agreement were powerful in their particular markets, and had market shares of some importance, their agreement considered in isolation might be caught under 85(1) so that the issue of exemption under the four conditions of 85(3) would become relevant. If, additional restrictions were placed upon the dealer, particularly those affecting the terms and conditions upon which he resold or his choice of customers, these were likely to weigh against the validity of the agreement. If, however, market entry by manufacturers through the opening up of new outlets (or finding alternative forms of distribution such as the use of hypermarkets) was in practice available, then clearly the exclusionary effect of the exclusive purchasing arrangements would be less pronounced.

By the time of the Seventh Annual Report, we find DG IV had taken a slightly less accommodating view of such agreements.[28] The Commission stressed that some benefit must arise for consumers from such arrangements; these benefits also must be large enough to justify the restrictions of competition which they involve. If, as the result of such an obligation of exclusivity, it became more difficult for other manufacturers to sell on the market, particularly if barriers to entry had been raised, exemption was likely to be refused because the parties would have opened the possibility of eliminating or at least substantially reducing competition in respect of a substantial part of the products. The first case referred to in the Seventh Report, the *Billiton/M & T Chemicals*[29] proposed agreement, had been decided on an informal basis; Billiton was the largest manufacturer in the Community of tin tetrachloride whilst M & T was the largest purchaser, normally taking half of Billiton's output. M & T had in the proposed arrangement agreed to purchase all its requirements from Billiton

[28] pp. 21–3. [29] 7th Annual Report, pp. 104–5.

so long as Billiton could guarantee supplies, and M & T had also agreed that it would not resell any such material. Following negotiations and under pressure from DG IV, the agreement was altered so that no obligations were placed on M & T to take any particular proportion of its requirements from Billiton. A restriction on resale was also removed.

In the other case, *Spices*,[30] a formal decision was actually issued by the Commission at the end of 1977. The case concerned spices distributed through supermarkets in Belgium. Brooke Bond Liebig was a spice producer with 39 per cent of the Belgian market. It entered into an arrangement with the three largest Belgian supermarket chains (having some 30 per cent of the food retail market) under which the latter agreed to stock only Liebig spices (apart from their own brands); arrangements were also made for special price rebates and for retail prices to be fixed on a uniform basis. Spices are apparently sold largely to 'impulse buyers' and are normally placed in large display units at eye level where customers can easily see them. Prices of Liebig products were apparently higher than most other spices, and the profit margin for the supermarkets from their agreements with Liebig correspondingly greater.

The Commission found that although benefits flowed from the agreement to both Liebig and the supermarkets, no part of this was passed on to the consumer. There were three main competitors to Liebig who would have been able to do more business if they too had had access to the major supermarkets, instead of being confined to a large number of smaller retailers. Liebig was already very strong in the Belgian market for spices, and the effect of the exclusive purchasing arrangement was not only to reduce the choice of the consumer, but to ensure that market prices were higher since the consumers were limited to choosing the store's own brands from these important supermarkets if they did not wish to buy Liebig products. The agreement was, therefore, not granted exemption.[31]

It was against this background that in the late 1970s DG IV, therefore, had to consider how best to provide a group exemption for exclusive purchasing. By this time it had reached the conclusion that any attempt to draw up a single Regulation covering exclusive purchasing and exclusive distribution would encounter insuperable difficulties given the different

[30] [1978] 2 CMLR 116.

[31] Exclusive purchasing arrangements are also sometimes condemned when forming an ancillary part of a wider horizontal arrangement. In cases which we have already dealt with such as *Atka A/S* v. *BP Kemi* [1979] 3 CMLR 684 and in *Rennet* [1981] ECR 851: [1982] 1 CMLR 240 (see Ch. 8) obligations to take either all the requirements or a substantial proportion of requirements, in the one case from the dominant supplier of ethanol in the Danish market, and in the other from the dominant Dutch producer of rennet, were both refused exemption as too restrictive on competition in their particular markets. We shall see in the treatment of Article 86 in Ch. 18 that the use of loyalty rebates by companies having a dominant position in their market, in order to ensure exclusive purchasing from their customers, has also been attacked by the Commission.

effects on competition brought about by the two types of agreement. In the event, the Commission chose to adopt a Regulation applicable to exclusive purchasing agreements solely relating to goods intended for resale and not applying to contracts that relate to goods purchased for the buyer's own use.[32] In the case of distribution agreements, no time limit on their permitted duration was included, either in the original 67/67 or in the later Regulation 1983/83. On the other hand, the potential exclusionary effect of purchasing agreements on manufacturers wishing to enter the market meant that some form of time limit was desirable. The original draft regulation limited exemption to agreements of only one year's duration with an upper limit of three years if the dealer was to receive some special financial or commercial terms in return for his commitment to the manufacturer. This was later extended following industry representations, to a term of five years with a complete ban on agreements that were for an indefinite period. A compromise proposal that exemption would be granted for agreements that were open-ended only if either party could terminate after five years at the latest by giving notice of no more than one year was later withdrawn, and the rather simpler final solution was that five years is the maximum period permitted.

The framework of exclusive purchasing agreements under Regulation 1984/83 can be considered as having five main elements. The main difference, of course, from exclusive distribution agreements is that there is no territory (T), merely a principal sales area which is not territorially defined in the agreement but operates naturally as the result of the geographical location of the purchaser. The five elements are that:

(1) The purchaser (P) agrees with the supplier (S) to purchase the contract goods as defined only from S for a period which cannot exceed five years. The contract goods may only be one type of goods or a range of goods that are connected with each other, but not two or more unconnected products.
(2) S agrees not to compete with P in P's principal sales area, e.g. by supplying direct to P's competitors or customers there.
(3) P agrees not to manufacture or distribute goods competing with the contract goods, during the period of the agreement.
(4) S places no restriction on P other than an obligation to purchase either a complete range or minimum quantities of goods and to sell them under the trade marks or in the packaging required by S, and to take the same measures for promotion of sales as can be imposed on exclusive distributors, e.g. advertising, maintenance of sales network, employment of technically qualified staff, etc.

[32] Contracts relating solely to goods purchased for the buyer's own use are often referred to as 'requirements contracts'.

(5) The group exemption only continues to apply so long as S and P are not manufacturers of competing goods giving each other purchasing rights on a reciprocal basis, or even on a non-reciprocal basis unless one of the parties has a total annual turnover of less than 100 million ECU.

The effect of the group regulation, therefore, is to allow a manufacturer to appoint several dealers in a Member State each of whom is bound by an exclusive purchasing obligation combined with a non-competition clause; it is not free, however, to require them to refrain from seeking customers for the contract goods outside the boundaries of their own Member State. No contract territory can be legally stipulated under the terms of the agreement, although naming a principal sales area is permissible. No market share criterion has been applied in spite of the wish of the Commission originally to limit such arrangements to those where the supplier held no more than 15 per cent of the total market (so as to reduce the extent of foreclosure of other competing manufacturers from suitable outlets).

5. Exclusive Purchasing Agreements: Special Rules for Beer and Petrol

There are, of course, several common features to these two sectors. Both involve liquid products, widely regarded as essential to the maintenance of normal life, which are consumed or utilized on a regular basis in relatively small quantities by individual consumers at any one time, but which have to be provided on a wide geographical basis from a large number of outlets. Brand preference is an important factor, so that heavy promotional expenditure is incurred in order to ensure brand loyalty from customers and to attract support for new products; product differentiation is a major element in the competitive struggle between manufacturers. In both cases outlets have to be regularly visited by the supplier's vehicles carrying fresh supplies, and tanks, pumps, and equipment have to be provided. Any interruption in these supplies (as the result of strikes or any other cause) may be commercially disastrous, particularly if local competitors are able to take advantage.

The cost of acquiring sites for retail outlets and of rendering them attractive to customers is substantial and often beyond the financial resources of those who will run them. Therefore, it is necessary for suppliers to ensure that these outlets can be let to tenants at rents ultimately linked with the level of turnover they can obtain or alternatively run by managers employed to run them effectively. Financial assistance to tenants of the outlets is often needed on a substantial scale for the fitting out of the premises to an appropriate standard. This assistance is also made available on favourable terms to owners of premises who are prepared to accept a partial or total tie to the lender's products in return for such loans.

These common factors have led in many cases to exclusive purchasing agreements being entered into by both the suppliers of beer (and other beverages) and of petrol (as well as of oil and related products). Of these, the Commission and the Court have had far more experience of agreements concerning the supply of beer and a number of these have featured in cases before the European Court[33] and which have already been referred to. The agreements referred to in these cases had all related to small cafés in Belgium which had entered into exclusive purchase requirements for lengthy periods of time covering the individual properties and their successive owners, agreements binding even after the original loans had been repaid. As in the case of all exclusive purchase agreements, there was no specified exclusive territory, but nevertheless as we have already seen in the *Concordia* case the Court held that Regulation 67/67 could apply by way of group exemption to such exclusive purchase agreements in those cases where the effects of the agreement (either individually or as the network) would have an effect on trade between Member States.

Since all these cases were merely requests for preliminary rulings on issues of EC competition law arising in the course of national Member State cases under Article 177, rather than requests for review of Commission refusals to accord exemption under Article 85(3) to such contracts after notification, the Court has not had to rule directly on the validity under Article 85 of the type of restrictions found in such agreements. The general view, however, of the European Court apparent from its judgments in those cases seems to have been that they were the kind of agreement that one could reasonably expect to find when breweries provided financial assistance to public houses or cafés. They were not automatically to be considered in breach of Article 85(1), any more than exclusive distribution agreements.

It had then become clear to the Commission that not only would exclusive purchasing need to be dealt with separately from exclusive distribution in the new group exemption, but that specific provisions would be needed to cover these two special sectors with their own particular problems. Amongst the relevant factors were the very large number of beer and petrol supply agreements within the Community containing some form of restriction (put by informed estimates at 250,000 and 150,000 respectively). The cumulative effect of these large numbers of networks of exclusive purchase

[33] See, e.g., the cases of *Brasserie de Haecht* v. *Wilkin-Janssens* (no. 1) (n. 27), and also *de Norre* (n. 22).

arrangements was to foreclose many of the best outlets from selling more than one brewery's or oil company's products; and in many cases the manufacturer was able to extend its foreclosure of outside suppliers into cider, wines, and soft drinks as well (in the one case), and tyres and batteries (in the other). This made it hard for independent manufacturers and suppliers to find adequate outlets for their products either within their own Member States or across national frontiers.[34] Nevertheless, as the Commission came to consider how to deal with such agreements in the framework of the group exemption, it was made aware of the substantial opposition which brewery interests in particular in some Member States would raise to prevent substantial alteration of the use of exclusive purchasing agreements.[35]

The Commission was concerned to increase interbrand competition between suppliers of beer and other drinks at the retail level, even if this involved the retention of the 'tied tenancy' and 'tied loan', possibly in an amended form. The various drafting changes through which the sections of the new group exemption passed between 1979 and 1983 reflected the wide differences of view both between and within Member States as to how far and in what way such agreements could be retained. The Twelfth Annual Report[36] refers to the Commission seeking to take action in both sectors 'where there have traditionally been vast networks of exclusive purchase agreements at national levels' and ultimately relevant Regulations were introduced (as Titles 2 and 3 respectively) on 1 July 1983.

Dealing first with Title 2 (Articles 6 to 9 relating to supplies of beer), the right of breweries to require exclusive purchase arrangements with those outlets in which their products are to be sold (not for off-licence sales) is recognized so long as some 'special commercial or financial advantage' is provided to the reseller. The tie itself can be a complete requirements obligation; if the requirements obligation is for beer only, the maximum period is for ten years but if it is for beer and other drink (which may include sort drinks, wine, and spirits) then five years is the maximum period. If, however, the brewery is also the landlord of the premises, the period of the agreement can be extended to the full length of the tenancy, even if this exceeds ten years. Once such an agreement comes to an end, there is nothing to prevent a new agreement being made on similar terms. There are special rules to cover purchase of non-beer drinks by a tenant, who is entitled to obtain these from an alterna-

[34] Almost the only method of entering a foreign market covered by these networks in the brewery sector was by making co-operation or licensing agreements of the kind considered in the *Carlsberg/Grand Metropolitan* Agreement (see Ch. 8, pp. 141–2).

[35] The exclusive purchase agreement or 'tie' in the brewery industry (whether linked to loans or to tenancies) is of most importance in Germany, Benelux, and the UK. It is of relatively minor importance in Mediterranean countries and is either substantially limited by law, or barred altogether, in France and Denmark.

[36] p. 27.

tive source if the landlord-brewery cannot match the terms which such a source offers.[37]

The correct interpretation of these Articles was considered by the European Court of Justice in 1991 in *Delimitis* v. *Henniger Bräu.* A dispute had arisen between Delimitis, a former tenant of a café in Frankfurt, and Henniger Bräu, the brewery which owned the premises, as to the sums owing between the parties. Delimitis claimed before the Oberlandesgericht that the exclusive purchase arrangements which his agreement with Henniger Bräu contained were not protected by the group exemption and were therefore void under Article 85(1). A number of important issues were raised in the consequent Article 177 reference.

The main elements in the judgment were as follows. First that there were advantages to both brewery and its tenant in providing an exclusive distribution agreement provided that it was not ever intended to nor had the effect of making it difficult for competitors to enter the beer market in competition with the brewery. The cumulative effect of the brewery's network was, as earlier case-law had established, one factor to take into account, looking at the number of the outlets so tied and the duration of that tie and the degree of loyalty of tenants to the breweries. In addition, however, the actual contribution of the disputed contract to the blocking effect on potential new entry had also to be weighed up. This had perhaps not been emphasized sufficiently in the earlier cases.

The Court also ruled that if a clause allowed the tenant retailer to purchase beer from outside the Member State, which actually gave him a real chance of buying competing products from abroad, then the terms of the agreement would be unlikely to affect trade between Member States. This clause, of course, is more common in Continental exclusive purchasing agreements than in the United Kingdom where the likelihood of importing substantial quantities of beer is for obvious reasons less attractive. The Court also ruled that the Group Exemption 1984/83 was to be strictly interpreted; if it did not expressly designate the types of beer and other drinks to be supplied exclusively by either brewery but merely referred to products listed on the brewer's price list (which could be unilaterally amended at any time) then the benefit of the Regulation would be lost. Moreover, its benefit would also be lost if a reseller renting premises from the brewer was not expressly authorized to purchase drinks (other than beer) from other suppliers offering better conditions than those given by the brewery. The final point made by the Court was that national courts cannot extend the benefit of the group exemption to beer supply agreements that are not completely in accordance with its terms. On the other hand, if national courts are certain that such agreements cannot qualify for group exemption and have no individual exemption then they are free to apply Article 85(2).

[37] Such a clause is known as an 'English clause'.

In June 1990 the Commission had published its review of the Community beer market.[38] The first conclusion was that no general change was required to the Community rules governing the tying arrangements between brewers and their outlets as set out in the group regulation which expires in 1997. Nevertheless, the Commission would evaluate whether further measures were needed in the United Kingdom market once the newly introduced national measures there had time to take effect. It noted that 62 per cent of all British beer sales had passed through the tied house system, a far higher proportion than any other Member State. The report also emphasizes that the exclusive purchase of beer from small brewers should not normally be covered by Community rules and the national rules would usually apply. The trend towards licensing agreements between major brewers would also be regularly examined by the Commission, in order to see if they were being used by way of market-sharing or for control of imports.

The provisions for service stations are contained in Title 3 (Articles 10, 11, and 12) and are similar, though slightly simpler owing to the smaller range of products supplied by oil companies as compared with breweries. Here the maximum length of agreement for the exclusive purchase arrangements is ten years, or again the period of the tenancy agreement if longer.[39] The only additional restrictions that can be placed on garages is that they may not use lubricants supplied by third parties if their supplier has itself provided the lubrication bay or other equipment, and there are other minor restrictions on advertising and servicing of equipment provided or financed by the oil company. For goods other than petroleum products, only non-exclusive purchase obligations can be imposed.

In both cases, the exemption can be withdrawn under Article 14 if the Commission finds agreements covered by the group exemption to have effects incompatible with the four conditions contained in Article 85(3). In particular this may occur if there is not effective competition to the contract goods from identical or equivalent goods or if there are difficulties of access into the distribution system in a substantial part of the Common Market for other suppliers or if the supplier itself without justification engages in unreasonable conduct to preserve its market position, e.g. by discrimination in price or refusal to supply.[40]

[38] [1990] 4 CMLR 588.

[39] Recital 19 of the group exemption makes clear that, in the case of service-station agreements, Member States remain free to impose maximum lengths of agreement less than those prescribed in the Regulation. Compare also the provisions of Recital 29 of the Motor Vehicle Distribution Group Exemption Reg. 123/85 discussed in Ch. 13.

[40] For a detailed textual analysis of the Regulations and Guidelines, see V. L. Korah and W. A. Rothnie, *Exclusive Distribution and the EEC Competition Rules* (London: Sweet & Maxwell, 1992).

13

Distribution: Selective Distribution and Franchising

Agency and exclusive distribution agreements are by no means the only method of establishing a satisfactory system for the marketing of goods in the Community. A number of other methods exist and in this chapter we shall consider two of those most commonly adopted, selective distribution systems and franchise agreements.

1. Selective Distribution Systems

Selective distribution systems differ from both open and exclusive distribution systems in that, whilst no specific territorial area is allocated to the dealer, a unique restriction is placed upon him, namely preventing the resale of contract goods to other dealers unless these other dealers have themselves already been admitted to the selective system. The essence of selectivity, therefore, is not simply that the supplier selects its main distributors or wholesalers, for if this were the only restrictive element, then it would be hard to see how Article 85 could apply at all to such a unilateral act of commercial policy. It is rather that the supplier limits the distribution of goods only to those wholesalers and retailers which satisfy appropriate criteria and as a result have been allowed to join the system. Wholesalers will normally sell to anyone except retail customers whereas retailers may normally sell to any level, i.e. to wholesalers, other retailers, and to consumers or users, provided that they do not sell to a dealer outside the approved list.

The establishment of a distribution system utilizing such restrictions might appear at first sight to raise problems under Article 85(1), if the other requirements of the Article were also present. In fact, however, both Commission and the Court have taken a tolerant view of such systems, which provide perhaps the only example of an important group of commercial agreements with specifically restrictive provisions being allowed to remain outside the scope of Article 85(1), for the sake of the advantages which they are said to confer for the distribution of certain types of product, always provided that the systems remain in accordance with the limits placed upon them in order so as to minimize their overall effect on competition. Selective distribution has developed in recent years particularly in respect of products that fall into one of two categories. The first category, and the most important, is of consumer products which require a high degree of after-sales service because of their inherent complexity and also the purchaser's reasonable requirements both for initial guarantee and later sales and service coverage. In this category come motor cars, electric and electronic equipment (such as high-fidelity equipment and personal computers), and (perhaps unexpectedly) even false teeth and other dental supplies. The second category comprises goods which are not necessarily highly complex but which are expensive and sold under prestige brand names which have been extensively promoted; their manufacturers believe continuing control over the environment in which the goods are sold is important to their goodwill and continuing success. Into this latter category come watches and clocks, jewellery, perfume, and some other borderline items such as porcelain and even possibly newspapers.

It was in the early 1970s that the Commission first began to give active consideration to the legal basis for the assessment of such systems of distribution for both technical and luxury products. The earliest cases arose in the perfume, automobile, and electrical goods markets. The most important of the early decisions was the *Metro/Saba* (no. 1) case[1] (European Court 1977). Saba was a German company specializing in electrical and electronic equipment including high-fidelity equipment and television sets. It had established a distribution network in Germany which involved the appointment of wholesalers who would purchase the Saba products and resell them to approved specialist dealers, whose turnover had to be obtained mainly from the sale of electric and electronic equipment of this kind. Department and discount stores would, therefore, normally have difficulties in obtaining appointment. Other qualifications related to the premises, the acceptance of substantial minimum supply figures, and the stocking of a full range of products, whilst distributors' employees had to have a suitable level of technical expertise. In other EC countries, sales by Saba were made direct to its sole distributor dealing also only with approved specialist dealers, who in turn served the public.

In the original forms of agreement, there were numerous prohibitions on the sole distributors (and wholesalers) and the specialist dealers. These included prohibition on export to other EC countries, prohibition on 'cross-supplies' (wholesaler to wholesaler or retailer to retailer), on 'return supplies' (retailer to wholesaler), and also on direct supplies by wholesalers or sole distributors to consumers. The Commission rejected these restrictions except for the last, which was accepted on the grounds that a separation of the function of wholesalers from those of retailers was appropriate to the multilevel distribution system that Saba operated.

[1] Case 26/76 [1977] ECR 1875: [1978] 2 CMLR 1.

The Commission's decision was challenged by Metro, a self-service wholesaler, running a cash and carry business for retailers which had been unable to obtain admission to the Saba system.[2] Metro challenged the decision of the Commission claiming that the prohibition on making supplies available to private customers should not be permitted under the exemption afforded by Article 85(3). It also challenged the requirement imposed by Saba that the turnover of wholesalers engaged in a number of different sectors and having a special department for such electrical goods must be on a level comparable with that of a specialized wholesaler. The decision of the Court, given in October 1977, provided the first authoritative judicial pronouncement on the criteria which the Commission had applied to selective distribution systems. The Court held that Metro, whilst entitled to challenge the ruling of the Commission, had failed to adduce evidence sufficient to overturn it. Pointing out that Saba's market share in relation to the various products concerned was in no case greater than approximately 10 per cent (and in some cases less) the Court stated that the nature and intensiveness of competition in such markets would vary to an extent dictated by the nature of the products as well as by the structure of the relevant market itself.

The Court emphasized that price competition was not the only form of competition for specialist wholesalers and retailers. The acceptance of a certain level of prices consistent with consumers' interests in retaining a network of specialist dealers alongside a parallel system of wholesalers operating self-service and other methods of distribution of a 'low price, no frills' nature formed a rational objective for the Commission. The availability of this choice for the consumer was considered appropriate in particular in sectors covering the production of high quality and technically advanced consumer durable goods, where a relatively small number of large and medium-sized sale producers offered a varied range of items readily interchangeable. While the Commission had exempted the various restrictions of the selective distribution system under Article 85(3), the Court took a slightly different approach and ruled that some of those restrictions did not fall under Article 85(1) at all, being normal requirements in the distribution of the sale of consumer durables. Any marketing system based on the selection of outlets necessarily involved the obligation on wholesalers to supply only appointed resellers; therefore, the right for Saba to verify that the wholesalers had carried out these obligations fell outside 85(1), as also the restriction on the sale only to dealers already approved as members of the system.[3]

[2] Metro was later admitted to the system on condition it complied with the restrictions applicable to wholesalers, which included a ban on sales to private customers and also to institutional customers such as schools and hospitals.

[3] Separation of functions as a part of the system also entitled Saba to prohibit wholesalers from selling direct to the large institutional customers which Metro had wished to serve.

This decision seemed to give a green light to the approval of such systems, provided that the choice of dealers within the system was based on the qualitative grounds adopted by Saba. On the other hand, the terms of the judgment suggested that a system based on other grounds, e.g. on limiting the total number of distributors within a given Member State or particular area, could only be justified if at all by way of individual exemption. To keep the agreement outside Article 85(1), the criteria applied had to relate to technical qualifications of the retailer and its staff and the suitability of its trading premises. All conditions had moreover to be laid down uniformly for all potential dealers and not in a discriminatory way, as subsequently confirmed by the Court both in an Article 177 reference in the *Perfumes* cases[4] (European Court 1980) and in the later *AEG/Telefunken* case[5] (European Court 1983) where the Court said:

> It is common ground that agreements constituting a selective system necessarily affect competition in the Common Market. However, it has always been recognised in the case law of the Court that there are legitimate requirements, such as the maintenance of a specialist trade capable of providing specific services as regards high-quality and high-technology products, which may justify a reduction of price competition in favour of competition relating to factors other than price. Systems of selective distribution, in so far as they aim at the attainment of a legitimate goal capable of improving competition in relation to factors other than price, therefore constitute an element of competition which is in conformity with Article 85(1). The limitations inherent in a selective distribution system are, however, aceptable only on condition that their aim is in fact an improvement in competition in the sense above mentioned. Otherwise they would have no justification inasmuch as their sole effect would be to reduce price competition.[6]

The application of these principles can be well illustrated by the subsequent (1984) decision of the Commission in the *IBM Personal Computers*[7] case. Here, IBM had created a selective distribution system for which it wished to recruit a large number of dealers capable of offering skilled pre- and post-sales service acting as independent undertakings to sell personal computers in direct competition with IBM's own sales force. Criteria had been published for such appointment and IBM had publicly announced that it would appoint any applicant who satisfied them. The distribution system would then consist of all these authorized dealers who would be required to sell the goods only to each other or to consumers, not to unauthorized dealers. IBM had also appointed wholesale distributors to operate alongside and in competition with its own wholesale subsidiary company. It stated that it would also appoint other wholesalers in such a capacity provided that they could also provide a service of the

[4] e.g. *Lancôme* v. *Etos* Case 99/79 [1980] ECR 2511: [1981] 2 CMLR 164.
[5] Case 107/82 [1983] ECR 3151: [1984] 3 CMLR 325.
[6] paras. 33, 34.
[7] [1984] 2 CMLR 342. See also the 14th Annual Report, pp. 60–1.

required standard. Given the high level of skill required from dealers, the criteria laid down for their appointment being matched appropriately to the nature of the product, and IBM's commitment to its uniform application without discrimination, the Commission gave negative clearance to the proposals.

The essential element in this system was that, although the restriction on selling to unauthorized dealers was retained, entry to the system was 'open-ended' for any qualified dealer, so that any restriction on competition flowing from the restrictions on resale outside the approved circle was kept to a minimum. If such agreements are to remain outside Article 85(1), it is essential that obligations imposed on the retailer cannot exceed those absolutely necessary to maintain the quality of the goods or to ensure they are sold under proper conditions. Obligations that go beyond this, e.g. to purchase complete ranges of the product or prescribing substantial minimum turnover figures over a particular period of time, would be regarded as restrictions of competition to the extent that, if they were to be included, individual application for exemption under Article 85(3) would be necessary. Likewise, no discrimination can be practised by a manufacturer between, on the one hand, specialized retailers and, on the other, department and discount stores, provided that each is able to meet the criterion of ability to sell the relevant products at the required standard. Nor may the manufacturer exercise its power of selection on the basis that a particular dealer has either exported or imported the goods outside official channels or has failed to comply with any particular pricing policy.[8]

It is important to see how these principles have been applied in a number of sectors, and we next review five sectors of particular interest. In the first four, we are dealing almost entirely with the case-law of the Commission and of the Court; in the fifth case, that of motor cars, we are dealing with a sector for which a specific group exemption has been issued, Regulation no. 123/85.[9]

Watches and Clocks

The first decision of the Commission relating to selective distribution concerned this sector, the decision being the *Omega Watches* case[10] (Commission 1970). Omega distributed its high quality watches throughout the EC selling through sole distributors in the individual Member States who in turn resold to distributors. The agreement with the distributor imposed on it the obligation to maintain a certain stock of various

[8] Such agreements will, however, possibly escape the reach of Article 85(1) altogether if the market share of the producers operating the system is itself very small.

[9] Effective from 1 July 1985.

[10] [1970] CMLR D49.

Omega models, to appoint for its relevant territory a limited number only of qualified retailers not selected simply on the grounds of their premises, staff, level of service, etc. (as in a qualitative system) but upon the calculated purchasing capacity of the particular area, so that not more than a certain maximum of retailers were allowed for each town and each region. The distributor could not sell in its own territory to retailers who were not part of the system though it was free to do so in other countries. Both distributors and retailers remained free to sell competitive products.

The theory behind this quantitative selection of retailers was that if all qualified applicants were allowed into the system, this would reduce to an unacceptable degree sales of such highly priced articles in each year by each retailer. This, in the view of Omega, would result in a deterioration in the services which its retailers supplied to the customer, so that they would not be able to stock a sufficient range of products and possibly have difficulty also in maintaining an appropriate standard of sales and after-sales service. Exemption was granted, therefore, in spite of both the quantitative limitations on the numbers of dealers chosen and also notwithstanding the ban imposed also on export sales outside the EC. The Commission took the view that this was unlikely to have any effect on trade between Member States since reimport of the watches into the Community was most unlikely because of the combined effect of relevant profit margins, transport costs, and the Common Customs Tariff (CCT) applicable on imports into the Market. The system was approved both with regard to the agreement with wholesalers and with the retailers, the agreements with the latter being merely 'the transposition to the resale stage of the selective distribution system organised by common agreement between Omega and its national agents',[11] the obligations on the retailers to be equivalent to those imposed on the wholesalers, namely that they could sell to any consumer and also to any retailer who was himself authorized to be an Omega concessionaire. It should be noted, however, that when at a later date Omega applied for renewal of the exemption, the Commission insisted in the light of changed conditions in the retail watch sector that the limitation of a maximum number of dealers in a country be removed.

In a subsequent case,[12] moreover, the Commission required substantially more changes in the system proposed before granting exemption. In this case, Junghans, a German company which was a subsidiary of Karl Diehl of Nuremberg, sold clocks and watches through wholesalers in Germany and through sole distributors in the other EC countries. Sole distributors also in turn dealt with a limited number of wholesalers and with

[11] By 'agents' the Commission meant its principal distributors in each country.

[12] *Junghans* [1977] 1 CMLR D82. See also Ch. 14, p. 274, in connection with clauses in the agreements relating to the maintenance of resale prices.

specialized retail dealers. The system of distribution was tightly controlled, although in this case open-ended without any preset limit on the numbers of dealers who could be admitted to the system. Qualitative criteria relating to staff, premises, levels of stock, and quality of service were laid down, and the dealer was required to show that it specialized in such products. Dealers were not allowed to solicit business outside their own territory nor to export outside the EC. Sole distributors (though not retail dealers) were restricted from selling competing products without specific consent of Junghans.

The proposed agreement contained a number of additional restrictions which were removed upon the insistence of the Commission, including:

(i) The sole distributor in Belgium was prohibited from supplying other Junghans wholesalers and from selling direct to trade customers.
(ii) Retailers in Belgium, France, Italy, and Luxembourg were prohibited from supplying wholesalers or sole distributors.
(iii) Some wholesalers and sole distributors were prohibited from making any direct or indirect exports to other EC countries.
(iv) Only undertakings having a proprietor or manager who was a specialist in clocks or watches were eligible for appointment as official retailers, thus denying supply of Junghans products to supermarkets and department stores.

The removal of such clauses from the system meant that wholesalers and sole distributors were able to sell Junghans and Diehl products not only to all other retailers in the Common Market but also to other wholesalers, sole distributors, and consumers throughout the EC, whilst specialist retailers could supply not only consumers in the EC but also other Junghans retailers, wholesalers, or sole distributors throughout the Common Market. Moreover, Junghans dealers at any level were entitled to deal in competing clocks and watches, save for certain restrictions only on sole distributors. Criteria for the appointment of sole distributors or retailers were objective. Prohibition on exporting outside the EC would only apply to countries not parties to a Free Trade Agreement with the EC.

The Commission agreed to exempt the selective system itself and also the obligation of the sole distributors in the individual Member States not to deal in competing goods. Obligation on sole distributors and wholesalers not to engage in active sales promotion in other EC countries was also accepted by analogy with the terms of the Exclusive Distribution group exemption which permitted a ban on active sales, although requiring that such dealers are entitled to respond to unsolicited requests for goods from outside their own territory.[13]

[13] Reg. 67/67, Article 2(1)(*b*).

Cameras and Film

Early consideration of the distribution system adopted by *Kodak*[14] is contained in the 1970 Commission decision. Kodak established a system within the EC under which sales by its distributing subsidiaries in the Member States were to be on terms that trade purchasers at the retail level should not themselves export Kodak products. If they intended to do so, sales by the subsidiary to them had to be made at the price prevailing in the EC country of destination. Products could also be sold only to dealers with premises suitably equipped for the sale of cameras. Not surprisingly, these standard conditions of sale were only accorded negative clearance after removal from them of the prohibitions on export both direct and indirect, so that the conditions themselves could not be used to impede parallel imports within the Community.

While accepting the principle that resale of goods such as cameras could be limited to dealers with the necessary qualifications, the Commission did not at this stage elaborate the basis upon which negative clearance could be given, the case occurring at an early stage in the development of policy and well before the *Metro* case mentioned above. The market for cameras was subsequently, however, considered in the *Hasselbad* case[15] (European Court 1984). This Swedish company sold professional photographic equipment using a distribution system not based on purely qualitative criteria, but which limited distribution to a certain restricted number of specialist distributors. The system originally came under the scrutiny of the Court and the Commission as the result of the application for interim relief made by Camera Care Ltd., a Belfast dealer, which had fallen into dispute with Hasselblad over its claim that Camera Care had sold products at below the resale prices which Hasselblad required to be charged. This dispute had led to Hasselblad cutting off supplies to Camera Care. The first (1980) decision of the Court to the effect that interim relief was available to protect Camera Care was later followed by Hasselblad's appeal to the Court against the Commission's decision that its distribution system was not capable of negative clearance under Article 85(1) or exemption under Article 85(3) unless certain clauses were removed. The Court in its second (1984) decision confirmed the finding of the Commission that the system was being used not only to maintain prices but also to impede parallel imports and protect the individual markets in Member States. At least three clauses were found actually to violate Article 85(1), namely a prohibition on supplies by dealers to other

[14] [1970] CMLR D19.

[15] *Hasselblad* v. *Commission* Case 86/82 [1984] ECR 883: 1 CMLR 559. The earlier application for interim relief by Camera Care to the Court (Case 792/79) is reported at [1980] ECR 119: 1 CMLR 334.

authorized dealers, and the right retained by Hasselblad to terminate an agreement summarily if a dealer failed to observe any term of the agreement or changed the location of its premises without prior consent. Furthermore, to deal with price cutting, Hasselblad reserved the right to control the content of advertising by its dealers, to prevent it placing undue emphasis on price cutting, and this was likewise confirmed as a breach of 85(1).

Perfume and Cosmetics

A common link between many of the sectors where selective distribution systems have been approved is that the goods are expensive, technically complex, and require sales and service facilities of a standard to ensure that customers obtain satisfaction from their purchases. To find the sale of perfume and cosmetics included within such categories comes as rather a surprise, and the legal basis for the utilization of selective distribution for goods that are expensive but lacking complexity or the need for after-sales service or repair requires explanation. The requirement for selectivity in retail outlets for such products is based on a claim by perfume manufacturers that their products can only be satisfactorily sold in a suitable atmosphere or ambience, where the luxurious quality of the products is not spoilt by association with more mundane and less expensive goods. The Commission's 1974 Annual Report[16] refers to the examination by DG IV of the distribution systems adopted by Christian Dior and Lancôme, which organized the sale both of perfume and other beauty and toiletry goods through a selective distribution network limited to a restricted number of approved retailers supervised by their general agent in each Member State. The Commission[17] insisted on the deletion from the agreement of several clauses imposing in its view severe restraints on competition. These included, for example, restrictions on resale prices to be charged by Dior and Lancôme dealers when supplying any other retail outlet, either inside or outside their own country. The retailers were also prevented from obtaining supplies from any source other than their national general agent. If such clauses had not been removed, the individual national market would have been partitioned quite securely, and their removal meant that the retailers were rendered able within their trade network to sell or purchase to or from any approved agent or retailer in any EC country, as well as to have freedom as to the resale prices to be

[16] pp. 60–1.
[17] The Commission proceeded in this case by way of granting negative clearance from 85(1) by informal settlement rather than by way of formal decision. Given the high prices being charged, and the apparent lack of consumer benefit from the systems, the granting of a formal exemption under 85(3) might well have presented difficulties to the Commission.

charged on goods obtained from any source. The Commission commented that if expected this would lead to an alignment in the prices for such products in the different Member States, and also that the market for these products is characterized by the existence of a fairly large number of competing firms, none of which held more than approximately 5 per cent of the market.

In the following year, the Fifth Annual Report referred also to further consideration of these distribution arrangements. A large number of other companies in the industry had been involved in negotiations with the Commission in order to obtain approval of distribution terms on a similar basis. We find the Commission here giving itself a complimentary accolade for having been able to apply a 'uniform general arrangement throughout an entire industry, the perfume industry, without having to issue formal decisions'.[18]

Litigation, however, over the selective distribution system did arise, and ultimately was referred by French Courts to the European Court for an Article 177 ruling. The case arose in the French Courts because of the fact that the Commission, though having approved the arrangements for the industry, had done so merely by a 'comfort letter' confirming that the system was not considered in breach of Article 85(1). Various perfume shops which had tried unsuccessfully to obtain supplies from Lancôme, Guerlain, and other leading suppliers brought action under the French law which forbids refusal to supply. The perfume manufacturers argued that the existence of the 'comfort letter' provided a complete defence to this claim, but the European Court rejected this argument.[19] It decided that national Courts are quite free to reach their own findings on the standing of such agreements. National Courts may, of course, act on the basis of all information available to them and are, therefore, not bound by the 'comfort letter', although they may take its contents into account.

In principle, a system which relies, as did the perfume manufacturers, not merely on qualitative criteria for admission to the system but upon a quantitative selection of retailers would fall within the prohibition of Article 85(1).[20] Nevertheless, in considering whether either an individual agreement or a set of agreements distorted competition, the Court ruled that it was necessary to consider the context within which they were established. Relevant factors here were the nature and quantity of the product, the position and importance of the parties on the relevant market, and whether the retail outlet was an isolated one or formed part of a

[18] p. 51.

[19] The *Perfume* Cases 253/78, 37, 99/79 [1980] ECR 2327, 2481, 2511: [1981] 2 CMLR 99, 143, 164.

[20] *Yves Saint Laurent* OJ C320/11 (decision dated 20 Dec. 1990). The individual exemption granted to YSL has been challenged by the Leclerc Group on appeal to the Court of First Instance.

large network. The judgment of the Court contains an interesting explanation of the rationale for allowing selective criteria in the choice of outlets for luxury goods. The Court indicated that a selective system could be accepted, however, provided it was laid down uniformly for all potential resellers, and that the objective criteria for selection were not applied in a discriminatory fashion. It was also necessary to examine whether the product in question required 'for the purpose of preserving its quality and ensuring that it is used correctly' such a system. National courts should also take into account whether national legislation had not already covered the requirements of this kind by specifying conditions of sale or qualifications for entry into the occupation of such a reseller.

A recent case raised the issue of whether the sale of cosmetics, often found in the same retail outlets as perfumes, can be limited to those shops where a qualified pharmacist is present. In its 1991 *Vichy* decision[21] the Commission found that the basis on which the French company operated was lacking in consistency. In France itself it supplied cosmetics through a variety of outlets, but outside France only to retail pharmacies run by a qualified pharmacist. Vichy argued nevertheless that this requirement was objective, necessary, and proportionate, that it improved quality control and the services given to customers, and enhanced interbrand competition with other cosmetic manufacturers. The Commission, however, rejected these arguments. Its finding was clearly assisted by the difference between the distribution systems adopted by Vichy in France and other Member States, and based principally on the grounds that the benefits claimed could be obtained by less restrictive means.

In this decision it followed the pattern established in the slightly earlier case of the *Association Pharmaceutique Belge* (APB) (Commission 1990).[22] Here the national association of pharmacists in Belgium, who were legally responsible for the products which they sold in their retail shops, gave manufacturers the right to affix the APB mark of guaranteed quality to their parapharmaceutical[23] products in return for the manufacturer's agreement to sell these products only through members of the association. Originally the manufacturers had agreed that they would not sell the goods (with or without the stamp) at all to other retail outlets but this clause had to be deleted; it would have not only limited the manufacturer's own freedom of choice as to outlet but would have restricted competition between the members of the APB and other distributors of such products. Negative clearance was granted to the agreement in its

[21] OJ L75/57 dated 21 Mar. 1991. This case was taken on appeal to the Court of First Instance, which upheld the Commission's decision on 27 February 1992. (Case T-19/91.)

[22] [1990] 4 CMLR 619.

[23] Parapharmaceutical products include such non-medicinal products as cosmetics, dietary products, and baby foods. They would not in the view of the Commission require to be sold only by a qualified pharmacist.

amended form under which the manufacturers could still supply such products to other outlets, though without the quality stamp.

Electrical and Electronic Equipment

This sector had in the *Metro/Saba* (no. 1) case already provided the first occasion for the European Court to lay down principles upon which selective distribution systems should be regulated. The application of these principles, however, continued to cause difficulties to the Commission. Some short time after the case, the Commission conducted an inquiry into the consumer electronic sector which indicated, as later disclosed in the Thirteenth Annual Report,[24] that where only the manufacturer or the exclusive distributor had the right to recognize dealers' qualifications, there was a risk that this might be used as a basis for excluding from the system either those retailers suspected of price cutting or equally exporters and importers thought capable of exploiting differences in the price of manufacturers' products in different Member States. The inquiry was also said to have revealed that approved dealers in the network who had acted in a way considered 'unworthy' had either been excluded from the system or threatened with exclusion or subjected to other pressures.

That these findings were based on solid evidence was later illustrated by two cases coming by way of review of Commission decisions to the European Court, the *Pioneer*[25] case which concerned pressure applied by this Japanese group to prevent parallel imports into France by its German and UK distributors, and the *AEG/Telefunken* case (n. 5) where the Court found that a selective distribution system already notified to and approved by the Commission had been utilized in a way to enforce resale price maintenance upon some of the dealers concerned. These cases show that selective distribution systems always need careful scrutiny and can affect the state of competition in the particular sector if their existence reduces the element of choice for the customer, a factor which was an important element in the original *Metro/Saba* (no. 1) 1977 decision. In granting a continuation in 1982 of the exemption for Saba, the Commission stated that it was satisfied that for technically sophisticated products the use of selectivity in the choice of distribution outlets was still justified and in the interests of the consumers, especially when competition was intensive as in the case of consumer electronic products.

The Commission stated that there were a large number both of European and Japanese suppliers in the Community with different distribution systems and trading techniques, but that some elements in the original distribution system could be used to exclude those dealers willing to cut prices from full

[24] pp. 39, 40. [25] Cases 100–103/80 [1983] ECR 1825; 3 CMLR 221.

continuing participation. Changes were, therefore, required to both the procedure for admitting dealers to the network and the grounds upon which appointed dealers could be excluded. Under these revised terms, Saba was required to admit all interested dealers to the system and had to decide on a dealer's application within four weeks. Wholesalers within the system were entitled to investigate on their own responsibility whether a retailer satisfied the criteria for appointment and to approve it on behalf of Saba if it did so. Exclusion from the network would only be possible if the dealer ceased to satisfy the criteria for appointment or was in breach of its contractual obligations in a way which would endanger the continued existence of the distribution system. The Commission felt that such amendments would ensure that selection criteria were applied uniformly and that qualified dealers would not fear discriminatory treatment because of pricing policy or because they chose to engage in parallel importing. The decision was, however, once again challenged by Metro[26] on the basis that there were, at least for this type of product, too many selective distribution systems which operating in parallel were collectively causing the price level in the sector to be too rigid. Their claim that Saba should not receive continuing exemption for its selective distribution system received support from Advocate General VerLoren van Themaat who in a long and impressively reasoned opinion argued that the Commission's decision was defective on both substantive and procedural grounds.

The Court, however, refused to follow his lead, confirming the exemption granted on these revised terms by the Commission on the grounds that the mere parallel existence of a number of such selective systems did not of itself lead to the conclusion that competition at the retail level was restricted. Whilst undoubtedly price competition itself was reduced to some extent by their existence, Metro had failed to establish that the degree to which these selective distribution systems dominated the market at the time of the further grant of exemption justified their prohibition, given that the consumer still retained a substantial range of choice for high-priced quality sales and service on the one hand and cheaper lower standard distribution outlets on the other. There were still sufficient competitors for Saba to ensure a workably competitive market. Moreover, Metro had failed to prove that it was unable to continue to obtain supplies of competitive products from other sources.

Cars

The motor car is probably the most complex consumer product of all, as well as being the most expensive purchase that many consumers ever

[26] *Metro/Saba* (no. 2) Case 75/84 [1986] ECR 3021: [1987] 1 CMLR 118.

make. These unique characteristics have been put forward by motor manufacturers as providing justification for special rules applying to the distribution of cars, and the issue first reached the Commission in the *BMW* case.[27] The importance of the Commission's decision is shown by the fact that some two and a half pages are devoted to it in the Fourth Annual Report.[28] The most notable feature of the exemption granted is that it permits a restriction on the number of dealers allowed in Germany, again a criterion of a quantitative not qualitative nature. The Commission justified its decision by saying that 'Motor vehicles, being products of limited life, high cost and complex technology, require regular maintenance by specially equipped garages or service depots because their use can be dangerous to life, health and property and can have a harmful effect on the environment'. It described the required co-operation between BMW and its dealers as going beyond the mere marketing of products and involving a form of co-operation which not only promoted economic progress and the distribution of the vehicles but also gave benefits to consumers in the form of improved service. The Commission felt that this close co-operation and specialization by the dealers would not be possible if BMW were required to admit to the system, without limit of numbers, any dealer required merely to show he had adequate premises, staff, and other qualitative attributes.

The Commission took into account that though dealers were primarily to concentrate their efforts on their own sales territories, they remained able to meet orders received from outside their territory. They were also free to obtain supplies of spare parts from other dealers, not only from BMW itself. They could use spare parts of other makes, if these matched BMW quality standards, even where the parts were of importance to the safety of the vehicle. They were also entirely free to purchase spares without a safety significance. A ban on export sales, however, within the EC was objected to and removed at the request of the Commission.

Whilst the BMW decision was a liberal exercise by the Commission of its discretion under 85(3), there was considerable pressure from motor trade sources that a group exemption should be issued to avoid the need for individual clearance of each manufacturer's arrangements. The first mention of this is found in the Ninth Annual Report[29] when the Commission mentioned that at the expiry of the BMW original exemption period of three years, an attempt would be made to try and use the experience gained in order to work out general principles to reflect the needs of the industry.

After a lengthy period of consultation, this culminated in the publication in June 1983 of the first official draft of a group exemption for car

[27] [1975] 1 CMLR D44, 56. [28] pp. 57–9. [29] p. 18.

distribution, which originally met with strong opposition from the motor industry. Amongst their complaints were that it sought to bring about an artificial realignment of vehicle prices within the EC, sought to interfere without justification in the detailed relationship between the manufacturer and the dealer, and failed to provide sufficient legal certainty for both, The content of the original draft actually owed a considerable amount to the principles already laid down in the *BMW* case. The Commission acknowledged that in the motor vehicle sector (covering cars, buses, and vans) many of the restrictions of competition approved in the *BMW* case fulfilled the requirements of 85(3). In particular, the benefit to the public of a system of authorized dealers was that it was claimed to provide more satisfactory maintenance and repair services; the degree of competition of an interbrand nature between the major car manufacturers made competition on an intrabrand basis between their dealers rather less important. The Regulation would grant exemption for the placing of restrictions on the motor dealer against:

(1) selling another manufacturer's vehicles or parts;
(2) actively seeking customers outside its allotted territory;
(3) sub-contracting distribution, servicing, or repairs without the manufacturer's consent; and
(4) selling the vehicles or parts to dealers outside the distribution network.

In return, the manufacturer would agree not to appoint other dealers in the allotted territory during the period of the contract.

Considerable concern, especially by consumers' organizations, had been expressed about difficulties placed in the way of purchasers desiring to make parallel imports of vehicles from those Member States where cars could be purchased more cheaply, and the draft contains clauses designed to give some protection to the consumer. Thus, consumers are entitled to have servicing or repairs done under a manufacturer's warranty anywhere in the Common Market and are entitled to order cars with specifications required at the place where they are to be registered from a dealer in another Member State, provided that the manufacturer or his importer sells the relevant model through the official distribution network at both places. Spare parts supplied by third parties, matching the quality of those supplied by the manufacturer, could also be purchased by dealers in the official network without breach of their agreement with the manufacturer.[30]

[30] It is interesting to compare this provision with the findings of the UK Monopolies and Mergers Commission in their Car Parts report (HC 318) published in May 1982, and subsequently implemented by S.1.1982/1146 issued under the Fair Trading Act 1973, prohibiting manufacturers from requiring their franchised dealers to obtain spare parts exclusively from those manufacturers.

The original draft dealt specifically with this sensitive issue of the freeing of parallel imports, Article 7 providing that if the difference between retail prices recommended in any two countries exceeded 12 per cent over a six-month period, the authorized dealers would be free to sell to dealers outside the network in that second country. This freedom would not apply if prices were artificially held at different levels as the result of very high Member State tax rates, or where there was a temporary legislative freeze on the fixing of prices or profit margins. Fierce debate had continued right up to the last moment as to whether overall the terms of the Regulation were unduly onerous upon the car manufacturers or upon dealers.

When Regulation 123/85 was finally issued at the end of 1984, some important changes had been made to the original draft. Dealers were now entitled to a minimum period of four years' appointment in order that they would not be made over-dependent on the manufacturer or importer. Additional safeguards were included to try to make it difficult for manufacturers or importers to prevent sales of right-hand-drive versions in the United Kingdom of cars bought through official channels on the Continent. The Commission, however, had removed the reference to the specific price differential of 12 per cent from the Regulation and transferred it to an accompanying Notice issued by way of interpretation of the Commission's attitude to the Regulation. The Commission indicated in the Notice that it would not seek to intervene to withdraw the exemption (under Article 10) if the recommended resale prices to final consumers did not vary between Member States by more than 18 per cent on the list price, or if the difference applied only to an insignificant number of motor vehicles within the dealer's contract programme. The Commission also attempted to deal with the problem of 'unfairly loaded charges' made in respect of cars destined for parallel importing. The Notice indicated that the Commission would not interfere if the supplier charged:

(a) an objectively justifiable extra charge in respect of special distribution costs as well as any differences in the specification of the model; and
(b) where a further supplement was charged, this would be permitted provided the final price of the vehicle did not exceed the price of the vehicle in that Member State in which the lowest price (net of tax) was recommended for sale.

It is inherent in the Regulation that whereas the principle of selectivity is necessarily unqualified (so that dealers are not allowed to sell cars for resale other than to another approved dealer or sub-dealer), the requirements of exclusivity are qualified. Thus the manufacturer may itself be allowed to appoint additional dealers in the dealer's original exclusive area of primary responsibility or to vary the size of the area if it can show objectively valid reasons for doing so. Equally the dealer can be released from its obligations of exclusive commitment to sales and servicing of that manufacturer's cars if in its turn it can show objective reasons for such a course. The Regulation also preserves the right of dealers to sell to intermediaries, as agents for individual customers, and this has been the subject of a recent Commission decision in *Peugeot* v. *Ecosystems*.[31] This decision in favour of the intermediary's right to act on behalf of customers has been accompanied by a Notice from the Commission limiting the number of vehicles which an individual dealer is entitled to dispose of annually in this way to 10 per cent of its total sales. The great majority of any dealer's sales will therefore still be made direct to individual purchasers and intermediary sales can remain only a relatively minor part of its total business.

In the Fourteenth Annual Report,[32] the Commission admitted that it was unable fully to harmonize car prices across the Common Market both because these partially reflected supply and demand in individual countries and also because of differences in national taxation. Nevertheless, it reserved the right to intervene if it suspected that 'national measures or private restrictive measures are behind' the differential; and if large and long-lasting these would lead to withdrawal of the group exemption.

It is noteworthy the degree to which the permitted 'white' clauses set out in great detail the obligations that may be placed upon individual dealers. To an extent (and probably more than any other group exemption), therefore, the Regulation itself, No. 123/85, provides a model form of agreement into which dealers and manufacturers will need more or less unchanged to enter if they are to secure the advantages of the Regulation even though there is no legal requirement on them to do so. This undoubtedly results from the fact that when draft Regulations relating to specialized sectors are drawn up, the tendency for the relevant trade associations of both manufacturers and dealers to press for their particular preferred (and sometimes opposing) clauses ultimately results in a compromise whereby many additional clauses covering such detailed issues are added to the 'white list' of the Regulation. This is one disadvantage, in the eyes of DG IV at least, of this type of Regulation, and may mean that in future it may be more hesitant before granting further group exemptions to individual sectors.[33]

Quite apart from this factor, Regulation 123/85, which comes up for review upon its expiration in 1995, has another uniquely unsatisfactory

[31] Commission Press Release IP (91) 1089 dated 4 Dec. 1991.
[32] pp. 43–4.
[33] DG IV encountered similar difficulties in the course of negotiations leading up to the preparation of Parts II and III of Regulation 1984/83 dealing with exclusive purchasing agreements in the beer and petrol sectors. See Ch. 12, pp. 236–40. The provisions and effect of Reg. 123/85 are discussed in some detail in the MMC Report on New Motor Cars (1992) Cm. 1808 ch. 9, pp. 149–62.

feature. The list of 'white' clauses contained in Article 3 is not expressed as exhaustive unlike, for example, the 'white' list in group exemptions such as 1983/83 and 1984/83 dealing with exclusive distribution and exclusive purchasing respectively. Equally there is no opposition procedure provided within the framework of the Regulation, by which additional clauses restricting competition can nevertheless be submitted to the Commission for clearance within a limited period. This means in the case of Regulation 123/85 that the status of certain common but restrictive clauses found in dealer agreements remains uncertain, particularly as there is arguably also ambiguity over the precise scope of some of the numerous restrictions which are permitted by Article 3.[34] Other group exemptions without an exhaustive list of 'white' clauses normally do contain such a provision for opposition and it is uncertain why such a clause does not appear in this Regulation.

This review of cases decided in some of the main sectors covered by selective distribution agreements leads to some interesting conclusions. Selective distribution systems seem to be treated in two main groups, which do not necessarily conform to the distinction already noted at the start of this chapter of 'technical' and 'luxury' products. The first group consists of those products where the supplier is able to impose selectivity upon outlets which can be restrained from trading freely with non-approved outlets so long as entry to this 'family circle' is open to all suitably qualified undertakings. These tests for suitability must relate to objective factors such as technical expertise, suitability of premises, adequacy of staff, and financial stability. The Commission tends now to require that these tests be applied quickly, i.e. within a few weeks, and not simply by the manufacturer but at a delegated level, i.e. at the level of the wholesaler or sole distributor, so as to weaken the ability of a supplier to find excuses to exclude those undertakings more inclined to aggressive promotional innovation or even price cutting. In this group, however, dealers must remain free to deal in competing goods but if these conditions are satisfied then the agreement falls outside Article 85(1) altogether, so that conditions cannot be imposed by the Commission on their operation nor close supervision maintained on their working.

The second group consists of those products where the technical requirements for pre- and post-sales services and guarantee work are such as to entitle additional restrictions being imposed by the supplier, namely quantitative limits on the number of appointed dealers within a Member State, and restraints imposed from dealing in competitive products so that dealers' energies may be focused on promoting only the goods of one particular supplier. Here, exemption is only possible if at all under Article

[34] For example, those permitted by Article 3(3).

85(3) on an individual basis and has so far been given on an extremely limited scale. On a group basis it has only been accorded to the special case of cars under the terms of Regulation 123/85. The combination of selectivity with a numerical limitation has ostensibly been permitted on the grounds of the nature of the product, and the consequent need for the closest co-operation between manufacturers and dealers, a commitment which, it is claimed, will not be obtained if the dealer is free to engage in the sale or servicing of competitive products. Moreover the manufacturer remains free to appoint additional dealers in the dealer's territory if there are objectively valid grounds for doing so, e.g. a failure by the original dealer to satisfy the full local demand.

Whilst the distinction thus drawn between the treatment accorded to different products under selective distribution is clear, its justification is less so. The main difficulty arises because of the Commission's acceptance, based on the judgment of the European Court in the *Perfume* cases, of the principle that selectivity can be justified equally by the surrounding circumstances in which goods are thought to have to be sold, as much as by their actual technical requirements. Thus, the privilege of classification as a selective distribution system is accorded to a variety of products not simply as the result of the nature of the product itself; instead justification is drawn from other extraneous circumstances and alleged commercial needs which many commentators feel are lacking in validity. In this category, for example, would fall the claim of newspapers, as in the *Salonia* v. *Poidomani*[35] (European Court 1981) and in the *Binon*[36] (European Court 1985) cases where it seemed that the justification for permitting selectivity turned largely on the problems of financing the return of unsold newspapers.[37] Whilst the Commission appears to have heeded the requirement imposed by the European Court in the *Perfume* cases that consideration should be given to whether national law itself provides sufficient safeguards so that selectivity of distribution outlets is unnecessary and has applied it in the *Grohe/Ideal Standard*[38] cases relating to bathroom fittings (and has moreover refused to accept that the sale of normal household furniture requires the setting up of such systems[39]), the criteria still appear unpredictable. The dividing line between the two permitted classes is itself indistinct since close examination of the case-law shows considerable ambiguity as to what is 'qualitative' and what is 'quantitative'. Thus,

[35] Case 126/80 [1981] ECR 1563: [1982] 1 CMLR 64.

[36] Case 243/83 [1985] ECR 2105: 3 CMLR 800.

[37] Another questionable example is the *Villeroy-Boch* case. where a selective distribution system relating to the sale of high-class porcelain was approved. 15th Annual Report, p. 66. OJ [1985] L376/15.

[38] 14th Annual Report, p. 61.

[39] 15th Annual Report, p. 66.

in the *Grundig*[40] case, a number of restrictions were described by the Commission as quantitative, which could as well have been classified as qualitative, and vice versa. It is in fact difficult to find a consistent thread of principle throughout the cases.

Difficulties arise also as to the extent to which exclusivity (i.e. restrictions on the dealers against dealing in competing products) can be combined with selectivity, in the way in which this is permitted for the distribution of cars by Regulation 123/85. In other words, is there any reason why other categories of products should not at the same time be eligible to fall into the category of selective distribution systems (thereby remaining outside Article 85(1) altogether), but also be covered by Group Exemption 1983/83 to the extent that restrictions are imposed on the dealer of the type permitted by that Regulation? The answer would seem to be in the affirmative, so long as the distribution agreement avoids the familiar prohibitions of the group exemption of restraints on parallel imports or the imposition of territorial or price restrictions. Article 2(3)(*c*) allows the distributor to be required to maintain a sales network within the framework of its obligation to take general measures for the promotion of the product. The mention of sale networks suggests that the supplier can lay down appropriate standards for inclusion in this network which may be the very criteria permitted in the first category of selective distribution systems so long as the dealer does not in setting up his network introduce the quantitative criteria that would bring him into the second category and so require individual exemption. Moreover, the Commission in the *Ivoclar*[41] case (1985) showed itself generous in granting an individual exemption for a selective distribution system for dental goods which operated through a number of exclusive distributors selling only to dentists, dental technicians, and medical institutions but not to other dealers. The apparent grounds for the decision were the complexity of the products and the level of training needed.

2. Franchise Agreements

Another method available for the large-scale distribution of goods or services is franchising. The basic idea here is that instead of appointing an independent distributor who takes on the risk of reselling the supplier's product within a particular area and who is responsible itself for determining both sales policy and choice of dealers within the territory, the manufacturer (franchisor) appoints an independent undertaking to operate

[40] 15th Annual Report, 66–7. Exemption was granted to the Grundig selective distribution system for consumer electronic goods.

[41] Ibid. 65. OJ [1985] L369/1.

in a manner far more closely integrated with the franchisor itself; the franchisor supplies the franchisee not just with a product or service for sale but with a standardized and highly detailed promotional framework within which the franchisee has to operate. This will probably include the following elements:

(*a*) a right for the franchisee to sell the relevant product or service from a location in a particular contract territory;
(*b*) the provision of trade mark licences and know-how licences to ensure that the presentation of the brand image of the product is identical to that of the franchisor and all other franchised outlets;
(*c*) a requirement that the required products or services are obtained through or from the franchisor, and if from some third party then in accordance with detailed specifications;
(*d*) the obligation on the franchisee to pay both a lump sum for the know-how provided and royalties in respect of all sales made, as well as to adhere strictly to the terms of the franchisor's package (in terms both of the layout of his retail premises and in all respects in which the business is conducted);
(*e*) a network of other franchise holders operating on an identical basis.

Franchises came into common use in the EC in the early 1970s, developing from business models adopted in the USA. By May 1982, there were 200 franchise systems in the Federal Republic of Germany involving 120,000 franchisees; in France by the same period there were nearly 400 franchise systems; in The Netherlands, 280 franchise systems in 1983. Similar developments took place in other Member States.[42]

There are three main varieties of business format franchising which apply to the retail distribution of goods and services. The first, and that which most concerns us here, is a *distribution* franchise where the function of the trade mark is to assist the effective distribution of the franchisor's goods in a recognizable way. Other varieties of the same concept of 'business format franchising' are *service* franchising, which is similar to the distribution franchise except that any goods provided are only ancillary to the service offered, and *production* franchising where the producer/franchisor arranges for the manufacture of the particular range of brand goods to be carried out by his franchisee. Within normal categories of service franchising can be found restaurants and hotels, car hire, accounting bureaux, and data processing services; in the field of production franchising, one often finds clothes, cosmetics, footwear, and other mass production items.

[42] These figures are taken from the opinion of Advocate General VerLoren van Themaat in the *Pronuptia* case (see n. 44) which itself provides a very full account of the economic role and development of the system of franchising.

Distribution franchising, however, can apply to almost every type of product. A normal franchise agreement for the distribution of goods is a lengthy document of up to 100 pages, presenting in great detail the way in which the parties are to carry out their mutual obligations. Inherent in most franchising arrangements are a number of clauses which restrict competition; and the issue immediately arises whether any of these clauses fall under Article 85(1) and if so whether exemption will be available. Since the essence of a franchise system, like that of exclusive distribution, is a network of agreements conferring partial or total territorial exclusivity on the franchisee in return for the substantial payments that he makes, the franchisor itself may be asked to accept some or all of the following restrictions, namely:

(1) not to appoint another franchisee in the territory allocated;
(2) not itself to be involved in the sale of the franchise goods within the territory, e.g. by opening a rival shop;
(3) not to sell, directly or indirectly, contract goods to persons who will resell them in the contract territory.

The franchisee on his part may have to accept a large number of restrictions including:

(1) the obligation not to sell franchise goods, except to customers of his or her specified shops at named locations, which have to be fitted out exactly in accordance with the franchisor's requirements;
(2) not to change the location of these shops nor to assign the agreement to third parties without the consent of the franchisor;
(3) not to sell franchise goods to dealers outside the franchise network;
(4) not to sell except at the prices (maximum or minimum) laid down or recommended by the franchisor;
(5) not to deal in any competing goods, or possibly to deal with them only to a limited extent;
(6) not for a period after the ending of the franchise agreement to enter into competition with the franchisor in respect of the contract product.

The Commission originally found itself in some difficulty in deciding how to classify franchise agreements since they exhibited in varying degrees features commonly found in no less than three other categories of agreement with which DG IV is familiar and for which it has, with the assistance of the European Court, developed detailed ground rules. These are exclusive dealing agreements, selective dealing agreements, and patent and trade mark licensing.

Although there are some similarities between each of these three classes of agreement and franchising, franchising nevertheless remains inherently distinct from any of them, by virtue of the close control maintained over the manner in which the franchisee conducts his business. In the case of exclusive dealing agreements by contrast, though the dealer can be required to sell goods under the supplier's trade mark and packaging under the terms of the group exemption, the dealer still retains considerable independence in its marketing policy and is normally free to sell from whatever outlets and in whatever way it chooses. Moreover, also under the terms of the group exemption, it must be left free to respond to orders outside its territory so long as it does not promote sales actively itself outside, and the supplier remains free to sell to other distributors or dealers outside the territory even if they will subsequently resell into the contract territory. Any licensing of know-how or trade marks is ancillary to these distribution rights.

The selective distribution systems discussed earlier in this chapter also showed considerable differences from franchising. Here again, the dealer remains free (except under Regulation 123/85 governing the distribution of cars and those other special cases for which individual exemption may be given under Article 85(3)) to sell competing goods in addition to the contract goods, and agreements normally do not prescribe specific territories beyond the allocation of territorial areas of primary responsibility. Nor did the patent licence Regulations or case-law on trade mark or know-how licences provide much assistance to the Commission. In such licences, the primary subject is the transfer of, e.g., know-how whereas in franchising these licences are merely ancillary to the whole 'business package', which the franchisee acquires usually for a substantial payment.

It was argued that the Commission should adopt a tolerant attitude towards franchise agreements, and a strong case can indeed be made out for claiming that it has some beneficial and pro-competitive aspects.[43] The close support received from the franchisor enables new retail outlets to compete strongly and immediately with other outlets, and it is also claimed that it contributes to the reduction of prices for manufacturers' goods without any sacrifice of quality. It can be argued to have a significant ability in the creation of an EC mass market and in breaking down frontiers, by allowing the coverage of large territories in a number of Member States without at the same time having to produce substantial product differentiation on a national level. The promotion of the 'brand image' enables recognition of the product to spread faster than would otherwise be the case, and thereby enables unit costs to be kept lower and prices possibly reduced. Whilst not all these arguments would necessarily

[43] A useful source of information on franchising, written from a US viewpoint before the Court delivered its *Pronuptia* judgment, and urging its liberal treatment on the grounds that it has a significant ability to make retail markets more competitive, is R. J. Goebel, 'The Uneasy Fate of Franchising under EEC Antitrust Laws'. 10 *E L Rev* 87 (1985).

be accepted, it was, therefore, widely expected that the European Court would take the opportunity, when it arose, to take a relatively positive attitude towards this form of distribution.

The opportunity arose in the well-known *Pronuptia* case[44] (European Court 1986). Mrs Schillgalis had entered into a franchise agreement in Germany several years previously to sell wedding dresses and other wedding items under the trade mark 'Pronuptia de Paris'. Her appointment covered three separate areas (Hamburg, Oldenburg, and Hanover) and contained restrictions on both her and Pronuptia. The main restrictions placed on Pronuptia were:

(1) that it would grant Mrs Schillgalis the exclusive right to use the trade name for the marketing of the contract goods in a specific territory;
(2) that it would not itself open any other Pronuptia shop in that territory or provide goods or services to third parties in the territory; and
(3) that it would give Mrs Schillgalis assistance with regard to all commercial aspects of her business including advertising, staff training, fashion purchasing, marketing, and indeed any other help needed to improve her turnover and profitability.

In return, Mrs Schillgalis accepted a large number of restrictions, which included:

(*a*) to sell the wedding gowns under the trade mark 'Pronuptia de Paris' only in the shops specified, which had to be equipped and decorated exactly as the franchisor required and could not be transferred to another location or altered without the franchisor's agreement;
(*b*) to purchase 80 per cent of wedding dresses and accessories to be sold from Pronuptia direct, and to purchase at least a proportion of other dresses from the franchisor or from suppliers approved by the franchisor, and to make the sale of bridal fashions her main concern;
(*c*) to pay an entry fee for the know-how of DM 15,000 and thereafter a royalty of 10 per cent on the total sales of Pronuptia products;
(*d*) to advertise only in a manner approved by the franchisor and in general to use the business methods prescribed by the franchisor, save that the fact that prices were recommended by Pronuptia did not affect her ultimate freedom to fix her own prices;
(*e*) to refrain during the contract and for one year afterwards from competing in any way with Pronuptia outside her own contract territory.

When later sued for substantial arrears of royalty by Pronuptia, she claimed that the agreement was void under Article 85(2) and that she was, therefore, not required to pay. Her argument succeeded before the

[44] Case 161/84 [1986] ECR 353, 1 CMLR 414. At the time when the relevant agreements were entered into, Reg. 67/67 was still in force.

Frankfurt Court of Appeal (Oberlandgericht); and the German Supreme Court (the Bundesgerichthof) referred to the European Court under Article 177 the issue of whether Article 85 applied to franchise agreements and if so whether the Group Exemption 67/67 could also have applied to it.

It is in the nature of such Article 177 references that the European Court does not review the case as a whole but merely answers the specific questions asked by the national Court. It is, therefore, less likely in such cases that the European Court will find it possible to pronounce in general terms on the legal principles involved, as compared say with the occasion when a review from a DG IV decision is brought under Article 173. Nevertheless, there have been occasions such as the *Perfumes* case when the Court had taken the opportunity of an Article 177 reference to state or restate basic principles relating to competition law, and it was hoped that it would also take this opportunity to provide assistance to the Commission in dealing with the increasing numbers and types of franchise agreements now being notified to it.

The Advocate-General, VerLoren van Themaat, took a relatively relaxed view of such agreements; after careful review of the earlier Court jurisprudence dealing with related areas (such as exclusive distribution agreements and licence agreements) he reached the general conclusion that franchising as a method of distribution had major advantages for both the franchisor and the franchisee, and in general should not fall within Article 85(1) unless:

(*a*) the agreement was made between a franchisor from one Member State or its subsidiary on the one hand and one or more franchisees in one or more other Member States on the other hand; and in addition
(*b*) the franchisor had through its subsidiaries and franchisees in one or more of those Member States or a significant part of their territory a substantial share of the market for that product; and
(*c*) the agreement had the effect or intention of restricting either parallel imports of the product into the contract territory or exports of those products by the franchisee to other Member States; or
(*d*) the agreement resulted in the establishment of unreasonably high resale prices, that is prices which could not be charged if effective competition existed for them.

The Advocate-General also considered that Regulation 67/67 did not apply to the Pronuptia franchises at issue, since no consideration had been given to the question of franchising when the Regulation was drawn up, and since in any case the issues of exclusive supply and exclusive purchasing covered by that Regulation are only minor factors in the normal franchise agreement.

The European Court, whilst agreeing that Regulation 67/67 was inapplicable, declined to give broad approval to franchise agreements in the way the Advocate-General had suggested. It accepted that such agreements had a substantial advantage both to franchisors and franchisees and allowed the franchisor to derive benefit financially from its expertise without having to invest its own capital. The franchisee benefited because it could receive, even without any past experience of the trade, access to trading methods which the franchisor had successfully utilized, and for which it had gained a reputation. In order for such a system to work, therefore, protecting the legitimate expectations of the franchisor, restrictive conditions could be accepted as falling outside Article 85(1) if they satisfied one of the following two conditions.

The first condition was that the franchisor must feel free to be able to pass on its know-how to franchisees and provide them with the necessary promotional and other assistance without running a risk that this might come to benefit its competitors. Under this heading, a clause preventing the franchisee both during and for a reasonable period after the agreement from opening a shop of the same or similar nature in another area (where he or she might compete with another member of the franchise network) was acceptable and outside 85(1), as also would be any restriction on proposed transfer of the shop to another party without the franchisor's approval.

The second basis for restrictive clauses was that the franchisor must be allowed to take the measures necessary for maintaining the identity and reputation of the franchise network, to the extent that it had to exercise control over it. This would cover obligations on the franchisee to apply the franchisor's business method and sell the goods covered by the contract only in premises laid out and decorated according to the franchisor's instructions which were intended to ensure uniform presentation. Likewise, the franchisor could impose requirements as to the location of the shop and its subsequent transfer and to prohibit the transfer by the franchisee of its rights and obligations without the franchisor's approval. In order that the goods supplied by each franchisee should be of the same quality, it was reasonable for the franchisor to require that the franchisee sell only products supplied by the franchisor or supplier selected by it, so that the franchisee must be allowed to purchase such goods from other franchisees.

The judgment, however, did indicate clearly that Article 85(1) would apply to a clause obliging the franchisee to sell goods covered by the contract only in specified premises, e.g. preventing him or her from opening up a second shop in his or her own territory when the franchisor is itself under obligation to ensure the franchisee exclusivity in that territory. For if such a restriction were applied throughout the network, it would mean that effectively a Member State would be divided up by the franchisor into a number of closed territories. If it was shown nevertheless by the franchisor that franchisees would not take the risk of entering the network and investing substantial sums on the purchase of the franchise without such an assured exclusive territory, then these considerations should be considered on an individual application under Article 85(3). The Court ruled that provisions which shared markets between the franchisor and franchisee or between the franchisees themselves were likely to affect trade between Member States, even if entered into solely by undertakings established in one Member State, because of their ability to prevent franchisees in one Member State from establishing themselves also in another Member State.

Finally, with regard to the question of the control over the resale prices the franchisor could exert, the Court took a more severe view than the Advocate-General. He had indicated that resale price maintenance should not be illegal unless one party is in a position of economic strength on the local market concerned where price maintenance was also applied by competitors or where the franchisor is in a position of a price leader on that particular market. The Court said that only price recommendations were permissible, but failed to distinguish between maximum and minimum price recommendation, and indicated that price recommendations themselves would only be acceptable so long as there was no concerted practice between the parties as to their actual application.

Pressure was now therefore growing for the unveiling of a draft group exemption, which was published finally in August 1987. Drawing substantially on the Pronuptia judgment, it adopted a generally favourable approach to the concept of both distribution and service franchises, without seeking to cover industrial or manufacturing franchising. Some aspects of this first draft were criticized. For example, the initial definition of 'know-how' was considered too narrow and technically demanding. In general, however, only a few changes were required before it was adopted officially as Regulation 4087/88 on 30 November 1988 coming into force on 1 February 1989 for a term of just over ten years.[45]

Its general shape followed the traditional pattern for group exemptions with a definition of its coverage and lists of exempted restrictions and prohibited restrictions, the 'white' and 'black' lists, a list of relevant conditions that had to continue to apply throughout the duration of the exemption, an opposition procedure, and grounds for withdrawal of the exemption. Article 2 contained a particularly long list of restrictions exempted including the exclusive right of the franchisee to operate his franchise in a particular territory from his contract premises only and without seeking outside that territory to solicit customers. Nor was the

[45] The annotated text of the Regulation can be found in [1989] 4 CMLR 387 and the explanatory notice of the Commission at [1989] 4 CMLR 93.

franchisee allowed to sell competitive goods or services. The 'white' list also included two main categories of clause which the Court had referred to in *Pronuptia*; those permitted in so far as they were needed to protect the franchisor's intellectual property rights and secondly those that maintained the common identity and reputation of the particular franchise network.

In the first category fell clauses requiring the franchisee to sell or use exclusively goods that met the franchisor's minimum quality specification, not to compete with the franchisor during the agreement (or for up to one year afterwards), and to sell to end users only. The second category included obligations to keep the licensed know-how secret and to use it only for the exploitation of the franchise, together with an obligation to communicate improvements back to the franchisor on a non-exclusive basis. If there were a network of franchisees, however, the individual franchisee was free to obtain the relevant goods from another franchisee as well as from the franchisor. Guarantees given by any one franchisee for goods must apply also to goods acquired from any other franchisee situate elsewhere in the European Community.

The 'black' list of clauses was comparatively brief. It included naturally any restriction imposed on the franchisee's freedom to determine sale prices, no-challenge clauses against intellectual property rights, and any attempt to limit competition between the franchisees or to impose restrictions on sources of supply for contract products. An opposition clause was included to enable additional restrictions not found in the lists to be presented to the Commission; and the 'safety valve' Article 8 provided a number of situations in which the Commission would be entitled to withdraw the benefit of the exemption from individual agreements, including any where franchisees engaged in horizontal price fixing or obstruction to parallel imports, where goods or services faced no effective competition in the Common Market, or where the franchisor used its contractual rights and considerable influence over the franchisee for purposes wider than the proper protection of its identity and reputation within the framework of the contract.

The comparatively liberal terms of the Regulation have led to their widespread use as a model by companies throughout the Community, in some cases to replace use of the Exclusive Dealing Group Exemption under Regulation 1983/83 with its far more limited list of permitted restrictions. Clearly this can only be done where the individual distributors receive a sufficient 'package' of intellectual property rights, know-how, and continuing technical assistance to be qualified as franchisees rather than simply dealers. The completion of the framework of the Regulation has moreover enabled the Commission to handle cases of request for individual exemption with greater confidence (some of which had been notified to the

Commission at an earlier date) if they did not fit the exact requirements of the Regulation itself.

Thus in *Servicemaster*[46] the franchise covered the supply of housekeeping, cleaning, and maintenance services for both commercial and domestic customers. Nevertheless the Commission pointed out that, although the Regulation only referred to the distribution of goods, franchises dealing with services alone should be treated similarly. Characteristic features of service franchises moreover include the relative importance of the know-how provided to the franchisee and the fact that his services were normally provided at the customer‘s premises creating a personal relationship with him. Individual exemption was granted here and also in another case, relating to the *Charles Jourdan*[47] franchise for the sale of a brand of French shoes and other leather goods. Charles Jourdan as franchisor reserved the right to approve individual managers who had been selected for the franchised shops and an exact territory was allocated to each franchisee. Franchisees all had to pay an entry fee either to run the business for particular premises within the territory or, alternatively, for being granted a 'franchise-corner', an individual area within a shop but with no exterior sign. The effect of the close control maintained by the franchisor was to give each franchisee relatively strong protection from competition within its territory against other franchisees. The Commission ruled, however, in granting the exemption that in this case unless territorial exclusivity were given few prospective franchisees would be willing to make the necessary investment in this purchase. The product itself only had 10 per cent of the French market for medium and top quality shoes and as little as 2 per cent of the whole Community market for such products. The decision illustrates that individual exemptions may be available for such arrangements even if the goods have no great technical complexity and are being sold alongside similar products not the subject of a franchise.

[46] [1989] 4 CMLR 581.

[47] [1989] 4 CMLR 591. The lesser restrictions placed on the 'franchise-corner' retailer were actually held to fall outside Article 85(1) altogether.

14

Distribution: Price Maintenance and Other Vertical Restraints

This chapter deals with vertical relationships, that is those that apply between a supplier and those who buy goods from it for distribution. It examines three separate situations where Article 85 may become relevant in such relationships. The first is where the supplier wishes to control the price at which the buyer resells the goods provided, a practice usually known as resale price maintenance. The second is where the seller is prepared to sell but only on terms that discriminate between different buyers; the third is where the seller refuses to deal with or supply a specific buyer at all. Article 86 applies also to the second and third situations, and its application in these situations is discussed in Chapter 18.

1. Resale Price Maintenance

Article 85 can apply only when there is some form of agreement or concerted practice or some decision of a trade association that satisfies all the conditions required under that Article as explained in Chapters 6 and 7. If there is a collective horizontal agreement between suppliers with regard to implementation of a resale price maintenance policy against their various purchasers, then this will be caught by the Article if it affects trade between Member States and has as its object or effect the prevention, restriction, or distortion of competition. If price maintenance is practised, whether or not there is in any particular sector a horizontal agreement or practice relating to resale price maintenance, there is almost inevitably a vertical agreement since the buyer will have been required to enter into an agreement with its supplier to observe its resale prices. The method of enforcement of these arrangements may itself be simply individual, or by way of collective enforcement with other manufacturers. Such an agreement is clearly one which 'directly or indirectly fixes purchase or selling prices . . .'. It may be difficult as a matter of proof to establish in certain cases whether the agreement actually exists but, once this has been established, then it is likely that Article 85 would be applicable to the vertical and any horizontal aspects of such an arrangement.

Resale price maintenance is a practice whose effects are variously regarded by economists.[1] A whole range of possible justifications for the practice have been put forward. The manufacturer or supplier may argue that if prices are cut, the prestige or luxury connotations of the product, e.g. perfume, may suggest to consumers that it is less to be valued.[2] A related argument is that if there is no resale price maintenance, dealers may sell popular brands at unreasonably low prices as loss leaders, even below cost (thereby in the manufacturer's view 'devaluing the brand') in order to attract customers to its premises who will then buy other goods at high prices. It is suggested that without resale price maintenance, the more aggressive retailer will drive smaller dealers out of business, depriving smaller communities of competition at the retail level and permitting the remaining dealers then to raise their prices to unreasonable levels.

Another argument in support of the practice is that resale price maintenance is necessary to guarantee dealers a generous profit margin, so that services can be provided which are of value both to customers and the cultural life of the community. In this category could be placed the profit required for electronic consumer goods, where extensive demonstration and advice from the dealer involves it in substantial expense, which may be incurred without any return if the customer can then purchase the goods from a discount store or department store offering no such services but merely the same goods at a lower price. This is an example of the 'free rider' argument referred to in Chapter 2. Alternatively, the manufacturer may believe that a large number of outlets are required for the best distribution of its product, and these can only be obtained by offering guaranteed margins. Resale price maintenance may be of value in helping new entrants to a particular trade obtain dealers who will be encouraged to promote their products if the margins are generous. There are, of course, contrary arguments available to most of these grounds for supporting resale price maintenance. The most important of these arguments is that legalization of the practice prevents retailers from competing with each other on price, and hence tends to raise the overall level of prices that consumers are required to pay. It tends also to have the effect that, if as its consequence inefficient dealers are protected against their inefficiency, efficient dealers are thereby prevented from taking the advantage to which they are entitled from their superior performance.

In general, the trend of the last two decades in Member States of the Community has been to reduce the extent to which resale price maintenance is accepted as a desirable economic practice,[3] and development of

[1] The literature on this topic is extensive. For citations see n. 7 of the opinion of Advocate-General VerLoren van Themaat in *Dutch Books (VBVB–VBBB)* cases 43–63/82 [1984] ECR 19: [1985] 1 CMLR 46.

[2] The same argument has been used to justify certain forms of selective distribution as mentioned in Ch. 13.

[3] It was largely abolished in the UK by the Resale Prices Act 1964, other than in respect of a limited number of agreements whose participants have been successful in satisfying the

Community competition policy over this period has reflected this tendency, as we find in the First Annual Report of DG IV the following comment:

> Purely national systems of resale price maintenance do not generally come under the Community law prohibiting cartels. To the extent that they are limited to compelling retailers in a Member State to respect certain prices for the resale within that state or products supplied by a manufacturer established on that market or by a concession holder appointed for that territory, trade between Member States will not, generally, be affected within the meaning of Article 85 . . . That is why the Commission considers that the question of vertical resale price maintenance is essentially a matter of national competition policy. The Commission ensures, however, that intermediaries and consumers are enabled to obtain supplies of the product concerned at the most favourable prices from wherever they choose within the Community.[4]

The sting in the tail, however, is to be found in the last sentence. The most perfectly watertight national system of RPM is of limited value if goods can enter that Member State from outside and can compete on price with the equivalent goods being distributed within the national RPM system. These imports may simply have come straight from a manufacturer in another Member State, free from tariffs or customs charges, or alternatively have originated in the local Member State and been the subject of reimportation. Efforts will undoubtedly be made to keep out such reimportations. This occurred in the *Deutsche Grammophon* v. *Metro* case[5] (European Court 1971). Metro had bought records in France produced by Deutsche Grammophon. Metro, having reimported them to Germany, sought to sell them there at prices well below the level at which the official German distributors (subject to German law which allowed RPM) were allowed to sell them. Deutsche Grammophon then sought to obtain an injunction against Metro preventing it from selling the records at a price lower than that fixed by the RPM provisions. The Court, agreeing on this point with Advocate-General Roemer, held that the holder of exclusive rights in sound recordings cannot, in order to achieve the aims of an agreement to fix selling prices which falls within the prohibition of Article 85, seek to exercise those rights by prohibiting the distribution of such recordings imported from another Member State, which it has itself placed on the market in that other Member State. This ruling of the Court of Justice effectively prevented Deutsche Grammophon from enforcing its

Restrictive Practices Court under the terms of the legislation (now contained in the Resale Prices Act 1976) that they come within the 'gateways' provided by Section 14 of that Act. One of these agreements was the Net Book Agreement which has subsequently been refused exemption under Article 85(3) by the Commission. See p. 279–80.

[4] p. 62.

[5] Case 78/70 [1971] ECR 487: CMLR 631. German law permitted individually enforceable price maintenance until the end of 1973, though collective price maintenance had been prohibited for a longer period.

RPM system in Germany, since its national legislation required that for the system to be valid it must be without any loopholes. In reinforcing this decision and hastening the end of resale price maintenance in Germany, the *Deutsche Phillips* case was of importance (Commission 1973).[6] Under the agreement made between Deutsche Phillips and its dealers, electric razors were to be offered and sold to customers only at the prices which Deutsche Phillips prescribed. This price restriction applied without distinction to products purchased direct from Deutsche Phillips in Germany and to such products obtained from foreign sources. The resale price restriction also applied to sales in other Member States; the Commission had little difficulty in establishing that these restrictions all fell within the prohibition of Article 85(1) and imposed a substantial fine.

Where resale price maintenance was in force on a widespread basis in a Member State through collective agreements, the Commission from its earliest days took active steps against it. We find in 1970 an early decision granting negative clearance to the rules of the *ASPA* (Association Syndicale Belge de la Parfumerie)[7] whose membership included manufacturers, general representatives, and exclusive distributors of perfumes and toiletries in Belgium. The original rules had contained stringent requirements on members to observe the fixed retail sales prices both for products manufactured by their members and for those imported from other Member States, coupled with an obligation to boycott any dealer or wholesaler who broke the association's rules relating to resale prices, discounts, or rebates. Apparently many other undertakings in this type of business, even though not members of the association, also adhered to these rules. Negative clearance was only granted on the basis that all such restrictions were removed. The Commission adopted a similar approach in the *Belgian Wallpapers* case[8] (1974). The Belgian wallpaper industry was comprehensively organized into a single marketing organization covering every level of the trade including manufacturers and wholesalers. Amongst the detailed rules were those involving collective resale price maintenance and the standardization of general selling conditions covering every aspect of the trade. The Commission found that the agreement infringed Article 85(1) and imposed heavy fines on the members of the association. Whilst the Commission's fine imposed for a collective boycott by the members was overruled on other grounds, the actual decision of the Commission, however, relating to the invalidity of the resale pricing scheme was not challenged on appeal.

Other cases followed shortly afterwards in which distribution arrangements extending beyond one Member State were granted clearance only on a condition that, *inter alia*, resale price maintenance clauses were

[6] [1973] CMLR D241. [7] [1970] CMLR D25.
[8] Case 73/74 [1975] ECR 1491: [1976] 1 CMLR 589.

removed. In the *Gerofabriek*[9] case of 1976, the largest of the Dutch cutlery manufacturers was refused exemption for its distribution arrangements in Holland and Belgium largely because of the extent to which these were based on price maintenance. It had some 60 per cent of the relevant markets in Holland in 1971 though this had fallen in subsequent years to approximately 50 per cent, and held approximately 15 per cent of the equivalent market in Belgium. It sold its cutlery both to retailers and direct to large buyers such as hotels and hospitals. The terms of trade for both resalers and wholesalers were that resales must only be at the listed prices. Moreover, retailers were debarred from selling on to other approved retailers so that it became difficult for dealers who had received the complete range of goods to supply other dealers who had not been allocated such a complete range. Moreover, different price levels in Belgium and Holland meant that dealers in Belgium were not allowed to sell to the Dutch dealers and vice versa. Altogether more than 2,000 dealers were affected.

In the other case of *Junghans*,[10] restrictions had been placed by the German clock manufacturer on French and Dutch sole distributors from dealing with dealers in Germany, unless these dealers had themselves entered into a resale price maintenance agreement with Junghans, nor were German retailers permitted to sell to customers in other EC countries except at the prices fixed for Germany. Moreover, the dealers were required to observe the fixed prices even when reselling in Germany goods which they had reimported from other Community countries. All these clauses were deleted before exemption was allowed under 85(3).

The *Metro/Saba* (no. 1)[11] European Court case and the *Spices*[12] Commission decision both decided in 1977 and already discussed are sometimes quoted as examples of setbacks to the Commission's policy in limiting strictly the extent to which resale price maintenance systems would be permitted. It would be wrong, however, to accept that either of these cases had the effect of weakening the Commission's policy. In the *Metro* case the European Court, in rejecting the claim that Saba's selective distribution system for consumer electronic goods of high quality was in breach of Article 85(1), certainly emphasized that price competition was not the only form of competition and stated that the desire of certain speciality wholesalers and retailers to maintain certain price levels which enabled additional service to be provided was not an objective which necessarily was prohibited by Article 85(1). The Court certainly did not, however, go on to say that competition in price was, therefore, an unimportant objective; it stressed rather that what was important was

[9] [1977] 1 CMLR D35. See Ch. 8, p. 140, for a more detailed account of the case.
[10] [1977] 1 CMLR D82. See also Ch. 13, pp. 246–7.
[11] Case 26/76 [1977] ECR 1875: [1978] 2 CMLR 1.
[12] [1978] 2 CMLR 116.

that consumers had the variety of choice of being enabled to choose such goods provided either through selective distributors at relatively higher prices or through more cut-price outlets offering lower standards of service and specialization. Unlike in the *AEG Telefunken*[13] case a few years later, however, there was no suggestion that Saba imposed resale prices on its selected dealers.

In the *Spices* case, the exclusive purchasing arrangements entered into by three major Belgian supermarket chains with Brooke Bond Liebig, a spice producer with 39 per cent of the Belgian market, was held unable to benefit from Article 85(3) for a variety of reasons. The strong market position of Brooke Bond Liebig, coupled with the restrictive nature of the arrangements which led to the total exclusion from a very large number of outlets in Belgium of all competitive spices, would probably have been sufficient on their own to justify the refusal of exemption. It is also clear from the terms of the decision that the Commission would have refused permission even for some less restrictive exclusive purchasing arrangements so long as they contained the resale price restrictions that Liebig had insisted upon. The existence of this resale price maintenance clause gave the supermarkets a very substantial profit margin but prevented them from passing on any benefit from the arrangements with Liebig to their customers.

Moreover, although the Commission had adopted, as we have seen, a generous attitude to selective distribution systems, these would not be approved if they contained any clauses which either restricted parallel imports or which had price restriction clauses of any kind. AEG/Telefunken, therefore, although it had obtained a negative clearance for its distribution system in Germany in 1976, subsequently lost the benefit of exemption because after an enquiry by the Commission it was found that the system was being utilized in order to enforce the resale price maintenance on a number of its dealers. The European Court in upholding the decision ruled that the Commission had received sufficient evidence to be entitled to decide that the distribution system had been improperly applied. The Court emphasized, however, that it had no objection to a system of the kind approved which by its nature would tend to support a higher level of price than one without the degree of selectivity. It assumed that specialist traders would necessarily charge prices within a much narrower range than that which might have existed had all competition been based simply on price and not also on the level of services

[13] Case 107/82 [1983] ECR 3151: [1984] 3 CMLR 325. The Court have also expressed strong disapproval of the practice in franchise agreements whilst accepting that non-binding recommendations as to price may be provided to franchisees (in the *Pronuptia* case). The Group Exemption for franchising agreements does not apply if the franchisor restricts the prices charged by the franchisee, whether directly or indirectly. Reg. 4087/88 Art. 5(*e*).

provided. The policy of AEG Telefunken, however, that a high profit margin was essential to survival of its specialist distributors, and that any of these which did not aim for or obtain a sufficiently high profit margin must automatically be regarded as incapable of providing the costly services expected from the specialist trade, was not acceptable; the maintenance simply of a minimum profit margin for traders could not itself be an object of the system. The fact that the specialist departments of discount stores were in fact able to provide the required services to consumers at a lower cost than other specialist shops proved that there was no justification for the attempts by AEG Telefunken to preserve minimum resale prices and profit margins.

Thus, in *Hennessy–Henkell*,[14] Hennessy, a major French cognac merchant, entered into a long-term arrangement for the distribution by Henkell of its cognac in Germany. It included the restriction that the consent of Hennessy was required if Henkell wanted to fix its resale prices either above cost plus 17 per cent or below cost plus 12 per cent. In return for this arrangement, Henkell had been guaranteed by Hennessy a price giving it a margin of 25 per cent (later reduced to 18 per cent), a deal intended to be flexible enough to protect it effectively against parallel imports. This arrangement was held to qualify neither for group exemption under 67/67 nor for individual exemption under Article 85(3) because the resale price controls were held neither of benefit to consumers nor indispensable to the establishment of the exclusive distribution arrangement.

In all these cases, the effect of the price restriction on trade between Member States was not seriously in doubt, so that the theoretical freedom for Member States still to operate their own national RPM legislation remained, In practice, the number of States which still treated resale price maintenance as lawful had shrunk by the start of the 1980s to Belgium, Holland, and Italy. France, Germany, and Luxembourg prohibited it, with minor exceptions, and there was very limited scope for its operation under the laws of the United Kingdom, Ireland, and Denmark. An interesting challenge to the Dutch law arose in the European Court case of *Officier van Justitie* v. *Van de Haar and Kaveka de Meern*.[15] On this Article 177 reference from the District Court of Utrecht, the defendant was a Dutch tobacco retailer charged with selling tobacco at prices below that fixed by the Tobacco Excise Act. Under this legislation, the price for tobacco is fixed by the manufacturer or importer and an official excise label showing this price is then affixed to the tobacco. The retailer is then required to sell the tobacco at only this figure. Competitive pricing is, therefore, impossible at the retail stage but remains possible at the intermediate

[14] [1981] 1 CMLR 601.
[15] Cases 177 and 178/82 [1984] ECR 1797: [1985] 2 CMLR 565.

stage, as a result of the possibility of obtaining various discounts and reductions, whilst market share can be sought competitively through promotion which emphasized product differentiation in quality and taste.

The argument raised by the retailer was that such legislation was in breach not only of Article 85 in that it fixed prices but also was contrary to the national obligation of The Netherlands under Articles 5 and 30. Issues raised by the latter allegation will be discussed in Chapter 23 but the Court gave a ruling relating to Article 85 which is relevant to the issue of resale prices; it held that, whilst a Member State was not entitled under the terms of the Treaty to enact legislation which would allow private undertakings to escape the constraints of Article 85, public legislation imposing fixed prices to be set by manufacturers or importers could not be attacked directly by individuals before national Courts, at least under the terms of this Article, although their position might be more favourable under other Articles.[16]

Books have always been considered as a unique product, and the arguments for according their retail pricing special treatment are at least better than those that can be put forward for most other products. The European Court has had an opportunity to rule on such arguments in two cases. The first, *VBVB and VBBB* v. *Commission* (n. 1), involved a horizontal agreement between two trade associations involved in the publishing of books in Dutch both in Holland and the Flemish-speaking part of Belgium. In both Belgium and Holland collective RPM was permitted under national law and the agreement under challenge by the Commission enabled each association to apply their rules directly against members of their own association and indirectly against members of the other association. Apart from rules compelling members of the association at any level to deal only with other members (upon penalty of exclusion from the trade), there were strict provisions for the collective enforcement of resale price maintenance. The Advocate-General, VerLoren van Themaat, gave a full review of the development of the competition policy of the Commission in relation to RPM. He characterized the agreement between the two associations as a corridor between the two national 'houses' or 'systems' of RPM. He accepted that the Commission had been wise in past years to adopt a 'wait and see' policy in respect of national systems of RPM, whilst it waited to see the effect of the elimination of transnational agreements restricting resale prices. He felt that it did not debar them now from proceeding against those transnational agreements which impeded imports or exports or from challenging clauses in solely national systems which could be shown, as in this case, to affect trade between Member States. This might involve restrictions on the freedom of

[16] On this issue see *Leclerc Books* case below. Case 229/83 [1985] ECR 1: 2 CMLR 286, discussed also in Ch. 23.

states to enact their own legislation on this topic, but this kind of restriction also existed in other areas of the Treaty, notably as the result of Articles 30 to 36 relating to the free movement of goods.

The European Court confirmed that the agreement between the associations violated Article 85(1) and was incapable of exemption by the Commission. It too emphasized that the Commission had challenged simply the transnational agreement, not the individual RPM systems adopted by Holland and Belgium. Neither national legislation nor judicial practices, even if common to many or all Member States, could prevail against the application of clear principles of EC competition policy. Even if the specific nature of books as an object of trade might justify special conditions for their distribution and pricing, the mere fact that these two large national associations of publishers and booksellers had extended the closely supervised rules and enforcement of RPM into a wider area constituted a sufficiently marked restriction of competition to justify the Community's refusal to give exemption. The fact that an RPM system might have the incidental effect of preventing unfair competition in the sale of books, particularly any sold as 'loss leaders', was an insufficient reason to refrain from applying Article 85 to a whole sector as the book trade. The Court did not in any case accept that the consequences to the book trade of preventing transnational agreements on RPM would be as serious as the two associations had suggested.

The actual right of a Member State to enact resale price maintenance laws relating to books was considered not many months later in the *Leclerc Books* case (European Court 1985). The French Book Prices Act of 1981 had required every publisher and importer of books to fix a public selling price for all the books which it published or imported. It also required retailers to charge an actual selling price of between 95 and 100 per cent of this fixed price.[17] If books were imported into France after being previously published in that country, the public selling price fixed by the importer must at least equal that fixed by the original publisher. Leclerc, retail establishments with a reputation for price cutting, were prevented by court action by other booksellers from selling books at prices below that set by this legislation; a question was referred to the European Court under Article 177 on whether the provisions of the Treaty, notably Articles 3(*f*), 5, and 85, should be interpreted as prohibiting the setting up in a Member State of an RPM system for books of the kind established in France. Objections were also raised under Article 30 whether the effect of the law was to restrict the free movement of goods. With regard to the retail price system, however, Advocate-General Darmon argued that the book sector undoubtedly constituted a special case, and that a Member

[17] There were limited exceptions for sales to schools and libraries.

State retained the right to enact legislation simply fixing a retail price of books by publisher or importer which would be binding on all retailers, provided that it did not conflict with any other of the specific provisions of the Treaty, particularly those relating to the free movement of goods.

On this basis, which was accepted by the Court, the basic law allowing RPM to books produced and sold in France was upheld, but some other portions of the law were struck down as contrary to Article 30 of the Treaty. These were provisions under which the importer responsible for carrying out the import formalities was responsible for fixing the retail price of the book, and also restrictions which required the selling price fixed by the publisher to be applied also to all books which had been first exported from France and then reimported. The law was, however, allowed to be applied to books exported in the opinion of the national court for the sole purpose of reimportation, in order to circumvent the legislation. The effect of the decision was to accept the right of a Member State to adopt a collective resale price maintenance system for books, which individual undertakings themselves would not be permitted to adopt by way of agreement or concerted practice.

The *Leclerc Books* decision raises some difficult questions so far unanswered.[18] The issue raised is whether the *Leclerc* decision indicates that a different test has to be applied under Article 85 to protect a purely, national RPM system as compared with the test to be applied under Article 30 under which (and without the benefit of any *de minimis* rule) quantitative restrictions on imports and all measures having equivalent effect are prohibited. Clearly, without the restriction on 'artificial reimporting', the French Prices Act would be substantially reduced in effect because of the ability of French publishers to export books and reimport books at lower prices. On the other hand, the judgment of the Court in *Leclerc* is apparently limited to the book publishing sector. For this sector at least it appears that the identity of the party implementing or enforcing the relevant measures is the critical factor in whether the Treaty permits the measures to operate. On the other hand, there do not appear to be such important reasons for defining 'imports' in Article 30 in a restricted fashion whereby such a necessary clearance for the French Book Prices Act effectively to cover 'artificial reimportation' is given, whereas 'trade' in Article 85 is defined so widely as to include reimports and thereby prevent such restrictions from being adopted by undertakings.

The Commission's concern, however, that resale price maintenance should not become an accepted practice in national markets even for books is shown by its subsequent refusal of exemption for the Net Book

[18] For analysis of other issues raised by the *Leclerc* cases, see Ch. 23, pp. 444–5.

Agreement in the United Kingdom.[19] This agreement is a collective agreement between publishers and booksellers under which, if a publisher gives a book a fixed resale price (known as its net price), the bookseller would resell it only at that price save in special circumstances.[20] The argument in favour of the maintenance of the agreement was the familiar one that it was necessary for the important cultural reasons referred to in the earlier cases, in particular for the protection of small bookshops unable to afford the discounting of 'bestseller' titles and to encourage the continued publishing of books with a small print run. It was claimed that there were practical difficulties and administrative burdens on both publisher and bookseller in making the same arrangements by way of a network of individual agreements between them. The Commission ruled, however, that a collective RPM system was unduly restrictive of competition even after taking into account the cultural objectives which the association relied upon. It referred to the *VBBB* case as authority for this approach and stated that administrative difficulties in implementing individual price maintenance could not justify the maintenance of such a highly restrictive collective system. In 1992 the Commission's decision was upheld by the Court of First Instance, which means that the agreement, insofar as it affects trade between Member States, can survive only if the European Court overrules it on appeal now pending.

2. Discrimination between Buyers

In vertical relationships, that is those between a supplier and buyer or potential buyer, Article 85 may, of course, be applicable if an agreement or concerted practice can be found, but as compared with RPM there may be real difficulty in establishing that such an agreement or concerted practice exists rather than simply a unilateral decision taken by the supplier. The situation usually arises when a single undertaking decides to use discrimination as a means of imposing different trading conditions (including pricing terms) on its distributors, wholesalers, and dealers in order to achieve total or partial insulation for individual national markets within the Community from each other. The methods used will include a variety of territorial and customer restrictions of the kinds already discussed in Chapter 9. Thus, for example, *Pittsburgh Corning Europe*[21] (Commission 1973) provides an example of an attempt by a manufacturer using customer restrictions to discriminate in price between two separate markets, namely the expensive German market and the low price Belgium

[19] *Re Publishers Association* [1989] 4 CMLR 825.

[20] For example sales to book agents, libraries, or in the annual sale of remaindered volumes.

[21] [1973] CMLR D2.

and Holland markets in respect of its supplies of cellular glass. The contract restrictions imposed on the Belgian and Dutch distributors requiring them to name the sites to which the glass was being delivered in order to receive rebate on the goods supplied illustrated the nature of the tactics that a supplier might adopt to enforce such a policy. We find numerous other examples in Commission and European Court cases of price discrimination as a method of keeping markets separate to protect profit margins in the higher price territory. It was clear, for example, in the *Kodak*[22] case that the agreements between Kodak and its various subsidiaries could not themselves be in violation of Article 85, since all the companies were part of the same large group; but the standard terms and conditions of sale attempting to maintain separate resale price levels in the different countries in which they operated were treated as agreements between each of the individual Kodak companies participating and their purchasers. Negative clearance for these conditions of sale was available, therefore, only when the discriminatory pricing clauses were removed, since clearly they would have had an effect on the prices paid by buyers in the various Member States. The Commission has not hesitated to apply Article 85 to such terms and conditions of sale, even though the view has been taken that Article 85 should apply only to contracts which limit one or more parties in their freedom to determine prices or trading conditions in other contracts. This would have meant that the terms upon which Kodak subsidiaries sold in Member States would not themselves have been subject to Article 85. Such an interpretation of Article 85, however, would not have assisted the Commission in its task of preventing individual discriminatory practices which had the effect of insulating the markets in individual Member States from each other.

Another reason for attempted discrimination in price in the supply of the same goods has not been simply the protection of profit margins in individual Member States which can be obtained either by reason of substantial demand or higher living standards, but rather the necessity of protecting a market that might otherwise cease to exist at all. The argument of the *Distillers* company in the 1980 European Court case[23] falls into this category; the argument raised by Distillers was that it could not continue to market Johnny Walker 'Red Label' whisky in Europe without being able to deduct from it the promotional costs that its sole distributors in Europe would incur. Its distributors in the United Kingdom operating in a relatively mature market did not need to incur heavy promotional costs, whereas in Europe whisky was alleged to compete against local spirits, such as various types of brandy, which received more favourable national fiscal treatment. This was the unsuccessful justification put forward for

[22] [1970] CMLR D19. See Ch. 13, p. 248.
[23] Case 30/78 [1980] ECR 2229: 3 CMLR 121.

requiring UK purchasers acquiring whisky for export to pay the additional sum of £5.20 per case.

Price discrimination has likewise played a major part in the important *Ford*[24] cases involving the pricing of Ford cars in Germany and other parts of the Community. The fact that the prices charged in the United Kingdom were substantially higher than those on the Continent prompted a number of British consumers to go to Germany and purchase cars there for import into the United Kingdom. Ford Europe, therefore, notified its German distributors that from 1 May 1982 it would no longer accept orders for right-hand-drive cars (except from military personnel serving in Germany) since all such cars in future would have to be purchased from Ford dealers established in the United Kingdom. The Commission having attempted unsuccessfully to have this circular withdrawn as an interim measure ultimately took a formal decision to withhold exemption under Article 85(3) for Ford's distribution system. This substantive decision was upheld by the Court largely on the basis that deliberate price discrimination between separate national markets invalidated selective distribution systems. The problem of finding in this situation an agreement upon which Article 85 could bite was surmounted by the finding that Ford dealers must be deemed to have accepted that, within the framework of the distribution agreement, Ford's policy on such matters as the choice of vehicles to be supplied was itself part of the distribution arrangements. Dealers must be deemed to have consented to all such policy decisions by Ford as if they had formed part of the original agreement.

Finally, we should note that Article 7 of the Treaty states that: '. . . any discrimination on grounds of nationality shall be prohibited.' This applies both to horizontal agreements to discriminate and to the vertical situation where goods are to be sold at different prices in different countries (after allowance for transport and other specific costs). This is supported by the wording of Article 85(1)(*d*) stating that the Article applies when undertakings 'apply dissimilar conditions to equivalent transactions with other trading parties thereby placing them at a competitive disadvantage'.

3. Refusal to Deal or to Supply

The individual refusal to deal in a vertical situation is dealt with under the cases decided on Article 86. A simple individual refusal to deal will

[24] Cases 25 and 26/84 [1985] ECR 2725: 3 CMLR 528, and also earlier citations to be found in that report, dealing with the Commission's unsuccessful attempt to impose an interim ban on Ford of Europe, Inc. and Ford Werke (its German subsidiary) from refusing to supply right-hand-drive cars to its German dealers for resale to UK purchasers for importation into the UK.

normally not, on the other hand, be easily brought within Article 85, again because of the problem of establishing an agreement or concerted practice to that effect when a single supplier is refusing supplies, regardless of the fact that the motivation may be anti-competitive. Article 85 does, of course, reach collective refusals to deal but these are more often found between members of the same trade association who bind themselves not to deal with certain classes of customers or with any individual customers who do not comply with the rules of the association.[25] The parties will often cite the need to maintain trading standards or quality of workmanship as the ground. The stringent examination by DG IV of the rules of such associations often indicates that these alleged justifications for the refusal to deal are extremely weak and could be achieved by means less restrictive of competition. A recent example of this attitude is the *Anseau-Navewa*[26] case where, in an agreement that had both horizontal and vertical aspects, washing machines were prevented from installation in Belgium unless they had a certificate of conformity. This certificate was available only from an association which restricted its membership to manufacturers of such machines and sole importers of foreign manufactured machines. The rules of the association, ostensibly concerned to protect public health, were in fact principally designed to make life more difficult for parallel importers, with whom no distributor of washing machines would deal unless the importer had first obtained an individual certificate of conformity for each machine involved. This, of course, meant that the position of parallel importers of foreign machines became uneconomic, as was intended; and the agreements in question, including the refusal to deal, were found by the Court in breach of Article 85.

Since, however, selective distribution systems are outside the range of Article 85 if they comply with the requirements set out in the *Metro/Saba* (no. 1) case involving qualitative criteria only in the selection of their dealers, the inherent nature of selectivity means that some refusals to deal are acceptable and will not invalidate the system so as to bring it back within the range of the Article. Inherent in the system is that a manufacturer or wholesaler or retailer cannot refuse to deal with qualified applicants and that the criteria must be applied without discrimination to all qualified applicants. Moreover, as we have seen, the criteria need not be applied in many cases by the manufacturer itself but can be applied at a lower level, namely by a wholesaler or sole distributor and normally within a short time limit as in the *IBM Personal Computers*[27] and

[25] As for example in the already mentioned *ASPA* case [1970] CMLR D12 and *Belgian Wallpapers* Case 73/74 [1975] ECR 1491: [1976] 1 CMLR 589.

[26] Cases 96–102/82 reported before the European Court as *IAZ International Belgium* v. *Commission* [1983] ECR 3369: [1984] 3 CMLR 276.

[27] [1984] 2 CMLR 342.

Grundig[28] cases. An interesting and not finally decided issue is the effect of a refusal to deal with a qualified applicant who has been admitted to the distribution system. If the refusal to deal is part of a deliberate policy by a manufacturer to introduce into the distribution system a discouragement to certain outlets which it feels are too inclined to cut prices or in other ways to depart from the manufacturer's policy, then if it is likely that the refusal to deal is more than an isolated incident the Commission may be justified in withdrawing any negative clearance or exemption given as in the *AEG/Telefunken* case. The case of *Demo-Studio Schmidt* v. *Commission*[29] (European Court 1983) deals with the rights of applicants to be appointed as selective distributors. The facts of the case were that Schmidt, a designer in a German machinery factory, decided to set up his own electronic equipment business as a sideline to his salaried activities. He opened a small shop in Wiesbaden with opening hours from 3.45 to 6 p.m. and Saturday mornings, selling a variety of consumer electronics made by Revox. Revox later discussed Schmidt's request to be supplied with its Series B products which were its more expensive items and afforded a higher profit margin and made it clear to him that any supply would be conditional on his sales premises remaining open for the whole of the working day. Schmidt said he would take a salesman on for this purpose. Later, however, Revox told Schmidt that they had decided that he could not be appointed a specialist retailer entitled to deal in Series B equipment because he did not satisfy the conditions laid down by them which related to the technical qualifications of his staff and the design of his sales area, as well as the observance of normal opening hours.

Revox having refused supplies of Series B equipment to him, Schmidt complained to the Commission who rejected his claim saying that it had no grounds under Article 85 to order Revox to supply him with products. The European Court was asked by Schmidt to annul this decision but likewise rejected his claim. Their view was that if a selective distribution system were properly adopted for highly technical electronic goods, it was a reasonable requirement of the manufacturer that retailers should be open for normal shopping hours and that supplies could be refused to a part-time establishment. Even if the selective distribution system had elements which made it incompatible with Article 85(1), this would not of itself entitle Schmidt to be appointed a dealer by Revox or to force Revox to supply any particular range of products to him. The effect of the case, therefore, does not seem to be to give a broad right of refusal to a manufacturer not to deal with any particular applicant (whether or not he satisfies the necessary criteria laid down for the relevant distribution system) but places a duty on the manufacturer (or distributor with delegated

[28] 15th Annual Report, pp. 66–7.
[29] Case 210/81 [1983] ECR 3045: [1984] 1 CMLR 63.

responsibility) fairly to assess the criteria in the case of each applicant, taking the facts as they were at the time when the application was made. What is, however, clear is that a manufacturer cannot impose an obligation on a dealer not to deal with an individual qualified applicant who for some reason has not been admitted to the distribution system, since that choice must be left to be made objectively by the dealer itself.

15

Intellectual Property Rights: Licensing

1. Patent Licences: Some Early Cases

The protection by national law of different types of intellectual property is common to all Member States. The range of that legislation, however, and the classification of the rights thus protected varies considerably between them: these rights include patents, registered designs, trade marks, and copyright together with various other minor or specialized categories such as plant breeders' rights. The methods of protection adopted themselves differ; some can be protected only by registration, after a more or less rigorous process of screening for acceptance, as with patents, trade marks, and registered designs. In other cases, rights can simply be acquired as the result of publication, as with artistic or literary copyright which is protected both by national legislation and a number of international conventions. Such proprietary rights can protect technical information of importance and value for commercial and industrial activities or, alternatively, may simply protect the identity and business individuality of an undertaking, as in the case of business names and trade marks.

There is an inherent conflict between, on the one hand, the existence and exercise of any such rights, which will necessarily give a degree of exclusivity and protection to their owners, and on the other the general aims of competition policy as embodied in Articles 85 and 86. In this chapter, we shall review the way in which the Commission in applying these Articles and Regulations made under the Treaty has dealt with agreements that relate to the use and exploitation of industrial property rights, in particular patent rights, to which most of its attention in this area has been devoted. There is, however, also a further limitation to the exploitation and enjoyment of industrial property rights, arising from the very nature of the European Community as described in the Treaty. This is the restraint placed upon owners of such rights in Member States to prevent their use in such a way that hinders the free movement of goods and services within the Community. Articles 30 and 34 of the Treaty set out the basic prohibitions on quantitative restrictions on imports and exports, and all measures having equivalent effect (a number of which directly arise as a consequence of the protection of industrial and commercial property), whilst Article 36 *inter alia* preserves the right of a Member State nevertheless to impose such prohibitions or restrictions on imports and exports required to protect industrial and commercial property, so long as those restrictions or prohibitions are neither a means of arbitrary discrimination nor a disguised restriction on trade between Member States. The exact interpretation of these concepts, which have been considered by the European Court in a large number of cases, will be considered in Chapter 16.

The owner of intellectual property rights naturally desires the greatest possible freedom to negotiate the basis upon which it can arrange their exploitation. Whilst, however, the manufacturer of goods will select an appropriate system of distribution through which its sales can be maximized, in the exploitation of intellectual property, on the other hand, the owner normally finds that it is best to part with only a limited interest in the asset, giving a licence to a third party for its use, whilst still retaining ownership of the asset.

The owner of such an intellectual property right has a wide variety of choices over the exploitation of its asset. A patent can, of course, be used simply to assist the owner's own manufacturing activities, a defensive shield preventing others from making use of the data comprised within its protection. Provided that the owner itself is making adequate use in its own processes of the information, then it is highly unlikely that under any national legislation licences for use of that right will be compulsorily issued, since this procedure can be adopted normally only when a patent owner is conspicuously failing to make adequate use of its patent.[1] On many occasions, however, merely the defensive use of the right will prove insufficiently profitable to its owner, who may then wish to obtain additional revenue by licensing one or more other undertakings to pay to utilize the information or process for their own purposes.

Three kinds of licence are found. The simplest is the *non-exclusive* licence, where none of the licensees receives any guarantee from the licensor as to the exclusivity of the rights provided which the owner remains free to grant to a number of other undertakings as well. Clearly, the value of each such licence is thereby reduced and the percentage payable by way of royalty normally less than for an equivalent exclusive licence. The *exclusive* licence, an equally common method of exploitation of industrial property, contains the same two-way bargain found in exclusive distribution arrangements. The owner agrees that it will not itself license any undertaking within the particular territory apart from the chosen exclusive licensee and also agrees that it will itself not exploit that particular right in that territory. An exclusive licence where the owner of

[1] For the legal position in the UK, see Sections 48–59 of the Patents Act 1977. Compulsory licensing of registered design rights and copyright may now also be required in certain circumstances. See sections 144 and 237–9 of the Copyright Designs and Patents Act 1988.

the right reserves for itself the right to compete in the licensed territory, but must refrain from licensing any third parties situate there, is known as a *sole* licence.

In spite of predictions from some sources that the national patent systems in Europe would become obsolete as a result of the ever quickening pace of new technical developments, arriving at both their high point of value and subsequent obsolescence more quickly than could be adequately coped with by these relatively cumbersome national patent systems, a large number of patents have continued to be granted both in Member States and since 1975 by the European Patent Office in Munich on behalf of the individual Member States. From the viewpoint of DG IV, therefore, a method of controlling and dealing with the large number of patent licences became an important element in early competition policy. Many of the problems which it faced were similar to those which arose as the result of the very large number of exclusive distribution arrangements notified. A substantial number of such individual agreements might fall within the terms of Article 85(1) and would, therefore, need consideration under Article 85(3) with a view either to the grant of exemption or its refusal. This would mean in turn an early demand from industrial sources for a group exemption in order to avoid the delay inherent in the need for individual consideration of a very large number of patent licence agreements, likely in number to exceed all other categories of agreement apart from exclusive distribution arrangements.

The same problem, however, confronted the Commission as with exclusive distribution. Group exemptions could not be provided, both as a matter of good administration and also under the express terms of Regulation 19/65,[2] until there was sufficient experience of individual case-law. This has meant in practice that the issuing of a group exemption covering patents has taken an extremely long time and was not implemented until well over twenty years after the adoption of Regulation 17. It would, however, be unfair to DG IV not to point out at this stage that, by comparison with exclusive distribution and purchasing, patents and other intellectual property rights and their licensing present considerably more complex issues. This is both because of their individual protection under national legislation and also because by their very nature a greater degree of exclusivity is involved, which in turn requires the consideration of a larger variety of restrictions to be placed upon licensees.[3] Given, therefore,

[2] The relevant recital reads 'Whereas it should be laid down under what conditions the Commission . . . may exercise such powers after sufficient experience has been gained in the light of individual decisions and it becomes possible to define categories of agreement and concerted practices in respect of which the conditions of Article 85(3) may be considered as being fulfilled'.

[3] This is clearly illustrated by the considerably greater detail and complexity of the Patent

the inherent tension between the exclusivity of the national laws on the one hand with the requirements on the other of Articles 85 and 86, it was not surprising that the problem of reconciliation took a considerable period of time.[4]

Amongst the very large number of agreements notified to the Commission at the end of 1962 and the beginning of 1963 under the provisions of Regulation 17 were many thousands of such licences. Considerable prior discussion had already taken place in DG IV as to the basis upon which patent licences would be dealt with. It was from the start accepted that a certain number of restrictions would inevitably be imposed by licensors on their licensees simply as the direct result of the inherent nature of patents, and of the fact that, in granting a licence, the licensor is not disposing of its entire asset but only parting with a limited interest whose bounds have to be carefully delineated. A difficult policy choice, which the Commission could not avoid, was the need to distinguish between those restrictions which merely reflected this right of the patentee to spell out the extent of the licence, as opposed to other restrictions which would actually restrict competition between the licensee and other licensees in respect of specific territories and in particular would interfere with the basic principle of the need for integration of the Common Market itself by the free movement of goods across national boundaries.

A few weeks before the closing date for the notification of bilateral agreements under Regulation 17 came the publication of the 'Christmas Message' on 24 December 1962, the first public statement by the Commission of the approach that it would adopt to patent licences. This Notice itself contained four sections. The first set out the list of clauses which it regarded as falling outside Article 85(1). These were:

— obligations imposed on the licensee having as their objects either limitation of the exploitation of the invention to manufacture, use, or sale, and limitations to certain fields of use;
— limitations imposed also to the quantity of products to be manufactured, the period and territorial area for which they were permitted to be used, and limitations on the licensee's powers of disposal, e.g. on sub-licensing;
— obligations as to marking of the product with a patent number and the observance of quality standards necessary for the technically perfect

Group Exemption Reg. 2349/84 compared with those on Exclusive Distribution (Reg. 1983/83) or Exclusive Purchasing (Reg. 1984/83).

[4] This does not mean, however, that the group exemption need ultimately have taken as long as it actually did, namely some eight years (1977–84) from the publication of the first draft to the implementation of the exemption.

exploitation of the patent, also non-exclusive improvement undertakings on a reciprocal basis between licensor and licensee;
— undertakings to ensure that the exclusivity of the grant to the licensee was protected.

The second section confirmed that all other clauses would be looked at on their individual merits and that in any case a general 'appraisal' was not possible for agreements relating to joint ownership of patents, reciprocal licensing, or multiple parallel licensing, or to any clauses which purported to extend beyond the period of validity of the relevant patent. The third section of the Notice confirmed that patent licences containing only those clauses referred to in the first part of the Notice did not require to be notified. The fourth stated that the basis of the Notice was that the clauses referred to 'entail only the partial maintenance of the right of prohibition contained in the patentee's exclusive right in relation to the licensee who in other respects is authorised to exploit the invention. The list . . . is not an exhaustive definition of the rights conferred by the patent'. After explaining the justification on this basis for a number of the other clauses referred to in the first part, the Notice concludes

By the undertaking . . . not to authorise anyone else to exploit the invention . . . the licensor forfeits the right to make agreements with other applicants for a licence. Leaving out of account the controversial question of whether such exclusive undertaking has the object or effect of restricting competition, they are not likely to affect trade between Member States as things stand in the Community at present.

When first published, this Notice received little adverse comment especially as the approach which it had adopted corresponded generally to current attitudes in Member States. The acceptance of the principle of exclusive licensing of patents was strongly influenced by the attitude of German law, whose statute against restraints of competition had adopted a relatively mild approach to such exclusivity. Moreover, the Commission had by this time only very limited case-law experience and was chiefly preoccupied with dealing with notifications relating to exclusive distribution agreements. If a more rigorous policy had been adopted from the outset towards patent licences, resources would have had to have been diverted from this other task. Whatever the reasons, no cases appear to have been dealt with formally until 1971, except for two cases in 1966 and 1970 respectively which are referred to in the First Annual Report.[5] The first of these involved the challenge by the Commission (but without formal decision) to clauses included in patent sub-licences granted by two undertakings which had cross-licensed each other, but on terms that sub-

[5] pp. 70–1. The relevant section of the German law (Gesetz gegen Wettwerbsbeschränkungen) is Section 20.

licensees would have to buy unpatented material from one or other of the sub-licensors. The Commission took the view that this was the case of an illicit extension of a patent monopoly in tying the unpatented material to the use of the patented process and not essential for the 'technically perfect application' of such patented processes as required under the 1962 Notice. In the 1970 case, the Commission had raised objections about the terms of a licensing agreement between a producer outside the Community and a licensee one part of whose territory was situate within it. The provisions of this licence agreement included restrictions on production and sale of patented products for a period beyond the term of the validity of the licensing agreement and of the underlying patent itself.

Nevertheless, in spite of the lack of actual decisions, a distinct change of policy occurred over the latter part of the 1960s. By the end of the decade, DG IV no longer regarded an exclusive patent licence as prima facie outside the scope of Article 85(1) but was already starting to take a stricter view, namely that exclusivity had to be justified by the particular circumstances of the licensing arrangements. A number of suggestions have been made as to the reason for this change in policy, including the influence of *Grundig–Consten*[6] decided by the European Court in 1966. Under this argument, the distinction drawn in that case between the existence of a right of intellectual property, which is protected by Article 36 of the Treaty, as opposed to the exercise of that right which is subject not only to Article 30 but also to Article 85, led inevitably to a more rigorous examination of the justification for exclusivity of licensing. The alternative argument advanced for the change in policy was that the Commission was influenced by cases decided by the European Court relating to the interpretation of Articles 30 to 36, as applied to the free movement of goods and the broad interpretation placed by the Court on Article 30, as compared with the narrower approach adopted to the limitations and exceptions set out in Article 36.[7]

Neither of the two suggestions is, however, completely convincing. Whilst undoubtedly the *Grundig–Consten* case, as the leading influence on policy within DG IV during the 1960s, influenced the thinking of officials and ensured the pre-eminence of the principle of market integration, more than this would have been necessary to effect the major shift apparent from the original Christmas Message of late 1962 to the position adopted in the cases decided in 1971. Moreover, nothing in the judgment leads inevitably to a conclusion that exclusive patent licences fall inherently within the scope of Article 85(1); the strictures of the Court apply specifically to such uses of those rights as may defeat the effectiveness of

[6] Cases 56 and 58/64 [1966] ECR 299: CMLR 418. For the facts of the case see p. 52.

[7] See L. Gormley, *Prohibiting Restrictions on Trade within the EEC* (Amsterdam: North-Holland, 1985), esp. pp. 184–9 and 230–3.

the Community law on restrictive practices. It can be argued that the limitation placed by the *Grundig* decision on the use of intellectual property rights, namely in this case trade mark rights, to make it hard for distributors to effect parallel imports was indeed covered by the terms of the Christmas Message, and that the case alone could not be the foundation for such a complete change of attitude to exclusivity in patent licences.[8]

On the other hand, the argument that the Commission was principally influenced towards a change of attitude by cases decided by the European Court under Articles 30 to 36 has its own weakness, namely that the major decisions of the Court in this area did not commence until the early 1970s by which time DG IV's change of policy had already occurred, as can be clearly discerned in the December 1971 decisions by the Commission in *Burroughs–Delplanque* and *Burroughs–Geha.*[9] It is clear that, despite the central place of the principle of the free movement of goods within the structure of the Treaty, it was not until the *Dassonville*[10] case of 1974 that the Court gave detailed consideration to the interpretation of Article 30 relating to measures having effect equivalent to quantitative restrictions, in which it delivered its seminal ruling that all trading rules enacted by Member States capable of hindering, directly or indirectly, actually or potentially, intra-Community trade were to be considered as measures having an effect equivalent to quantitative restriction.

Although the influence of this and later cases on such issues would undoubtedly themselves have had an effect on the development of DG IV's policy relating to intellectual property rights, this can have applied only after the date of publication of the decisions in such cases and could not have had relevance to the initial change of policy at the very end of the 1960s. We are, therefore, thrown back upon a need to find other explanations for the change of policy. It would surely be reasonable to conclude that it occurred primarily because of the weight of experience accumulating within DG IV as the result of the examination of a large number of patent agreement notifications during the six or seven years following Regulation 17's early implementation at the start of 1963. Some evidence of this is the very full treatment given to this topic in the First Annual

[8] While the subsequent European Court decision in *Parke, Davis* v. *Prober* Case 24/67 [1968] ECR 55: CMLR 47, confirmed the application of *Grundig–Consten* to patents, it took the issue of exclusivity no further. See Ch. 16, p. 325.

[9] *Burroughs–Delplanque* [1972] CMLR D67 and *Burroughs–Geha* [1972] CMLR D72. It is, of course, also true that the Court had in June 1971 in *DGG* v. *Metro* Case 78/70 [1971] ECR 487: CMLR 631, a case involving copyright, given some indication as to its interpretation of Articles 30 to 36 as applied to the exclusivity involved in intellectual property rights but the significance of the judgment in the context of Article 85 was not fully appreciated for some time afterwards. See Ch. 16, pp. 326–7.

[10] Case 8/74 [1974] ECR 837: 2 CMLR 436. This principle was subsequently elaborated in many other Court decisions on Articles 30 to 36, notably *Rewe* v. *Bundesmonopolverwaltung* (*Cassis de Dijon*) Case 120/78 [1979] ECR 649: 3 CMLR 494.

Report of the Commission.[11] It was clearly felt at that time that an explanation was needed of the course of development of policy relating to patent licences since the Christmas Message, and an account is given of the examination of some 500 of such licences. The Commission explained that these agreements were analysed in order to determine which clauses most often occurred, also to ascertain which of the restrictions imposed were made essential either by the inherent nature conferred by national patent rights, or in order to maintain the secrecy of related know-how, or which for other reasons did not have the effect of restricting competition. This analysis was stated to be in the process of utilization for the possible preparation of a group exemption covering two-party agreements such as patent licences, since this was permitted by the terms of the Council Regulation 19/65 which had also given the Commission the necessary powers to introduce the existing group exemption for exclusive distribution agreements.

This change of approach to exclusivity of patent rights, however, had been found initially in the *Burroughs* cases. Burroughs, a US company, controlled patents for the manufacture of plasticized carbon paper, which is more expensive than ordinary carbon paper. Other manufacturers were also in this market in Europe, and there was considerable competition in its production. The Swiss subsidiary of the US parent licensed Delplanque as the exclusive licensee to manufacture in France and to sell this product non-exclusively throughout the European Community and in various other territories and to use on a non-exclusive basis certain trade marks relating to the product in all the territories licensed, except for France. The agreement was to be for a ten-year period, and Delplanque was also required to keep secret the confidential know-how supplied under it for a period of ten years after its termination. Delplanque had no right to grant sub-licences, and its market share of the French carbon paper market at that time was only approximately 10 per cent. The Commission, however, in the decision included the following statement which indicated its shift on the position of exclusive rights as such:

> A patent confers on its holder the exclusive right to manufacture the products which are the subject of the invention. The holder may cede, by licences, for a given territory, the use of the rights derived from its patent. However, if it undertakes to limit the exploitation of its exclusive right to a single undertaking in a territory and thus confers on that single undertaking the right to exploit the invention and to prevent other undertakings from using it, it thus loses the power to contract with other applicants for a licence. In certain cases, the exclusive character of a manufacturing licence may restrict competition and be covered by the prohibition set out in Article 85(1).[12]

[11] This report appeared in 1972 and would cover the work of the Directorate up to the end of 1971. The relevant pages are 65–74. [12] [1972] CMLR D70.

Further clarification of the Commission's policy came the following year in the *Davidson Rubber*[13] case. Davidson was a United States company which had patented a method of manufacturing padded elbow rests and seat cushions for cars and had also acquired considerable know-how about the most effective and economical way of applying these techniques in large-scale production. It granted licences covering the relevant patents and know-how to several countries within the Common Market. All these licences were on an exclusive basis, and the licensees agreed to exchange information with each other and with Davidson that might assist in the manufacturing process and also to cross-licence patents covering any improvements which they might discover that were of a patentable nature. Sub-licensing was allowed only with the consent of Davidson, but prohibitions on exporting the products were removed during negotiations at the request of the Commission. Licensees delivered the products direct to the car industry and the total aggregate output of the licensees represented approximately 20 per cent and 40 per cent of the total production in Germany and France respectively.

The Commission took the view that such exclusive manufacturing and sales rights conferred by Davidson on its licensees were caught by the terms of Article 85(1), since curtailing the number of suppliers produced a noticeable effect on the competitive process within the Common Market and the development of trade between Member States. The substantial market share which Davidson's licensees accounted for within the EC made such a conclusion inevitable. The Commission did, however, accept that the restraint of competition which derived from Davidson's policy of exploiting its inventions through exclusive licensees in those countries which had large motor industries was a restriction without which the commercial objectives of patent licences and know-how agreements could not be achieved. The licensees would not have agreed to invest substantially in the new process if they could not have been sure of exclusivity within their particular territory. The fact that the car manufacturers themselves accounted for a substantial part of their needs from their own resources and had a strong bargaining position as Davidson's licensees, therefore, meant that the agreement did not have the effect of eliminating full competition from the relevant market and could, therefore, fall within the terms of the exemptions contained in Article 85(3). Moreover, the licensees themselves were entitled to compete with each other, not being limited to making sales in their own territory.

The growing importance of intellectual property rights to DG IV was marked in 1973 by the establishment of a new division that specialized in this subject. Although the number of published decisions remained small,

[13] [1972] CMLR D52.

work now began on developing detailed lines of policy which would eventually lead to the issuing of the first drafts of group exemptions for patent licences. In the *Raymond/Nagoya Rubber Company*[14] case, also in 1972, negative clearance had been granted to an exclusive patent licence given by the German subsidiary of the French company, Raymond, to the Japanese company, Nagoya. The terms of the licence included the right to manufacture and sell the plastic attachments used in the construction of cars, but in this case the Commission concluded that the export ban imposed on Nagoya against selling within the Common Market had no effect on competition within the Market; because of the nature of the items concerned they were manufactured only to special order by the car manufacturers, and it was most unlikely that such special orders would be placed for supply in Europe. Therefore, the exclusive licence permitting Nagoya to incorporate their works into Japanese vehicles and their spare parts for export anywhere in the world could not affect the competitive structure in the Community.

This development of policy continued with an increasingly critical assessment being given to claims of justification for exclusivity and in a more rigorous examination of ancillary restrictions placed on the parties in such cases. In *Kabel -und Metallwerke Neumeyer (KMN)/Luchaire.*[15] KMN was a German company which had developed a range of special techniques for the machining of steel parts, protected by patents and know-how, and used for a wide variety of items including pistons, axles, spindles, and cogwheels. Luchaire had been granted an exclusive licence for France, under which KMN provided all the necessary technical assistance in return for royalties on the product produced with the use of the information. Luchaire agreed not to sell products made under the agreement outside the EC and Spain and Portugal, though the restriction did not apply to exports that merely incorporated parts made under the agreement, as components in a larger assembly. KMN agreed not to grant any licences on more favourable terms to any other undertaking elsewhere in the world. From the agreement were deleted obligations on Luchaire to transfer to KMN ownership of improvements whether patentable or not which it might develop over the period of the agreement, an obligation not to sell the goods in question within the EC, and an obligation not to contest the validity of the patented processes; with these deletions, an exemption was given under Article 85(3). The Commission stressed in its decision, however, that the restrictions on KMN not to appoint any other licensees in France, not being themselves of the essence of a patent, fell clearly within the restrictions on competition covered by Article 85(1).

[14] [1972] CMLR D45. Nagoya, controlled by Toyota, was at the time the largest subcontractor in the Japanese motor industry. See also the 2nd Annual Report, pp. 52, 53.

[15] [1975] 2 CMLR D40.

In another decision issued only a week later,[16] the Commission actually refused exemption to a patent licence agreement between a Dutch company covering the installation of a well-point drainage system of importance to major drainage projects by its method of linking filter tubes to pumps. Three licensees had been authorized and a further company, ZN Bronbemaling, then applied for its own licence but was refused because of opposition by a majority of the existing licensees whose consent was essential under the terms of their agreement. The significance of the process was that, being highly capital intensive, it was particularly suitable for major drainage projects by public authorities. The effect of refusing the licence was to limit the number of companies capable of tendering for such contracts. The Commission found that the clause did not satisfy the necessary condition of promoting technical or economic progress, and indeed possibly impeded it by restricting the number of companies able to tender for public contracts, and in particular companies from outside Holland.

The growing awareness of the Commission, however, of the scope for application of Article 85(1) to exclusive patent licensing is perhaps most clearly shown by its decision in the *AOIP* v. *Beyrard*[17] case. Beyrard was a French inventor resident in Paris who granted an exclusive patent licence to AOIP to manufacture and market in France and certain foreign French territories rheostats and speed changers for electric motors. In the decision, the principle of exclusivity itself was challenged. The clauses struck down by the Commission as in breach of 85(1), and not capable of exemption under Article 85(3), included the exclusivity of the rights granted and also an export prohibition applying to any country where Beyrard had either licensed his patent or assigned his rights to third parties. Among other clauses struck down were a no-challenge clause and an obligation to pay royalties during the lifetime of the most recent original or improvement patent whether or not the patent was being exploited by the licensee. Finally, there was an obligation on the licensee to refrain from any kind of competition with the licensor in the field covered by the agreement. If the licensor invested new processes or devices in the field (even if based on different principles but which could be used for the purpose of the same end) use such processes and devices automatically fell within the scope of the agreement. If the licensee on the other hand made any such devices by processes other than those of the licensor, he still had to pay royalties on them. All these restrictions were treated as inherently incapable of receiving exemption.

[16] *Zuid-Nederlandsche Bronbemaling en Grondboringen* v. *Heidemaatschappij Beheer* [1975] 2 CMLR D67.

[17] [1976] 1 CMLR D14.

2. The Evolution of the Group Exemption (1976 to 1985)

The distance which Commission policy had now travelled from that set out in the 1962 Christmas Message was so great that some official identification of the content of patent licences acceptable to the Commission had clearly become necessary. The principal point of difficulty was the extent to which the licensor could establish a group of exclusive licensees within the Common Market with territorial protection from each other and whether this territorial protection could be made complete or would be declared invalid to the extent that it impeded the movement of parallel imports under the *Grundig–Consten* principle. In the view of the Commission, all such arrangements involving even a limited degree of exclusivity had to be considered under Article 85(3), thus raising the possibility in every case that a decision rejecting the claim for exemption would be given.[18]

This unsatisfactory uncertainty about the status of exclusive licences could not continue indefinitely and, from 1976 onwards, one finds reference in the Annual Reports of the Commission to the gradual progress of draft group exemptions on this subject. The process, however, proved even more protracted than in the case of other group exemptions. The Sixth Annual Report refers to a conference held between the Commission and governmental experts in December 1974 to discuss the main problems arising in this area.[19] The draft Regulation was presented to the Advisory Committee on restrictive practices and dominant positions for discussion in December 1976; these discussions continued throughout 1977. An amended draft Regulation was then submitted at the end of that year for consideration by the Member States, which was made public for consultative purposes. In this early draft, the inclusion of any of the following clauses, amongst others, are indicated as likely to prevent a group exemption from being available:

(1) no-challenge clauses debarring the licensee from challenging the relevant patent;

(2) non-competition clauses between the licensor and licensee in respect of matters not directly covered by the relevant patents and know-how;

[18] A problem raised for the Commission was that the 1975 Luxembourg Patent Convention included an express provision (Article 43, para. 2) that the exclusive territorial licence for a part of the Community formed part of the Community's patent regime and was, therefore, immune from Articles 85 and 86. The Commission never accepted that this was a correct statement of the legal position. The Convention, however, which would enable the granting of a truly Community-wide patent is not yet in force, owing to the fact that it has not yet been ratified by all the Member States of the Community. See Ch. 16, p. 322–3.

[19] At p. 17.

(3) extension of agreements beyond the life of the most recent patents covered by the agreement unless the agreement is terminable by the parties on reasonably short notice; also forbidden would be a requirement to pay royalties after the expiry of the last patent;

(4) agreements requiring purchase of unpatented supplies for use with the patented process;

(5) restrictions on the manner in which the product made with the licensed patent is sold, e.g. in small tubes or large bottles and in what kinds of packaging;

(6) exclusivity of sales and export restrictions exceeding the period necessary for a new product having to penetrate its market. The amended draft had introduced the condition that the combined licensed territory for individual licensees (or licensees within the same group) was of a maximum population of one hundred million. Conditions imposed for allowing prohibitions on export of licensed products outside the licensed territory were that the patented products are available for sale by independent third parties on the open market by way of parallel imports and that the licensee is itself either a manufacturer or sub-contracts out such manufacturing. Such a condition prevents the exemption from applying in the case of a licence applying to sales only without manufacturing responsibilities being taken on, on the basis that a licensee only needs protection in connection with a substantial investment in manufacturing facilities.

Nevertheless, criticism of the draft as too stringent continued from industrial sources; objection was taken to the limit on the extent to which field of use restrictions can be imposed, which would, it was claimed, make the proposed Regulation unattractive to a majority of potential licensors. The restriction of one hundred million population for a territory within which exclusive licensing would be allowed incurred particular criticism as unnecessarily arbitrary;[20] and in response the Commission indicated that it was prepared to consider replacing it by a more generous qualification.

Further indications of the views of the Commission on the preferred content of the draft Regulation were shown by its decision in the *Maize Seed* case.[21] This case arose out of the assignment by a French State agency for agricultural research (Inra) of plant breeders' rights in a variety of maize seed to Eisele, a German resident, who proceeded to register them in Germany. Eisele and Inra also later concluded an arrangement

[20] Similar objections had also been raised to the proposed application of such limitation on the areas for which exclusive distribution agreements could be permitted.

[21] The report of the original Commission decision is at [1978] 3 CMLR 434. The European Court reports referring to the cases *Nungesser* v. *Commission* no. 258/8 are at [1982] ECR 2015 and [1983] 1 CMLR 278. The rights originally granted to Eisele were later transferred to Nungesser KG, a business controlled by him.

under which Eisele received also the exclusive right to distribute Inra's maize seed varieties in Germany, and under which Inra accepted an obligation to prevent other imports of its own maize seed into Germany. Subsequently Inra assigned to a French company, Frasema, an exclusive licence for commercial exploitation of its maize seed and vested all its own rights in that company. Meanwhile, Eisele agreed to purchase two-thirds of the maize seed which he sold in Germany from Inra, whilst in respect of the remaining third he was free either to produce it or to arrange for its production under his supervision. Eisele agreed also not to sell competing seeds and to set prices in agreement with Inra, although in practice apparently this last arrangement was not enforced. Eisele had enforced his rights in the German Court against at least one parallel importer and had obtained a settlement under which the importer agreed not to sell Inra's maize seed in Germany without Eisele's permission.

The decision of the Commission in 1978 was that all the restrictive terms referred to above and also those incorporated in the settlement in the German Court were in violation of 85(1). The claim for exemption under Article 85(3) was rejected. Eisele then proceeded to appeal to the Court of Justice against all the findings of the Commission except those relating to the requirements clause, the non-competition clause, and the resale price maintenance requirement. The Commission's viewpoint can be summed up by the following extract taken from the subsequent Court decision:[22]

> The Commission considered that, as in the case of a patent, exclusive propagation rights granted by the owner of breeders' rights to a licensee within the Common Market are, in principle, capable of being considered to have satisfied the tests for exemption under Article 85(3). There are even circumstances in which exclusive selling rights linked with prohibitions against exporting might also be exempted, for example when the exclusivity is needed to protect small or medium-sized undertakings in their attempt to penetrate a new market or promoting a new product provided that parallel imports are not restricted at the same time.

In *Maize Seed*, however, the Commission felt that the absolute territorial protection afforded to Eisele did not satisfy these tests especially as the prices paid for the seed in Germany were as much as 70 per cent greater than were paid in France.

The first official publication of the Patent Licensing group exemption, already considerably revised from the draft first shown to the Advisory Committee, appeared in 1979. Extensive consultations had already taken place, and oral hearings were held in Brussels on three days during October 1979 attended by international and national trade associations,

[22] [1982] ECR 2023–4.

and by a variety of companies, academic and practising lawyers, and patent agents, where both legal and economic aspects of the draft Regulations were discussed. The general framework of this draft had similarities with that of Regulation 67/67 relating to exclusive distribution agreements. Article 1 specified that exclusive rights to manufacture and distribute specified products under licence were exempted altogether from 85(1). Dealing with the key issue of exclusivity, the recitals to the draft stated:

> The territorial protection that arises from exclusive sales rights and related export bans can only be allowed if it is requisite for ensuring the expansion of technical progress. The Commission accepts that this protection is necessary for the majority of undertakings as a determining factor to facilitate decisions on investments relating to the development and marketing of new technologies. For undertakings with very high turnovers this protection would not on the other hand seem appropriate having regard to their extensive financial resources . . . the turnover limit set in the Regulation will ensure that most independent undertakings in the Community that grant or take licences will be able to qualify for the exemption. But the exemption will not be available for a number of firms which have particularly large financial resources and which, moreover, hold the bulk of the patents in force in the Common Market. Subject to this restriction, territorial protection may be allowed for the full duration of all patents extant at the time of the licensing agreement . . . the Regulation assumes that the licensee himself undertakes investments for the manufacture of the licensed product. It does not, therefore, apply to mere sales licences.

The qualification referred to was, of course, the proposal that the annual turnover of the licensor or licensee did not exceed one hundred million units of account, a level which would have excluded from benefit most multinationals. All the subsidiaries within an individual group would have to be taken into account in calculating this aggregate figure.

Article 2 listed a number of ancillary restrictions that would not be held to debar exemption. In addition to the clauses already mentioned in the earlier draft, these include field of use restrictions (provided that the relevant products in each of the fields from which the licensee is excluded differed materially from the products for which the licence was granted), an obligation to respect the licensor's specifications relating to minimum standards of quality, and an obligation on a non-exclusive basis to give information to the licensor as to improvements made whilst working the patent.

Article 3 contained the so-called 'black list', fourteen restrictions, the presence of any one of those being sufficient to preclude the application of the group exemption. These included no-challenge and no-competition clauses, obligations to pay royalties after the expiration of the last relevant patent,[23] maximum quantity restrictions, and restrictions relating to the price or condition upon which the licensee should sell including restrictions on classes of customer. Also prohibited were clauses preventing the licensee using his own trade mark in parallel with that of the licensor; if any of these clauses appeared, then only individual exemption could be sought.

In spite of the publication of the draft Regulation, further progress was apparently blocked by the need for the Commission's procedures to be underpinned by the Court's judgment on the appeal pending from the Commission's *Maize Seed* decision. In the meantime, however, interesting light on the attitude of the Commission to the important and related topic of the treatment of know-how after the termination of the agreement was also to be given by the publication in December 1978 of a Notice on subcontracting agreements. This topic had been in discussion for some four or five years with interested industrial groups, trade associations, and Member States. Under such agreements, a main contractor, often but not necessarily as the result of an order which it had itself received from a third party, would place with a sub-contractor an order for supply of goods or services that had to be integrated closely together with its own manufacturing or service operations. A typical example would be the supply by a sub-contractor of components for assembly and incorporation in complex mechanical and electrical machinery or apparatus such as an aircraft, motor car, or computer. Some similarities exist between such subcontracts and patent/know-how licences, although the sub-contract would normally be for a more limited period of time than a patent licence, so that the main contractor would, therefore, be particularly concerned to place restrictions on the sub-contractor as to the future use which the latter made of the patented or unpatented information provided to him for the purposes of the sub-contract. It was also important to ensure the close integration of the sub-contractors' manufacturing processes with those of the main contractor. Article 85 has relevance to the restrictions under which the main contractor seeks to protect his interests by limiting the use of this information by the sub-contractor once the contract has been terminated.

The original Commission working paper produced in 1974 was criticized by industry groups on the grounds that it placed too great a limitation on the ability of main contractors to protect their interests by imposing consequential restrictions on sub-contractors, in other words that the contract was being treated more as one simply for the exclusive

[23] It is clear, however, if the licence can be terminated by the licensee on reasonable notice, that such a restriction cannot be a breach of Article 85(1). *Ottung* v. *Klee* 320/87 [1989] ECR 1194: [1990] 4 CMLR 915.

supply of goods to the main contractor rather than one for the exclusive supply of goods closely linked to the licensing of relevant industrial property rights and know-how. Suggestions were made that a group exemption should be issued on this subject, but it did not apparently claim sufficiently high priority over the other competing claims on the time of DG IV for this to occur. The solution eventually adopted was, however, a compromise, the publishing in 1978 of a Notice. This, of course, had less value in terms of legal certainty than a group exemption. It at least indicated, however, DG IV's views on the restrictions which it regarded as available for placing within a sub-contract and enabled greater scope to be given to a main contractor's desire to restrict the subsequent use of know-how by its previous sub-contractors.

The Notice can be summarized as follows:

1. Article 85 would not apply to a sub-contracting agreement where the sub-contractor received either industrial property rights or unpatented secret data (know-how) or documents, dies, or tools belonging to the contractor and which permitted the manufacture of goods in a manner which differed in form, function, or composition from other goods on the market, unless the sub-contractor was able to obtain access to the relevant technology and equipment needed to produce the relevant goods from sources in the public domain.

2. If the relevant conditions were satisfied, then Article 85 would not apply to a clause whereby the technology or equipment provided by the contractor, and necessary for the sub-contractor to carry out its obligations, could not be used except for the purposes of the sub-contracting agreement or made available to third parties, and whereby goods or services resulting from its use should be supplied only to the contractor.

3. Further ancillary restrictions would be permitted, namely that during the period of the sub-contract the parties would not disclose confidential information, and that the sub-contractor might enter into an obligation not to reveal the relevant know-how of a secret nature even after the expiration of the agreement, and to pass on to the main contractor improvements which the sub-contractor has discovered, on a non-exclusive basis. On the other hand, the mere previous existence of a sub-contracting agreement should not be allowed to displace the sub-contractor's ability to compete in the future with the main contractor, utilizing independently its own technical information and know-how even if this had developed from experience gained during the main contract.

The Notice attempted to strike a fair balance between the interests of the main contractor in preventing unfair activity at a subsequent date by the sub-contractor in utilizing the main contractor's information, whilst relieving the sub-contractor from onerous restrictions placed upon it after the termination of the contract, that would restrict it from utilizing the benefits of the research and development which it may have carried out, and which in turn may have substantially contributed to its own specialized input to the contract. The difficulty of balancing these equally legitimate interests in the sub-contracting area well illustrated the difficulties that would face the Commission in preparing the framework of model agreements acceptable as group exemptions in the still more involved field of exclusive patent licences.

The complexity of the whole subject of patent licences and the different interests involved was inevitably to cause a certain degree of delay in its resolution; and this was contributed to by procedural difficulties encountered by the European Court of Justice in finalizing the hearings on the *Maize Seed* case, where its final decision was not rendered eventually until the summer of 1982. The Court's ultimate decision, however, reversed some parts of the Commission's 1978 decision, ruling that the following restrictions did not *per se* amount to a violation of Article 85(1):

(*a*) the obligation upon Inra (or those deriving rights through it) to refrain from producing or selling relevant seeds in Germany to other licensees; and
(*b*) the obligation upon Inra (or those deriving rights through it) to refrain from producing or selling the relevant seeds in Germany themselves.

On the other hand, the Court did uphold the remainder of the Commission's decision, namely that a violation of 85(1) was involved by the obligation upon Inra or those deriving rights from it to prevent third parties from exporting the relevant seeds into Germany without Eisele's authorization, as also by Eisele's concurrent use of his own contractual and proprietary rights to prevent all imports into Germany of the seed. These violations resulted from the fact that the licensee had obtained absolute territorial protection not only from the licensor, but from fellow licensees in other territories, as a result of the combined effect of the licensor's undertaking to prevent exports by third parties into Germany and by Eisele's own use of his exclusive rights to challenge parallel imports in the German Courts. The Court also confirmed that an agreement which did permit such 'blanket' protection would not receive exemption under 85(3). The Court also indicated that an earlier German Court approved settlement between the licensee and David, a parallel importer, requiring David to obtain Eisele's consent before marketing seed in Germany, violated Article 85(1).

The Court in its decision drew an important distinction between on the one hand an 'open' exclusive licence in which the licensor only agreed not to grant rights to other licensees for the same territories without affecting the ability of third parties, such as parallel importers and licensees for other territories, to compete, and on the other hand exclusive licences accompanied by absolute territorial protection aimed at eliminating any competition from third parties, and especially parallel importers. Whilst absolute territorial protection could never receive negative clearance under Article 85(1), the grant of exclusivity under an open licence had to be considered in the light of a number of factors, including the nature of the product, the novelty and importance of the relevant technology, the investment risks assumed by the licensee, and the development of interbrand competition. In particular these last two factors deserved to carry most weight.

Commentators on the case subsequently indicated that they felt its effect would be very considerably to widen the scope for argument by the owners of intellectual property rights, and that the Commission should show a more generous attitude towards restrictive clauses necessary to protect the interests of licensors and licensees in exclusivity. The Court had after all made it clear that it regarded its decision as relevant not only to the narrower field of plant breeders' right but to the more general application of patent licences. In late 1983, a further draft of the group exemption appeared incorporating a number of amendments, to be followed in due course by a 1984 version containing still more alterations mainly extending the number of restrictions that could be imposed. It also introduced for the first idea the concept, already familiar in the group exemption of exclusive distribution, that a licensee could be prevented from active sales outside the territory whilst keeping ability to respond if unsolicited sales orders happen to be received from outside its territory.

Even the 1984 draft turned out not to be the final version, since a further draft was produced subsequently including for the first time the right for the licensor to impose an absolute ban for a period of five years on the licensee against the making even of 'passive sales' in respect of orders received from outside its territory, a concession widely welcomed by industrial and commercial interests. Finally, with effect from 1 January 1985, the Regulation came into force as no. 2349/84.

Whilst the analysis of the process by which the final form of the Regulation was reached is important in showing the gradual development of the Commission's thinking in this area, ultimately it is of course the final form of the Regulation that is of greatest interest. Space does, however, not allow a fully detailed analysis of the complex provisions of this final version which is accessible in specialized monographs.[24] In terms of the legislative process, however, it is important, in addition to noting the painstakingly detailed process of consultation with both all the Community institutions themselves (including the Parliament) and with

[24] In particular see V. L. Korah, *Patent Licensing and EEC Competition Rules: Regulation 2349/84* (Oxford: ESC Publishing Ltd., 1985).

Member States and official bodies within them, to be aware of the substantial extent to which critical comment received especially from individual sources within industry was taken into account. In particular this comment played a part in ensuring that the final version of the Regulation reflected a more accommodating approach on the part of the Commission on the question of exclusivity than had the original drafts.

The group exemption did not apply to a licence simply to sell or distribute products; the licensee must be engaged either directly or through subcontractors in the manufacture of products (so that a substantial investment is almost certainly being made by it). Provided this qualification was satisfied, the territorial protection available was substantial. Thus the licensor itself could grant exclusivity of manufacture to its licensees on an unconditional basis, and thus might prevent any competition for sales with the exclusive licensee within its territory. So far as the protection of the licensee from licensees in other territories is concerned, absolute territorial protection could be given for a maximum of five years calculated from the date when the patented product was first marketed in the EC. After this period the licensee would be open to competition on a 'passive' basis, the assumption being that by this time it would have been able to obtain a market for its product sufficient to justify the original investment, and by which time it should be well enough established to cope with competition from other licensees.

The Regulation commenced with very extensive recitals referring both to the legislative history of the document and to the policy considerations which had governed the Commission's decision to accept some restrictions as worthy of exemption on a group basis but to reject others. The Regulation applied to agreements covering the entire Common Market or an area wider than the Common Market but having effect also within the Common Market. The Regulation applied not only to patent licences but to agreements covering both patents and non-patented technical knowledge (know-how), provided that this know-how was secret and permitted a better exploitation of the licensed patent. The know-how need not be merely ancillary to the patent, but could involve any combination of the two so long as at least one patent remains in force. Article 1 in the final version listing permitted 'white' clauses both includes the restrictions which the Court of Justice had deemed permissible in the *Maize Seed* case, and also other obligations to give a certain degree of territorial protection to the licensees as already described. A further obligation exempted under this Article was an obligation on the licensee to use only the licensor's trade mark (if the licensor required it) provided that the licensee was also entitled to identify itself as manufacturer of the licensed product. Underlying the basis for all these restrictions, however, was the continuing need for at least one of the licensed patents to remain in force.

Article 2 contained a list of eleven other restrictions permitted in addition to those contained in Article 1. These permitted exemptions were, however, themselves restricted to some extent by the content of the 'black' list in the following Article 3 listing those clauses banned from any licence benefiting from the group exemption and the two Articles have, therefore, to be read carefully together.

Article 4 was of considerable importance in that it contained an opposition procedure, first to be found in the 1983 draft. If an exclusive patent licence contained certain restrictions not falling within the existing 'black' list (Article 3) it could be notified, and if the Commission did not raise a formal objection to the relevant clauses within six months, the agreement was deemed to have been accepted, as if these restrictions were covered by Articles 1 and 2 of the Regulation. This procedure is described in the relevant recital as a simple means of benefiting clauses in exclusive licensing agreements not strictly within the terms of Articles 1 and 2, but which did not entail any of the restrictive effects of the clauses listed in Article 3, by giving them a method of benefiting from the legal certainty that the group exemption was intended to provide. At the same time, it gave the Commission an opportunity to check whether the clause was in breach of the spirit of the Regulation, if not its letter.[25]

Under Article 9, the Commission could withdraw the benefit of the exemption if it found that an agreement exempted nevertheless had effects incompatible with the conditions of Article 85(3), in particular those where the licensor did not have the right to terminate exclusivity after five years at the latest from the date when the agreement was entered into and annually thereafter if the licensee failed to exploit the patent adequately. The benefit could also be withdrawn if the licensee refused without 'any objectively valid reason' to meet unsolicited demands from users or resellers in the territory of other licensees or made it difficult for users or resellers to obtain products from other resellers within the Common Market whether by the use of industrial or commercial property rights or by other measures.

Since the implementation of the Patent Regulation, the number of notifications of individual agreements has substantially been reduced, as DG IV had hoped. Nevertheless, there will still always be a number of patent licences which do not fall within the framework of the group exemption. An indication of the likely attitude to such agreements was given by the later Court decision in the *Windsurfing International* v.

[25] There is nevertheless always an uneasy tension between this form of procedure (to which the safeguards provided by Reg. 17 applicable to an application for individual exemption under Article 85(3) do not apply) and the continuing need for a relatively speedy procedure by which objection can be raised to the application of the opposition procedure by competitors or consumers, having the necessary standing to apply to the Commission for such a purpose.

Commission case[26] (European Court 1986). The importance of this case lies in the fact that it was the first occasion since the introduction of the group exemption when the Court had an opportunity of laying down rules for the assessment of individual patent licences. The Commission had imposed a fine, albeit a small one, on a patent licensor, Windsurfing International. The licences under challenge were those between Windsurfing International and its German licensees. The Court accepted the Commission's finding that the coverage of the German patent obtained by Windsurfing International was not for a complete sailboard but only for the rig (mast, sail, and spars) attached to the board. There was a market for the rig sold separately, though usually boards and rigs were sold as a complete integrated unit. Several restrictions in the licence were stated by the Commission to be in breach of Article 85(1) as follows:

(1) the licensed product having been specified as a complete sailboard, the licensor's approval was nevertheless needed for any type of board on which the licensee intended to place the rig;
(2) the obligation on the licensee to sell only complete sailboards, i.e. the patented rig and an approved board;
(3) the licensee's obligation to pay royalties on the entire selling price of the complete sailboard (rather than solely on that of the patented rig);
(4) the requirement on the licensee to fix a notice on the board stating that it was 'licensed' by the licensor;
(5) no-challenge clauses with respect to the licensor's trade mark and patent;[27]
(6) a restriction on the licensee against manufacturing the sailboard except at certain factories in Germany.

Advocate-General Lenz disagreed with a number of findings made by the Commission and himself undertook a very thorough investigation of the facts. In his lengthy opinion, he concluded that only three of the restrictions could be held to be in breach of Article 85(1). These were numbers (1), (2), and (3) above, and with regard to the others his opinion was either that the restrictions could not have any effect on competition or in practice would have insufficient effect on trade between Member States. He, therefore, recommended that the fine be reduced.

The Court, however, took a stricter view, concurring with the views of the Commission on all except item (3) above and reduced the fine slightly. The reasons given by the Court have importance for the future. With

[26] Case 193/83 [1986] 3 CMLR 489. The founder of the company, Mr Hoyle Schweitzer, was one of the leading figures in the development of the sailboard.

[27] On the other hand the Court has subsequently held that a 'no-challenge' clause cannot restrict competition if the licence itself is free since then the licensee suffers no competitive disadvantage by any payment of royalties. *Bayer* v. *Süllhofer* Case 65/86 [1988] ECR 5249: [1990] 4 CMLR 182.

regard to restriction (1), the Court found that Windsurfing International had failed to show that their control over the manufacture of non-patented components, such as boards, related to any objective criteria relating to the patented rig as no technical method of verifying the quality of the board was provided by the agreement. The Court found that its real interest lay 'in ensuring that there was sufficient product differentiation between its licensee's sailboards to cover the widest possible spectrum of market demand'. If any passing off of boards as official Windsurfing International boards by the licensees was found, this could have been dealt with by an action in the national Courts. The distinction drawn by the Court in respect of item (3) was that the method of calculating royalties simply on the entire sailboard might have a small effect in restricting competition with regard to the board but no effect on the sale of rigs, especially as it turned out that the royalties levied on the sale of complete sailboards proved to have been no higher than that laid down for the sale of separate rigs.[28] With regard to restriction (6), the Court rejected the argument by Windsurfing International that the restriction of manufacturers to one location was required by quality control. The Commission's argument to the contrary was accepted, that in fact numerous licensees throughout Germany were manufacturing components for the rig, and that Windsurfing International in any case would have had no control over the conditions of quality control of such manufacturers. The Court also gave an important general ruling relating to the interpretation of Article 85(1); it said there was no need for the Commission to examine each clause restricting competition to see if individually they could be shown to have an effect on trade between Member States. Provided that the agreement itself, taken as a whole, had such effect, Article 85(1) would be applicable to it. The Court appears to have found the necessary effect on trade from the levels of imports of both complete sailboards into the Community (about 20 per cent of national markets) and of their individual components (where the corresponding figure was about 10 per cent).

This case is also important as illustrating that the Court will generally support the Commission in its analysis of restrictions imposed by the holders of an intellectual property right (here a patent, but equally applicable to other forms of such rights). It will divide them into those (comparatively limited in number) which flow from the inherent subject matter of the right (and can, therefore, be said to form part of the right itself) and those which on the other hand represent an attempt to extend the holder's rights beyond, even by a short distance. It will also respect the Commission's findings as to the scope of the right itself (provided that

[28] The licensee acknowledged that it would be equitable to accept a higher rate of royalty if it were calculated on the price of the rig alone.

the findings appear reasonable in the light of all the evidence presented); crucial in this case, of course, was its finding that the patent extended only to the rig and not the complete sailboard itself, a factor that influenced the majority of the adverse rulings against Windsurfing International.

3. Know-how Licences

Much technical information exists that is either essential, or at least of substantial importance, to manufacturing operations or other industrial activities. Although undoubtedly a form of industrial property, it cannot be protected by registration, notification, or symbol; it exists in its own right or can form part of a 'package' of technical information which includes patents and other intellectual property rights. It is well defined by Hawk as 'a catch all expression embracing a broad and ill-defined spectrum of unpatented and unpatentable inventions, techniques, processes, formulae, devices, blueprints, technical and production skills, etc.'[29] Since no time limit applies to the period of protection, it has been argued that it should be more strictly treated by the Commission than patents. It could also be argued that a great deal of what starts as technical know-how will reach the public domain relatively quickly so that in practice its commercial value may be available for exploitation for a considerably shorter period than patents, trade marks, or copyright. These factors would argue in favour of a shorter period of protection for know-how than in respect of such rights.

Case-law in this area took some time to develop. The first decision was *Schlegel Corporation and CPIO*[30] where a know-how licence granted in connection with an exclusive distribution agreement for automatic weather-seals was approved for a seven-year term by the Commission. A more important case was *Boussois* v. *Interpane*[31] where a German company (Interpane) sold to a French company (Boussois) an entire factory in France for making fine thermal insulation coatings for flat glass, for use particularly with double glazing units. Most of the technology involved was unpatented, though both secret and important. The contract provided full details of how the plant was to work using Interpane's own current technology, and Interpane was also obliged to provide full details of all improvements in development over the next five years. Boussois was to be allowed to make the product in France to the exclusion of any other licensee for five years and after that indefinitely on a non-exclusive basis.

[29] B. E. Hawk, *United States, Common Market and International Antitrust: A Comparative Guide*, 2nd edn., ii (New York: Harcourt Brace Jovanovitch, 1990 Supplement), 691.

[30] [1984] 2 CMLR 179.

[31] [1988] 4 CMLR 124.

Interpane would nevertheless be free to build its own plant after two years had elapsed. Boussois had to agree not to manufacture in any other Member State but had an exclusive right to sell the products made with the technology in France for five years, thereafter non-exclusively both inside and outside France; it remained free to sell competing products.

This was not, of course, an open exclusive licence because of the barriers for five years on other licensees selling into the territory reserved for Boussois. The licensing covered a large package of information in which, although patents were involved, a detailed body of manufacturing know-how was the dominant element. The Commission held that the contract failed to meet the requirements of the patent block exemption 2349/84 for apparently patents protected only one of the various coatings required and even then in only five of the twelve Member States.

Nevertheless, an individual exemption under 85(3) was given because the protection given to Boussois was regarded as no more than necessary to encourage it to invest and produce; it was reasonable in the Commission's opinion to assume from the substantial investment being made by Boussois that the consumer would benefit from it. The exemption was, therefore, granted for a five-year term and certain ancillary restrictions, such as that improvements were to be on a reciprocal non-exclusive basis, were held to fall outside 85(1) altogether.

The decision showed, however, that the existing Regulation 2349/84 covering patent licences was not an appropriate group exemption for a large number of licensing agreements in which, although patents were included, they were not a dominant factor. Considerable pressure had already built up, therefore, over the 1980s from UNICE and other industry sources for a separate group exemption to cover this situation; it was claimed with some justification that such know-how was actually more crucial to the development of innovative new processes of importance to the Community than patents themselves. After some three years of internal discussion, DG IV produced a first draft of a group exemption in 1987. As with all group exemptions, the Commission required, for both legal and practical reasons, further case-law experience prior to the making of the final Regulation, in order that the legal basis for the Regulation was firmly based on case experience. During this interim period of 1987 and 1988 two further cases were decided which indicated that the Commission was prepared to be relatively liberal in the degree of protection against sales activities by other licensees that licensees would be allowed in return for their investment in the purchase of the know-how. These two cases were *Rich Products/Jus-rol*[32] and *Delta-chemie.*[33]

In *Jus-rol* individual clearance under Article 85(3) was given for a

[32] [1988] 4 CMLR 527. [33] [1989] 4 CMLR 535.

licence from Rich Products to a licensee, Jus-rol, relating to a process for making frozen yeast dough products. Jus-rol received an exclusive licence to manufacture in the United Kingdom as well as a non-exclusive right to sell the product in all Member States. This ban on manufacturing outside the United Kingdom was held to constitute a breach of Article 85(1), but exemption was granted under 85(3) because the exclusivity was regarded as necessary to induce the licensee to make the necessary investment. Various ancillary restrictions including an obligation on the licensees to buy a secret pre-mix essential to the process only from the licensor were also exempted. In *Delta* individual exemption was granted to an agreement between a German company, Delta, and a UK company, DDD, for the manufacture and sale of a series of stain removers and bleaching agents. DDD received exclusive rights to manufacture the products in the licensed territory (UK, Ireland, and Greece) for twenty years, though its exclusive selling rights at the Commission's request were limited to ten years. DDD were not protected, however, against parallel imports, and this was probably the main reason why the Commission were prepared to give the exemption for as long as the twenty-year duration of the agreement.

The Know-how Regulation itself—no. 556/89—came into force on 1 April 1989 and does not expire until the end of 1999. Its recitals follow the normal pattern of setting out at great length the policy considerations underlying its introduction. After emphasizing the growing importance of know-how within the Community, they stress the importance of limiting with precision what is to be properly covered by this very general phrase. Not all useful information is to be benefited; it must be secret, substantial, and identified. It also originally included the requirement that it must be of 'decisive importance' to the technology involved, but this additional requirement was later discarded. None of the three definitions is totally straightforward. 'Secret' means not that all the information is totally unknown (save to the licensor), but that, taken as a whole, it is not generally known or easily accessible so that its use can give the licensee lead time once communicated to it. 'Substantial' means that it has important value for the whole or a significant part of a manufacturing process or a product or service or its development. In practical terms it must be useful to the licensee in its commercial operations. 'Identified' means that the package of information is described or recorded in sufficient detail for the other obligations of substantiality and secrecy to be verified and to enable the licensee to be clear as to where its own existing knowledge stops and the new know-how package begins. The group exemption does not cover licences connected to franchising or distribution (as opposed to production technology).[34]

[34] Apart from transitional arrangements when a licensee acts temporarily as a distributor prior to commencing manufacture with the aid of the know-how.

Article 1 confirms that it applies both to pure know-how agreements and to 'mixed' agreements which may also cover patents but do not fall within the terms of Regulation 2349/84. The Commission in practice has adopted a relaxed view in the interpretation of the recitals and this Article, and accepts that the benefit of the group exemption can be obtained by nearly all 'mixed' agreements covering both patent and know-how without any investigation of their relative importance. The exemption applies to two-party agreements which include any one or more of eight particular obligations, namely (i) an obligation on the licensor not to license other undertakings to exploit the know-how in the licensed territory; (ii) an obligation on the licensor not to exploit the know-how itself in that territory; (iii) an obligation on the licensee not to exploit the know-how in other Member States which are reserved for the licensor; (iv) an obligation on the licensee not to manufacture or use the know-how in Member States licensed to others; (v) an obligation on the licensee not actively to market the know-how within other Member States licensed to others; (vi) an obligation on the licensee not to put the product produced with the know-how on the market in Member States licensed to others (even by way of 'passive sales' requested from him); (vii) an obligation on the licensee to use only the licensor's trade mark or 'get up' to distinguish the product produced with the know-how provided that the licensee is not prevented from identifying himself as their manufacturer; (viii) an obligation on the licensee to limit the quantities of production made with the know-how to the amount required for manufacturing its own products and to sell the licensed products only as an integral part or replacement part for his own products, though in quantities which it determines for itself.

It has been suggested that a licensor who sought to avail himself of the protection of all these restrictions might indeed be receiving protection to an excessive degree.[35] It has to be remembered, however, that compared with the licensing of other intellectual property, once know-how has been licensed it is almost impossible to terminate the licence in a manner which prevents the licensee in practice from making substantial future use of the know-how provided.

The structure of the remainder of the Regulation follows the familiar pattern of 'white' and 'black' clauses supported by an opposition procedure for 'grey' clauses. The 'white' clauses include those making the licence exclusive to the licensee and requiring confidentiality on its part (extending for secret material even after the end of the agreement), prohibiting sub-licensing, and limiting the use of the know-how to specific technical applications (field of use). It also covers payment of a minimum

[35] T. Frazer, 'The Commission's Policy on Know-how Agreements' (1989) 9 *YEL* 1.

level for royalties and production of minimum quantities. The licensee may insist on a most favoured licensee clause to ensure no other licensee receives better terms. The 'black' clauses in general follow the pattern of those prohibited in the patent licence group exemption, Regulation 2349/84. Thus no-challenge and non-competition clauses are prohibited; royalties cannot be charged on production of goods for which the know-how is irrelevant, nor can licensee or licensor be restricted to particular categories of customers, maximum production quantities, or in connection with prices to be charged or discount given. The licensee further cannot be required to assign or give an exclusive licence to the licensor for improvements which it develops over the life of the licence, though it may be required to give the licensor a non-exclusive licence during that period.

The degree to which the licensor is protected against the licensee failing to make full use of the know-how is significant.[36] The licensee can be required to use his best endeavours to exploit the licence process and its technology; the licensee's right to compete with the licensor in research, development, and production or use of competing products is limited by the licensor's rights if this occurs to withdraw the exclusivity granted and to refuse to give details of future improvements. The only exception would be that the licensee itself can prove that the know-how which it has received has not been used in the production of any goods and services other than those actually licensed, and proving a negative in this way is probably a difficult task.

In general, the periods of protection for licensed territories tend to be potentially shorter than those applicable under the patent group exemption, especially when it is remembered that all the maximum periods set out in the Regulation will be further shortened if the information provided to the licensee ceases at any time to be secret or substantial, for example by becoming public knowledge. The main time periods set out in the Regulation are:

(*a*) protection for the licensor against his licensee and vice versa: *ten years* from the date when the licensor first makes an agreement covering licence technology for that territory;

(*b*) protection for the licensee against other licensees actively selling into its territory: *ten years* from the date when the licensor first makes an agreement covering the particular technology within the Community, i.e. a period that begins once the first licence has been granted in any Member State;

(*c*) protection for the licensee against other licensees passively selling into his territory: *five years* from the same date as in (*b*), also therefore a period that reduces once the first licence has been granted.

[36] Article 3(9).

The only exception to these time periods is where the same technology is also protected in a Member State by licensed patents, in which case the licensee can continue to benefit from them during the life of the patent.[37]

Article 4 contains the opposition procedure which is in similar terms to that found in the patent group exemption. The period available for the Commission to indicate its opposition is six months from the date of notification. The terms upon which the Commission may withdraw the benefit of the Regulation are also similar to those set out in the patent group exemption. They apply in particular to the situation where the effect of the agreement is shown to prevent licensed products from being exposed to effective competition from identical or equivalent products, or where the licensee is found without an objectively valid reason to refuse to meet unsolicited demand from users or resellers in the territory of other licensees once the five-year total exclusivity provision has expired. The benefit of the Regulation equally does not apply to technology pools or horizontal joint ventures or cross-licences.

4. Trade Mark and Copyright Licences

Both trade marks and copyright are familiar as types of intellectual property rights that often carry considerable commercial value: their development and protection are justified by the time, effort, and expenditure that their owners have incurred in their creation. Both, of course, have potentially much longer lives than patents; indeed while the legal protection of copyright does ultimately expire, the use of trade marks can continue on an indefinite basis. In some early European Court jurisprudence a feeling can be detected (perhaps notably in *Sirena* v. *Eda* mentioned below) that perhaps trade marks are not as worthy of protection as other forms of intellectual property. In more recent cases, however, notably the *Hag* (no. 2) case,[38] a firm basis has been provided for treating trade marks as no less important than any other form of intellectual property. Their chief importance lies in the ability they provide to a manufacturer who consistently produces high quality goods to extend his reputation from one market to another without having to start from scratch again in every case. Earlier suggestions by the Commission that it was considering the issue of a draft group exemption for trade marks have not, however, been followed up.

Early European Court cases concerned themselves mainly with horizontal agreements relating to trade mark rights. The early case of *Sirena*

[37] Article 1(4).
[38] [1990] 3 CMLR 571, See Ch. 16. pp. 332–3, for a detailed description of this case.

v. *Eda* (European Court of Justice 1971)[39] shows the width of application of Article 85 in this context; in this case, an Italian Court sought under an Article 177 reference to know whether Articles 85 and 86 prevented an Italian company (Sirena) from asserting its registered trade mark 'Prep', so as to prevent the sale in Italy under the same trade mark of similar hairdressing products made in Germany by another manufacturer. Both the Italian and the German undertaking claimed to have rights to the use of this trade mark, by assignments from an American company as long as thirty years previously. The problem had arisen because the German product had been imported into Italy for resale at prices lower than those at which the Italian product was sold, and Sirena wished to prevent these imports by relying on its registration of the mark in Italy. The Italian importer of the German product naturally contested such action on the basis that such a finding would be in breach of Article 85, claiming that the previous assignment of the two licences itself had the effect of constituting an agreement between the United States owner and the licensees in their two separate markets, even though these assignments had taken place a considerable number of years previously.

As in all Article 177 cases, the Court was not called upon to rule as to the actual merits of the case between the parties, but only to provide the reply to certain questions on legal issues submitted to it by the national Court responsible for deciding the case. The terms of the opinion given, however, were substantially in favour of the importer of the German product. The Court agreed with its argument that the effect of licences or assignments relating to trade marks can be that of preventing imports, notably through the allocation of particular watertight areas to individual undertakings, and that it would equally be a breach of Article 85 if such arrangements actually had such an effect even if not intended to do so. The result of the *Sirena* v. *Eda* case was initially felt alarming, since Article 85 seemed to be applicable without reference to the length of time that had passed since the date at which the original trade mark rights had been allocated. Nevertheless, the potentially wide effects of the decision were substantially reduced by the later case of *EMI Records Ltd.* v. *CBS United Kingdom Ltd.*[40] (also considered in the next Chapter) where the Court confirmed that in order to bring Article 85 into play, the original allocation of trade marks would need to be accompanied by a continuing relationship between the original owner of the trade mark and its licensees or assignees.

[39] Case 40/70 [1971] ECR 69: CMLR 260. Had the case reached the European Court at a later date, after the development of the extensive case-law considered in Ch. 16 relating to the effect upon national intellectual property rights of Articles 30–3, it is likely that it would have been decided (with much the same ultimate result) by reference to those Articles rather than to Article 85.

[40] Case 51/75 [1976] ECR 811: 2 CMLR 235.

A more recent case in this category which reached the European Court in 1985[41] related to a trade mark agreement between BAT (the German subsidiary of the multinational British-American Tobacco Ltd.) and a small Dutch tobacco manufacturer, Segers. The agreement limited the use of trade marks which were alleged, rather unconvincingly, to be of a confusingly close nature. BAT owned a German trade mark 'Dorcet', which had never been utilized for any particular tobacco product; Segers owned a mark in Holland 'Toltecs Special' in respect of fine cut tobacco. Segers had applied to have its trade mark registered in Germany (as well as in some other countries) and BAT had opposed this on the grounds of potential confusion. The compromise set out in the relevant agreement involved withdrawal of opposition by BAT, if Segers restricted the application of 'Toltecs' to the sale of his fine cut tobacco and agreed not to market his tobacco under the Toltecs mark in Germany except through BAT's approved importers. Moreover, Segers was not in any case to exercise his trade mark rights against BAT even if the Dorcet mark remained unused for a period of over five years.[42] The Commission had found that in reality there was no serious risk of confusion between the two trade marks and imposed a fine on BAT for extending the no-challenge clause beyond the five-year period. The Court annulled the fine on technical grounds, ruling that the restriction on Segers restricted only the use of the mark, not trade in goods or services, but confirmed that BAT had, by attempting to control the distribution of Segers's products in this way, misused their trade mark rights available under national German law.

While the trade mark agreement with horizontal elements is thus treated with relative severity, vertical agreements containing restrictive clauses have been treated more leniently, in a way more comparable with that accorded to distribution agreements. This is appropriate as such trade mark arrangements of a vertical nature are often themselves ancillary to distribution systems as was, for example, the case in *Campari*.[43] Here a network of exclusive licences to a variety of undertakings from the original manufacturer of that well-known aperitif was granted exemption once tying clauses and export bans affecting other Member States in the Community had been removed from the documents.

Trade mark licensing may also play an ancillary part in other forms of licensing. Thus trade marks usually form an important part of the package of information and intellectual property rights which any franchisee

[41] *BAT Cigaretten-Fabrieken* v. *Commission* Case 35/83 [1985] ECR 363: 2 CMLR 470. The original Commission decision is reported as *Toltecs-Dorcet Trade Marks* [1983] 1 CMLR 412.

[42] This period of continuous disuse of the trade mark would normally under German law have led to its cancellation. Not all settlements of trade mark disputes, however, involving an allocation of territories between competitors will necessarily be regarded as in breach of Article 85(1). See e.g. *Penney's Trade Mark* [1978] 2 CMLR 100.

[43] [1978] 2 CMLR 397.

receives from its franchisor: moreover, a patent licensee may well be required to market the goods produced under the licence showing only the licensor's trade mark provided that the licensee is not prevented from identifying himself as the manufacturer of the products (Article 1(7)). Although a similar clause appears in the know-how licensing clause, the know-how group exemption cannot be applied to agreements that include the licensing of trade marks where these marks are not required for achieving the object of the licensed technology, or where the agreement contains additional restrictions on competition apart from those permitted to be attached to the licensed know-how itself. This has the effect that some 'mixed' trade mark and know-how licences will require individual exemption particularly where the main value of the licence lies in the trade mark rather than the know-how.

This is well illustrated by the *Moosehead* case.[44] Here Moosehead, a Canadian brewery, wished to sell its beer in the United Kingdom but could not justify the capital costs of setting up its own production and distribution facilities there. By entering into a licence, however, with Whitbread, a well-established UK public company with its own brewing and distribution facilities, Moosehead could obtain access to the UK market. Under the arrangements some know-how was to be provided to Whitbread to enable them to produce and sell Moosehead beer. Nevertheless the principal object of the agreement was the exploitation of the Moosehead trade mark rather than this know-how. The Commission granted individual exemption to a number of restrictive clauses including both the exclusive use of the trade mark in the United Kingdom and prohibitions on Whitbread from actively selling the product outside the UK or from dealing in competing brands of lager.[45]

The importance of copyright in protecting a wide variety of commercial rights continues to increase. It applies not only to books, records, films, and television programmes but also now to computer software, following the enactment of directive number 91/250[46] adopted on 14 May 1991, which requires Member States to have changed their domestic laws to comply by 1 January 1993. It applies also to certain other specialized rights such as plant breeder's rights. The Court's willingness to look at the special characteristics of copyright were shown early in the *Coditel* cases. In *Coditel* (no. 1),[47] the European Court had ruled that Article 59, protecting the freedom

[44] [1991] 4 CMLR 391.

[45] The Commission also ruled that a 'no-challenge' clause to the validity of a trade mark may constitute a restriction of competition within the terms of 85(1), as a successful challenge could bring the mark into the public domain for use by everyone, but only if the mark were already well known (not the case with Moosehead in the UK) and its use an important advantage to a company competing in a particular market, so that its unavailability would be a barrier to entry.

[46] OJ [1991] L122/ 42.

[47] *Coditel* v. *Cine Vog Films* (no. 1) Case 62/79 [1980] ECR 881: [1981] 2 CMLR 362, discussed in more detail in Ch. 16, pp. 331–2.

to provide services, did not prevent the granting of a seven-year exclusive copyright licence in a film by the French producer to a Belgian distributor (Cine Vog) thus entitling Cine Vog to prevent Coditel, a Belgian cable television company, from intercepting the transmission of the film and reshowing it to its subscribers, even with the consent of the original French producer. In the sequel case (which is in this context more important), *Coditel* no. 2,[48] the Court went on to rule that an exclusive grant of copyright in a film did not automatically bring the licence within Article 85(1) since there were many special circumstances relating to the film industry; the difficulties in obtaining a return on an investment in a film, which did not apply to other goods and services, entitled the Court to make a full economic analysis of the effects of the particular agreements before reaching a conclusion.

The Court did state that the exercise of exclusive rights would, therefore, fall within Article 85(1) only if after such an economic analysis it was shown either to give the parties to the licence the opportunity of creating artificial barriers to the market, or of charging fees exceeding a fair return on the investment, of providing exclusivity for a disproportionate period of time or, on a more general note, enabling the parties in a particular Member State to prevent or restrict competition in that country. The Court seemed to accept, reflecting its earlier conclusions in the *Maize Seed*[49] case referred to above, that original investment in films is far more likely if accompanied by the possibility of an exclusive licence being granted. Given also the comparatively short life of many films, it would seem reasonable to accord them greater protection for that short period than in the case of many other products, including those, for example, protected by plant breeders' rights as in the *Maize Seed* case.

In all such cases the Commission and Court will take into account not only the geographic and product markets concerned but also the duration of the licence being granted and the particular characteristics of the industry involved. In the German TV films case (*Commission* v. *Degeto Film*[50]) the proposed licence related to a large number of feature films owned by MGM/United Artists Corporation. The original intention was to have licensed these exclusively to Degeto, a subsidiary of the association of public broadcasting companies in Germany. The exclusivity apparently covered some 1,350 of the total of 3,000 films owned by MGM/UAC and would have lasted for terms of approximately fifteen years starting on a number of different dates. During these exclusive periods, third parties would have no access to the films. Moreover, Degeto would also have had rights over new films produced by MGM/UAC during the period of the agreement. The Commission found that feature films played an important

[48] *Coditel* v. *Cine Vog Films* (no. 2) Case 262/81 [1982] ECR 3381: [1983] 1 CMLR 49.

[49] At p. 298.

[50] [1990] 4 CMLR 841.

part in television programming as they were very popular with audiences and often of a higher artistic standard than those films produced specifically for television. The licence period of fifteen years, however, was at least five years longer than the normal permitted period for such agreements and the total number of films covered extremely large; both these factors caused concern to the Commission.

The licence was eventually given Article 85(3) exemption following substantial amendments to it; 'windows' were introduced to the agreement, permitting the licensing of the films to third parties for particular periods during its lifetime. These 'windows' would last for a period of from two to six years, thereby considerably reducing the degree of exclusivity. Exemption was granted because with these amendments some advantages could be obtained for German viewers who would have greater access to a larger number of films. The package deal provided would reduce the price per film licensed and thereby reduce the operating costs of public broadcasting organizations. The remaining degree of exclusivity was accepted as necessary in order to allow MGM/UAC a fair return for their original investment.[51]

One final general point needs to be made which applies to all forms of licensing of intellectual property. The terms of the licence cannot be used so as to extend the inherent right of the owner of the particular form of intellectual property. The Commission has in recent years been attempting to harmonize these rights and the software directive referred to above is one attempt to lay down common standards for not only these rights but also the terms upon which software licences may be granted. The Council's First Directive on trade mark licences (dated 21 December 1988) has also sought to specify the rights which the owner of a trade mark may regard as his under Community principles. These include a definition of trade marks, enumerate cases when they cannot be accepted for registration, and define the rights which they are capable of conferring. They also provide that non-use for a five-year period after registration can lead to revocation of a proprietor's rights.

Special rights may exist for other types of property. For example, it is clear from the *Erauw-Jacquery* v. *La Hesbignonne*[52] case (European Court 1988) that the rights of the owners of basic seeds may well extend to obtaining protection against improper handling and the control of propagation by those establishments regarded as suitable. For this reason, and to this extent, a clause prohibiting a licensee from selling or exporting

[51] This decision is apparently under appeal on rather unusual grounds. One of the companies benefiting from the exemption granted is objecting to some of its terms on the grounds that they are still too anti-competitive! In the past the Commission has assumed that it has power to grant an individual exemption even over the objection of one of the parties.

[52] Case 27/87 [1988] ECR 1919; 4 CMLR 576.

basic seeds would not come within the prohibition of Article 85(1) even though an equivalent clause would not be permitted in a copyright or know-how licence. The Court pointed out that the breeder of basic seeds is in some respects similar to a franchisor rather than simply the owner of a copyright and that its control over use of the seeds was essential to prevent its know-how being made too easily available to its competitors. The validity of licence terms in novel types of intellectual property yet to be developed may likewise well be influenced substantially by the nature of the right which is being licensed.

16

Intellectual Property Rights: The Effects of Articles 30–36

It is not only under the terms of Articles 85 and 86 that intellectual property rights fall to be considered. Articles 30 and 36 of the Treaty contain basic rules protecting the free movement of goods; Article 30 itself provides the central principle and Article 36 sets out the limited exceptions. The wording of Article 30 is concise: 'Quantitative restrictions on imports and all measures having equivalent effect shall, without prejudice to the following provisions, be prohibited between Member States'. Article 34 sets out similar provisions applying to exports, and Article 36 contains the qualifications to those basic rules, namely:

> The provisions of Articles 30 to 34 shall not preclude prohibitions or restrictions on imports, exports or goods in transit justified on grounds of public morality, public policy or public security; protection of health and life of humans, animals or plants; the protection of national treasures possessing artistic, historic or archaeological value; or the *protection of industrial and commercial property*. Such prohibitions or restrictions shall not, however, constitute a means of arbitrary discrimination or a disguised restriction on trade between Member States.

This chapter will, therefore, consider the relationship between, on the one hand, Articles 85 and 86 with their prohibition on agreements and concerted practices between undertakings and, on the other hand, Articles 30 to 36 addressed to Member States including their legislative, judicial, and administrative authorities responsible for implementation and enforcement of national intellectual property rights.

Before doing so, however, it is essential to understand the place of Articles 30 to 36 within the framework of the Treaty. Within any common market, one of the essential freedoms is the free movement of goods. The removal of internal tariff barriers and of straightforward and quantitative limits, such as quotas, on the import and export of goods between Member States occurred at an early stage in the development of the Common Market. It is, however, non-tariff barriers including national legislation and administrative practices relating to the regulation of prices, indications of national origin and other labelling requirements, public health and consumer legislation, which have presented, and still provide, far greater difficulties. As a result, the expression 'all measures having equivalent effect' in Article 30 has been considered in a large number of cases since 1974 by the European Court of Justice. The vast majority of these cases have been references made by national Courts to the European Court of Justice for a preliminary ruling under Article 177, rather than appeals against specific decisions of the Commission under Article 173. A considerable body of case-law on this subject has built up since the landmark decisions of *Dassonville* and *Cassis de Dijon* on the interpretation of Articles 30 and 36 and much of this has had an important influence on the competition law relating to intellectual property.[1]

If the founders of the Community were to be allowed today to redraw both the Treaty of Rome and Member State legislation in the way required to preserve the essential characteristics of the original Treaty, whilst providing a completely fresh basis for the framing of the rules relating to national rights of intellectual property, the experience of the recent decades would mean that their objectives would clearly have to include the definition of such rights and their essential characteristics in a way which could be accepted as valid throughout the Community, rather than allowing the perpetuation by individual Member States of such national rights varying in both classification and content, and which through their very diversity raise major problems for the free movement of goods and commercial integration. While Article 36 includes the 'protection of industrial and commercial property' as one of the exceptions to the basic principle of free movement of goods, there is no doubt that the attainment of an integrated market will remain difficult, if not impossible, so long as such varied exceptions to the basic rules are preserved. The harmonization, therefore, of the law relating to intellectual property within the Community, in particular that relating to patents and trade marks, as well as the gradual harmonization of the classification and definition of all the other various intellectual property rights found to exist within Member States covering rights as varied as copyright, registered design, and rights against unfair competition, remains an objective of great importance.

Some progress has been made towards providing uniform rules for the entire Community in the case of patent law. The European Patent Convention of 1973 has been in operation for some time and enables applicants to obtain from the European Patent Office in Munich a set of patents in respect of each of the States which have ratified the Convention. This at present comprises, *inter alia*, all the Member States of the Community apart from Denmark and Ireland. The Luxembourg Patent Convention by contrast signed in 1975 actually provides for the grant of a Community patent. The Convention was first signed in 1975, but never

[1] See L. Gormley, *Prohibiting Restrictions on Trade within the EEC* (Amsterdam: North-Holland, 1985), *passim*: *Procureur du Roi* v. *B & G Dassonville* Case 8/74 [1974] ECR 837: 2 CMLR 436: *Rewe-Zentral* v. *Bundesmonopolverwaltung für Branntwein* Case 120/78 [1979] ECR 649: 3 CMLR 494 (the *Cassis de Dijon* case).

entered into force because of insufficient ratifications. In December of 1989, all the Member States signed an updated version of the Convention,[2] including protocols on the Common Appeal Court and on the settlement of litigation. An additional protocol provides for the possibility for a new intergovernmental conference to limit the number of ratifications required for the entry into force of the Convention. In the meantime, the institutions of the Community and undertakings operating within it have to operate as best as they can in the transitional period, making use of the provisions of the Treaty as interpreted by the European Court. The primary issue which has presented itself to the Court has been whether to give a narrow or broad interpretation of the words of qualification in Article 36. If the interpretation given had been broad, the effect might have been that individual national intellectual property rights could be asserted, even where they might have the effect or purpose of preventing free movement of goods, so long as they did not conflict with the express prohibition in that Article of being either 'a means of arbitrary discrimination' or 'a disguised restriction on trade' between Member States.

Fortunately, from the viewpoint of those concerned with the effectiveness of competition law (and in particular the officials of DG IV), the choice of the Court has been, with very few exceptions, to confine claims based on Article 36 under the reference to 'the protection of industrial and commercial property' to a narrow range; in striking a balance between the requirements of Community law and national legislation, the needs of the Community have been given priority. The often repeated (and criticized) distinction which the Court has made between the 'existence' and 'exercise' of such rights is itself no more than the necessary application of the express provisions of the Treaty.[3] The 'existence' of the rights is protected apparently by Article 222: ('The Treaty shall in no-way prejudice the rules in Member States governing the system of property ownership'),[4] but any existing rights, if they are to have any commercial value, must also be capable of being exercised and the issue that has concerned the Court primarily is how far that exercise can be allowed, whilst at the same time preserving freedom of movement of goods between Member States including the freedom of parallel imports. Whilst debate as to the way in which the Court may have chosen to interpret the effect of national legislation in any particular case is essential and necessary, to criticize (as some commentators have done) the very distinction the Court

[2] OJ no. L 241/1 (30 Dec. 1989).

[3] See *Hag* (no. 2) Case C-10/89 [1990] 3 CMLR 571 especially Jacobs A.-G. at pp. 581–2, though the Court itself in its judgment appeared to lay less stress on the existence/exercise dichotomy.

[4] It is often claimed that it is wrong to attach too much importance in this context to the general words of this Article, which does little else but preserve the right of Member States to nationalize or denationalize individual sectors.

has made between 'existence' and 'exercise' is to miss the point that this distinction is part of the very framework of the Treaty itself. The distinction is not simply a 'gloss' read into the Treaty by the Court. More crucial, however, than that distinction is the issue of how the Court would define those special rights regarded as so essential and inherent to the very existence of a national intellectual property right that they had to be protected under the terms of Article 36 even where their effect on trade between Member States could be observed.

The term 'industrial and commercial property' is not necessarily the exact equivalent to 'intellectual property rights' or 'industrial property rights'. From the Court's case-law, there is certainly no doubt that there is a wide area of overlap between the two concepts; industrial and commercial property cover patents, registered designs, trade marks, copyrights, and plant breeders' rights. The Court tends in borderline cases to look for an analogy between special types of intellectual property for which protection is claimed and the well-established categories, by seeing if there are common features, for example, rights of exclusivity following registration or prior use, and also by examining the provisions of relevant international conventions to see which definition they accord in the particular area. On the wrong side of the borderline have fallen a number of cases such as the *Prantl Bocksbeutel* case[5] (European Court of Justice 1983) where the Court refused to accord the protection requested by the German Government to the producers of wine from Franconia for a bottle of distinctive shape against imports of a different variety of wine in similar bottles from Italy. The decision was, however, taken not on the basis that legislation conferring rights relating to certificates of national origin could never constitute industrial and commercial property, but because the Court felt that the particular German legislation was too general in its application to qualify under such a heading, since it protected descriptions even of only a very general or generic nature. The honest concurrent use of similar bottles by Italian producers could be permitted without damage to consumer interests, by simply requiring adequate labelling indicating national origin.

The prohibition of Article 30 applies to a wide range of actions. It goes far beyond coverage of formal governmental action or legislative acts and applies to government agencies, nationalized bodies, and indeed any public bodies wholly or partially financed by government or as a result of charges levied under legislative authority, whether direct or by way of secondary legislation, as well as to acts of regional or local government both by way of administrative action, bylaws, and to judicial decisions. The acts involved may be formal or informal and cover practices and cus-

[5] Case 16/83 [1984] ECR 1299: [1985] 2 CMLR 238.

toms as well as legal rules or administrative systems having the required effect on trade. Even if the effect on trade is slight, Article 30 will still apply since the principle of *de minimis* established by *Völk* v. *Vervaecke*[6] has no application in view of the absence in this Article of any reference to 'having an effect on trade between Member States'.

Grundig–Consten provided early evidence of the approach that the Court would adopt to the potential conflict between Articles 36 and 85. After confirming that Article 36 cannot limit the field of application of Article 85, the Court continued:

> Article 222 confines itself to stating that 'the Treaty shall in no way prejudice the rules in Member States governing the system of property ownership'. The injunction in . . . the contested decision to refrain from using rights under national trade mark law in order to set an obstacle in the way of parallel imports, does not affect the grant of those rights, but only limits their exercise to the extent necessary to give effect to the prohibition under Article 85(1). The power of the Commission to issue an injunction, for which provision is made in Article 3 of Regulation 17 . . . is in harmony with the nature of the Community rules on competition, which have immediate effect and are directly binding on individuals. Such a body of rules . . . does not allow the improper use of rights under any national trademark law in order to frustrate the Community's law on cartels.[7]

Some two years later, the Court confirmed that similar principles apply to patents. In *Parke, Davis* v. *Probel*,[8] the Court held that the normal use of a patent cannot be treated as abuse of a dominant position under Article 86, whereas its use for some other purpose would be capable of constituting such an abuse, for example, if used in a way equivalent to the use of the 'GINT' trade mark in the *Grundig* case. The Court here ruled that a patent holder, the owner of a Dutch patent for an antibiotic process, was entitled to prevent the marketing of this product in Holland by the defendant, who had obtained it from a third party who had manufactured it in Italy, where patent protection was at that time totally unavailable for drugs. The Court held that the use of the Dutch patent to prevent importation of the unpatented Italian product was a normal use of the right of exclusivity under the patent, since the patent holder had not itself placed the product on the Italian market nor indeed was it possible for itself to have had its own product patented in that country. This principle applied whether or not the price charged in Holland exceeded that payable for the drug as imported by Centrafarm.

This case was, of course, not concerned with the case of parallel imports from Member States where the products had been placed originally on the market either by the plaintiff or by its own licensee, and

[6] Case 5/69 [1969] ECR 295: CMLR 773.
[7] Cases 56 and 58/64 [1966] ECR 299, 345: CMLR 418.
[8] Case 24/67 [1968] ECR 55: CMLR 47.

revolved, therefore, around the interpretation of Article 86 rather than Articles 30 to 36. The terms of the judgment, however, gave clear indication that once a case was referred to the Court involving an alleged abuse of an intellectual property right in order to prevent parallel imports, the Court was likely to look sympathetically on a defendant who relied on the argument that marketing of the patented product acquired in the normal course of business from the plaintiff would be protected by Article 30. In the early 1970s, cases were referred to the Court dealing with copyright, patents, and trade marks raising this very point, and decisions followed predictable lines in the light of the earlier case-law. The first case in time was the *Deutsche Grammophon* v. *Metro*[9] case, a reference from a German Court under Article 177 which we have already considered in relationship to the development of law on resale price maintenance. The Court, affirming Advocate-General Roemer, also held that the right of the holder of exclusive rights, namely copyrights, in sound recordings to prohibit the import of records into its own country which have been marketed by its owner in another country is not any part of the 'industrial and commercial property' for which protection is available under Article 36. The owners of the right to the copyright are limited to protection afforded to enable the first sale to be made either in its own country or in a foreign country. Once the goods have been placed on the market, however, it is a 'measure having equivalent effect' for the Court of the country where the original copyright has been granted to seek to prevent the import by a purchaser in the normal course of trade from seeking to market the goods, as parallel imports, in Germany.

The application of the same principles to patent and trade mark cases was dealt with in the well-known *Centrafarm* cases of 1974 at the Court which involved the Sterling-Winthrop group, a drug manufacturer in the United Kingdom, and Centrafarm whose business involved the reselling of drugs purchased in the United Kingdom and Germany in The Netherlands where for a variety of reasons a higher price level prevailed. In the first case (*Centrafarm* v. *Sterling Drug Inc.*),[10] Centrafarm had bought various drugs from subsidiaries of Sterling-Winthrop under the trade name of 'Negram' in both the United Kingdom and Germany for resale in Holland. Sterling tried to prevent Centrafarm from selling the drugs at a profit in Holland, claiming exclusivity under Dutch patent law. The same basic facts applied in the second case, *Centrafarm* v. *Winthrop BV*,[11] when

[9] Case 78/70 [1971] ECR 487: CMLR 631. See Ch. 14, p. 272.

[10] Case 15/74 [1974] ECR 1147: 2 CMLR 480. For a more detailed analysis of this and other cases referred to in this Chapter, see G. Leigh and D. Guy, *The EEC and Intellectual Property* (London: Sweet & Maxwell, 1981), ch. 8.

[11] Case 16/74 [1974] ECR 1183: 2 CMLR 480. Other characteristics of the 'specific object' of intellectual property rights are considered in *Parke, Davis* v. *Probel* and *Hag* (no. 2)

Centrafarm marketed in Holland under the trade mark, Negram, drugs previously purchased in the United Kingdom. Winthrop relied on the fact that it was the proprietary owner in Holland of the trade mark, Negram, covering this particular drug, to stop Centrafarm from selling in the Dutch market. The Court was able to apply precisely the same principles in both cases. It held that the effect of Article 36 is to permit exceptions to the principle of free movement of goods only when the restrictions can be justified for the purpose of safeguarding rights constituting the specific object of the intellectual property right. This phrase had been used by the European Court in the *Metro* case; but now it is defined in its application to patents and trade marks. Dealing with patents, the Court identified the characteristics specific to them as: first the right to exploit an invention for the purpose of making and then selling a product and second the corresponding right to prevent an 'infringement' by a third party, seeking either to make or sell the product without the patent owner's consent. What the phrase did not include was any right to prevent imported products from coming into its territory, which had been marketed in another territory either by it or by a third party with its consent, merely because it thereby sustained economic loss, possibly because of Member State intervention in pricing levels. Applying the same criteria to trade marks, the court suggested that the 'specific object' covered the protection of the owner against competitors by the exclusive use of the mark for the purpose of first putting the product into circulation, but did not extend as far as preventing the import of goods bearing the mark which legitimately have been marketed into or in another Member State.

The limitation thus placed on the holders of both patents and trade marks was necessary because of the express words of Articles 30 to 36, which were incompatible with any wider view of their rights, regardless of the differences in prices between the respective countries. Sterling-Winthrop was, therefore, unable to prevent Centrafarm from marketing drugs purchased in a conventional manner on the United Kingdom market and reselling them in Holland. By reason of the first sale in the United Kingdom, the rights of the holder of the intellectual property were 'exhausted', and thereafter the purchaser of the products which are the subject of the intellectual property rights was free to market them as it might think fit.

Some four years later, Centrafarm was involved in a further pair of cases involving similar facts but where the product, a tranquillizer, was not resold exactly as when originally purchased in the United Kingdom but in a repackaged form. The issue presented under Article 177 to the

Whilst the relevant French word used in this context is 'objet', a more exact English equivalent may be 'subject matter'.

European Court in the first case of *Hoffmann-la Roche* v. *Centrafarm*[12] was whether this slight change of circumstances did give the owner of the relevant trade mark the right to prevent the sale, unless the goods had been purchased from the original patent holder just as in the earlier cases. The Court's ruling was that once again Centrafarm remained free to sell the product, even after repackaging, provided that the same simple conditions were satisfied. Thus, the repackaging itself must not adversely affect the original condition of the product, and prior notice had to be given to the owners of the trade mark of the proposed marketing of the repackaged product; the repackaging must state by whom it had been carried out. The final requirement was that it must be shown that enforcement of the trade mark rights against the proposed repackager would contribute to the artificial partitioning of the Common Market. The national court must be satisfied that each of these conditions has been met, if it is to be able to conclude that the enforcement of the trade mark right does amount to a disguised restriction on trade within the meaning of the second sentence of Article 36.

A similar issue was raised in the companion case of *Centrafarm* v. *American Home Products Corporation.*[13] The defendant here was the registered proprietor of trade marks 'Seresta' and 'Serenid' used to describe particular sedatives in the Netherlands and the United Kingdom respectively. The effect of the drug was similar, but the composition and taste mildly different. Centrafarm bought the drug in the United Kingdom under the Serenid D mark and repackaged it without the consent of the defendant with the 'Seresta' mark reselling it under that mark in The Netherlands. Although there were the slight differences in the characteristics of the product, they were effectively identical and the Court dealt with the case on that basis.[14] The Court ruled that the use by the defendant, the registered proprietors of their trade mark, of the national right to prohibit the marketing of a product to which their trade mark had been affixed in an unauthorized manner was part of the specific subject matter of the trade mark and, therefore, justified under the first sentence of Article 36. The action, however, would also have to be justified by the criteria contained in the second sentence of Article 36, namely that it would not artificially result in the application of restrictions on trade between Member States.

In the case both of patents and trade marks, therefore the Court's jurisprudence has placed substantial limits on the ability of the owner of

[12] Case 102/77 [1978] ECR 1139: 3 CMLR 217.

[13] Case 3/78 [1978] ECR 1823: [1979] 1 CMLR 326.

[14] Whether different trade marks were used in different Member States for such purposes was a question for the national Court. National Courts in at least some jurisdictions, including Germany, have held that the use of different trade marks in these circumstances is not an artificial partitioning of markets and is, therefore, valid.

the right to prevent purchasers who have lawfully acquired the goods in one country from marketing them in a second Member State where the owner's rights are likewise protected by parallel patent or trade mark.[15] The broad principles behind the interpretation of Article 36 that the Court would apply had now been clarified, but on those principles were to be grafted several important exceptions to be established by the Court in later cases. In the 1976 case of *EMI Records* v. *CBS United Kingdom*,[16] the issue was whether the principle of exhaustion of rights of trade marks would enable EMI which owned the relevant mark 'Columbia' throughout the EC to prevent CBS from manufacturing records under that trade mark through subsidiaries established in the various Member States. The original trade mark had been owned solely by US companies. By a series of agreements over a period of thirty years prior to the expiry of the Second World War, ownership was then divided so that CBS owned the mark in the United States, whilst in other parts of the world EMI owned it, including the whole of the European Community. Although there had been some co-operation subsequent to the end of the Second World War between the parties, ownership of the trade mark rights had been clearly divided for a substantial period. The Court held that the Treaty did not prevent EMI as proprietor of a trade mark throughout the Community from exercising its right to prevent the import of similar products carrying the same mark and arriving from a third country outside the EC. The rationale of this decision was that exercise of such rights could not affect trade between Member States, even if it were classifiable as a measure equivalent to a quantitative restriction under Article 30, and thus it could not pose any threat to the integration of trading within the Common Market which that Article was designed to bring about. In the light of the decision in the *Hag* (no. 2)[17] case, however, which is discussed below, it is unclear whether EMI would have been in the same fortunate position had CBS owned the trade mark rights even in only one of the Member States.

A similar result was achieved in the case of *Polydor Records* v. *Harlequin Record Shops*,[18] a case referred under Article 177 to the European Court by a United Kingdom Court. Polydor, the plaintiff, owned the copyright in recording of some songs made by the Bee Gees group. The defendants imported copies of the records from Portugal where they had been manufactured by a licensee of the plaintiff. Both the importation into and sale in the United Kingdom were without the consent of the plaintiff. When sued for copyright infringement, the defendants referred to the Free Trade

[15] It is clear from *Keurkoop* v. *Nancy Kean Gifts* (Case 144/81 [1982] ECR 2853: [1983] 2 CMLR 47) that similar principles will likewise be applied to registered designs, though the Court refrained from explaining what it considered the inherent elements of this form of intellectual property right.

[16] Case 51/75 [1976] ECR 811: 2 CMLR 235.

[17] See below at p. 332.

[18] Case 270/80 [1982] ECR 329: 1 CMLR 677.

Agreement between the Community and Portugal, which contained clauses equivalent to those set out in Articles 30 and 36 of the Treaty. Although the Court agreed that the language was similar, it declined to allow the defendant the benefit of the principle of exhaustion of rights. Its explanation for this decision was that the Free Trade Area provisions relating to free movement of goods did not have the same purpose as the Treaty of Rome. The Free Trade Area agreement between the Community and a non-Member State did not purport to create a single market reproducing as closely as possible the conditions of a domestic market, because the instruments which the Community had at its disposal to achieve uniform application of Community law and the progressive abolition of legislative disparities had no equivalent in the context of a relationship simply between the EC and Portugal. The principle of strict territoriality for copyright items no longer existed between each of the Member States, but could continue to exist in the relationship of the Community with non-member countries, even those with whom it had a free trade agreement.

The second important exception to the basic rule of exhaustion of rights is derived from the case of *Parke, Davis* v. *Probel*. In that early case the Court had held that the exhaustion principle did not apply where goods had been placed on the market originally in a country where no patent protection was available, and without the consent of Parke, Davis who held the patent protection in The Netherlands where the defendants sought to sell the goods. This decision had been based on the principle that the principle of exhaustion could not apply when Parke, Davis had not had the opportunity of obtaining a reward for the investment in the invention of the relevant drug, because of the failure of Italian law to provide patent protection. In the more recent case of *Merck* v. *Stephar*,[19] the facts were similar to *Parke, Davis* except that Merck had themselves marketed the drug known as 'Moduretic' in Italy, even though unable to obtain a patent either for the drug or the manufacturing process. Merck had obtained some return for their investment through their sales in Italy, even though the lack of patent protection possibly reduced the price which they were able to obtain; the Court felt that the goods had been placed in commerce within the Common Market. Merck could not expect to guarantee that simply by placing the product on the market the full reward potentially available for patented products would likewise be available for its non-patented products. By contrast, however, the original placing on the market of a patented process not by the voluntary act of its proprietor but because a mandatory licence has been granted under the provisions of national patent law to a licensee would lack the essential

[19] Case 187/80 [1981] ECR 2063: 3 CMLR 463.

element of the proprietor's consent, and the patentee would be entitled to invoke his patent against parallel imports into a second Member State where its patent protection existed.[20] Similar principles would apply in the case of licences of right also available under the national patent law of certain Member States.

On the other hand the proprietor cannot prevent the actual import of products subject to the licence of right provided that the importer undertakes to pay the licence fee laid down by agreement or, in default of agreement, as determined by the national patent authorities. The importer has to be treated in the same way as the domestic manufacturer of such products under the licence of right. This means that the importer cannot be required for example to give security in advance for royalties payable or to delay distribution of the product until satisfactory health and safety clearance has been given. The fact that the relevant product was a pharmaceutical product imported from a Member State where patenting of such products is not possible again did not affect the principles involved.[21]

Another exception to the rule of exhaustion is where the property right concerned relates not to goods but to services covered by the provisions of Articles 59 to 66. The Belgian court made two successive references for preliminary rulings under Article 177 in the *Coditel* cases.[22] In the first case the European Court held that the provisions of Article 59 did not prevent the owner of copyright in a film under national law from invoking its rights against cable TV companies. The facts of the *Coditel* (no. 1) case (European Court 1980) were that Ciné Vog had acquired the exclusive right to show the film 'Le Boucher' in Belgium for a period of seven years provided that it did not show it on television for a minimum of forty months after the first screening in Belgium. The film was sold by the owners for screening on German television, but when shown the transmission was picked up by Coditel which operated the cable television service in Belgium and then distributed to its own cable subscribers in Belgium. Ciné Vog then claimed that Coditel had breached the copyright vested in it under the original seven-year agreement.

In the second reference (*Coditel* no. 2) the issue before the Court was

[20] This principle was established in *Pharmon* v. *Hoechst* Case 19/84 [1985] 3 CMLR 775. In *EMI Electrola* v. *Patricia* (European Court 1989) (Case 341/87 [1989] ECR 79: [1989] 2 CMLR 413) the same principle was applied in the context of copyright. The owner of a Cliff Richard record, for which copyright in Denmark had expired, was allowed to continue to make use of its copyright in Germany so as to exclude imports from Denmark. It had given no consent to the manufacture of the record in Denmark and therefore, by analogy with *Pharmon* v. *Hoechst*, Article 30 had no application.

[21] *Allen and Hanburys Ltd.* v. *Generic (UK) Ltd.* Case 434/85 [1988] ECR 1245: 1 CMLR 701: [1989] 2 CMLR 325.

[22] *Coditel* v. *Ciné Vog Films* (no. 1) Case 62/79 [1980] ECR 881: [1981] 2 CMLR 362. *Coditel* v. *Ciné Vog Films* (no. 2) Case 267/81 [1982] ECR 3381: [1983] 1 CMLR 49.

whether an exclusive licence to show a film would be subject to the prohibitions of Article 85; in the course of its judgment the Court indicated that the distinction drawn in the earlier cases between the existence and exercise of industrial property rights under Article 36 could also apply when such rights were being exercised in the framework of the provision of services. In both cases it held that cinema films were made available to the public by performances capable of being repeated without limit, so that the owners of copyright had legitimate interests in calculating the fees due in respect of an authorization to show the film 'on the basis of the actual or probable number of performances and in authorising a television broadcast of the film only after it had been exhibited in cinemas for a certain period of time'. It treated in other words the nature of copyright in a film as inherently different from that in either a book or gramophone record. It is noteworthy that, in his opinion in the first *Coditel* case, Advocate-General Warner indicated that in his view the omission from Articles 59 to 66 of the Treaty dealing with services of any provision for the protection of industrial property was more likely the result of oversight than of deliberate intention. Given the opportunist nature of the action of Coditel, the Court's decision has commanded general support, and especially the finding after appropriate economic analysis that an exclusive copyright licence does not necessarily violate Article 85.

For a long time it was believed that the fact that a particular trade mark came from a 'common origin' or 'common source' also protected it from challenge by the proprietor of a related mark. This belief arose from the case in 1974 of *Van Zuylen Freres* v. *Hag*[23] where identical trade marks had come from a common source. Van Zuylen had become the assignee of the Benelux trade mark in 'Hag' coffee, having acquired it by purchase from the Custodian of Enemy Property following the confiscation of the assets of the former Belgian subsidiary of Hag AG of Germany at the end of the war. The trade mark was registered in Benelux, and Van Zuylen sold coffee at the retail level. In the meantime, the original German company, Hag AG of Bremen, had recommenced the marketing of coffee in the Benelux area under its own identical trade mark, and Van Zuylen tried to stop Hag AG from importing coffee into Luxembourg, claiming it to be an infringement of the mark owned and registered by it. The Court rejected the claim by Van Zuylen, repeating the distinction already made in earlier cases between the existence and exercise of rights. It held that it could not allow the holder of a trade mark effective in one Member State to prohibit the marketing in that Member State of goods legally produced in another Member State under an identical trade mark that had the same origins.

[23] Case 192/73 [1974] ECR 731: 2 CMLR 127.

This decision was, however, overruled in the *Hag* (no. 2)[24] case towards the end of 1990. Here in his opinion Jacobs A.-G. convincingly argued that the original *Hag* decision was flawed in that the consent of the owners of the Benelux mark to the use of the similar German trade mark in Benelux had been wrongly presumed, from the mere fact of the original common source. The facts of the *Hag* no. 2 case, however, presented the reverse situation from the earlier case, as it was now the German Hag company (which in the first case had successfully defended its right to sell coffee in Benelux under its German trade mark) which sought to prevent the owner of the Benelux trade mark (successor in title to Van Zuylen) from selling its coffee in Germany.

In upholding the right of the German company to prevent the Belgian company from selling its coffee in Germany, Jacobs A.-G. pointed out that the doctrine of common origin is nowhere referred to in the Treaty and had no rational basis. The earlier judgment had failed to analyse satisfactorily the inherent subject matter of trade marks and had, therefore, underestimated their importance as property rights. They were necessary to protect the manufacturer who consistently produced high quality goods and should not be regarded as a secondary form of intellectual property rights since they enable existing quality to be recognized in the new markets. To remove the exclusive right to the mark from a manufacturer takes from him the power to influence and capitalize on the goodwill associated with it. The specific subject-matter of the trade mark is the guarantee given by its owner that it has the exclusive right to use it in order to put into production and circulation a quality product, which can be protected against competitors who later seek to acquire unfairly the benefit of the original manufacturer's goodwill.

Whether the decision applies also to the voluntary (as opposed to compulsory) division of trade marks is less clear. Some commentators have indicated that under Community law the voluntary assignment of trade mark rights should not affect the rights of undertakings under national law to prevent imports into their respective territories of goods manufactured by the other. They warn, however, that agreements of this nature may still be caught by Article 85(1) in so far as they had as their object or effect a partitioning of Member State markets within the Community.[25]

Concern had been expressed as a result of *Hag* (no. 1) that the European Court would also show hostility to attempts by owners of trade marks in one Member State to prevent the use in that Member State of another trade mark originating from a second Member State that might be considered confusingly identical, even where there was no element of

[24] Case C-10/89 [1990] 3 CMLR 571.

[25] See on this subject R. Joliet, 'Trade Mark Law and the Free Movement of Goods' (1991) *IIC* 303.

common origin and the relationship of the parties was totally at arm's length. This fear, however, was laid to rest by the *Terrapin (Overseas)* v. Terranova[26] case two years later. Terranova was a German company which had manufactured plaster for façades of buildings for a considerable period of time and was the registered proprietor in Germany of a number of trade marks including the words 'Terra', 'Terranova', and 'Terra-fabrikate' in respect of various building materials. Terrapin was an English company which manufactured prefabricated houses and components under the trade mark 'Terrapin'. The English company operated in Germany directly and through a subsidiary and had applied to register their trade mark there. A number of legal actions arose between the parties because of the similarity of their respective trade marks until finally the German Supreme Court referred the issue to the European Court under Article 177: the issue was whether the registration of the United Kingdom mark in Germany would give rise to confusion with the mark owned by the German company, a legitimate proprietor of its own trade mark in its own country; and if so whether the German company could then legitimately within the terms of Article 36 oppose use of the name 'Terrapin' to describe the English product in Germany.

Had the Court taken an excessively purist view of industry property rights in the Community under the regime of Articles 30 to 36, one might have expected that the doctrine of honest concurrent use which the Court did apply in other cases might have prevailed. Nevertheless, the Court, clearly feeling that the risk of confusion was a genuine one owing to the similarity of the respective companies' business and trade marks, found that a refusal by German authorities to register 'Terrapin' in Germany would not be in breach of Article 30. The situation was very different from that in the *Hag* (no. 1) case; the trade mark had arisen quite separately under the laws of different Member States and there were no links of a legal or economic nature between the companies. The desirability of free movement of goods as against the legitimate interest of the proprietor of the threatened trade mark had to be reconciled in such a way as would both protect legitimate use of trade mark rights whilst preventing their abuse in a manner enabling segmentation of the territories within the Common Market. The Court concluded, therefore, that the German Court could restrain the use of the imported mark, since otherwise, if the principle of free movement of goods was nevertheless to prevail in such a case, even a specific and important object of the system of industrial and commercial property rights, the differentiation of competing products, would be seriously undermined. It was essential, however, in such cir-

[26] Case 119/75 [1976] ECR 1039: 2 CMLR 482.

cumstances that the national Court be satisfied both that there were no links between the respective companies and that the origin of the marks was completely independent.[27]

The wide effect of Article 30 on the exercise of intellectual property rights will by now be apparent. It is also important, however, to understand the relationship of the rules relating to free movement of goods (Articles 30 to 36) with the rules relating to freedom of competition (Articles 85 and 86). Although their objectives are not necessarily in conflict, there is equally no inherent harmony between them since each group of rules operates in its own particular way and largely without overlap. Articles 30 and 36, as we have seen, are addressed to Member States and their executive, legislative, and judicial authorities at all levels, to prevent the use of legislation, judicial decision, or administrative practice from maintaining non-tariff barriers, save to the limited extent allowed by the narrow interpretation placed by the European Court of Justice on Article 36. By contrast, Articles 85 and 86 are addressed only to undertakings, not to Member States, and for their operation require either an agreement or concerted practice between at least two undertakings or the existence of the abuse of a dominant position. A sanction for entering into an agreement in contravention of Article 85 is that an undertaking may incur fines and penalties from the Commission; no such sanction is available if a government or its agency breaches Article 30. On the other hand, Article 85(2) provides that the offending clauses of such an agreement are void and, therefore, without legal effect. By contrast a Court in which an Article 30 defence is raised, by a defendant accused for example of infringing the patent or trade mark rights of a plaintiff seeking to exercise such rights in a situation where the European Court has ruled it would be in breach of Article 30, is required to uphold its defence. It must moreover reject the plaintiff's claim in such a way so as to ensure that neither civil nor criminal sanctions are imposed upon the defendant.

In general terms, both sets of rules have similar objectives, the freeing of trade between Member States from restraints imposed by either private or public sources. The European Court, however, has made it clear, in a number of cases that it will not allow one set of rules to be used as justification for freedom from the other set. Thus, as far back as the *Grundig*[28] case in 1966, these provisions later being embodied in Regulation 67/67, the European Court had laid down that protection

[27] It should not be assumed that the use in national courts of copyright or trade mark law to prevent the import of goods lawfully marketed under a distinctive name in another country is in any way facilitated by this decision; this was made clear by *Musik-Vertrieb Membran* v. *GEMA* Cases 55 and 57/80 [1981] ECR 147: 2 CMLR 44 and *Dansk Supermarked* v. *Imerco* Case 58/80 [1981] ECR 831: 3 CMLR 590.

[28] Cases 56 and 58/64 [1966] ECR 299: CMLR 418. See also Ch. 5, pp. 52–5.

given by Article 36 to industrial and commercial property should not enable those rights to be used in a way which supported a distribution system that sought to exclude the possibility of parallel imports; and the same principle underlies subsequent decisions of the Court in the case of *Sirena* v. *Eda*[29] and in the *DGG* v. *Metro* case referring to copyright and the several *Centrafarm* cases in the years immediately following. On the other hand, since the freedom of trade rules protect existence of industrial and commercial property, the mere exercise of rights under national law which do not conflict with the principle of exhaustion of rights would not be in breach of Article 86, a principle established by both the *Probel* and *Terrapin* cases.

A similarity between both sets of rules is, of course, that Member States are under a duty not to introduce legislation which would cause a breach of them. This obligation can be specifically seen in Articles 30 and 34, but has been stated clearly also to apply to the competition rules in the case of *Inno* v. *ATAB*[30] (European Court 1977). The European Court in that case spelled out the duties of Member States not to enact measures enabling private undertakings to escape from the constraints of the rules against competition even if in practice such measures might also well offend Articles 30 or 34.

Whilst only Articles 85 and 86 have a *de minimis* provision which is inherent from the wording of the Article ('. . . and which may affect trade between Member States . . .'), it is arguable that Articles 30/34 are to be interpreted as if containing an inherent 'rule of reason'. In *IDG* v. *Beele*[31] (European Court 1982), the Court concluded that, in a straightforward case of 'passing off' involving exact imitations of cable ducts, Article 30 should be interpreted so not as to prevent a rule of national law (applying to domestic and imported products alike) from allowing a trader to obtain an injunction against another trader to prevent him from continuing to market a product lawfully marketed, but which 'for no compelling reason' was sold in a form almost identical to the first named product, therefore causing needless confusion between the two.

Exceptions to the prohibitions on restrictions of competition contained in or effected by agreements and concerted practices are based on grounds that are mainly economic in nature, namely the two negative and the two positive conditions of Article 85(3), whereas the better view in relation to Article 36 appears to be that arguments of an essentially economic

[29] Case 40/70 [1971] ECR 69: CMLR 260.

[30] Case 13/77 [1977] ECR 2115: [1978] 1 CMLR 283. The case involved a Belgian statute requiring all tobacco products sold at retail to have a label affixed showing the maximum price. This was claimed by the Belgian Government to have as its purpose the prevention of fraud on the tax authorities, but the Court ruled it had mainly been introduced so as to protect a resale price maintenance system. The case is further considered in Ch. 23.

[31] Case 6/81 [1982] ECR 707: 3 CMLR 102.

nature cannot be used to justify claims for exemption; on the other hand, the mere fact that governmental measures introduced on other grounds referred to in Article 36, e.g. national security, may also have some economic advantages will not automatically mean that the measure will be held to violate Article 30. The precise relief, however, which such measures may be permitted to give to government will be strictly limited to those proportionate to the national security interest required to be protected.[32]

The development of competition policy by DG IV in respect of patent licences as discussed in Chapter 15 might possibly cause difficulties in the future under Article 30. While case-law has made clear that intellectual property rights cannot be invoked as a method of preventing or restricting parallel imports, i.e. goods imported by persons who have obtained them lawfully directly or indirectly from licensees in other Member States, the position is less clear with regard to the exercise of such rights directly or indirectly either by the original licensor against one of his own licensees or by one licensee against another licensee of the same licensor. Under the terms of the patent licence group exemption permitted under Regulation 2349/84, the restrictions that can be placed on licensees include a total ban on 'active' sales by licensees outside their area, and a five-year ban even on responding to requests for sale received from outside the area on a passive basis. If, in infringement of such agreements, sales of patented products were made by a licensee into the territory of another licensee, an action to restrain the licensee from doing so might be met by a claim that Article 30 itself prevented courts from approving it. The established precedents of *Centrafarm* could be quoted in support of this view notwithstanding that the patent licence terms themselves clearly were covered by Article 85 and the group exemption, so that the rights were not being used in an abusive way to avoid Article 85 as in the *Grundig* case.

In other words, while it has long been clear that the application of Article 85(1) to an agreement will not be restricted because the agreement might also involve a breach of Article 30, it is far less clear whether the fact that Article 85(1) will *not* be held to apply to a particular agreement, e.g. as a result of a group exemption, means that Article 30 cannot then be raised by a defendant against whom the agreement is pleaded.

The European Court of Justice might well seek, as a matter of policy, to limit the application of Article 30 to situations other than those where the Commission itself had, after protracted consultations and negotiations,

[32] This is illustrated by the *Campus Oil* v. *Ministry of Industry and Energy* Case 72/83 [1984] ECR 2727: 3 CMLR 544, where the Irish Government had introduced measures to require various oil companies to acquire a certain minimum quantity of oil from its only State refinery. The Court ruled that the national Court had the jurisdiction to approve such measures to the extent that they applied only to the minimum supply requirements without which the State's public security would be affected, or at which the level of production had to be maintained to keep the refinery's production capacity available in the event of a crisis.

established a model for patent licensing arrangements within the Community which it followed would not in its view restrain competition within the terms of Article 85. The fact that the restriction permitted went beyond the reach of that open exclusive licence permitted in the *Maize Seed* case (even though far from comprising absolute territorial protection of a kind clearly unacceptable under Article 85) should not of itself permit Article 30 to be successfully raised as a defence by a licensee sued in these circumstances. The Court would be likely to seek a solution which harmonized the approaches of the two Articles rather than emphasized the differences between them.

17

Article 86: The Concept of Dominance

The preceding chapters in this book have dealt almost exclusively with the development of the law relating to Article 85, in its application to both horizontal and vertical agreements and concerted practices. Although the resources devoted by DG IV to the administration of Article 85 have far exceeded those spent on Article 86, this latter Article, the second of the twin pillars of the competition policy established under the Treaty, has nevertheless given rise to some decisions of great significance, both from the Commission itself and from the European Court of Justice. The Article remains of increasing importance to the Commission in its continuing enforcement of competition policy; it retains substantial potential for further development and application by way of control, in a great variety of circumstances, of the problem of the abuse by undertakings of dominant positions in specific markets.

This Chapter examines the concept of dominance, which in turn leads necessarily on to consideration of the process of market definition, both of products (goods and services) and the relevant geographical areas in which they are sold or provided. The following Chapter will be concerned with the concepts of abuse and abusive conduct and exploitation.

It is an abstract word, but used in a commercial context it refers to a position of power for an undertaking in relation to a specific product market and within a relevant geographical market both of which have to be defined. Fortunately dominance was considered in detail by the European Court in one of its early cases arising out of this Article, *United Brands.*[1] The market power of this company (engaged in the large-scale international fruit business) derived substantially from the degree to which it had integrated its various activities. Although it also had a market share in the four relevant Member States of between 40 and 45 per cent, its strength and dominance derived from the fact that at each stage of the production and distribution process it was able from its own resources to accept and react to variations in demand. It alone was capable of carrying over two-thirds of its production in its own fleet of specially designed ships. It alone was able to advertise the 'Chiquita' brand name and to revolutionize the commercial exploitation of the banana. Its large capital investment in the purchase and equipment of its plantations gave it an

[1] 27/76 [1978] ECR 207: 1 CMLR 429.

ability to increase its source of supply, so as to overcome any effects from disease or bad weather. It was enabled by reasons of the scale of its capital investment to introduce a distribution system of a highly perishable commodity through the extensive use of refrigeration facilities, giving it a strategic advantage over all its competitors. In particular, the ability to control the timing of the ripening process at close quarters to the ultimate retail market within the Community gave it an important element of strategic leverage, placing all its competitors at a disadvantage. The combined effect of these measures led the Court to characterize the concept of dominance as follows: 'The position of economic strength enjoyed by an undertaking enabling it . . . to behave to an appreciable extent independently of its competitors and customers and ultimately of its consumers. In general, it derives from a combination of several factors which taken separately are not determinative.'[2]

In the *Hoffmann-la Roche* case, the Court gave a rather more extended definition to the concept of dominance:

> The dominant position . . . relates to a position of economic strength enjoyed by an undertaking which enables it to prevent effective competition being maintained on the relevant market by affording it the power to behave to an appreciable extent independently of its competitors, its customers and ultimately of the consumers. Such a position does not preclude some competition which it does where there is a monopoly or quasi-monopoly but enables the undertaking which profits by it, if not to determine, at least to have an appreciable influence on the conditions under which that competition will develop, and in any case to act largely in disregard of it so long as such conduct does not operate to its detriment.
>
> . . . The existence of a dominant position may derive from several factors which taken separately are not necessarily determinative but among these factors a highly important one is the existence of very large market shares.[3]

Such comments by the Court need to be illustrated by reference to specific markets in cases heard by it. In *Hoffmann-la Roche*, the Court found that Roche held the following market shares within the Community:

Vitamin A — 47 per cent
Vitamin B_2 — 86 per cent
Vitamin B_3 — 64 per cent
Vitamin B_6 — 95 per cent
Vitamin C — 68 per cent
Vitamin E — 70 per cent
Vitamin H — 95 per cent

[2] [1978] 1 CMLR 486, 487.

[3] Case 85/76 [1979] ECR 461, 520: 3 CMLR 211, 274. This case is often popularly called the *Vitamins* case.

In 1974, the turnover of the company in the Common Market comprised 65 per cent of the total sale of vitamins world-wide. The group had altogether some 5,000 customers, many of whom were large multinational companies and who purchased in large orders the entire range of vitamins produced, particularly under pressure of the fidelity rebate scheme already discussed. On the basis of these figures, for all except Vitamin A there was virtually an overwhelming presumption of dominance because of the market share alone, which gave Hoffman-la Roche a position of strength, allowing it a certain freedom of action in responding to changes in market conditions. In the case of Vitamin A, where the market share was lower, the factors upon which the Court placed heavy reliance were the great disparity between its market share and that of its next largest competitors, and the wide technological lead which it enjoyed over them. The absence of real challenge in many of the markets, and the extent to which the Roche sales network was more highly developed than any of its competitors, were also factors that could be taken into account.

On the other hand, whilst the Commission had placed weight on the pure size of the company and the volume of turnover which it provided, the Court did not accept that size and turnover alone could be indicative of dominance. The Court also placed far less reliance than the Commission on the range of products produced by Roche as well as the fact that it had retained market share over a continuous period of time, pointing out that this could be explained by a large number of other factors, including simply its ability to compete effectively in a normal manner.

The concept of dominance, however, is not limited to large-scale markets but may be found also in markets that are narrow and where one undertaking has either by virtue of legislation or other circumstances obtained either a monopoly or very strong position which completely or substantially reduces competition. The earliest example of this was the *General Motors Continental* case.[4] The alleged breaches of Article 86 involved the charging of excessive prices for granting type approval on five cars manufactured in Germany but which needed such official approval before use in Belgium. GM Continental had carried this out as an authorized agent of the original manufacturer under Belgian law. The relevant market was the provision of such approved services for General Motors' cars in that country; and the five cars in respect of which the complaint arose came into Belgium not through official channels but by way of parallel imports. Type approval could only under Belgian law be carried out by the manufacturer or its officially approved agent, and the price was fixed by the manufacturer. The abuse lay, according to the Commission, in the imposition of an excessive price relative to the

[4] Case 26/75 [1975] ECR 1367: [1976] 1 CMLR 95.

economic value of the service provided; it had the effect of making it less profitable to import cars into Belgium outside official distribution channels by imposing an additional financial charge. General Motors claimed that the five transactions, for which it admitted charging an excessive price, were the first transactions after it took over testing responsibility from the Belgian Government, and that the prices were those which would normally have been applicable on the import of cars from the United States. They also pointed out that they had, after receiving complaints, reduced their charges to actual cost and had refunded the excess to the five owners. All this had indeed taken place before the European Commission's investigation.

On appeal, the Court not surprisingly indicated that it felt that the Commission had been unjustified in raising what appeared so trivial a case under Article 86, especially as General Motors had corrected the overcharge made at an early stage of the inquiry, and the decision was annulled. It has nevertheless a certain value as a precedent, as pointed out in the Fifth Annual Report.[5] Any undertaking which holds an exclusive legal right for the performance of a statutory duty, delegated by a State or public authority, is subject in carrying out its duties not to abuse the powers conferred in a way that may enhance other objectives, such as rendering less attractive the possibility of parallel imports.

A direct descendant of the *General Motors Continental* case is the more recent *British Leyland PLC* v. *Commission* case[6] (European Court 1986), where the Commission's fine on British Leyland was upheld. The actions about which the Commission complained here were BL's refusal to issue national type approval certificates for Leyland cars imported into the United Kingdom other than through official channels, and for charging £150 for certificates of conformity to those national type approvals for those dealers and individuals who sought to import Metros with left-hand drive for the Continent.[7]

The ownership of intellectual property rights often provides a dominant position in a product covered by the relevant patent or other intellectual right. In *Volvo* v. *Eric Veng (UK) Ltd.*[8] the car makers Volvo sued Veng for breach of its registered designs covering a front wing panel for a particu-

[5] p. 29. An interesting development of this principle is to be found in *Hilti* v. *Commission* (Case T 30/89 [1992] 4 CMLR 16, where the Court of First Instance ruled that it was an abuse of dominant position for a company owning a patent, subject under national legislation to grant a licence of right, to demand a fee six times higher than that ultimately awarded by the Comptroller of Patents, thereby prolonging the proceedings needlessly.

[6] Case 226/84 [1987] 1 CMLR 185.

[7] *Hugin* v. *Commission* Case 22/78 [1979] ECR 1869: 3 CMLR 345 raised similar issues. The market here was that of spare parts for Hugin cash registers. The Court's decision, however, to find no breach of Article 86 turned on its findings that Hugin's practices in refusing to supply Lipton with such spare parts had no effect on trade between Member States.

[8] Case 238/87 [1988] ECR 6039: [1989] 4 CMLR 122.

lar Volvo model. The issue in the case (an Article 177 reference from the High Court) was primarily whether Volvo's refusal to grant a licence to Veng to manufacture such panels was an abuse by Volvo of its undoubted dominant position with regard to that part; in a parallel case decided on the same day the Court also rendered its opinion under an Article 177 reference in the *Renault*[9] case. Here, Renault had claimed it had monopoly rights under Italian law for its ornamental body panels. It was accepted by the Court in these cases that if under national law Volvo or Renault could validly secure the exclusive right to manufacture these panels and thereby obtain complete dominance of that product, this by itself did not justify an adverse finding under Article 86. Dominance alone, even as to 100 per cent, is insufficient as it is not the monopoly or quasi-monopoly on its own that can be challenged but its abuse.[10]

One hundred per cent control of a particular market may arise under other national legislation; thus, in the *Magill* cases decided by the Court of First Instance,[11] it was the law of copyright which gave the BBC and ITP in the United Kingdom, and RTE in the Republic of Ireland, complete control of the entire compilation of programme schedules for radio and television which were not, therefore, available (absent the grant of an appropriate licence by their owners) to those entrepreneurs who wished to publish a weekly magazine including such details. National legislation conferring a monopoly may also place a business in such a position of dominance; a notable example has been the granting of sole rights in The Netherlands and Spain for certain categories of letter delivery to the national postal authorities to the exclusion of commercial organizations who wished to compete for this business.[12]

At the other end of the scale the Court in the *Metro*[13] case made it clear that no question of dominance could arise when a supermarket chain had only less than 10 per cent of the general electronic equipment market for leisure purposes and less than 7 per cent of the colour television market. In practice, it seems unlikely that any undertaking having a market share of less than 25 per cent will be held to have a dominant position. Whilst neither the Court nor the Commission has made a firm statement to this effect, it is noteworthy that a recital to the Merger

[9] *Maxicar and Others* v. *Renault* Case 53/87 [1988] ECR 6211: [1990] 4 CMLR 265.

[10] For an example of the same principle applied by the English High Court, see *Pitney Bowes* v. *Francotype* [1989] 4 CMLR 466.

[11] *RTE* v. *Commission* Case T 69/89 [1991] 4 CMLR 586: *BBC* v. *Commission* Case T. 70/89 [1991] 4 CMLR 669: *ITP* v. *Commission* Case T 76/89 [1991] 4 CMLR 745.

[12] See *Dutch Courier Services* [1990] 4 CMLR 947 and *Spanish Courier Services* [1991] 4 CMLR 560. The European Court, however, has subsequently on procedural grounds quashed the Commission's decision in the *Dutch Courier Services* case. (Cases 48 and 66/90, a judgment dated 12 Feb. 1992.)

[13] Case 75/84 [1986] ECR 3021: [1987] 1 CMLR 118.

Regulation 4064/89 sets out a presumption that a concentration significantly impeding competition cannot be created if the parties' joint market shares do not exceed 25 per cent. Once over that percentage, however, dominance can become an issue and from then on the larger the percentage share becomes the more difficult it is to avoid the presumption that the undertaking is starting to benefit from a lesser degree of control from normal competitive pressures. Notwithstanding the lesser emphasis placed by the Court than the Commission on levels of market share in *United Brands*, the Commission itself continues to place great emphasis on such percentages and once these can be shown to reach 45 per cent it becomes almost impossible to claim that such an undertaking lacks power unless there is another undertaking in the same market with a share of equivalent size.[14] Once the market shares exceed 65 per cent, the presumption becomes almost impossible to displace, especially if the undertakings in competition are themselves all of relatively minor significance.

But we have not yet answered a central question, namely of what markets are these percentages. It is always one of the most difficult issues in any Article 86 case to establish both the product and geographic markets which are relevant.[15] The search for the relevant geographic market begins with the words of the Article '. . . within the Common Market or in a substantial part of it . . .'. Thus a Member State's territory will often provide a natural geographic setting equivalent to the commercial area in which competition is taking place, and in earlier cases the assumption was made rather easily that geographic markets would be, if not the entire area of the Community, at least no smaller than an individual Member State. This tendency was, however, first challenged in the *Sugar Cartel* case[16] when the Court accepted the Commission's contention that a substantial part or region in a Member State could likewise constitute a market given a sufficient volume of sales.

In each case the parties concerned and the Commission will advance relevant economic arguments to justify their particular contention, including any existing barriers to the free movement of goods, consumer buying habits, and costs of transporting goods in from neighbouring areas. Obviously the easier it is to bring goods in from a distance, using the

[14] In *AKZO* v. *Commission* (see p. 357 below) AKZO's share of the relevant product market was found to be some 50 per cent and this was regarded as clearly sufficient to justify a presumption of dominant position.

[15] For a more extended account of the problems of defining both geographic and product markets, see B. E. Hawk, *United States, Common Market and International Antitrust: A Comparative Guide*, 2nd edn., ii (1990 Supplement) (Englewood Cliffs, NJ: Prentice-Hall Law and Business, 1990), 748/88.

[16] The Court here accepted southern Germany as a geographic market for sales of sugar. See [1975] ECR 1991–4: [1976] 1 CMLR 463–5.

improved road and rail systems now to be found in many parts of the Community, the wider the likely market will be. It is usually in the interests of the undertakings to suggest that a market is a relatively wide area so that their individual shares are thereby reduced. The Commission will normally by contrast argue that narrower areas form natural individual markets. The Commission will often reach a conclusion that, since undertakings organize their distribution arrangements on a national basis, this choice itself implies that they regard such national territories as separate markets, an approach adopted, for example, in both *United Brands* and *Michelin*.[17] If the effect of national legislation is also to create different market conditions in individual Member States, this too will tend to establish them as separate markets.

The Court in at least one recent Article 177 opinion has shown that in this context it is prepared to look beyond the geographical focus of the immediate case and to define the geographic market more broadly if circumstances require. In *Alsatel* v. *Novasam*[18] (European Court 1988) the original dispute arose when the Alsace Telephone Authority tried to enforce payment from Novasam, an employment agency, by way of compensation for Novasam's early termination of a long-term telephone rental installation. Under the terms of an admittedly onerous contract, Novasam as the price of early cancellation had to pay some three-quarters of the remaining payments still outstanding over the original duration of the agreement. Moreover, the agreement stated that if the equipment provided was at any time supplemented so as to increase Novasam's annual rental by more than 25 per cent, then this automatically extended the contract for a further fifteen years. Alsatel alone were entitled to make any extension or modification to the equipment. The making and enforcement of an onerous contract in these terms was found to be potentially abusive by the Court but the relevant geographic market was considered to be not simply Alsace (with which the immediate case was concerned) but the whole of France. Businesses such as Novasam were not limited to obtaining telephone equipment from Alsatel but could acquire it by purchase or rental from any part of France. In this wider market Alsatel had a market share well below that required for a finding of dominance.

Another French case referred to the Court under Article 177 which involved this type of issue was the *Bodson* v. *Pompes Funèbres*[19] case (European Court 1988). The Court did not reach a finding on the facts as to whether Pompes Funèbres had a dominant position, this being left for the national Court to decide. Nevertheless, it pointed out that Pompes Funèbres held the exclusive concession for conducting funerals (apart

[17] Case 322/81 [1983] ECR 3461: [1985] 1 CMLR 282.
[18] Case 247/86 [1988] ECR 5987: [1990] 4 CMLR 434.
[19] Case 30/87 [1988] ECR 2479: [1989] 4 CMLR 984.

from the provision of church services, flowers, and monuments) for some 2,800 of those 5,000 local communes in France in which the performance of these external aspects of conducting funerals had been delegated exclusively to one specified private firm. These 5,000 communes covered 45 per cent of the entire French population. The contrast between this and the *Alsatel* case is that the individual purchaser of funeral services could clearly not in practice make a choice from a wide variety of enterprises but was obliged to use the firm or firms authorized by the local commune. The Court stressed that in assessing dominance (as well as the effect on trade between Member States) it was the share of the population affected by the exclusive allocation that was crucial rather than simply the total percentage of communes affected.

The definition of product markets is, if anything, even more elusive. The Commission has however recently in the course of its published Telecommunications (Antitrust) Guidelines[20] itself provided a useful working definition of the product market. This states

a product market comprises the totality of the products which, with respect to their characteristics, are particularly suitable for satisfying constant needs and are only to a limited extent interchangeable with other products in terms of price, usage and consumer preference. An examination limited to the objective characteristics only of the relevant products cannot be sufficient: the competitive conditions and the structure of supply and demand on the market must also be taken into consideration.

There are few products for which there are not substitutes of some kind, and the interrelationship of quality, price, and availability is in nearly all cases difficult to analyse with exactness. Cross-elasticity of demand[21] between substitute products has been considered in a number of cases, notably *United Brands*, in which the banana's unique characteristics were considered in meticulous detail in the context of an ultimately unsuccessful argument that it should be considered as part of a wider product market involving fruit of different kinds. The Court upheld the Commission's view that in terms of year round availability, price, suitability for particular types of consumer, notably the very young and very old, and other characteristics, the product market could not be said to include any other kind of fruit, a decision which had an effect in increasing the market shares held by United Brands. Similar consideration of the interchangeability of products was given by the Court in *Hoffmann-la Roche* where among the Court's findings was one to the effect that a simple product could be considered as belonging to separate product markets if applied for more than one end use.

[20] [1991] OJ C-233/2 (6 Sept. 1991), also at [1991] 4 CMLR 946 at para. 26.

[21] The concept of 'cross-elasticity' between substitute products was considered in detail in the well-known *Cellophane* case in the US Supreme Court *US* v. *du Pont* 351 US 377 [1956].

There are, of course, a number of other factors which, as a matter both of common sense and economic theory, should be taken into account. The attitudes of buyers to any closely matching substitutes are clearly relevant, especially when the purchase is not one of a basic necessity but forms part of a potentially wide range of consumer goods items. Technical qualities, composition, specification, and price range will also be part of this issue. The existence of barriers to entry to the market must be considered (these may include some of the practices forming part of the alleged abuse) and in particular whether companies can at fairly short notice switch production from one type of goods to another without substantial capital expenditure or delay. It is normally easier for consumers to change their buying habits at short notice than for businesses to do so whose production facilities may be committed to the particular type of product which they have previously been utilizing.

The recent approach of the European Court to product market definition appears from the *Ahmed Saeed*[22] case to be becoming more flexible. Here, two travel firms in West Germany had sold airline tickets at prices which substantially undercut the official Federal Government approved tariffs, by exploiting an anomaly in the pricing system. The tickets they purchased were for a journey from Portugal via a German airport to a third country at rates cheaper than the shorter journey from that German airport to the same foreign destination. Under German law such opportunism is not encouraged but is a criminal offence. On an Article 177 reference arising out of an appeal by the travel firm against their convictions for such ticket sales the Court ruled that the product market in Article 86 cases was not limited to different forms of air travel, namely charter and scheduled flights, but could take into account all other possible methods of travelling including rail and coach services. Every case, however, will turn on its own particular facts.

In the *Tetrapak*[23] (no. 1) case before the Court of First Instance, the Commission had treated two different types of milk cartons as representing distinct product markets as well as the separate machinery utilized for producing them. The cartons were acquired respectively for fresh pasteurized milk and for sterile UHT (ultra high temperature) treated milk. The Commission's market findings were not ultimately challenged by Tetrapak before the Court; they were based on the evidence that the two types of milk were quite different in taste and other characteristics so that consumers may not have regarded them as normal substitutes for each other. Moreover, the element of packaging in the total cost of the milk filled carton was relatively small, and an increase of even some 10 per cent in the

[22] Case 66/86 [1989] ECR 803: [1990] 4 CMLR 102.

[23] [1991] 4 CMLR 334. The machines for packaging fresh pasteurized milk are less complicated than for UHT milk.

cost of packaging was unlikely to have more than a 1 per cent effect on the price of the filled carton; these factors were significant even though on a superficial analysis the different types of milk might have been considered as reasonable alternatives.

In another case,[24] *Hilti* v. *Commission*, by contrast the Commission's market definition was challenged before the Court of First Instance, as it was regarded as crucial to the eventual outcome of the case. Here Hilti, a Luxembourg company, had patent protection for both its nail guns and cartridge strips inserted into the gun but not for the individual nails which the gun then fired from that strip. Hilti argued that the product market here could not be limited to powder-actuated fastening systems (the technical description of the particular type of nail gun) but covered a variety of other fastening methods used on building sites. It pointed out that such fastenings could alternatively be done in certain circumstances with hand or power drills, spot welding, screws, bolts, and nuts, The choice of the particular method to be used would depend on a variety of factors including the technical problems of the particular materials to be fastened, the load which it would be required to bear, the skill of the operators, and the time available for carrying out work, quite apart from the cost of the fixing materials.

The Commission regarded powder-actuated fastening systems in certain situations as giving builders particular advantages. These included versatility of use, easy portability, and speed of operations, even if there were some materials which these systems could not fix and some loads which such fixings could not sustain. Moreover, the nails and cartridges used were often considerably more expensive than the equivalents, such as screws and bolts. Nevertheless, in its view powder-actuated fastening systems could be properly regarded as a separate product market from the other methods because the cost of fixing represented only a very small element in the total cost of building; the relatively minor variation in cost attributable to the use of that system meant it was unlikely that a small change in the price of Hilti's own guns, cartridge strips or nails would cause a shift by customers to an alternative method. The Court upheld the Commission's decision and the fine imposed of six million ECUs. Hilti's market share in the United Kingdom for nails was found to be between 70 and 80 per cent. The Court also confirmed that not only were guns, cartridge strips, and nails all separate product markets but that the powder-actuated fastening system which in combination they comprised was sufficiently different from other fastening systems to justify a finding that the degree of substitutability between them was relatively low.

Another Commission decision on product markets taken on appeal to

[24] Case T. 30/89 [1992] 4 CMLR 677.

the Court of First Instance is that of *British Plasterboard Industries PLC* v. *Commission.*[25] This company was fined under a Commission decision for giving special rebates to customers for plasterboard in return for dealing exclusively with it, in order allegedly that the business of smaller rivals could be curtailed. British Plasterboard argued that plasterboard itself was not the relevant product market since in some cases wet plaster could be used as an alternative (a product in which it had a smaller market share than in plasterboard). The Commission rejected this argument on the grounds that there were many kinds of partition and building where wet plastering was not technically possible. Moreover, it had always to be laid by skilled men who had to receive a long period of training, whereas plasterboard could simply be placed in position by employees after a far shorter instruction period. The supply too of the dry powder utilized in the making of wet plaster was also under the control of British Plasterboard.

In summary, therefore, neither the importance nor the difficulty of the task of analysis of both geographic and product markets can be overemphasized, and such analysis remains a central issue in the great majority of Article 86 cases. Similar problems of market definition arise of course also in merger and joint venture cases under the Merger Regulation considered in Chapter 20.

[25] Case T. 85/89 [1990] 4 CMLR 464.

18

Article 86: Abuse and Abusive Exploitation

1. Introduction

Article 86 itself contains a number of examples of abuse, but these examples give only a general indication of the very wide variety of practices found to be in breach of the Article in Commission and Court decisions. As with Article 85, the development of the law has derived mainly through the use of two different procedures. The Commission itself, after investigations by DG IV, will issue decisions on individual cases, which are then often challenged by the parties by way of review to the European Court of Justice under Article 173 of the Treaty. The European Court of Justice will also, however, frequently be asked to give its ruling on cases sent to it by national courts under the provisions of Article 177, and have provided through this process invaluable guidance as to the correct interpretation of Article 86.

The range of cases decided under this Article shows that the concept of abuse is wide, capable of covering a variety of policies and actions that depart from normal competitive and commercial practice. It is a concept which still contains considerable possibilities for growth, and represents a potent weapon in the hands both of the Commission and of complainants. It would for example clearly cover any action taken by a dominant company to make parallel imports more difficult and thus to promote the effectiveness of an export ban. It is well suited to the control of leverage sought to be utilized by a single undertaking with a monopoly or a dominant position in one market, who seek to use that to obtain an unfair advantage in another. Some of the main types of abusive conduct found in the cases are considered in the remainder of this chapter.

2. Refusal to Deal and Discriminatory Pricing

The first case involving a refusal to deal was the *Commercial Solvents* case[1] in 1974. Nitropropane is a compound that results from the nitration of paraffin and is the base product for the industrial production of another

[1] *Istituto Chemioterapico Italiano and Commercial Solvents Corporation* v. *Commission* Cases 6–7/73 [1974] ECR 223: 1 CMLR 309.

substance called aminobutanol. This in turn is the base product for the industrial production of ethambutol, a compound used in the treatment of pulmonary tuberculosis. Commercial Solvents Corporation (CSC) had a world monopoly in the manufacture of products derived from the nitration of paraffin, including both nitropropane and aminobutanol. Although the relevant patents had largely expired, it was difficult for other enterprises to enter the market because of the difficulty of finding outlets for other products that could also be derived from the nitration of paraffin, and because of the high cost of research and development. The necessary manufacturing installations were also of considerable complexity and cost.

CSC had originally been perfectly willing to supply ethambutol to a smaller Italian company, Zoja, through its Italian subsidiary, Istituto Chemioterapico Italiano (ICI) over which it exercised complete control. After merger talks between ICI and Zoja had broken down, ICI on the instructions of CSC raised its prices for ethambutol to Zoja and sought gradually to reduce the level of its supplies. Originally, aminobutanol remained available from other sources, but these gradually dried up. As a result, by the start of 1971 Zoja was unable to manufacture any further ethambutol and complained to DG IV that CSC and ICI were abusing their dominant position as world leaders in production both of nitropropane and aminobutanol. It claimed the object of this conduct was to eliminate Zoja from the market place in Europe as a producer of ethambutol. Ethambutol was of importance for the treatment of tuberculosis in conjunction with other drugs, and there was no valid reason why Zoja could not be supplied by CSC, which had adequate supplies. On the basis of these facts, the Commission imposed a fine of 200,000 units of account on CSC and ordered an immediate resumption of supplies by it to Zoja at a price no higher than the maximum price which it had previously charged for both nitropropane and aminobutanol.

The Advocate-General, although finding that the facts established by the Commission were largely correct, would have cancelled the fine because prior to its decision CSC had offered to resume sufficient supplies to Zoja to carry on business within the Community. The Court, however, upheld the imposition of a fine, though reduced to 100,000 units of account, and in other respects upheld the findings of the Commission.

The Court in particular agreed with the Commission that CSC and ICI between them held a dominant position in the relevant market of nitropropane and aminobutanol. By abusing a dominant position in these markets, they were able to have the effect of restricting competition in the market for the derivative obtained from those products, namely ethambutol. Such effect must be taken into account in considering the effect of CSC's action even if the market for the derivatives was itself not self-contained but could be met from other sources. The refusal to supply Zoja involved an abuse of its market power, particularly as it had already at one time supplied it and then later discontinued supply without adequate commercial reason. The case would, of course, have been less clear had CSC never supplied Zoja, and even more so if Zoja had merely been a potential entrant to the market rather than an actual competitor for in such situations it is doubtful if CSC would have had an obligation to supply.[2] Finally, the Court ruled that any undertaking that abuses its dominant position within a substantial part of the Common Market in such a way as is likely to eliminate a competitor is in breach of Article 86, regardless of whether the conduct by the dominant company relates to its exports or to trade within the Common Market so long as it can be established that the elimination of the competitor would itself have repercussions on the competitive structure of relevant markets within the Community. Whilst the case was also concerned with the detriment to consumers that would flow as the result of the action taken by CSC, the Court placed greatest emphasis on the effect on competitive structure which would be likely to come about as the consequence of the conduct complained of.

The most important case, however, to come so far before the European Court under this Article was a case referred to already in Chapter 17, the well-known *United Brands*[3] case (European Court 1978). This United States multinational exported bananas from its own plantations in South America and had a substantial proportion of the banana market in the Community. The majority of bananas were shipped in the company's own fleet to Rotterdam where they were resold to national distributors in the various Member States, who purchased bananas on a 'free on rail basis' from Rotterdam at prices which varied quite substantially. United Brands held a substantial market share, amounting to some 40–45 per cent in each of the four geographic markets in issue in the case, namely Germany, Benelux, Ireland, and Denmark, and held a degree of dominance in these markets even greater than might have been expected simply as the result of its market share. The particular conduct of which the Commission complained involved several separate practices. The first three practices were:

(*a*) the prohibition of the resale by distributors of green, i.e. unripe bananas;
(*b*) the refusal of supplies on certain occasions to a Danish distributor

[2] Whilst the later *ABG* case (pp. 355–6) established that regular customers have a right to equitable treatment from a supplier with a dominant market position, it still left unclear the rights of occasional customers. A dominant company would not appear to have any obligation, however, to supply an undertaking with which it had never previously dealt, even if it were an actual or potential competitor.

[3] Case 27/76 [1978] ECR 207: 1 CMLR 429.

which had offended United Brands by promoting a competitive brand of banana;

(*c*) pricing that showed discrimination between the markets of different Member States.

The facts established by the Commission showed that United Brands' policy was to try to sell its product at prices which, both by their absolute level and by their variation as between Member States, sought to extract the highest possible profit from all these countries. Its distributors were left with a relatively small margin for resale which nevertheless had to be accepted because of the market strength possessed by United Brands.

The Court had little difficulty in establishing abusive conduct under the first two heads. The prohibition of trade in green bananas had the effect of an export ban. The bananas arrived in the green, unripened form in Rotterdam, and only ripened at the time and at the speed required by ripeners in order to bring them into the right conditions of sale to suit local markets. In a freer market, the various distributors might well have exchanged supplies with each other in the face of the varying demands of different national markets. The limiting of such resales of unripened bananas meant that in practice the short periods between the attainment of ripeness and the subsequent onset of decay ensured that no exchange of supplies was possible. The Court acknowledged that a restriction of this kind, if imposed simply to maintain a certain quality level, would have been justifiable. A complete restriction, however, on the transfer of all green bananas went far beyond such a legitimate aim, because its clear real purpose was to weaken the bargaining position of the distributors with United Brands.

The second commercial practice found to violate Article 86 was a refusal to continue to supply Olesen, a Danish distributor. The alleged basis for this was that Olesen had been active in promoting sales of competitors' fruit. The Court found that the reaction of United Brands to Olesen's conduct was out of proportion to its alleged 'disloyalty', and ruled that it was not permissible for an undertaking with a dominant position to cut off supplies to a long-standing distributor, so long as its orders placed on the supplier had remained within the normal range.[4]

The third charge against United Brands was that it had allegedly charged discriminatory prices to its distributors in different Member States in direct breach of Article 6(*c*): 'Applying dissimilar conditions to

[4] An interesting issue with which the Court did not deal is whether it would have been abusive conduct for United Brands to have insisted on the distribution of fruit through exclusive dealers or distributors: in practice apparently exclusivity was unlikely to have been accepted by dealers and distributors in perishable products, whose handling raised many problems not encountered in selling consumer products with a reasonable or indefinite shelf life.

equivalent transactions with other trading parties thereby placing them at a competitive disadvantage'. It was shown that there were substantial differences in prices between, on the one hand, Denmark and Germany where prices were relatively high, and in the Benelux countries and Ireland on the other hand, where they were much lower. It was apparently the practice of the Rotterdam management of United Brands to fix the prices of the sale to the distributor so as to reflect as closely as possible the anticipated price that would be obtainable in its territory for ripe (yellow) bananas in the following week. The ability of United Brands to obtain such prices from its different distributors and so, therefore, disadvantage those distributors who purchased at the higher prices served to underline the strength of the company in Western Europe. All the bananas were sold 'free on rail' (f.o.r.) from Rotterdam and costs were virtually identical in each case. United Brands claimed that the difference in the pricing resulted from a commercial decision as to what each part of the market could bear, but the Commission and the Court concluded that this was not an objective justification for the differences imposed.

It is not absolutely clear from the terms of the Court's decision to what degree such a company when determining its prices may take into account market forces of supply and demand. It appeared to suggest that this would be permitted only in cases where the seller takes a substantial risk, having involved itself in local production and distribution to an extent that market movements in that country will be likely to affect it directly. The facts in this case were different, United Brands not being involved in the local markets as it was selling to all the distributors at Rotterdam, and in practice it was the distributors who bore the risks of the individual national markets. The Court's judgment left uncertain exactly what was meant by 'involvement' in local markets, but gave the impression that United Brands might have had a stronger case, had the discrimination in pricing reflected pressures resulting from market conditions in the individual countries, i.e. 'market lead pressures', rather than appearing to arise from United Brands' own ability to impose higher prices on its distributors in those markets because of the absence of effective competition. Had United Brands itself been carrying out the distribution function in the Member States, then it would have been taking a greater risk and having a closer involvement with the national markets; to that extent its freedom to have adjusted its prices to local dealers would have been greater.

Use of the Article is relatively straightforward when the recipient of the abusive behaviour or treatment is a well-established customer, either as part of a distribution network (as in *United Brands*) or by way of a series of individual transactions (as in the *Commercial Solvents* case). Useful pointers on the permitted response of a dominant company to the requirements of customers falling within neither of these categories is found in the *ABG* case considered by the European Court in 1978.[5] This case arose as the result of the oil shortages during the crisis period that started in November 1973. The Commission reached a decision that BP had breached Article 86 by substantially reducing supplies of petrol during this period to a Dutch independent central buying organization, ABG. The figures on which the Commission relied were that whilst BP had reduced its supplies at this time to all its customers in The Netherlands, the percentage reduction to ABG was of the order of 73 per cent while the reduction on average for its other customers was only some 12.7 per cent. The result of this apparent discrimination was that ABG had to buy its remaining needs on the open market or through the official marketing scheme established for the crisis period by the Dutch Government. The Commission defined abuse in this context as 'any action which reduces supplies to comparable purchasers in different ways without objective justification, and thereby puts certain of them at a competitive disadvantage to others, especially when such action can result in change in structures of the particular market'. The Commission pointed out that BP had not supplied any objective reasons for the proportionately greater reduction in supplies to ABG and, after taking into account the extent, regularity, and continuity of commercial relationships between the parties, the conduct was ruled as abusive.

The European Court, however, took a radically different view of the commercial relationships between BP and ABG. Accepting the findings of the Advocate-General, it held that ABG was far from being a regular customer either by way of long-term contractual obligations or by reason of having made regular spot purchases. ABG on the contrary was simply an occasional customer and BP was able to show that it had, owing to a reduction in its own crude oil supplies from Libya and Kuwait, reduced its contractual commitment to ABG twelve months before the oil crisis had occurred. It had actually requested ABG that in the future it should obtain the greater part of its crude oil from other sources, though BP remained willing to refine this crude oil. Temporarily, in the earlier part of 1973, BP had 'lent' ABG some crude oil to be repaid at a later date. The Court ruled that it was clearly justifiable to remove these figures from the total oil sales between the parties; once this had been done, the percentage reductions in supplies to ABG during the crisis period became too small to justify any finding of abuse by BP of its market position. Nevertheless, in spite of the reversal of the Commission's decision on the particular facts, the duty of the dominant supplier to deal equitably with

[5] Reported as *British Petroleum* v. *Commission* Case 77/77 [1978] ECR 1513: 3 CMLR 174. The Commission's decision is reported at OJ [1977] L117/1 [1977] 2 CMLR D1, as *ABG* v. *Esso and others.*

its regular customers is now clearly established, even though individual application of this rule will clearly raise difficulties in individual cases.

A more recent example of refusal to supply is *Napier Brown & Co. Ltd.* v. *British Sugar PLC*[6] (1988, Commission). In this case British Sugar was fined three million units of account for abuse of dominant position for refusing to supply industrial sugar to Napier Brown. It had given the excuse that it had insufficient supplies but the true commercial reason was that it wished to make it more difficult for Napier Brown to compete with it in the market for retail consumers purchasing one-kilogram bags. British Sugar was found also to have engaged in other practices to make life difficult for Napier Brown, for example maintaining a margin between the price it charged Napier Brown for raw materials supplied and the charge which it made for its own consumer sales such that Napier Brown were unable profitably to transform the industrial sugar into consumer retail packets. Moreover, it prevented customers from collecting industrial sugar from its factory and denied supplies of raw material of particular types and origin to Napier Brown, whilst remaining willing to supply these to other companies. It also offered rebates to other buyers on condition that they all bought from it exclusively in future.

3. Excessive Pricing

The fourth charge of abusive conduct in *United Brands* was simply that the prices charged by the company were excessive. Here, the Commission had relied on the prices charged for Ireland, which were the lowest, as a base line from which to claim that the higher prices in the other countries were demonstrably excessive. As in *Continental Can*[7] and some other cases, the Court refused to accept the Commission's economic analysis on the basis that it lacked the depth and adequacy to substantiate the claims made. In particular, it indicated that it felt that the basis for the Irish prices had not been sufficiently investigated, to see if in fact they were loss-making prices. The Court confirmed that Article 86 could apply when the dominant undertaking directly or indirectly applied unfair prices to its customers. The Court stated that prices are excessive when they have no reasonable relation to the economic value of the product supplied, and that Article 86 could be breached if the consumer suffered as the result of such pricing policies, even if no effect on competition could be shown. It would perhaps have been too much to expect that the Court would indicate exactly how an individual price or range of prices could be assessed as excessive, a problem only too familiar to economists. The Court did,

[6] [1990] 4 CMLR 196. [7] See Chapter 20, p. 387.

however, indicate that in every case a detailed cost analysis was an essential preliminary, and that once this had been established, the next step would be to compare the prices charged with those charged by competitors for the same product. The finding of the Court on this issue was that the difference between the Chiquita bananas and those of its principal competitors was only 7 per cent, and this evidence alone was felt insufficient to justify a finding of excessive prices. Hence, on this aspect of the case, United Brands were not adjudged to have acted in breach of Article 86.

The *United Brands* case shows, however, that Article 86 can be effective as a means of controlling the power of dominant companies able to insulate themselves to some extent from competitive pressures. It also illustrates that the task of DG IV in establishing the necessary underlying evidence for such a case will always be difficult, and that it will have to produce well-researched economic evidence showing both that the margin between prices charged and costs involved is not within the normal commercial range of profit and cannot be considered reasonable in comparison with prices charged for identical goods. In default of this evidence, the case is likely to fail.

Article 86 can, of course, equally be applied to pricing by a dominant company which is predatory, that is at a figure below its production costs. An important example of this is the *AKZO* case (European Court 1991).[8] This case concerned an appeal by AKZO against a decision of the Commission taken as long ago as 1985 following a dispute with the original complainant, ECS, which was a small producer of benzoyl peroxide in the United Kingdom; ECS originally supplied its product for use as an additive to flour, a small market found only in the United Kingdom and Eire. It planned to expand its sales within the Community by marketing its product also to the plastics industry, a much larger potential market, and AKZO became unhappy about the prospect of this competition in Europe from ECS. According to the complaint by ECS, AKZO had threatened over a period of several years to take reprisals against it by way of selected price cuts unless ECS abandoned its intention of selling to the plastics market. AKZO, as a much larger company to whom its trade in the flour additive field was of little importance compared to its trade with the chemical companies, could afford to undercut ECS substantially in the flour additives field, even to the extent of putting it out of business altogether, and appeared prepared to do so to protect its position in the more important market.

Interim measures were applied for and granted to protect ECS, and the

[8] Case C-62/86. The delays in this case were extreme: the Commission's decision was dated 14 Dec. 1985, the Advocate-General's opinion was published on 19 Apr. 1989, and the decision of the Court announced on 3 July 1991.

Commission issued a decision that AKZO's predatory pricing and use of threats to induce ECS to withdraw from the plastics market were indeed breaches of Article 86. A substantial fine of 10 million ECUs was imposed on AKZO. The Commission refused, however, to prescribe any specific pricing rules linked to costs, or to define the precise stage at which price cutting by a dominant firm became abusive. It concentrated rather on the objectives of a dominant company which were shown (from the evidence obtained by the Commission through the use of its investigatory powers under Regulation 17) to have been intended to eliminate or damage ECS as a competitor in the relevant markets.

The Court in its decision, however, went into considerable detail as to the basis upon which the finding of abuse was to be confirmed. It stated in its judgment that, in applying the *Hoffmann-la Roche* principles to AKZO's pricing, it was to be presumed that prices charged that were lower than average variable costs alone were intended to eliminate competitors since they involved sacrificing recovery of all relevant fixed costs.[9] Moreover, pricing even at a higher level above average variable cost would still be presumed abusive if it did not succeed in recovering average total costs[10] and was part of a deliberate plan to eliminate a competitor with smaller financial resources.

The Court also confirmed the Commission's findings about the use of threats against ECS and that offers of both benzoyl peroxide and other related chemicals had been made (at prices below average total costs) by AKZO to customers which normally dealt with ECS. AKZO was also found to have sold its products cheaply in the flour market to some major customers in return for exclusivity of purchase by them. The fine, however, on AKZO was reduced to 7,500,000 ECUs, mainly because of the novelty of the pricing issues and because ECS had not actually lost substantial market share as the result of AKZO's conduct.

4. Discounts and Rebates

The European Court has also been concerned with the basis upon which dominant companies grant rebates to their customers. The normal function of a rebate is the encouragement of a customer to do business with a supplier on a regular basis by offering it a retrospective cash payment calculated on its purchases over the year or other fixed period. It has the same commercial purpose over a longer period as does the offer of an individual discount on a single particular transaction, the object of encouraging the buyer to do business with that seller rather than with its

[9] i.e. those which remained constant regardless of the quantities produced.
[10] i.e. variable plus fixed costs.

competitors. A rebate, although sometimes offered in addition to ordinary discounts, is in effect an aggregation of discounts seeking to bind customers to a particular supplier.

A rebate may often be used quite legitimately, particularly in markets where there is active price competition, and also in cases where the seller may reasonably claim that its costs can be reduced if customers show a degree of loyalty by placing a high volume of orders with it. Rebates can, however, be used by a dominant company in a way which makes it difficult for customers themselves without substantial countervailing market power to switch purchases between it and other suppliers, either to take advantage at particular moments of price fluctuations or other changes in market conditions or because some of the products of the other suppliers are considered of better quality or suitability. The use of the rebate by a dominant company is often, therefore, linked with its desire to ensure 'fidelity' from its buyers, using the leverage of one particular product to compel the buyers to acquire all the remainder of their needs from it, involving thereby a form of 'tying' on the buyer.

In a notice of a settlement made with the Coca Cola Export Company in 1988,[11] the Commission recorded that the use of fidelity rebates by a Coca Cola subsidiary having a dominant position on the Italian Cola market had been discontinued, it being accepted that the fidelity rebates were an abuse of its position in that market. In the text of the settlement, however, a distinction was drawn between certain types of unacceptable rebate, namely those conditional to targets linked to volume of purchases in the previous year or linked to targets which relate to a group of products in different markets. On the other hand a rebate conditional on a purchase of a series of different sizes of product or of participation in advertising promotions was acceptable. In *British Plasterboard* referred to in Chapter 17 the abuse was found to consist of the payment of rebates in return for not dealing with competitor suppliers of plasterboard; on the other hand discounts given to customers within a limited geographical area in return for taking particularly large loads were held not to constitute an abuse.

The treatment of such practices under Article 86 is shown by two European Court cases, *Hoffmann-la Roche* (1979) and *Michelin* (1983).[12] Hoffmann-la Roche is a multinational group whose parent company is based in Switzerland. As already discussed, the Commission had itself decided that this company had abused its dominant position for seven categories of vitamins. A particular abuse was the entering into of exclusive or preferential supply contracts with major industrial purchasers, contain-

[11] Dated 13 Oct. 1988: IP (88) 615.

[12] *Hoffmann-la Roche* Case 85/76 [1979] ECR 461, 541: 3 CMLR 211, 290. *Michelin* Case 322/81 [1983] ECR 3461 [1985] CMLR 282.

ing 'fidelity rebates' which placed considerable pressure on the buyers to obtain virtually all their vitamin requirements from Roche. A very large fine had been imposed and, although the Court reduced the fine by one-third, it substantially upheld the main elements in the Commission's decision. The Court for the first time took the opportunity to give a general definition of what was meant by 'abusive conduct'; it stated that this was the conduct of an undertaking which, having a dominant position and thereby able to influence market structure to the point where the degree of competition was weakened then 'through recourse to methods differing from those which condition normal competition in products or services on the basis of the transactions of commercial operators, has the effect of hindering the maintenance of the degree of competition still existing in the market or the growth of that competition'.

Pressure thus applied by a dominant undertaking to its purchasers to obtain all or most of their requirements exclusively from itself is an abuse, whether or not it is accompanied by an appropriate rebate payable to the customer so long as the customer remains loyal for all (or at least a large proportion of) its requirements. The Court felt that the application of such pressure is itself incompatible with the normal working of competition, since it has the effect that orders are placed not because the buyer has made a simple commercial choice in its own interests, but because the buyer, given his dependence on the dominant supplier, simply cannot afford to spread his orders over a variety of suppliers and thus risk losing the rebate offered by the dominant company.

Roche had placed considerable reliance in their argument on the price alignment clause which it included in its agreements widely known as the 'English clause'.[13] Under this, a customer who obtained a better price quotation from another producer was entitled to ask Roche to align its prices to those prices. If Roche failed to do so, the customer was then able to buy from the other producers without losing the benefit of the fidelity rebate. The Court felt that this clause, rather than enabling competitive pressure to work effectively for the benefit of the buyer, actually had the effect of providing Roche with a great deal of information about the state of the relevant vitamin markets, and the pricing and marketing strategies open to it in the light of actions being adopted by its competitors. The information which it received in this way further enhanced its ability to dominate, and the effect of the 'English clause', therefore, was still further to weaken the competitive structure of the market, a factor which the Court regarded as of some importance.

The *Michelin* case had also concerned itself largely with fidelity rebates. Again the initial decision of the Commission was upheld by the Court,

[13] Though the use of the expression is widespread, its precise origins are disputed.

although the fine was reduced in this case by more than half. Michelin held a strong position in the Dutch replacement tyre market for heavy vehicles including trucks and buses. The Commission's decision had been based on a number of grounds and had focused particularly on the fact that the rebates fixed by Michelin in return for the achievement of sales targets of individual distributors were not sufficiently objective, and the rebate schemes characterized by a general lack of transparency. There were also findings that the relevant targets and rebates given for 1977 had linked two different tyre markets (heavy vehicle and light vehicle markets respectively) and then used Michelin's market dominance for heavy vehicle tyres to promote sales of light vehicle tyres. The Commission had ruled that for the use of rebates to be acceptable as normal commercial practice for the purpose of Article 86, they must be both clearly known to the purchaser and defined on an objective basis, the sums payable to dealers being commensurate with the task they performed and the services which they actually provided.

Fears had been expressed by manufacturers following the Commission's decision that its effect would be considerably to reduce the extent to which manufacturers with a strong market position would feel able to use any form of rebate or bonus scheme. The subsequent decision of the Court has to some extent reduced that concern. The Court did, however, concur with the Commission's view that the system adopted by Michelin for variable discounts linked to the attainment of individual sales targets, and the method of administering the scheme through frequent personal visits by Michelin representatives applying pressure to its distributors to attain those targets (particularly towards the end of the sales year) constituted an abuse, in view of the undue pressure placed on dealers. Michelin set out neither the rebate scale nor sales targets in writing; and this would increase the disinclination of the dealers to deal with Michelin's competitors particularly towards the end of that relevant year. All these factors contributed to strengthening Michelin's market position and increasing the barriers to entry in the Dutch tyre market by its competitors.

These findings alone were sufficient to justify confirmation by the Court of the Commission's decision that an abuse had occurred, even though in certain respects the factual findings of the Commission were rejected. The Court did not, for example, accept that resellers of tyres would have hesitated to complain about the lack of written confirmation of sales targets, and also rejected allegations that both targets and rebates were related to the percentage that Michelin sales made by the client bore to the total sales of all makes by it, a formula known as 'temperature Michelin'. It also rejected the finding by the Commission that different rates of rebate, representing differences in treatment between dealers, were the result of the application of unequal criteria and had no basis in ordinary commercial practice.

Overall, however, the result of the *Michelin and Hoffmann-la Roche* cases would seem to be that the use of rebates by dominant companies, although not prohibited outright, requires very careful analysis of the services for which they are granted. It is not possible to tie them simply to exclusive purchasing obligations, but they can be linked to the achievement of objective sales targets provided that the operation of the system is sufficiently objective and transparent so as not to impose undue pressure on dealers, and does not discriminate between different categories of dealers on a basis that does not reflect cost savings to the supplier. The supplier will also be in a stronger position to withstand a charge under Article 86 if its rebate system corresponds to those generally adopted by its competitors in the same markets.

5. Abuses by Individual Owners of Intellectual Property Rights

The first ever case to reach the Court on interpretation of the Article came in 1968 under an Article 177 reference from a Dutch court, the *Parke, Davis* v. *Probel* case,[14] and concerned the rights of an individual patent owner.

Parke, Davis was the owner of a Dutch drug patent for an antibiotic process called chloramphenicol and sought to use its ownership of this patent to exclude drugs from Holland which had been produced in Italy, where patent protection was not available at that time for drugs or medicine. In so far as the Court's judgment reveals, Parke, Davis had had nothing to do with the production or marketing of the drugs in Italy.[15] Such a normal use of patent protection under national law was held by the Court to be a right protected by Article 36, and by itself could not constitute abuse; abuse necessarily involved an element of impropriety in the way in which the patent right was exercised. Such impropriety could not be said to occur simply because Parke, Davis utilized its patent rights to exclude from Holland drugs which it had neither placed on the market in Italy nor allowed to be manufactured there by a third party. The mere fact that prices were higher in Holland for this particular drug than on the Italian market had no effect on the legal position; indeed, the likely reason for the ability of the Italian manufacturer to undercut the price

[14] Case 24/67 [1968] ECR 55: CMLR 47. See also Ch. 16, p. 325.

[15] Some commentators have subsequently suggested that Parke, Davis had in fact provided some assistance in connection with the production of the drug but nothing in the Court's decision refers to this.

charged in Holland was that the cost of the research and development of the drug had been incurred largely by Parke, Davis rather than the potential importers. The Court did, however, leave open for the future an important issue, namely the extent to which the existence of the higher price level in one Member State might in some cases be relevant and should be taken into account, in determining the existence on other occasions of improper exploitation of market power. The authority of *Parke, Davis* has, however, after some twenty years been narrowed by a remarkable group of cases in the 1980s relating to copyright and industrial design.

Whilst, of course, it is decided cases from the European Court and Court of First Instance which carry the greatest authority in this area, the next important pointers for the future interpretation of this aspect of the Article by the Commission were found in a case which never reached formal decision. This was the widely publicized settlement involving International Business Machines (IBM) (Commission 1984). IBM was originally challenged with four practices claimed to be in violation of Article 86, though two of these charges were later withdrawn after IBM had modified its marketing policies. The charges which were the subject of the final settlement involved (*a*) the practice of IBM in including the main memory function in the price of a central processing unit and refusing to supply it separately, a practice known as 'memory bundling', and (*b*) its policy of not disclosing important information about the interface of IBM computer systems until they had actually been marketed, thereby placing at a disadvantage the manufacturers of equipment that would have been compatible with the IBM hardware, who were unable to obtain the necessary 'interface disclosure'. The ultimate terms of settlement were that the Commission suspended its proceedings under Article 86 in return for an undertaking by IBM to amend its practices relating both to memory bundling and interface disclosure. The Commission retained the right to reopen legal proceedings against IBM should the relevant undertakings not prove in its view satisfactory; IBM itself agreed to observe the terms of the undertaking at least until 1990 and to give twelve months' notice of any intention to terminate the settlement.

IBM were alleged to have obtained market dominance in large main frame computers and also in the submarket of the IBM Systems 370; the issue of 'memory bundling' ultimately appeared to be less important than that of interface disclosure because by the time of settlement memory no longer formed a large part of the central processing unit. IBM indicated that it was willing to offer the EEC System-370 without any main memory capacity or at least only with such capacity as was required for reasonable tests to be carried out. The more difficult issue of interface disclosure began with an allegation by the Commission as to the foreclosure from competition of independent manufacturers of peripheral equipment that would be compatible with IBM computers; but this original complaint was later extended to alleged non-disclosure of the IBM System-370 source code permitting a company to design and prepare software programs that could work in conjunction with IBM software, and also to practices adopted regarding its communications standard known as Systems Network Architecture (SNA). The Commission had alleged in particular that the non-disclosure of interface information relating to such Architecture could conceivably have adverse effects on the data communication industry's efforts to develop its own standard computer networking procedures, since such non-disclosure would prevent other brands of computer equipment from being able to be combined in a simple communications system with IBM products.

Under the terms of the complex settlement, IBM agreed to make interface information available within 120 days following the announcement of a new System-370 product intended for sale within the Community and to make source code interface information available to competitors as soon after the announcement of a new product as the interface had been technically proved. IBM would also disclose adequate information to competitors to enable them to interconnect with System-370, using SNA once the relevant format and protocols were stable. The undertakings given, however, did not extend to interfaces between two distinct products of a subsystem, from which the design of the products could be detected.

This highly complex and technically demanding inquiry[16] may have been satisfactorily resolved from the viewpoint of the Commission as well as from that of IBM, for the completion of any litigation involved in resolving the case would probably have occupied several years and a disproportionate amount of the resources of DG IV. It is clear that the Commission will place great importance on any activity of a dominant company which involves foreclosure of competitors from a particular market, either by insisting that two related products can only be purchased together, e.g. the CPU and its memory, or in cases of what is often called 'implicit tying' where the smaller firms within an industry have adapted their products to those of the dominant firm and subsequently suffer a commercial disadvantage when the dominant firm redesigns its products so that competitors' complementary products cease to be compatible.

If implicit tying is likely to lead to an allegation of abusive conduct, it seems to indicate that companies with a dominant position may not only have a negative obligation to refrain from action that places their smaller competitors at a disadvantage in this way, but even as a result of their dominant position may assume a responsibility for maintaining a com-

[16] The 14th Annual Report contains details of the decision at pp. 77–9.

petitve market structure regardless of whether that commercial practice can be shown to have other justifications. This argument would seem to be of particular relevance with regard to the issue of non-disclosure of interface information relating to SNA, which was considered likely to restrict the development of markets for communication and area networks. Its use, however, may well be of value in future to the Commission in other areas of advanced technology where a powerful company appears likely to be able, if left unchecked, to use its technical superiority to dampen the level of competition in its chosen markets.

The right of the owner of patent or other intellectual property right in a particular jurisdiction to exercise them simply by excluding goods that infringed his territorial monopoly had been clearly confirmed in *Parke Davis* v. *Probel* as not constituting an abuse. By analogy it appeared almost certain that such an owner would be entitled also to refuse to license his property to third parties.[17] In *Volvo* v. *Eric Veng*[18] and its companion case *Renault* v. *Maxicar*[19] (two similar cases decided on the same day in October 1988), this principle was accepted but subject to a qualification so substantial as to render the right itself of limited value. The Court agreed that the owners of protected design, in these cases for car body panels, had the right to prevent third parties from manufacturing, selling, or importing products which infringed the design and that this right to prevent third parties from entering the market comprised the very subject matter of the exclusive right which national law had granted. If, therefore, Volvo and Renault were obliged as a matter of law to grant third parties a licence to supply products (even in return for reasonable royalties) then this would in effect deprive them of the substance of those exclusive rights. Hence a refusal to grant licences could not in itself constitute an abuse.

The qualification which the Court then stated reflected the underlying relationship between car makers on the one hand and on the other the owners of cars, who will need inevitably to have them maintained and repaired from time to time if they are to be kept in use. The Court stated that where a manufacturer arbitrarily refused to supply spare parts to independent repairers or fixed prices of spare parts at an unfair level or made a decision no longer to produce spare parts for particular models, even though many cars of that model were still in circulation, this would be an abuse of the dominant position which each manufacturer necessarily has in respect of many kinds of spare parts for its range of vehicles. The Court seems to be saying that there are situations where the relationship

[17] Though there would of course always be the risk that the grant of a licence of right would be ordered.

[18] Case 238/87 [1988] ECR 6039: [1989] 4 CMLR 122.

[19] Case 53/87 [1988] ECR 6211: [1990] 4 CMLR 265.

between the manufacturer-owner of intellectual property rights and the seller of the product which it has manufactured is such that the rights themselves cannot be used in such a way as effectively to reduce the value of the product supplied. The Court's final comment, reminiscent of the judgment in *Parke, Davis*, was however that the mere fact that prices charged for spare parts by such manufacturers which enjoyed national protective right for their designs were higher than those charged by independent producers did not necessarily indicate an abuse, since the higher prices might be necessary to provide 'a return on the amounts invested in order to perfect the protected design'.

The line of reasoning adopted in these cases was, however, taken a stage further in the subsequent Court of First Instance decision in *Magill* v. *Radio Telefis Eireann (RTE)*.[20] Here the intellectual property right concerned was the copyright in the compilation of radio and television schedules for seven days; RTE, the Irish radio and television authority, had consistently refused to license Magill, an organization which wished to publish a weekly magazine containing the schedule of these programmes. Basing its refusal on the grounds of copyright, it was prepared merely to license the publication of a 24-hours schedule in newspapers (and for 48 hours at weekends) with some additional provision at public holidays. Magill claimed that this refusal to license the weekly programme details was an abuse of the monopoly which RTE held in them by virtue of its copyright. The Court of First Instance, largely adopting the reasoning of the Commission but after extensive citation of the earlier case-law of the Court of Justice (both from cases concerning Article 86 and those involved with Articles 30 to 36 of the Treaty), confirmed that decision. After defining the relevant product market as RTE's own weekly programme listings and also the guide in which it published them (which of course had a monopoly) it pointed out that it was only weekly guides with their comprehensive listing for seven days ahead that would enable viewers and listeners to decide in advance which programmes they wished to follow and so to arrange their leisure and domestic activities accordingly. The existence, moreover, of a demand for a comprehensive weekly guide was shown by the fact that they were provided in nearly all other Member States irrespective of other available sources of programme information. RTE clearly had a monopoly in the listings themselves and in the right to reproduce and market them as it thought fit through its RTE guide. Third parties like Magill who wished to publish a general television magazine were in a position of economic dependence upon RTE, which could thus hinder the emergence of any effective competition.

[20] Case no. T. 69/89 [1991] 4 CMLR 586. Similar judgments were given in parallel cases involving the BBC and ITP in respect of their own UK programme schedules. For detailed references see Ch. 17, n. 11.

The most interesting section of the decision, however, concerned the definition of abusive conduct. The Court of First Instance drew attention to the existing extensive case-law, which balanced the need for free movement of goods against the protection of subject-matter of intellectual property right including copyright. It pointed out that the cases showed that the principle of the supremacy of Community law, particularly in regards to principles as fundamental as those of free movement of goods and competition, normally prevailed over any use of a rule of national intellectual property law which was contrary to those principles. It referred to the cases of *Volvo* and *Renault*, drawing an analogy between the hypothetical refusal there to supply spare parts to independent repairers or to provide parts at all for cars still in current use with RTE's actual refusal in this case to license a weekly magazine when there was a clear demand for it. By this action RTE was preventing the emergence on the market of a new product, namely a general television magazine likely to compete with its own programme magazine. Copyright was being used in the listings of the programmes which RTE produced as part of its broadcasting activities in order to secure a monopoly in the derivative market of weekly television guides and the provision of the information for 24 hours or 48 hours alone did not meet that need.

The Court found that to prevent the production and marketing of a new product for which there is a potential consumer demand and excluding competition from the market simply to secure a monopoly went well beyond the essential function of copyright as permitted under Community law. RTE's refusal to authorize such publications was justified neither by the specific needs of the broadcasting sector nor of the technical requirements of producing magazines containing the schedules. This decision is likely to be highly contentious and the analogy drawn between the *Volvo/Renault* case and this case may not be generally accepted. Interestingly enough two potentially powerful additional arguments that could have been used to justify the Commission's decision were not referred to by the Court. The first is that in considering the concept of abuse, the normal practice to be found within the Community may be of considerable relevance. Here the facts found were that in all Member States apart from the United Kingdom and Ireland publications containing a full week's programmes of television and radio were commonly found and that competition existed for such publications. The fact that these publications existed not only proved that they represented a separate product market for which there was substantial demand but also illustrated that there could be no overwhelming commercial or technical reason for refusing listeners and viewers the opportunity of planning their activities at least a week ahead. In other words, within the concept of abuse is included an element of departure from norms of commercial conduct commonly found throughout the Community.[21]

The second distinguishing feature of the *RTE* v. *Magill* case is that the copyright in the programme material only arose in the first place because of the existence of the statutory privilege or monopoly conferred on RTE, namely to act as a broadcasting authority. It could, therefore, have been appropriately argued that RTE had not simply a right to prepare programme schedules in which it had copyright but a duty to publicize in every reasonable manner the details of the programmes that it was putting out pursuant to this monopoly. In other words, the existence of a copyright which came into being simply in order to enable the proper performance of a public function may be treated more severely by the Court than a copyright voluntarily acquired by the expenditure of creative effort. This point was not expressly taken up by the Court but may nevertheless have played an important underlying part in its thinking.[22]

It has been suggested that the consequence of this decision (which is on appeal to the European Court by RTE and ITP) may be far reaching in other sectors since it could be used as a justification for requiring owners of a wide variety of intellectual property rights in, for example, software to grant licences on reasonable terms if it can be shown that there is an existing demand for the material covered by the licences and no convincing technical or commercial reason for refusing to grant it on reasonable terms. It could, therefore, be used as a means of removing from the owners of a variety of intellectual property rights even that limited protection from obligations to grant licences of relevant material to the extent that the *Renault* and *Volvo* cases had appeared to provide. The special features, however, present in the *Magill* cases suggest that the Court will not be enthusiastic about applying such principles freely to other types of copyright such as computer programs or patents where such factors may be absent.

It is of course true that the recitals to the Software Directive 91/250 confirm that its terms are without prejudice to the application of Article 86 when a dominant supplier refuses to provide information necessary for the interoperability of equipment, as in the *IBM* case. Moreover, Article 6 of the Directive allows a person authorized or licensed to use a program to

[21] An interesting and comparable approach is found in the case of *Höffner and Elser* v. *Macroton*. Case C. 41/90 decided by the European Court on 23 Apr. 1991 where the practice of the German Government in permitting only public authorities to engage in employment agency business was treated as a breach of Article 90 leading to an abuse under Article 86 partly because it was found to be a practice not followed in any other Member State. See also *Merci Convenzionali Porto di Genoa* v. *Gabrielli*, a Court of Justice decision dated 10 Dec. 1991 Case C.179/90 [1991] ECR I 5889.

[22] Another distinction between the copyright in the programme information and other forms of copyright, e.g. books or artistic works, is its extremely ephemeral value, worthless once the relevant programmes have been transmitted.

'decompile' it so that it can communicate with another program which the licensee has created. This right of decompilation, however, which is achieved through reverse engineering, may not be used in such a way as to unreasonably prejudice the legitimate interests of that program's owner nor so as to conflict with his normal exploitation of his program. In the past, however, such an owner has been free to refuse to license it to persons whom he suspects may seek to carry out 'decompilation'; a possible outcome of the *Magill* case may be that (unless it is reversed by the European Court on appeal) such a refusal on those grounds alone may be considerably harder to maintain. On the other hand, whilst second-hand cars cannot be operated without the availability of a full range of spare parts and a television service without adequate programme schedules is of limited value, the potential range and function of new software programs that licensees may wish to produce to communicate with (and expand the use of) their existing software is unlimited. The analogy with the *Magill* case may not therefore be a straightforward one.

The abuse of power provided by intellectual property rights continued to be a major theme in two later cases, namely *Hilti* (Court of First Instance 1991)[23] and *Tetrapak* (Court of First Instance 1990).[24] The facts presented in *Hilti* were essentially unremarkable; they involved an attempt by Hilti to use its intellectual property rights which it possessed through patent and copyright respectively in nail guns and cartridge strips (into which the nails were fixed before firing) in order to discourage other companies from manufacturing or buying the nails themselves (which enjoyed no such protection) from third parties. Its methods included tying the sale of such nails to the purchase of the cartridges and upon sales of cartridges made without correspondingly large purchase of nails to reduce the discounts provided. The Commission rejected Hilti's argument that it was primarily concerned with preventing damage to users of its guns and cartridges if it made use of nails from other sources. There was no evidence that the use of non-Hilti nails had led to accidents and a complete absence indeed of any current written complaints from Hilti about the quality of nails produced by competitors for use with Hilti's guns and cartridges.

The facts of the *Tetrapak* judgment by the Court of First Instance were, however, altogether more unusual and the decision itself a landmark in the development of case-law relating to abuse.[25] Milk can be provided for household consumption in two main ways. It can be sold fresh, normally

[23] Case T. 30/89 [1992] 4 CMLR 16. The issues of market definition in the case have been considered in Ch. 17 at p. 348.

[24] Case T. 51/89 [1991] 4 CMLR 334.

[25] It is noteworthy that in its first major case the Court took the opportunity to lay down broad principles of the relationship between Articles 85 and 86 rather than deciding the case on relatively narrow technical grounds.

in familiar 'gable shaped' cartons, or it can come as UHT milk, which is aseptic (after being sterilized by hydrogen peroxide) and is sold normally in 'brick shaped' cartons. Consumers find it much easier to open the gable shaped carton and for some years Liquipak, a United States company, had been trying to develop technology to produce machinery for manufacturing gable shaped cartons in which UHT milk could be sold (thus eliminating the need for brick shaped cartons). It had collaborated with, and had as its Community licensee, a Norwegian company, Elopak, and with this company it had carried out numerous trials in the development of the new product which it believed had been brought to the point of commercial full-scale development. In doing so it had had the help of an exclusive patent licence from the British Technology Group, BTG, which had been granted in 1981 but in terms which complied with the subsequent Patent Group Exemption (Regulation 2349/84) introduced in 1984. By 1986 Elopak believed it was in a position to challenge Tetrapak, which had at that time some 90 per cent of the Community market for both the machinery for manufacturing the cartons and the individual cartons themselves.

In that same year, however, Tetrapak acquired the Liquipak group of companies and with them the exclusive licence for developing the 'gable shaped' aseptic milk carton and necessary filling machinery. Elopak raised a complaint under Article 86 with the Commission and after proceedings had been instituted against Tetrapak, Tetrapak abandoned any claim to exclusivity under the licence, which would have meant that BTG could then have licensed other would-be entrants (had it wished to do so) to the market for 'gable end' aseptic milk cartons. In spite of this, however, the Commission continued with the case because it raised important issues of principle. Tetrapak's acquisition of Liquipak had left only one other active competitor capable of producing aseptic filling machinery and cartons enjoying only a mere 10 per cent of the market; moreover, the technology of this company, PKL, was inferior to that of Tetrapak since it could supply only individually flattened blanks and not cartons produced in continuous rolls. The market for the filling machinery was limited, as dairies did not change their machines often as they had a life of at least ten years; milk moreover was not a growing market. All these factors meant that the barriers to entry by new competitors were high.

The issue raised by the actions of Tetrapak were whether the acquisition of the exclusive licence previously owned by its main competitor could itself constitute an abuse; Tetrapak argued that this could not be so because of the fact that the agreement itself benefited from an existing group exemption under Regulation 2349/84. It is clear from the Court's decision that the mere existence of the group exemption covering the licence agreement did not automatically prevent an abuse coming into being. In his lengthy review of the earlier cases on abuse Advocate-General Kirschner pointed out that undertakings with a dominant position were entitled to act in a profit-orientated way in order to expand their business activities. They were not required to act in a way which made no economic sense or against their own legitimate interests. On the other hand certain commercial steps which might be open to undertakings without a dominant position were, when effected by a company in that position, to be treated as an abuse of its power; nor was it necessary for an Article 86 finding for the dominant company to have used that market power in order to obtain the benefit of the acquisition or licence agreement concerned. He referred back to the judgment in *Hoffmann-la Roche*[26] and pointing to the concept of proportionality underlying the various kinds of abuse dealt with in earlier cases including *United Brands*[27] and *BRT* v. *SABAM*[28] and other performing rights cases. Here a company with some 90 per cent of the market in the aseptic filling machines and the cartons made for use with them acquired an exclusive licence which belonged to its main potential competitor endeavouring to break into that market. The effect of the acquisition of the Liquipak group was that the alternative technology protected by that patent was taken out of reach not only from Liquipak but all Tetrapak's potential competitors. The acquisition of this exclusive licence was a disproportionate method of carrying on business since the acquisition of a non-exclusive licence would also have enabled Tetrapak to use its patented process to improve its own products, but without at the same time impeding the access of competitors to the market over which Tetrapak enjoyed so dominant a position.

The Court speedily made clear that the principles laid down in *Parke, Davis* as later qualified in the *Renault/Volvo* cases would not serve to help Tetrapak. A distinction was drawn between the development in those cases of technology by an individual company which it was then entitled to protect or refuse to license (other than in exceptional circumstances) whereas in this case the acquisition was by Tetrapak of technology developed by others. The Court, dealing quite briefly with the concept of abuse since during the hearing of the case Tetrapak had withdrawn their appeal against the Commission's decision on this aspect (though not on other aspects)[29] of the case, found that the mere fact that an undertaking in a

[26] Case 85/76 [1979] ECR 461, 520: 3 CMLR 211, 274. This case is often popularly called the *Vitamins* case.

[27] Case 27/76 [1978] ECR 207: 1 CMLR 429.

[28] Case 127/73 [1974] ECR 51 (the Court's judgment dealing with the direct effect in national courts of Article 86) and 313 (the Court's judgment dealing with the issue of whether SABAM was exploiting its dominant position as a copyright authority in Belgium): 2 CMLR 238 is a consolidated report of both judgments.

[29] The Court's important ruling on the relationship between Articles 85 and 86 is considered in detail in Ch. 19, pp. 383–5.

dominant position acquired an exclusive licence did not *per se* constitute abuse; on the other hand in the specific circumstances of the acquisition the exclusivity of the licence acquired had not only strengthened Tetrapak's already considerable domination of the market but had the effect of preventing or at the very least considerably delaying the entry of new competitors into the market.[30]

6. Abuses by Collective Owners of Intellectual Property Rights

The application of Article 86 to performing rights societies within the Community raises some novel issues, and cases decided by the Court of Justice have illustrated the potential flexibility of the Article as applied to situations outside normal industrial and commercial markets. The role of performing rights societies found in most Member States is the collection and administration of very large numbers of copyrights for music to be publicly performed, representing the creative work of composers and librettists, so as to enable the licensing of these works (often by way of a block licence) to all those undertakings which may require their use. These may range from broadcasting authorities and cable TV companies on the one hand to individual proprietors of businesses such as discothèques and public halls on the other. The existence of such bodies has a dual advantage; it enables the owner of the individual copyrights (which are often assigned to music publishers for exploitation) to obtain and monitor a greater number of performances of his or her work with a corresponding increase in income, while enabling those paying the licence fees to obtain by a single payment a far greater range of copyright material than would be administrably possible if individual negotiations were required for each copyright.

Nevertheless, performing rights societies have on occasion sought to introduce into their rules restrictions going beyond those reasonably necessary for the proper protection and promotion of their members' interests. Such restrictions have been challenged under the provisions of Article 86, involving performing rights societies and similar undertakings in Germany, France, and Belgium. The leading case is that of *BRT* v. *SABAM* (European Court 1974).[31] Two Belgian radio employees decided to prove that even a 'nonsense song' could become a hit and wrote a

[30] It is noteworthy that within 12 months after the rendering of this decision the Commission had in 1991 imposed on Tetrapak for a range of other breaches of this Article the largest ever fine imposed, namely 75 million units of account. In respect of this case the range of actions found to constitute abuse included sales below cost, tying arrangements preventing dairies from using its machinery unless they also used its cartons, and other exclusionary tactics.

[31] See Ch. 19, n. 1.

song entitled 'Asparagus Beans' assigning the copyright to Belgian Radio and Television (BRT). BRT then arranged for it to be recorded and for the discs to be sold. SABAM claimed rights in the song on the basis that under its standard form of agreement it obtained rights over the work of its former members (who included the BRT employees who had written the song) for five years after they had ceased to be members. BRT claimed that SABAM's claim was invalid because SABAM was a *de facto* monopoly, and the five-year restriction went so far beyond its reasonable requirements as to constitute an abuse of its dominant position. An Article 177 reference by a Belgian Court resulted in a ruling from the European Court that any undertaking entrusted with the exploitation of copyrights was likely to hold a dominant position, and that if it imposed on its members obligations not absolutely requisite for the attainment of its legitimate objects, it would be in breach of Article 86. Protection could not be afforded to that organization by reason of Article 90(2) since the responsibilities taken on by SABAM had not been assigned to it by the State, but merely involved the management of private interests, namely intellectual property rights protected by Belgian law. The extent, however, to which the validity of any contract in dispute before a national court was thereby affected was for the national court to decide.

There had already been earlier cases heard and decisions issued by the Commission in connection with the rules of such organizations. In 1971, the *GEMA* no. 1 case involved a complex set of rules relating to the German Performing Rights Society which were challenged. The Commission decided that the rules in a number of respects violated Article 86; they discriminated against nationals of other Member States who were denied credit for royalties which they received from other similar associations in the Community, when their status and eligibility for ordinary membership in GEMA were being considered.[32] Moreover, foreign nationals were unable to become members of GEMA's council, nor could music publishing houses outside Germany be admitted to membership. The rules also required higher royalties on records imported or reimported from other Member States than on records produced in Germany alone.

GEMA was also found to have applied over-wide obligations to all its individual members, amongst other obligations requiring assignments of their rights in all categories of creative work, and for the whole world; it was made extremely difficult to withdraw from the association by requiring that membership be for a minimum period of six years at one time.

[32] This was a more favourable category of membership than the extraordinary membership normally accorded to foreign nationals. [1971] CMLR D 35.

GEMA was, therefore, required to amend its rules and practices to remove the discriminatory and over-wide provisions.[33]

Subsequently, the *Greenwich Film Productions* v. *SACEM* case[34] raised further issues arising out of a dispute between the French Performing Rights Society (SACEM) and the producers of a film for which two members of SACEM had composed music. Under the terms of their membership, the copyrights in their work had been assigned to SACEM, though some performances would take place in territories over which SACEM had no direct jurisdiction, outside the Common Market. SACEM sued Greenwich for the royalties in respect of these performances, and Greenwich put forward Article 86 as the ground for non-payment. The issue before the Court was whether Article 86 could apply to the performance in non-EC countries of contracts entered into within the territory of a Member State by parties within the jurisdiction of that State. On an earlier occasion, SACEM had had to alter its rules to comply with Commission requirements that it reduce the restrictions which it imposed on the same basis as the findings in the *GEMA* no. 1 case. SACEM had, therefore, already eliminated from its rules discrimination against nationals of other Member States and made a reduction in the period for which a member had to bind himself or herself to SACEM.

The results of the case ultimately proved unsatisfactory to the composers as well as to the Commission. The Court held that if Article 86 abuses were found, it was for French courts to decide to which extent the interests of the composers would be affected. The rules of the association were still, however, in breach of Article 86, and the fact that such an abuse affected only the performance in non-member countries of contracts entered into in the territory of a Member State would not preclude the application of that Article. In the event, however, the French Court de Cassation avoided any application of the ruling by the European Court to the particular case, claiming that the original membership of the two composers in SACEM antedated the application of Article 86 to the case, so that no effect on trade between Member States could be considered attributable to the conduct of SACEM. This unsatisfactory ruling (ignoring the direct effect in national law of Article 86 which was in force at the date when both composers had joined SACEM) offers an example, fortunately rare, of a national court giving a ruling which in effect paid scant attention to the terms of the decisions of the European Court.

[33] In a later Commission decision ([1982] 2 CMLR 482) there has been further review of provisions of the GEMA rules, and a negative clearance granted to an amendment aimed at preventing broadcasting companies from obtaining unfair advantages through special royalty-sharing arrangements with composers in return for 'song-plugging'. The Commission emphasized that practices which might in other contents be an abuse of a dominant position would not necessarily be so if 'indispensable to carrying out the essential purposes of a collecting society'.

[34] Case 22/79 [1979] ECR 3275: [1980] 1 CMLR 629.

Article 86 has also been applied[35] to an organization established in Germany for the protection of copyright in order to manage the rights of performers to royalties in respect of visual or sound recordings of their performances which are then shared on an equitable basis with the manufacturers of these recordings. This type of right is normally called 'secondary exploitation' and is, of course, quite separate from the primary exploitation of the copyright in the music and words of the individual works. The defendant, GVL, was the only organization of its kind in Germany, and until November 1980 refused to conclude agreements with performers who were not resident in Germany. The relevant market was found by the Commission and Court to be the commercial management of the secondary exploitation rights of performers, and for this the Court ruled that residence was not a relevant factor. Therefore, the refusal of GVL to act for foreign performers was itself an abuse of their *de facto* monopoly position. Although no fine was imposed because GVL's internal decision had been reversed before the date of the Commission's decision, a valuable ruling had been obtained; in view of the express provisions of Article 7 of the Treaty forbidding discrimination on the grounds of nationality, however, the decision itself caused no surprise.

The question of the level of charges made by such organizations reached the European Court in four linked cases[36] involving the rules of SACEM referred under Article 177 from French courts. In the first three cases SACEM sought payment of royalties from some French discothèques in the Poitiers region which had refused to take out a licence from it; in the other case a disco owner sued Tournier as managing director of SACEM in respect of alleged unfair trading practices under French law. Since 1978 there has been a long-running dispute between SACEM and a large number of French discothèque owners who thought that its rates were too high, discriminated unfairly between categories of discothèque, and refused without objective justification licences for the only category of music in which they were really interested, dance music of predominantly Anglo-American origin such as rock. As it was an Article 177 reference the Court was not asked to resolve the specific disputes between the parties but simply to reply to those questions which the French courts had raised. These questioned the compatibility of the practices of SACEM with respect to this kind of musical material, in the light of its undoubted dominance as a performing rights society within France and the various complaints about the manner in which this was carried out which the cases had raised.

[35] In *GVL* v. *Commission* Case 7/82 [1983] ECR 483: 3 CMLR 645.

[36] The cases were *Ministère Public* v. *Tournier* Case 395/87: [1989] ECR 2521: [1991] 4 CMLR 248. *SACEM* v. *Lucazeau* (Case 110/88). *SACEM* v. *Debelle* (Case 241/88). *SACEM* v. *Soumagnac* (Case 242/88): all at [1989] ECR 2811: [1991] 4 CMLR 248.

In one aspect of the case the Court's ruling would have disappointed the discothèque owners. It found that SACEM's refusal to grant them licences simply for the foreign repertoire likely to be preferred by its patrons would not be a restriction of competition unless providing access to a part only of the repertoire (at presumably reduced royalties) could also entirely safeguard the interests of composers without thereby increasing the costs of managing and monitoring the use of such copyright works. In other ways, however, the Court judgment was more favourable to the discothèque owners. It pointed out that if an undertaking such as SACEM imposed special fees for its services which were appreciably higher than those in other Member States, that difference might itself be indicative of an abuse of SACEM's dominant position. The burden of proof would then be on SACEM to justify that difference by showing that there were objective differences between its own position and that of equivalent organizations in other Member States. The Court mentioned that the level of operating expenses in SACEM appeared particularly high and wondered if this was attributable to lack of competition. It refused, however, to rule on whether the amount charged to discothèques themselves as compared to the rates charged to other large-scale users of recorded music such as TV and radio stations were themselves unfair discrimination, leaving that awkward question to the national court. The Court decision followed the spirit, though not all the detail, of Jacobs A.-G.'s opinion; he emphasized the particularly dominant position of this kind of organization which had no need to fear competition from foreign performing rights societies and no likelihood of substantial regulation of its affairs by public authority, so that it had a near absolute freedom of action, coupled with a clear inequality of bargaining power between it and discothèques who were completely dependent upon it.

The general attitude of the Court in such cases, therefore, has been robust. It has swept aside a number of the arguments raised by the societies, including protests that they are not 'undertakings', and that the writing of songs and music is a cultural rather than economic activity and that individual composers or publishers are, therefore, not entitled to the protection of Article 86. The Court has, moreover, refused to allow the rules of such societies to limit the free transfer of copyright items between Member States which have been placed in commercial circulation in a single Member State. Articles 30 to 36 do, of course, also here come into the reckoning as illustrated by the case of *Musik-Vertrieb Membran and K-tel International* v. *GEMA* (European Court 1981).[37] This involved

[37] Cases 55 and 57/81 [1982] ECR 147: 2 CMLR 44. This case follows closely the earlier precedents of the Court in respect of patents and trade marks. See Ch. 16. The distinction between this and *EMI Electrola* v. *Patricia* 341/87 [1989] ECR 79: 2 CMLR 413 is that in the latter case the owner of the copyright had not given consent to the marketing of the records.

the question of whether the German collecting society owning the copy-right in musical works reproduced on gramophone records was entitled to claim from a parallel importer a fee equal to the royalties ordinarily paid for marketing such records in Germany, less the lower royalties already paid in the United Kingdom, where the records were originally put into circulation by the copyright owner. The Court gave a negative answer emphasizing that sound recordings even if incorporated in copyright material are covered by the system of free movement of goods, so that it was contrary to Article 30 for the owner of the copyright in the records to rely on copyright legislation to prevent the import of a product lawfully marketed in another Member State by the owner or with his consent.

19

Article 86: Its Relationship with Article 85

The case-law of the European Court over the last thirty years has clarified the meanings of both these Articles in their application to a wealth of differing circumstances and industrial sectors. The great majority of cases have been concerned with one rather than both Articles, but in the course of certain decisions the relationship of the Articles to each other has become a central feature. No account of EC competition law can, therefore, disregard the important interaction of these two Articles with each other.

At first sight the possibility that such interaction could be significant might seem remote. Although their drafting has some shared characteristics, their individual scope and purpose would appear mutually exclusive; whereas in Article 85 cases the essentially prohibited element is the agreement or concerted practice between two or more undertakings, the primary target under Article 86 is abusive action by a single undertaking of its market dominance, without the need for proving any agreement or concerted practice involving any other undertaking. One of the main problems indeed encountered by the Commission in enforcing Article 85 is that of proving the existence of the relevant agreement or concerted practice in the absence of unequivocal written evidence. This problem of identification does not arise under the terms of Article 86.

Four further important distinctions may be noted between the Articles:

1. Since it is necessary under the terms of Article 86 to establish that the abuse occurred in a particular market comprising a substantial part of the Common Market, the chief difficulty that arises in such cases (rather than relating to proof of an agreement or concerted practice) is the need to define accurately the relevant geographic and product markets to which the dominance applies. Such dominance has to be found either within the entire Common Market or within a single Member State or group of Member States or possibly even within a substantial part of a single Member State. By contrast, under Article 85, although the Commission must still carry out economic analysis to identify the relevant markets where the agreements or practices have their effect, their delineation is normally less acutely critical to the outcome of the case.

2. Article 86 itself provides no sanctions of voidness for illegality against the prohibited conduct, for there is no equivalent of Article 85(2).

On the other hand, since Article 86 is directly enforceable in the courts of Member States (a point not in dispute since the 1974 European Court decision in *Belgian Radio and Television* v. *SABAM*)[1] the offending clauses in any contractual provision imposed by way of abuse may by national courts be declared void and unenforceable against contracting parties. Moreover, damages or injunctions can be applied for in the courts of Member States.[2]

3. Since the prohibited action is an 'abuse' of existing market power, there can logically be no exceptions provided of the kind available under the 'four conditions' contained in Article 85(3). If the dominant company wishes to raise arguments of the same kind as those considered by DG IV when reviewing the application of these four conditions, it will have to do so by contesting the Commission's claims that the acts which it admits to performing are an 'abuse', by establishing on the contrary that they are really no more than the normal use of its economic strength within the boundaries of legitimate competition or to be justified on other grounds.

4. Whilst a finding of a breach of Article 85 necessarily involves some restriction of competition, a breach of Article 86 can occur when the abuse of dominance is simply 'exploitative', i.e. an advantage obtained for the dominant company at the expense of the consumer without any effect on the competitive process or the structure of the market in which the dominant company operates.

Notwithstanding these basic differences between the two Articles, their similarities in both language and position within the overall scheme of the Treaty are significant. Article 87 (which provides the legal basis for Regulation 17/62) requires the Council to adopt regulations or directives to give effect to the principles set out in both Articles; and Article 88, dealing with the interim obligations of Member States pending adoption of such Regulation, also refers to both Articles. Moreover, Article 89 obliged the Commission as soon as it took up its duties to ensure the application of the principles laid down in both Articles. Regulation 17 applies, therefore, to both Articles and even though exemption is not available from Article 86 it is possible to lodge details of a specific practice or practices, especially if embodied in written documents, in order to seek a negative clearance. Such a negative clearance is, of course, simply a confirmation that Article 86 does not apply. As the same procedural powers are available to the Commission under Article 86 cases as with Article 85 cases, its powers to conduct investigations under Article 14 of Regulation 17/62

[1] Case 127/73 [1974] ECR 51 (the Court's judgment dealing with the direct effect in national courts of Article 86) and 313 (the Courts judgment dealing with the issue of whether SABAM was exploiting its dominant position as a copyright authority in Belgium): 2 CMLR 238 is a consolidated report of both judgments.

[2] See Ch. 22.

and its range of sanctions including penalties, fines, and termination of prohibited conduct are equally available.

Two other important similarities should be noted:

1. In both cases the conduct challenged must be such as to 'affect trade between Member States'. There may be marginal differences in the application of this condition between the two Articles. In the *Bodson*[3] case particular emphasis was placed by the European Court that the national court in investigating this element should actively take into account the effect on imports of the allegedly abusive conduct, a point not so clearly made in earlier Article 86 cases at the Court.

2. The examples of abusive conduct given in 86(*a*) to (*d*) include examples identical, or at least very similar, to those referred to under Article 85 as the subject matter of prohibited agreements or concerted practices. Only (*c*) in Article 85 ('the sharing of markets or sources of supply') is omitted from Article 86 since for that 'sharing' it is necessary that there must be at least two or more undertakings involved. The types of conduct referred to are simply examples and the width of the definition of those agreements or concerted practices which have as their object or effect the prevention, restriction, or distortion of competition on the one hand and the definition of abusive conduct on the other is not adversely limited in either case by the nature of the examples set out.

The first opportunity for the European Court to comment on the relationship between the Articles came in the *Continental Can* case. As the summary of the case in Chapter 20 indicates,[4] the Court took trouble to emphasize that both Articles had been included in the Treaty for the same reason, to ensure that both Article 2 and Article 3(*f*) could be implemented. They should be interpreted in a purposive way that ensured that the implementation of one Article did not frustrate the purposes of the other. The important result of the approach in this case was to provide the Commission with jurisdiction over those mergers entered into by companies already holding a dominant position. This jurisdiction has not been of great value to the Commission in practice and as explained in Chapter 20 has now been rendered less important following the introduction of the Merger Regulation 4064/89 with effect from September 1990. The long-term result of the *Continental Can* case, however, has been to make it more likely that Article 85 and Article 86 will be treated as two parts of a whole (or possibly two sides of the same coin) rather than as largely unrelated.

Whilst in the great majority of cases, proceedings brought by the

[3] *Bodson* v. *Pompes Funèbres Générales* Case 30/87 [1988] ECR 2479: [1989] 4 CMLR 984.

[4] p. 387.

European Commission or in national courts relate to one or other of the Articles. Article 86 does expressly refer to the possibility of abuses being practised by more than one undertaking often referred to as 'joint' or 'collective' dominance. It this occurs, and there is also some form of agreement or concerted practices between the parties, then clearly there could be a breach of both Articles. In the majority of cases where the Commission have alleged a horizontal cartel against a group of companies with an aggregate market share of a sufficient level for market dominance to be alleged, the Commission has not sought to raise Article 86 arguments in addition, contenting itself with proceedings under Article 85. However, in the *Italian Flat Glass*[5] cases (European Commission 1988), the Commission has sought to link the agreements and concerted practices referred to in Article 85 to the joint abuse of dominant position covered by Article 86. The reason for this appears to be that a decision of the Court on joint dominance may provide assistance to the Commission in dealing with problems of oligopolistic markets. In such markets the parties may be able, through closely parallel conduct (without the necessity for agreements or concerted practices even within the European Court's broad definition) to eliminate some degree of competition between them. The Commission is, therefore, anxious to be able to establish the principle that joint dominance and joint abuse under Article 86 can also arise from the same facts as would justify an Article 85 decision. In the case where an Article 85 agreement or concerted practice can be shown, then the ability to bring this Article into operation is of minor importance, but in the more borderline case where the parties have been scrupulous to avoid the appearance of Article 85 relationships the extent of acceptance of the joint abuse doctrine may well become important. Moreover, the Commission would be entitled, if an Article 86 violation could be shown, to impose a heavier fine than would be possible for a violation of Article 85(1) alone. Still more important to it is the potential scope for increasing the jurisdiction of the Merger Regulation, if joint dominance of particular markets were to become an accepted part of the Court's jurisprudence to supplement existing case-law on single firm dominance.[6]

The traditional approach of the European Court to this issue has been that Article 86 only applies to joint abuses in cases where the undertakings concerned are themselves linked within corporate group or at least there is some kind of permanent structural relationship between them.

[5] Cases T. 68/89 and T. 77 and 78/89: [1990] 4 CMLR 535. The decision of the Court of First Instance was rendered on 10 March 1992, quashing the Commission's decision against the three companies under Article 86, for failing to show that they had presented themselves on the market as a single entity and also because of deficiencies in market definition. The Court did not however disagree with the concept of joint dominance under Art. 86.

[6] See Ch. 20, p. 400.

This approach was adopted in both *Züchner*[7] and *Bodson*.[8] It fits logically with its requirements that Article 85 can only apply if there are at least two independent entities (not being members of the same corporate group) between which the agreement or concerted practice exists because it is only between such independent entities that competition exists and, therefore, can be restricted. The essence of the Commission's decision in *Italian Flat Glass* is that the three major companies who held between them 79 per cent and 95 per cent of the non-automotive and automotive flat glass markets respectively in Italy made arrangements for the exchange of different categories of product of such scope and permanence that they constituted established structural links rather than simply the existence of agreements and concerted practices between them.

There is little in the jurisprudence of the European Court to date which indicates that the Commission's argument on Article 86 will be successful in these particular circumstances even if the doctrine of joint dominance is given support in principle. The underlying principle of *Continental Can*, that the two Articles are to be read together in a purposive way to ensure that implementation of one Article does not frustrate the purpose of the other, does not seem to go so far as to permit a considerably extended interpretation of Article 86 simply because of the inherent difficulty of applying Article 85 to oligopolistic markets. The Court's own approach may be illustrated by the *Ahmed Saeed*[9] case already referred to. On that occasion the Court emphasized that there was only one situation in which a party to an agreement capable of challenge under Article 85 was also at risk of being challenged under Article 86, namely where the party to that agreement had already at the time of making it obtained a dominant position so that the effect of the agreement was simply to 'set the seal' on the position of power which it had already obtained. Moreover, in such circumstances it would only be the dominant company, not the other subordinate parties to the agreement, which would be challenged under Article 86. One could of course imagine a situation where a particular market is shared more or less equally between two or more companies, where each had sufficient market share and power before any agreement had been made to qualify potentially as dominant, notwithstanding the countervailing power of the other. In those highly unusual circumstances it might be conceivable that both parties having entered into an agreement that effectively eliminated competition between them, might be challenged under both Article 85 and Article 86. Given, however, the presumptive equality of the two parties in that case, it seems far more likely that Article 85 alone could and would be utilized. Nevertheless the

[7] 172/80 [1981] ECR 2021: [1982] 1 CMLR 313.

[8] Cases 29 and 30/83 [1984] ECR 1679 [1985] 1 CMLR 688. See Ch. 6, p. 86.

[9] Case 66/86 [1989] ECR 803: [1990] 4 CMLR 102.

Commission's continuing incentive to strengthen its armoury against leading companies in duopolistic or oligopolistic markets under both Articles 85 and 86 and the Merger Regulation mean that the principle of joint dominance is likely to be raised in other suitable cases.

The mere existence, however, of an agreement which prima facie falls within the prohibition of 85(1) and to which a dominant company is party is clearly, therefore, not going to be allowed to prevent the utilization of Article 86 by the Commission in a suitable case. An example of this situation was raised by the *Tetrapak*[10] case where the Court of First Instance had to decide whether to confirm a Commission decision which raised the question, namely what should happen where the dominant company had entered into an agreement subject to Article 85(1) but within the terms of a later group exemption regulation that effectively removed the prohibition. The resolution of this potential conflict between Articles 85 and 86 was ultimately the only issue before the Court, other matters dealt with in the decision of the Commission relating to the existence of dominance and relevant markets having been conceded by Tetrapak during the hearing.

Tetrapak's argument ran along the following lines. Whilst it was admittedly dominant in relevant markets, namely the production of machinery for filling aseptic UHT milk cartons and the cartons themselves, by acquiring the benefit of an exclusive licence in favour of its competitor Liquipak through takeover it had not breached Article 86 because the agreement itself was in terms permitted by the Patent Group Exemption (Reg. 2349/84). Such an exemption was a 'positive' action of the Commission in terms of the *Walt Wilhem*[11] judgment and should not, therefore, lead to any liability under Article 86, since the Commission had an alternative remedy being free at its discretion to withdraw the exemption under the terms of the group exemption regulation.

The Court of First Instance nevertheless upheld the Commission's decision that the acquisition of the licence was a breach of Article 86. It was pointed out that the withdrawal of the block exemption could only operate in a prospective way, not in respect of the past. The two Articles were independent legal instruments that within the overall scheme of the Treaty addressed different situations. The acquisition by a company with a dominant position of a licence that was exclusive was not *per se* an Article 86 abuse, but the circumstances of the acquisition and the effect it might have on the structure of competition in the relevant market must be taken into account. The Court supported the Commission's conclusion that in this case the acquisition of the particular licence for the packaging

[10] Case T. 51/89 [1991] 4 CMLR 334: [1990] 4 CMLR 47. The facts are set out in Ch. 18, pp. 369–70.

[11] Case 14/68 [1969] ECR 1: CMLR 100.

of the aseptic milk containers not only strengthened Tetrapak's substantial existing dominance but prevented or at least delayed the entry of new competition into the market. The takeover of Liquipak was simply the means by which Tetrapak had acquired the Liquipak licences and the mere fact that the licences themselves had received a group exemption could not render inapplicable the prohibition of Article 86. The Court also pointed out that group exemptions are merely secondary legislation being a specific application of the grounds for exemption set out in Article 85(3) to particular categories of agreement whereas Article 86 is primary legislation in the Treaty, which takes precedence over such secondary legislation.

The position would, however, differ if an individual exemption had been granted to a particular licence agreement because, if exemption had been sought for an individual agreement in this way, the Commission would have to take account of the particular circumstances in which it was made. The second negative condition of 85(3) itself refers to the requirement that the parties to such individual agreements shall not be afforded by it the 'possibility' even of eliminating competition for a substantial part of the relevant products. Unless these circumstances had altered by the time an Article 86 allegation was later made the earlier findings of the Commission would be relevant and taken into account. By contrast, with a group exemption there is by definition no case-by-case examination of the circumstances of the parties to those agreements and no positive assessment of the overall market situation. The benefit or group exemption is, therefore, different from obtaining any form of negative clearance from Article 86, which had a binding effect upon the Commission. The benefit of group exemption is not automatically held back from companies which have a dominant position and, therefore, cannot take into account in advance their position on relevant markets in respect of any particular agreement.[12]

An appeal to the European Court from this decision is, of course, possible but it will be surprising indeed if it were successful in overturning the essential point of principle. The Court of First Instance decision, its first of a substantive nature, makes very clear what perhaps had been only implicit in earlier Court decisions, that the grant of a group exemption under Article 85(3) cannot be utilized in such a way as to defeat the objectives of Article 86. After all, group exemptions are not permanent legislation but temporary legislative and administrative arrangements giving exemption from notification and fines, and the effects of 85(2), to a broad category of agreements and excluding others from their coverage without reference to the individual market position or power of the parties

[12] The key paragraphs of the judgment on this important issue are 25 to 29 at [1991] 4 CMLR 385–6.

concerned. Group exemptions are very different in all these ways from the individual exemptions granted under 85(1) when the detailed and specific circumstances of the case can be fully investigated.[13]

In understanding the relationship therefore between these two Articles, the *Tetrapak* judgment is of central importance. The Court of First Instance has made it very clear that Article 85(3) cannot be utilized in any way (whether by way of group or individual exemption) to reduce the scope of Article 86. Nevertheless it also emphasized that these two Articles remain 'independent legal instruments addressing different situations'.[14] This suggests that the Commission should proceed with caution, in trying to make up for any perceived deficiencies in Article 85 in dealing with closely knit oligopolies simply by seeking a corresponding extension to the application of Article 86, however great the incentive it may feel for such a change.

[13] The Court, however, stresses also that this does not mean in any way that the grant of an individual exemption under 85(3) can operate as any form of 'exemption' from 86 (para. 25 of the judgment).

[14] para. 22.

20

Mergers

1. Events Leading up to the 1989 Regulation

Competition authorities tend to regard control over mergers between large and powerful companies as one of their most important concerns; for authorities are concerned not only with the conduct of undertakings but also with the effect that their conduct has on market structures and especially on the degree of concentration (and consequent increase in market power) which mergers may bring about in particular product and geographic markets. They naturally prefer, therefore, that such mergers shall not be allowed to occur without having undergone careful scrutiny as to their likely effects on the competitive process: this should take place preferably before they occur, rather than afterwards. In the absence, however, of any rules or mechanism for ensuring that this scrutiny occurs, two possibly undesirable consequences may occur. Undertakings which are prevented from making anti-competitive agreements with each other because of the prohibitions contained in Article 85(1) may simply determine that to achieve the objectives of such agreements, they will merely merge their business operations into a single unit, thereby avoiding the reach of that Article; they may also, by their merger, increase the market power of their combined undertaking in a particular product or geographic market and thus reduce the scope for competition within it. Any competition authority, therefore, with inadequate substantive and procedural control over major merger proposals operates under a severe disadvantage.

Nevertheless, there seems little doubt that those responsible for the drafting of Article 86 did not intend that it should give control over mergers to the Commission. Neither the actual wording of the Article nor the evidence of those who participated in the negotiations leading up to the Treaty or in its early administration support any contrary argument. Nevertheless, concern within DG IV over this apparent gap in its powers was felt even during its first decade of active operation when, as we have seen, it had a variety of other pressing problems to which prior attention had to be given, in particular those relating to vertical agreements. In a memorandum in 1966,[1] it admitted that the concentration of under-

[1] 'Le problème de la concentration dans le marché Commun', *Etudes CEE* (Series

takings following transfer of shares or assets was not covered by Article 85. As we have seen under the joint venture cases, Article 85 ceases to apply once the integration of two or more undertakings has gone beyond the stage of temporary assistance and co-operation to a stage of full-scale merger where both policy direction and management have been centralized in a single body. The possible utilization, therefore, of Article 86 in a suitable merger case was not overlooked by the Commission and, in spite of considerable doubts even within DG IV itself, such an opportunity occurred in the early 1970s. It was then decided to seek a ruling from the European Court that Article 86 did enable the Commission to issue a decision prohibiting a merger where the merger itself strengthened pre-existing dominance in a particular market.

The case chosen for this attempt became known as the *Continental Can* case.[2] The United States company of that name had acquired an 85.8 per cent holding in a German company, SLW of Brunswick. In the following year, Continental Can had formed a Delaware corporation known as Europemballage Corporation which opened an office in Brussels and soon afterwards with financial help from its parent acquired 91.7 per cent of the shares of a Dutch company, TDV. SLW was the largest producer in Germany of packaging and metal closures whilst TDV was a leading manufacturer of packing material in Benelux. In its decision[3] prohibiting acquisitions by SLW of TDV, the Commission stated that it took into account not only the share of the market which the Continental Can group would as the result of the mergers control in Germany and Benelux, but also the group's advantages over its competitors resulting from its size and economic, financial, and technological importance. In the view of the Commission, all these factors gave the company room for independent action, conferring on it a dominant position at least on the German market for lightweight containers for preserved meats, fish, and shellfish as well as on the market for metal closures.

So far as the actual facts of the case are concerned, the Court held in a relatively brief judgment that the factual evidence failed to support the Commission's conclusion that Continental Can held a dominant position in the German market. In simply assigning tin cans for fish and tin cans for meat to separate markets and assuming that a high market share within each area involved market power, DG IV had failed to consider the possibility and ability of producers shifting from the manufacture of one

Concurrence no. 3, 24 (1966)), quoted by Hawk, *Antitrust*, ii, 659–60. As a statement of principle this has now been overruled by the *Philip Morris* decision (n. 5) and inevitably therefore this part of Chapter 20 deals with Art. 85 as well as with Art. 86.

[2] *Europemballage and Continental Can* v. *Commission* Case 6/72 [1973] ECR 215: CMLR 199. On the question of the basis for the Commission's jurisdiction over this case, see Ch. 24, p. 463.

[3] [1972] CMLR D11.

product to another, e.g. from fish cans to meat cans or vice versa. It also failed to consider whether large buyers would decide to manufacture their own cans if the price of Continental's cans was too high. The Commission had not prepared adequate evidence relating to the suitability of other types of containers or the probable entry by other companies into the markets, if Continental Can utilized their market dominance to raise prices. Its market analysis was simply insufficient to support its conclusion.

Whilst on the facts the Commission suffered a severe defeat, however, on the main issue of principle they gained a rather surprising victory. The Court of Justice chose to interpret Article 86 in a purposive way to give effect to the spirit of the Treaty of Rome and to ensure that consistency applied between Articles 85 and 86. The Court referred to the terms of Article 3(*f*) and concluded that, as the result of it providing for the institution of a 'system' to protect competition within the Common Market from distortion, there was an underlying principle to be found there that competition must not be eliminated. Its existence was also supported by the principles contained in Article 2 relating to the harmonious development of economic activities. Articles 85 to 90 had, therefore, been included in order to preserve the principles set out in these two Articles. It was not to be supposed (said the Court) that the Treaty should in Article 85 prohibit certain agreements, decisions, and concerted practices that might restrict but would not eliminate competition, but at the same time would allow Article 86 to permit already dominant undertakings simply by making further acquisitions to strengthen their position to the extent that any serious possibility of competition in the relevant market would be eliminated. Such a diversity of legal treatment could open a breach in the whole structure of competition law that would jeopardize the proper functioning of the Common Market. Articles 85 and 86 had to be interpreted in a manner which reflected the fact that they were both inserted to achieve the same ends, those set out in Articles 2 and 3(*f*).

The Court went on to say that Article 86 applied not only to practices having a direct prejudice for consumers but also to those that caused a prejudice to consumers indirectly through interference in the structure of markets. It referred to the risk that, if individual undertakings obtained a sufficiently powerful position in markets, all those who remained would become dependent on the dominant undertaking with regard to their market behaviour. There was no need for the Commission to show any cause or connection between the dominant position and the abuse of it. An increase in market share might alone be sufficient to constitute the abuse.

Perhaps no better example of the difference of approach of the European Court to the interpretation of the Treaty of Rome from that which would have been adopted by any English Court can be found in this decision, whose outcome was a surprise to many, although received gratefully by the Commission. Unfortunately, though the Commission won the victory relating to the applicability of Article 86 to mergers by companies already with a dominant position which, therefore, gave it in principle better standing for the future on such issues, the task of winning the subsequent broader campaign required to obtain effective substantive and procedural control over mergers proved considerably more difficult. The relationship between the *Continental Can* decision and the scope and content of the draft of the first Merger Regulation issued some few months after the decision of the Court (at the request of Member States) was, therefore, more psychological than substantive. A Council Regulation was essential to merger control because the use of Article 86 simply on the basis of the *Continental Can* decision would not have provided the Commission with the desired ability to control the creation of a new dominant position as opposed to the ability to prevent an undertaking with an existing dominant position from strengthening it still further, as under the rather unusual facts of the *Continental Can* case. Such a Regulation was also essential to ensure that pre-notification of major mergers became mandatory, so that any necessary powers of the Commission could be exercised before the merger had taken place rather than being employed in the always more difficult and sensitive task of unscrambling it after completion. Moreover, the use of interim measures which have been adopted in some Article 85 and 86 cases subsequently are themselves no real substitute for a regulation containing full procedural powers to control mergers equivalent to those provided by Regulation 17/62 for other competition cases.

The original 1973 draft that appeared shortly after the *Continental Can* decision provided a basic scheme of control, though significantly it refers not to Article 86 but to Article 235 of the Treaty as the authority for its enactment.[4] The lengthy recitals refer to the importance of preventing concentration in markets within the Community. For this reason, prior notification of major mergers over a specified threshold was essential. The only mergers to be excluded from its control would be those where undertakings had a joint turnover of less than two hundred million units of account and where the market share of goods or services affected would in no Member State exceed 25 per cent. This was, of course, an extremely low threshold. Article 1 of the Regulations, drawn in terms strongly reminiscent of Article 85 itself, stated that control would apply over all

[4] Article 235 reads 'If action by the Community should prove necessary to attain . . . one of the objectives of the Community and this Treaty has not provided the necessary powers, the Council shall, acting unanimously on a proposal from the Commission and after consulting the Assembly, take the appropriate measures.'

transactions that had the direct or indirect effect of bringing about competition in the Common Market if they conferred power on the combined undertaking to hinder effective competition. Mention is also made that consideration should be given to exempting those mergers 'indispensable to the attainment of an objective which is given priority treatment in the common interest of the Community'. The recitals referred in general terms to the factors to be taken into account in assessment, including the economic and financial power of the undertakings, market structures, and the scope for choice available to suppliers and consumers. The Regulations spelled out the detailed mechanics of control, following in many respects the terms of Regulation 17, which is expressly stated in the draft to have thereafter no application to mergers. The threshold for compulsory notification was set at one thousand million units of account calculated on the joint turnover of the undertakings, unless the target undertaking itself had a turnover of less than thirty million units of account. Decisions had to be rendered within a total time limit of 12 months and were subject to review by the Court, just as any other decision of the Commission. The Regulation would apply to both agreed and to contested mergers, since it applied to 'transactions' as well as to 'agreements'.

With hindsight, the scope of the Regulation was undoubtedly drawn too broadly. It would have brought within its scope a very large number of mergers, far more indeed than the staff of DG IV either in 1973 or even in their slightly increased numbers of later years could have dealt with within the time limits proposed, save perhaps on what would have been a demonstrably inadequate basis. In any case, however, the political opposition raised by Member States in the Council ensured that the original 1973 Regulation was not adopted. Subsequent realization within the Commission that the scope of the draft Regulations would need to be modified so as to affect only a smaller proportion of mergers led to new drafts of the Regulations being put forward to the Council in 1982, 1984, and 1986, but even these modifications initially failed to overcome the political deadlock. The more important changes to the draft were found in Article 1(1) with the express reference to the fact that competition must be shown to be limited at Community level (rather than simple Member State level) before the Commission can intervene and a market share requirement had now been added to the jurisdictional threshold. Unless the market share in the goods or services covered by the merger exceeded 20 per cent either in the entire Common Market or a substantial part of it, the merger was to be deemed compatible with the provisions of the Treaty, it being presumed that the merger is incapable of hindering effective competition. The joint turnover jurisdictional threshold was raised from 200 million to 500 million and subsequently to 750 million ECU.

Much of this increase would, of course, be accounted for by changes in the value of money since 1973. In the case of mergers with a lesser turnover, they were now to be totally exempt from control unless the merged undertakings had a 50 per cent market share in a substantial part of the Common Market.

Commissioner Sutherland indicated on a number of occasions from 1985 onwards that failure by Member States to break the deadlock would leave the Commission no alternative but to rely on Articles 85 and 86 for purposes of merger control. Whilst initial reactions by Member States to this suggestion were sceptical, given the many substantive and procedural difficulties involved, the subsequent effect of the *Philip Morris*[5] judgment of the European Court in November 1987 was to bring home to them the realization that control based on a new regulation might be far better than if linked to the unpredictable consequences of that case.

The case involved an agreement by Philip Morris to acquire a 30 per cent interest in its competitor, cigarette manufacturer Rothmans, from its South African owners, Rembrandt, though the voting rights to be obtained by Philip Morris were limited to 24.9 per cent. Both groups kept the right of first refusal should the other wish to sell, and there were other arrangements to ensure that Philip Morris had neither board representation nor managerial influence over Rothmans. The Commission had given exemption to these proposals, and the Court upheld its decision in the face of a challenge by two other tobacco companies, BAT and RJ Reynolds. It ruled that the acquisition of an equity interest in a competitor did not of itself restrict competition, but might serve as an instrument to that end. Article 85 would then become applicable to the relevant agreement.

The central section of the Court's judgment on this legal issue read as follows:

> That will be true in particular where, by the acquisition of a shareholding or through subsidiary clauses in the agreement, the investing company obtains legal or de facto control of the commercial conduct or the other company of where the agreement provides for commercial co-operation between the companies or creates a structure likely to be used for such co-operation. That may also be the case where the agreement gives the investing company the possibility of reinforcing its position at a later stage and taking effective control of the other company.[6]

The decision nevertheless left over many problems for the Commission. These included the effect of Article 85(2) on the agreement, the lack of time limits or jurisdictional thresholds, and uncertainty as to the exact

[5] *BAT and RJ Reynolds* v. *Commission* Cases 142 and 156/84 (decided by the Court on 17 Nov. 1987) [1987] ECR 4487; [1988] 4 CMLR 24.

[6] paras. 38–9.

range of mergers affected. Possibly, however, the most important outcome of the case from the Commission's viewpoint was its likely effect in concentrating the minds of the more unwilling Member States, fearful of the use of the *Philip Morris* precedent to attack mergers under Article 85, on the terms of the draft Regulation.

All this led to the submission to the Council in March 1988 of an amended draft containing some important new elements, dealing both with substance and procedure. The test for determining whether a merger had a 'Community dimension' was considerably expanded so that it included both a qualitative and a quantitative test. The qualitative test was that each of the undertakings concerned had its main field of activities in a different Member State or, even if this condition were not satisfied, one of the enterprises involved had substantial operations in at least one other state through its subsidiaries' direct sales or purchases. Alternatively one or more of the undertakings could have their main field of activities outside the Community but one at least must have substantial operations in a number of Member States through its subsidiaries' direct sales or purchases. In all cases, however, there was deemed to be no Community dimension if at least three-quarters of the aggregate turnover of two companies derived from the same Member State. The quantitative test was that the combined world turnover of the undertakings involved was one thousand million ECU with a minimum of fifty million ECU Community turnover for the target company. The criteria for assessment for the first time included language familiar from both Articles 85 and 86, of 'compatibility' with the Common Market. A merger that gave rise to or strengthened a dominant position either in the Common Market as a whole or a substantial part of it would not be compatible and could not be allowed, and a long list of factors relating to the competitive structure of the relevant markets were declared to be relevant to the assessment of this compatibility. The procedural framework for the operation of the Regulation was tightened and prior notification of all qualifying mergers made compulsory. Two months were to be allowed for the initial decision as to whether proceedings should be opened and a further four months for completion of any such full investigations.

Although the negotiating process still had some time to run, these new proposals seemed to break the deadlock, and the arrival of Sir Leon Brittan as the new Commissioner with responsibility (*inter alia*) for competition brought new impetus. Early in 1989 the Commission responded to Member State criticisms by raising the thresholds for jurisdiction considerably, namely to five billion ECU for world turnover and to two hundred and fifty million ECU aggregate Community turnover for each of at least two companies involved (with the intention that both figures should be reduced at a later stage after experience of the operation of the

Regulation). In addition, the exception for undertakings both having a substantial proportion of their Community turnover in the same Member State was reduced from a level of three-quarters to two-thirds.

These changes very considerably reduced the number of mergers likely initially to fall to be within the reach of the Regulation. When France took over presidency of the Council in June 1989 it made clear that it regarded agreement on the final text of the Regulation as having high priority; the succeeding six months of its presidency saw the final difficult stage of the negotiations edge forward to completion. An important difference lay between those Member States that wanted the criteria expressly limited to competitive issues (especially Germany and the UK) and those, such as France, which wanted social and industrial policy issues to remain among the criteria that could be applied. Both Germany and the United Kingdom also sought to retain the right of national competition authorities to retain jurisdiction in those cases where major national interests and markets were involved. On the other hand the majority of Member States, as well as industrial and commercial interests, were considerably attracted by the possibility of a 'one-stop shop' for large-scale mergers with the Community dimension.

Agreement on the text of the Regulation was finally reached at a Council meeting on 21 December 1989 following last minute concessions by nearly all Member States. The resulting text is contained in Council Regulation 4064/89, though there are also a number of other relevant documents published by the Council and the Commission at the same time which relate to its interpretation and application.[7] The Regulation came into effect some nine months later, on 21 September 1990. Its recitals are naturally extensive; after pointing out the inadequacy of Articles 85 and 86 to protect the system of undistorted competition envisaged in the Treaty, and the consequent need to introduce a new Regulation based not only on Article 87 but principally on Article 235, it restates the essential principle behind the new instrument; this is that any concentration with a Community dimension creating or strengthening a position as a result of which effective competition in the Common Market or a substantial part of it is significantly impeded is to be declared incompatible with the Common Market. Jurisdiction over such concentrations is to rest exclusively with the Commission save only for some narrowly

[7] These are conveniently collected in a Community publication, 'Community Merger Control Law', Supplement 2/90. This contains not only the texts of both Regulations 4064/89 and 2367/90 containing the relevant substantive and procedural rules but also two Notices on the definition of concentrative and co-operative joint ventures and the principle upon which contractual restrictions ancillary to concentrations can be cleared under the Regulation. These can include non-competition clauses, licences of industrial and commercial property rights, and purchase and supply agreements.

defined exceptions to be found in Articles 9(3) and 21(3), which are discussed below.

The Merger Regulation 4064/89

Jurisdiction—the Community dimension

The Commission's jurisdiction is over a range of concentrations which are both of a very substantial size and take effect wider than simply within an individual Member State. It is not normally, therefore, concerned either with companies of a relatively small size nor those whose main activities are outside the Community; on the other hand if large multinationals, even those whose primary activities are, say, in Japan or the USA, have sufficient turnover within more than one Member State, their acquisitions or joint ventures may still fall within the jurisdiction of the Regulation even if at first sight they have little material connection with the Community. This objective is ensured by the minimum jurisdictional figures first proposed in early 1989 and finally adopted:

(i) at least five billion ECU (£3.6 billion) combined world-wide turnover for all the undertakings, and
(ii) at least two hundred and fifty million ECU (£180 million) Community turnover for at least two undertakings involved,
(iii) but *no* jurisdiction if this Community turnover for each of the participants is achieved as to at least two-thirds from the same Member State.

Article 1 lays down this basic jurisdictional principle, supported by Article 5, which provides some necessarily detailed provisions in support to allow the calculations to be made in individual cases. Turnover for ordinary trading companies is defined so as to cover all sales of goods and services but excluding value added tax and sales tax; for insurance companies and banks different methods of calculation relating to gross premiums and total assets are provided. To the turnover both of the bidding and target company are to be added the turnover of both its parent and subsidiary companies forming part of the same group.[8] If, however, the acquisition is simply of part of a company's business, for example where a division manufacturing a particular product is being sold rather than the entire

[8] See the approach of the Commission to this point illustrated in the *Arjomari-Prioux/Wiggins Teape* case (IV/M025 dated 10 Dec. 1990) decided in the early months of the new Regulation. The decision that a company (GSL) having a 43% holding in Arjomari-Prioux was not able to control the composition of more than half of Arjomari-Prioux's board of directors shows that difficult issues can arise on the interpretation of Article 5(4).

company, then the relevant turnover in calculating the seller's world-wide turnover is only that of the particular division which it is selling, not the entire company (Article 5(2)). In joint ventures, however, extremely small transactions may be caught since in all cases the turnover of both parents is relevant. In order to prevent, however, the artificial splitting up of a major purchase into a series of such transactions each of which remains below the jurisdictional threshold, the Commission may aggregate all transactions made between the parties over a two-year period. It is important to note that before the end of the first four years of the Regulation, i.e. before 1994, the jurisdictional thresholds may be amended by qualified majority of the Council so as to extend to a wider range of transactions.[9]

Jurisdiction—the meaning of 'concentrations'

Whilst we have used the general expression 'merger' to refer to the normal situation where one undertaking (a bidder) acquires control of another (the target), the actual term used in the Regulation is a 'concentration', as defined in Article 3. It is, of course, an essential element of any scheme of merger control (in order to provide legal certainty) that an exact definition of those transactions and situations within its jurisdiction must be given. The expression 'concentration' covers the case where two separate undertakings merge into a single body and also the more common situation (at least in the United Kingdom) where one person or undertaking acquires 'direct or indirect control' of the whole or part of another. 'Control' in its turn is defined as 'rights, contracts or any other means which either separately or jointly . . . confer the possibility of exercising *decisive influence* on an undertaking', whether this occurs through share ownership, voting, or management agreements or through any other form of rights operating on the undertaking. Attention has to be paid in each case to the extent to which the acquiring company has rights which affect the commercial strategy and detailed financial management of the company acquired as opposed to those which relate simply to the protection of its investment. It can include an acquisition either of 100 per cent of the shares in an undertaking or merely a lesser percentage sufficient in practice to enable such 'decisive influence' to be exercised. It can also cover cross-shareholdings, cross-directorships, and indeed a variety of other means of exercising control.[10]

[9] The 19th Annual Report (para. 16) indicates that the Commission's objective is to reduce the world-wide turnover figure to two billion ECU and the Community threshold by a similar proportion.

[10] See also the *Arjomari* case referred to above for the Commission's initial approach to the concept of decisive influence under Art. 3(3). A shareholding of 39% where no other shareholder held more than 4% was held to confer such influence. It is, however, clear from

A joint venture may also fall within the definition provided it is properly described as 'concentrative' rather than 'co-operative'. Although the Regulation provides a detailed definition for the two categories, its application to the often complex facts of large-scale multinational operations will undoubtedly prove difficult as early experience of the operation of the Regulation has shown. Concentrative joint ventures are those which Article 3(2) describes as 'Forming on a lasting basis all the functions of an autonomous economic entity, which does not give rise to co-ordination of competitive behaviour of the parties among themselves or between them and the joint venture'. Whilst the co-operative joint venture is, by contrast, one whose object or effect is co-ordination of the competitive behaviour of undertakings which remain independent, cases remain where the characterization of the transaction in one or other of these categories is difficult. The terms of the 'Notice on Joint Ventures' already referred to[11] are an attempt by the Commission to assist in this task. An important element in the concentrative joint venture is that the parent companies involved have cut off their own ability to compete with or trade with the newly created joint venture, which has itself been provided with sufficient resources to compete in its own right in its chosen markets without a continuing dependence for financial or technical resources on its parents. Moreover, the joint venture must not have as either object or effect the co-ordination of the parents' activities with those of the new joint venture. Following the publication of the Notice it was criticized by a number of commentators on the grounds that the degree of concern expressed on this ground[12] was in economic terms overstated. The Commission's subsequent response in the initial cases has been to develop a new doctrine, not to be found in the original Notice. Under this, if one of the parent companies is effectively the 'industrial leader' of the joint venture, in particular by being able to designate the managing director, and the other parent company accepts this, then any product overlap or apparently even explicit co-ordination between the first parent and the joint venture will not be regarded as anti-competitive within the terms of the Regulation because in any case no competition could reasonably be expected in these circumstances between them.[13]

The contrast between this concept and that of the joint venture which primarily facilitates the co-ordination of the parents' continuing activities

this case that the requirements for a finding of decisive influence are less than those needed for determining under Art. 5(4) whether an undertaking should be considered as forming part of the corporate group whose aggregate turnover has to be calculated for the purpose of establishing a 'Community dimension' for the concentration.

[11] OJ C203 (14 Aug. 1990). See also Ch. 11, p. 209.

[12] In paras. 20–36 in particular.

[13] See *Thomson/Pilkington* Case IV/M086 dated 23 Oct. 1991 and *UAP/Transatlantic/Sun Life* Case IV/M.141 dated 11 Nov. 1991.

in a market was almost immediately raised by one of the first cases decided by the Commission after the Regulation was brought into force, the *Renault/Volvo* case.[14] Both these large companies based in France and Sweden respectively manufactured cars, buses, and trucks. They planned to take cross-shareholdings of 25 per cent in each other's car operations but a larger percentage of 45 per cent in the remaining operations. The joint venture was to operate on the basis that a number of joint committees would have the power to make relevant decisions on all commercial issues, whether research and development or more closely associated with manufacturing problems, purchasing of components, and marketing. In the situation where the parties could not reach an agreement through the committees, they would retain the right to make independent decisions.

With regard to the coach, bus, and truck operations, the Commission ruled that, since the parties would end up sharing profits and losses on an almost equal basis, this would create a strong situation of common interests; in practice this would mean that the parties would normally reach joint decisions as a result of the committee structure and would not need to act independently. With the much lower shareholdings held in the other's car manufacturing operations, however, the Commission did not feel that either of the parties would feel under the same practical compulsion to reach joint decisions on a permanent basis. Moreover, in the case of the car operations the companies had entered into far fewer binding commitments with the other with regard to the integration of the manufacturing businesses and purchase of common components. The car joint venture was adjudged therefore co-operative, the other joint venture as concentrative and subject to the Regulation.

The Notice on joint ventures also emphasizes the importance to be placed on the nature of the continuing interests retained by the parents. These may be retained in those territories where the joint venture will not operate, or by contrast still in related product markets in the same geographical territories. Alternatively, they may be in other markets upstream or downstream, for example, where the joint venture deals exclusively or at least very often in buying or selling to or from its parent company in respect of raw materials or components. Also relevant would be any continuing involvement by parent companies through their representatives in the commercial business of the joint venture to an extent greater than would be normal for shareholders merely making long-term strategic decisions at reasonably extended intervals.

[14] IV M.004 decision dated 7 Nov. 1990.

Criteria for assessment of individual concentrations

Once rules have been established for deciding whether any particular transaction falls within the Regulation, the substantive criteria for assessment of that transaction become of central importance. As mentioned, the formula chosen has been whether the concentration is 'compatible' with the Common Market. Criteria of this kind can be in principle defined either in a broad manner, including not only competition issues but social and industrial policy, or they can be specific and limited to particular issues, for example the effect on competition in the relevant product and geographic markets. Contrary to what may have been expected, the Regulation in its final form does not reflect a compromise between the two approaches; its character is firmly based on competitive issues, even though the relevant Article 2 does apparently contain a suggestion that other factors might be taken into account in a particularly evenly balanced case. The Article sets out the rules to be applied by the Commission by way of three separate subclauses whose relationship to each other is not immediately obvious. Article 2(1) lists several factors to be considered in deciding upon 'compatibility':

— the need to maintain and develop effective competition within the Common Market in view of, among other things, market structures and actual (or potential) competition from other undertakings;
— the economic and financial power and market position of the parties and also of their suppliers and consumers (and their access to alternative supplies);
— legal and other barriers to entry;
— trends of supply and demand in relevant markets;
— development of technical and economic progress provided that it is to consumers' advantage and does not form an obstacle to competition.

The Commission is not, however, simply to assess these factors and then decide, after weighing them up, whether the concentration should be permitted, rather as it might do in deciding whether to grant an exemption to a particular agreement under Article 85(3). The Commission has rather to make its findings in the light of these factors and then see into which of the alternative categories provided by Article 2(2) or 2(3) the concentration will fit. If the findings of the Commission are that a dominant position has not been created or strengthened or in the considerably more unlikely event that, even if this has happened, it will not significantly impede effective competition in the Common Market, then the concentration is to be declared compatible and cleared. If, on the other hand, the effect of the concentration is to create such a dominant position with the necessary effect on competition, then it is to be declared incompatible and blocked or at least its implementation made conditional on the divestment by one or more of the participants of certain parts of their undertakings so as to eliminate the anticipated detriment to effective competition.

The terms of this Article 2 raise a number of difficult issues, which will only be resolved once there have been authoritative court rulings upon them. The concept of dominant position has, of course, been considered in a number of Article 86 cases by the European Court.[15] It is far from clear, however, whether the meaning of the expression as used in Article 2 will be held to mean precisely there what it has been held to mean in these Article 86 cases. If it were to be interpreted in exactly the same manner, the consequence would be that before any concentration could be refused clearance, a finding of a relevant market share of the combined aggregates of the bidding and target company would have to be at least 40 per cent, this being consistent with the rulings of the Court in a number of earlier cases such as *United Brands*.[16] This percentage would have to be enjoyed in at least one relevant market consisting either of the Community itself or one or more Member States or a relatively large part of one of the main Member States. The same analysis of both product and geographic markets will be needed as in normal Article 86 cases, though under far greater time constraints. Whilst the terms of Article 2 in this context are imprecise and reflect continuing differences of emphasis between Member States, it is unlikely that the Article would be interpreted in such a restrictive way as to prevent the Commission in all cases from blocking the merger unless such a substantial market share were the result. The European Court is more likely to approach the Regulation in its normal teleological manner, i.e. asking its main purpose. The purpose of the Regulation is clearly the prevention of either the creation or strengthening of a dominant position which might significantly impede effective competition in one or more Member States; whilst the recitals contain a presumption that effective competition is not impeded by a concentration where the parties have in combination less than 25 per cent (and even this as a presumption is capable of rebuttal by sufficiently strong evidence) it contains no reference to any other quantitative share or percentage. Equally of course the possession of such a market share in either a Member State or the Community itself is not an automatic guarantee that the undertakings concerned do hold a dominant position there (see *Volvo/Renault* where the national market shares for buses exceeded 60 per cent for Volvo in Sweden and Renault in France, and yet the merger was found compatible).

Moreover, the assessment of the dominant position in this Regulation,

[15] See Ch. 17. The wording of Articles 2(2) and 2(3) is notably similar to wording in the merger control articles of the German Law against Restraints on Competition.

[16] (27/76) [1978] ECR 207, 345: [1978] 1 CMLR 429: 3 CMLR 83.

which is concerned with the prediction of what may happen in the future to the competitive structure of a market as the result of a particular transaction, is inherently different from its assessment under Article 86 cases, where Commission and Court are concerned with past actions and whether they have been abusive of the market position of the parties. Where the Court's concern is with the past and with the punishment of firms for proven abusive conduct (leading possibly to substantial fines), considerations of legal certainty strongly support the establishment by the European Court of a relatively high threshold for a finding of dominant position. It would be unreasonable to impose such a fine in any case where the market share of the parties was manifestly only on the borderline of dominance. By contrast, in the course of the analysis of the future effect of a proposed concentration, when no question of punishment for past abuse arises, it would be both realistic and reasonable to adopt a more flexible approach. In suitable cases therefore the Commission may accept that when the combined market share of target and bidder is between 25 and 40 per cent such an aggregation may also be capable of constituting a dominant position. On the other hand it is difficult within the terms of Article 2 to see how the merger of the second and third companies in a market can ever be prevented where the largest company itself has a market share of over 40 per cent (absent some strong likelihood of joint dominance between them all in the future).

The possible interpretation of Article 86 to cover cases of joint or collective dominance has been mentioned in Chapter 19, particularly in connection with the *Italian Flat Glass* case. Any development of the Court's jurisprudence in this area would be also of major significance to the interpretation of 2(2) and 2(3). It would permit an important extension of the jurisdiction of the Commission in that in oligopolistic markets the 'dominant position' referred to in those Articles need not be simply that of a single powerful undertaking. It could also be applied to two or more undertakings which by parallel action (falling short of agreement or concerted practices within the terms of Article 85) are able to impede effective competition in a relevant product or geographic market. Whilst the Merger Task Force has made no secret of its desire to be able to utilize the concept of collective dominance, it would be premature to assume that this is necessarily a correct interpretation of Article 2 or indeed Article 86.

A flexible approach, therefore, to a definition of dominant position has also the advantage that it enables the criteria contained in Article 2(1)(*a*) and (*b*) to be read in a manner which gives greater rather than lesser meaning to them, so that it is primarily through their application to the individual facts of a case that the Commission will be able to assess the effects of a merger or joint venture on the state of competition in that market. An appraisal under Article 2(1) might, for example, show that market shares in a particular sector fluctuated substantially from year to year, perhaps as the result of changes in consumer taste, so that a 40 per cent share held in one year could rise to a 60 per cent share in the next year reducing to 20 per cent in the third year. In other product markets by contrast shares might vary by only a point or two over a five- or even ten-year period. Barriers to entry may in some cases mean that market shares are likely to remain substantially unchanged as between the existing participants in the market, whereas in other situations where barriers to entry are much lower the chance of challenge would be far greater. All these factors could be determinative in deciding whether a dominant position 'significantly impeding' effective competition will be created or strengthened by the merger, and it is to be hoped that full effect will be given to them in the application of Article 2(1) for this purpose.

Whereas some earlier versions of the Merger Regulation[17] had given more weight to social and industrial policy criteria, the criteria as set out in Article 2 are now primarily concerned with competitive issues except only for the final phrase 'development of technical and economic progress', which of course appears also in Article 85(3). It is of course possible that this phrase could be used by the Commission as justification for permitting a merger (otherwise objectionable on competition grounds) because it allowed development of a European champion to expand its research and development capability so as to compete more effectively with Japanese or United States multinationals; it is hard, however, to see the legal basis for such an approach, particularly in view of the structural relationship between the different subclauses of Article 2.

Whilst it might be supposed from the strict wording of Article 2 that concentrations would either be cleared or refused in their entirety, this is not so, for Article 8(2) makes clear that the decision of the Commission may be to declare the concentration compatible, but to attach to its decisions conditions and obligations which are intended to ensure the original merger plan is modified so as to remove any concerns which the Commission may have felt as a result of particular aspects of its investigation. In two of the first full investigations under Article 6(1)(*c*), decisions to approve were made conditionally.[18]

Intervention by Member States

In the course of the final negotiations two concessions were made to Member States who felt strongly the need for involvement of national merger authorities in cases which were particularly sensitive to their

[17] See R. Kovar, 'The EEC Merger Control Regulation' 10 *YEL* (1990) 55.

[18] *See Alcatel/Teletra* Case IV/M.042 dated 12 Apr. 1991, 1991 OJ L22/42, and *Magneti-Marelli/CEAC* Case IV/M.043 dated 29 May 1991, 1991 OJ L 222/38.

national interests. The principle of the 'one stop shop' is to this extent compromised, though in practice it may prove a limited breach as the Commission has shown only occasional willingness to allow Member States to invoke exceptions to this principle. The two concessions are, however, quite different in substance from each other and need careful analysis in order to show the degree to which they do actually represent exceptions to the Regulation's basic principles.

Article 9[19] sets out a procedure under which a Member State may inform the Commission that a particular concentration threatens to create or strengthen a dominant position as a result of which effective competition will be significantly impeded 'on a market within that Member State' which presents all the characteristics of a distinct market. There is no requirement that this distinct market is a substantial part of the Common Market. If such an application is made, it is for the Commission itself (not the Member State) to apply the terms of the Article in order to decide whether the case should be referred to the Member State or dealt with in the normal way by the Commission. If it is referred to the Member State (and the reference may be of only those aspects of the concentration that directly relate to the distinct local market) then the Member State will apply its own national competition law. When a Member State is permitted to 'take back' a merger in this way it has four months in which to assess the case and publish its report and findings. The action which it may take is limited by Article 9(8) to 'measures strictly necessary to safeguard or restore effective competition on the market in question'. It is noteworthy, however, that the Commission is not required to yield a jurisdiction over any concentration which meets the requirements of the Article and may retain the case for itself 'in order to maintain or restore effective competition'.[20]

Quite different is the operation of Article 21(3), which allows a Member State to intervene in connection with a concentration that has a Community dimension on non-competition grounds. It may in certain circumstances allow the Member State to block a merger that the Commission would under normal principles decide to be compatible with the Common Market, because the Member State can show it has a legiti-

[19] Sometimes called the 'German clause' because Germany pressed most strongly for it. An important justification for the exception is that while under Article 2 the Commission has jurisdiction over all concentrations of a Community dimension it can make a declaration of incompatibility only in regard to those which have the required effects in the Common Market or a substantial part of it, so that it cannot investigate their effect on local geographic markets. C. Bright, 12 *ECLR* (1991) 139–47.

[20] See e.g. the published notes on this Article of the Regulation, which state the views of both Council and Commission that 'the referral procedure provided for in Article 9 should only be applied in exceptional cases . . . confined to cases in which the interests in competition of the Member State concerned could not be adequately protected in any other way'. The first such reference was made to the UK in Case IV/M/80 Steetly/Tarmac on 12 Feb. 1992.

mate interest in preventing the merger which must fall under one of the following categories—public security, the plurality of the media, or prudential rules relating to financial or fiscal issues. It does not apply to mergers that the Member State may wish to block on economic or social grounds such as employment or regional considerations and would not permit a Member State to allow a merger simply because of the existence of such a legitimate interest. Member States may, however, request the Commission to recognize other public interests within the framework of this Article, though the Commission must be satisfied that the class of interest is in accordance with existing principles and provisions of Community law before permitting a Member State the right to prevent a merger on this ground. The framework of this Article does (unlike Article 9) provide in principle for the possibility of parallel proceedings at Community and Member State level; uniquely the result of such proceedings may be that a concentration declared compatible by the Community may nevertheless be legitimately prevented by the individual Member State. The Member State, however, cannot clear under this Article a concentration that the Commission has declared incompatible.

A third situation can possibly arise outside the normal rules for jurisdiction. If under Article 22(3) a Member State invites the Community to investigate a concentration that does not meet the turnover requirement of the Regulation and is, therefore, not strictly of a 'Community dimension', the Member State may also ask the Community to make a finding as to whether the concentration creates or strengthens a dominant position as a result of which effective competition would be significantly impeded within that Member State's territory. In these circumstances which, of course, can only arise where a Member State decides to initiate the procedure, the Community may 'insofar as the concentration affects trade between Member States' adopt a decision and take measures that are strictly necessary to maintain or restore effective competition within the territory of that Member State. One can imagine that this Article might be called into play by a Member State without its own competition authority or one with resources insufficient to deal with a merger that had a significant effect on its economy. If thresholds for Community dimension, however, have been reviewed downwards at the end of the fourth year, this option for Member States will cease to be available.

The relationship of Articles 85 and 86 to the Regulation

The Regulation makes clear in Article 21(1) and (2) that no Member State may apply its national legislation on competition to any concentration with a Community dimension, for it is the Commission alone that now has jurisdiction to take decisions in such cases. Ideally, therefore, this new

Regulation would have rendered unnecessary the consideration of whether any residual jurisdiction remained in the Commission in respect of mergers under either Article 85 or 86. Legally, however, these Articles still retain in principle some relevance to mergers and, of course, as primary provisions of the Treaty, cannot be overruled by any subsequent Regulation. Before the new Regulation came into force in September 1990, it had been Regulation 17 that had enabled the implementation of Articles 85 and 86 without constricting the rights of national courts to apply both Article 85(1) and Article 86. The arrival of the new Regulation with its express disapplication of Regulation 17 (and certain other Regulations) to concentrations of Community dimension[21] automatically caused the basic initial rules of Article 88 and Article 89 to re-enter into force. These Articles provided the initial basis for the control by the Member States (Article 88) and by the Commission (Article 89) of both restrictive agreements and abuse of a dominant position. In practice it is unlikely that in respect of concentrations with a Community dimension the Commission would take action other than under the terms of the Regulation. Nevertheless, the possibility of private actions in respect of such concentrations under Article 86 may still continue to be available, though a court would be unlikely to grant any form of injunctive relief while the Commission is still carrying out an investigation under the Regulation. Concentrative joint ventures of the relevant size are, of course, now fully covered by the Regulation, although co-operative joint ventures are still treated under Regulation 17 applying Article 85.

The position is more complex for those concentrations which, for some reason (normally because of inadequate world or Community turnover), do not have a Community dimension. So far as Articles 85 and 86 are concerned the European Commission has expressly reserved its right to take action under Article 89. This may be simply a defensive move to prevent a claim by anyone that the Commission is neglecting its duties under the Treaty following the introduction of the new Regulation to cover the largest mergers. The Commission's published statement at the time of the introduction of 4064/89 was to the effect that it would not normally regard concentrations below a particular level as having any significant effect on trade between Member States. The level fixed was that of a world turnover of two billion ECU and where also at least two parties have a turnover in the European Commission of at least one hundred million ECU, provided that two-thirds of the Community turnover of each company is not in the same Member State.

So far as the operation of national courts are concerned in relation to such smaller mergers, it is unlikely that the parties will find them willing

[21] Article 22(2).

to accept jurisdiction on the basis that an agreed merger is claimed to constitute an agreement under Article 85 but use of Article 86 still remains an option theoretically available. Since the Commission is less likely to intervene itself national courts might even feel emboldened to issue injunctions more readily in such cases.

Procedural control by the Commission

Of all major commercial transactions mergers and takeovers are particularly sensitive to even quite short delays. The volatility of stock exchange prices and the difficult relationships that may prevail between bidder and target (whether the bid is hostile or amicable) mean that in many cases even a delay of a few months can be fatal to the completion of a proposed concentration. For this reason it is now normal in national merger control systems for fairly strict time limits to be imposed within which decisions have to be rendered by regulatory authorities. In the case of the Commission the time limit originally suggested in 1973 of twelve months has by the Regulation been reduced to at most approximately five months for the entire procedure, which has two stages.

The first stage starts with the prenotification of the concentration within seven days from the making of the relevant agreement or acquisition of a controlling interest or announcement of a public bid. This notification involves provision to the relevant department of DG IV, known as Merger Task Force (MTF), of a substantial dossier of information about the transaction. This will include details of the method proposed for effecting the concentration, the details of the relevant corporate groups, their accounts for the last three years, and full explanation of this ownership and control and of any relevant links between them. A full analysis of all the affected markets including both product and geographic dimensions will also be needed, with detailed assessment of the shares held by the parties and information as to actual or potential barriers to entry. Further information will need in appropriate cases to be provided on the degree of vertical integration, and distribution and service systems; research and development expenditures and the existence of co-operation agreements, including the licensing of patents and know-how, may also be relevant. In practice notification on a formal basis is normally preceded by a number of meetings between the parties and the MTF appointed team which enables the framework of the actual notification to be more easily agreed in the light of the particular circumstances of the case.

Once official notification has been made the MTF team begin an intensive three weeks of analysis of all the information provided and may also seek additional information from a variety of sources, including suppliers, customers, trade associations, and public authorities. The MTF will also consult widely within the Commission itself and with Member States. Quick visits or surveys may be made to or in relevant geographic markets to obtain additional information, and urgent information requests under Article 11 of the Regulation sent out to interested parties. Implementation of the transaction is normally suspended for a period of 21 days but this is often extended by agreement between the parties and the Commission until after the end of the first stage at the end of one month. The great majority of concentrations notified have been cleared at the end of this first stage and given an appropriate ruling under Article 6(1)(*b*) to the effect that the concentration fell within the scope of the Regulation but did not raise serious doubts as to their compatibility with the Common Market.[22] It is only in respect of the other small percentage that the second phase of the investigation then begins under Article 6(1)(*c*), where the concentration is believed to raise serious doubts as to its compatibility with the Common Market. Both the initial notification and subsequent clearance, or commencement of the second stage will be announced in the Official Journal, often accompanied by a press release.

The investigation stage lasts for no longer than four months after the first month has elapsed. In practice most of this more detailed investigation will have to take place mainly during the first six weeks of the second stage whilst all the relevant information is assessed and a variety of issues explored. Consultations will be continuing more or less continuously with the other relevant departments of the Commission, with the Legal Services Division and with the Cabinet of the Commissioner for Competition. If, after this period, a conditional or negative decision is thought likely, then a formal statement of objections setting out the arguments against the concentration in its original form will be prepared. The parties will be given fourteen days to respond in writing and, if they wish, to attend an oral hearing. Other interested parties such as trade unions, employees' associations, or competitors will have a right to be heard under Article 18 and the parties will have full access to the Commission's file of relevant documents apart from those defined as business secrets.

After the hearing an initial draft decision will be prepared; this will have to be translated into all the nine official languages of the Commission and then sent to Member States so that their representatives can consider it at the Advisory Committee. Unusually, under this Regulation, the Advisory Committee can recommend publication of its opinion and publication has now occurred on several occasions. After the Advisory Committee meeting has been held, a draft decision is prepared and approved by the various divisions of the Commission already mentioned. Finally, it is sent (in only the three major languages, French,

[22] Or alternatively that it fell outside the terms of the Regulation altogether perhaps because the turnover thresholds were not met.

English, and German) to the Commission for adoption by the written procedure if uncontroversial, but by formal review if of a controversial nature.

During the operation of the Regulation up to the end of January 1992 the actual number of notifications made and their outcome had been as follows:

— decisions under Article 6(1)(*a*)–8, concentrations not of Community dimension, e.g. because not sufficient world or Community turnover.
— decisions under Article 6(1)(*b*)–66, concentrations having a Community dimension but compatible with the Common Market.
— decisions under Article 6(1)(*c*)–6, referred to full second stage investigation as raising serious doubts about compatibility of these having ultimately been cleared unconditionally and after conditions had been imposed on the proposals. Of these only one proposed concentration had been rejected as incompatible with the Common Market.[23]

The procedural machinery for the new Regulation has generally been regarded as operating successfully. Careful preparation of a notification form (form CO) and skilful management of the administrative procedures and personnel, with special attention being given to pre-notification discussions and co-operation with the parties and their lawyers, have all meant that time limits have been adhered to and the majority of cases which present little or only moderate legal difficulty under the Regulation smoothly dealt with. Undertakings and their advisers have been surprised by the flexibility and imaginative approach of the MTF. The very success of these procedures, however, has caused important questions to be raised about the relationship of the MTF and its administration of the Regulation to the remaining departments of DG IV with their general responsibility for Articles 85 and 86. Comment on these features as well as on a number of possible future problems arising under the substantive criteria of the Regulation and its overall framework will be considered in Part III.

[23] This was the Aerospatiale/Alenia bid for De Havilland, a Canadian division of Boeing which manufactured small (20–70 seat) commuter aircraft. (Case IV/M053, decision dated 2 Oct. 1991), 1991 OJ 334/42.

21

Trade Associations

Trade associations play an important part in the commercial life of the Community and its Member States. Article 85(1) recognizes this with its prohibition of 'decisions by associations of undertakings' which have the required effect on trade between Member States and on competition within the Common Market. A similar wording was included also in Article 65(1) of the Treaty of Paris.

The normal use of the term 'association' referred to in the Article is to describe a separate legal entity or person formed in order to pursue particular objectives for its members. Since these objectives in the context of Article 85 are almost invariably commercial, it is normal to refer to them as 'trade associations'.[1] It is important, however, whilst using this term to remember that Article 85 applies to any kind of association which has some form of economic or commercial purpose, and that the association need not have a separate legal personality (or even any formal constitution) to be reached by the prohibitions of Article 85. If the association is unincorporated, the prohibitions of that Article fall in effect upon its participating members even though a decision of the Commission may be expressed to cover the association as well. If the association is incorporated the decision will be directed both to it and its participating members.

There are, however, a number of organizations which may loosely be described as 'trade associations' but which do not normally have to be considered in the context of Article 85 or indeed or Article 86. Thus, we find there are *employers' organizations* which are associations whose primary function is to bargain collectively for their members in a particular sector over wages and conditions of service for employees in particular industries. *Chambers of Commerce and Trade* exist in virtually all Member States with local and regional organizations which reflect the interests in a particular geographical area of manufacturers, wholesalers and retailers, and providers of professional and other services. In some countries, such organizations have standing in public law and indeed in France they have responsibility for the operation of public facilities including ports and air-

[1] A convenient definition for this purpose is found in Section 43(1) of the 1976 Restrictive Trade Practices Act, namely 'a body of persons (whether incorporated or not) which is formed for the purpose of furthering the trade interests of its members or of persons represented by its members'.

ports. There are also *national business organizations* like the Confederation of British Industry (CBI) representing manufacturing and commerce generally who will often include representatives not only from employers' organizations and Chambers of Commerce and Trade as defined above but also from trade associations proper; the chief responsibility for such national organizations is the maintenance of good relations with government and other official bodies over proposed legislation and other policy issues including on occasions negotiations with the European Commission itself.

A trade association proper, however, of the kind with which competition law is normally concerned, limits its membership to those with a common involvement at one level of a manufacturing or service industry rather than attempting to cover a wider span of interest. Further, it normally operates on a national basis rather than regionally, although within its organization it may have sections or classes of specialists of separate categories of membership. Whilst differing from Chambers of Commerce and Trade in that they are nationally rather than regionally based, they differ also from employers' organizations in that their activities cover a wider range of interests affecting undertakings in the same industry. The normal objectives, therefore, of such a trade association would include:

(*a*) representation of the views of members to government and official bodies on topics of current interest, especially in connection with proposed legislation, (national or within the Community) and likewise passing back the views of Government and authorities to the industry as to how it should order its affairs either on specific topics, e.g. environmental protection, or more generally by promulgating codes of conduct;[2]

(*b*) providing a range of services for members including information on matters relating to their businesses, the performance of the sector, and the provision of legal and financial guidance on general problems;

(*c*) in handling the public relations of the industry with its consumers and the public as a whole, and including utilization of press, television, and other media.

It is important to emphasize that many, indeed probably a majority of trade associations, raise no problem for authorities such as the Commission, required to enforce competition law. The constitutions of such bodies will be carefully drawn to limit their authorized range of activities so as to avoid any suggestion of facilitating agreements between their members that would themselves offend the provisions of competition law;

[2] In the UK these may include codes officially approved by the Director General of Fair Trading under the Fair Trading Act 1973, Section 124(3).

precautions will also be taken to try and refute any suggestion that (regardless of the constitution of the association) in practice the association may provide a useful cover for co-ordination of business activities between competitors. Legal advisers to such associations are aware that the boundary line between the legitimate and the illegitimate in this context, though recognizable, needs careful and regular monitoring.

It is only in a minority of cases that the so-called trade association is essentially no more than a cartel seeking to dress itself in respectable clothes. It is with such cases, however, that this chapter will necessarily be primarily concerned. The Commission has in many cases had little difficulty in penetrating the disguise, either because the actual rules or practices of the association are blatantly restrictive or because they clearly create a market situation where its members are encouraged to regard the unity of the industry as more important (as a matter of 'solidarity') than the advantages each might gain from strenuous competitive effort. It is rare in this kind of case, therefore, for the Commission to have to perform the kind of 'balancing' test that we have considered in earlier chapters as when dealing, for example, with systems of distribution or joint ventures. With the possible exception of the regulation of trade fairs and exhibitions,[3] the likely outcome of a Commission investigation of a trade association's activity is either a negative clearance or a prohibition under Article 85 of specific activities. The first answer is likely, though not of course certain,[4] if the association's activities can be shown to be limited to issues such as technical and quality standards, safety and environmental protection; the second answer will be given when the activities of the association are less carefully limited and have the necessary effect of making competition between the members less attractive or essential to them.[5]

The trade association will normally have a written constitution and relatively predictable rules. Though constitutions vary, as a minimum they normally contain provisions on the following points:

(1) *The objectives* of the association and the definition of the interests which it represents whether these be simply those, e.g. of manufacturers of, say, perfume or writing paper, or whether it seeks to represent more than a single class of undertaking within the industry, e.g. manufacturers, wholesalers, and retailers. Reference will also be made to any ancillary

[3] See p. 423 below.

[4] Even agreement between members on apparently technical issues may have important effects on the level of competition between them.

[5] The case-law from the US is particularly rich in this area providing a wide variety of cases to illustrate the type of arrangement associations have indulged in to reduce the rigours of unrestrained competition. See A. Neale and D. Goyder, *The Antitrust Laws of the United States*, 3rd edn. (Cambridge University Press, 1980), pp. 43–50.

objectives which the association may have such as the exchange of technical data and the promoting of education and training.

(2) *Qualifications* for membership. These refer not only to the particular type of business it carries on but to minimum standards for eligibility including number of staff employed, premises, location, and financial status. In a trade association membership is usually open to all qualified applicants. In some cases, however, it may be hard to tell if a body is such an association or is simply a group of companies joined together for their own purposes.[6]

(3) *Classes* of membership may be indicated, possibly simply full and associate status, or a number of classes of different categories and requiring different qualifications may be found. Restrictive conditions may be found as to eligibility for admission to particular classes (as in the GEMA[7] case).

(4) *The rights* of members will be indicated including the right to make nominations for election of members of the governing body of the association, which may be called a management committee, council, or board of managers or trustees.

(5) *The rules* for conduct of business by the governing body will be set forth and the extent of the power conferred by the members. An important point is whether the governing body has the power to make decisions which can be binding upon the members or whether they can only make recommendations of a non-binding character.[8]

(6) *The rules* for termination of membership will often be set out: the majority of associations reserve the right to expel members believed to have acted against the interests of the association.

(7) *Provision* for winding up of the association and disposal of any surplus assets.

There may also be by-laws which will set out detailed rules for the procedural conduct of the affairs of the association.

A temptation for a trade association, especially in a sector where there is little product differentiation and an oligopolistic market structure, is to use the association itself to eliminate as many as possible of the uncertainties inherent in the competitive situation, especially those relating to pricing. This temptation is present even when the association represents

[6] As in the *Roofing Felt* case: *Belasco* v. *Commission* (246/86): [1989] EC 2117: [1991] 4 CMLR 96.

[7] [1971] CMLR D35.

[8] As we shall see later in the chapter, the fact that members of the association may not officially be bound by any such decision may not prevent the application of Article 85(1). Likewise the Restrictive Trade Practices Act 1976 specifically makes recommendations of trade associations upon some matters, such as prices to be charged or conditions of contract, registrable even if under the rules of the association they do not bind its members (Section 8(1) and (2)).

merely a single interest group within the industry, e.g. all manufacturers or wholesalers or all retailers. If information can, however, be passed to the membership about the way in which each member establishes its costs, sets its prices, obtains its orders, and achieves a particular level of profit and turnover, this will considerably assist the business planning of its fellow members and reduce the uncertainties that will otherwise exist. A member's uncertainty is not simply as to the factual nature of the business of its actual competitors (about which normally it will be fully informed) but also as to the likely reaction of those competitors to any change of policy which the individual member might decide to initiate, a far less easily acquired type of information. The form in which this data is disseminated will, as some recently decided cases illustrate, make a considerable difference to the extent of competition to be found between the members.

What, however, has been made very clear by such cases, decided by both the Commission and the Court, is that those associations which have the greatest potential for anti-competitive activities are not those where just a single interest is represented in the membership. The potential is substantial where one association includes a number of persons or undertakings that have different roles within the same industry or, alternatively, where an association of associations has been created to provide a regulated basis for a particular industry which almost totally eliminates the element of uncertainty and arm's length negotiation that would otherwise exist between the different levels. These arrangements, of which particularly notable examples have been found in cases involving Belgian and Dutch trade associations discussed in this chapter, normally provide for 'collective exclusive dealing'; another appropriate description for such entities might be 'closed circuit associations', where a high percentage of the market is insulated from outside entry by rules notable for their rigidity and complexity. On the other hand, trade associations which allow membership not only to producers but also to 'consumers' of the relevant product at different levels may possibly enable the 'consumers' to monitor anti-competitive restrictions by the producers.

Having analysed the basic structure of the trade association, it becomes possible to define more exactly what is meant by the expression 'decision' found in Article 85. Within the context of such an association, a 'decision' normally means a resolution by the authorized organ of the association (usually as we have seen a board of management or council with authority delegated by the members under the terms of its constitution) to impose restrictions and obligations on its members, which will have the effect of reducing their freedom of action. These obligations and restrictions may cover the list of persons with whom members may deal, and may also control to a greater or lesser extent the terms upon which any business can be carried out, A wide variety of examples can be found among cases decided by the Commission and Court.

Decisions of associations implemented by agreements between more than one association will, of course, fall within the category of 'agreements'. An early example is the 1972 decision of *GISA*, the Dutch trade association for wholesale sanitaryware.[9] Under the rules which were binding upon the members, with fines imposed for any breach, lists were drawn up by the association to cover all relevant transactions by its members which represented some 75 to 85 per cent of the total Dutch sales of sanitary fittings. The wholesalers were not only required to sell these at the approved prices but were also required to purchase a large proportion of their requirements from another association, that of Dutch sanitaryware manufacturers, who themselves applied a range of minimum prices and agreed not to sell its output to other wholesalers at lower prices. In addition, members were required to add a heavy surcharge to manufacturers' selling prices before resale in the case of goods categorized by GISA itself (not the individual members) as luxury items.

Similar price restrictions were found in a more recent Dutch case, that of *Vimpoltu*,[10] where the trade association was one of importers of farm tractors. Sources of purchase here were not restricted, but all members were required to observe decisions of the association. All Dutch tractors were imported from abroad and local demand was gradually declining, although prices remained higher in Holland than in most parts of the Community. For these reasons, the association was anxious to negotiate with the association of dealers in agricultural machinery that the maximum discount that would be offered by dealers to farmers would not exceed 25 per cent. It was agreed that fines would be imposed on any dealer breaking this limit. Importers were required to exchange price lists with each other so that none were able to 'cheat' by importing goods at lower prices than others,[11] and standard terms of business were adopted. In both cases, the rules of the association were held in breach of Article 85(1), and the lack of any benefit to consumers from the arrangements meant that no exemption was available under Article 85(3).

It is not necessary, however, for decisions of a trade association to be binding in a formal sense on its members before Article 85(1) can apply. An informal decision of a trade association, even one made outside its rules altogether, may be sufficient. There will, however, have to be at least some evidence that the conduct of members has been or might in

[9] [1973] CMLR D125.

[10] [1983] 3 CMLR 619.

[11] The Commission suggested in its decision that even the exchange of recommended prices between importers might serve to reduce competition between them, since each of them would derive from the list a better idea of the likely pricing policy of other importers (para. 38 of the Decision).

the future be influenced by the information passed by the association. A clear example of this was the *IFTRA* case[12] (Commission 1975) where European manufacturers of virgin aluminium had adopted a standard contract which contained fair trading rules, also known as the IFTRA rules. The rules purported to set out principles of trading to be adopted by the members which could prevent 'unfairness', but in practice gave the members an opportunity to take joint action against normal competitive actions and reactions by individual undertakings. A further set of 'international free trade rules' is found in the case of *IFTRA Rules of European Manufacturers of Glass Containers.*[13] The obligation here was for producers to communicate to their competitors the essential elements of their home sales policy including price lists, discounts and terms of trade, and any concessions given to particular customers. It also obliged the members to deal on the basis of supplying the containers on a delivered basis, with no option for the customer to reduce costs by purchasing ex works and paying for his own transport. The elimination in this way of a potential variety of transport charges clearly rendered the market more 'transparent' in a way that impeded rather than aided competition. A finding of a breach of Article 85(1) was made by the Commission even though the code of conduct did not form part of the constitution of the association. The European Court has itself held that even a recommendation by an association which is described as 'non-binding' can be a 'decision' if it is really an accurate expression of the association's wish to co-ordinate the commercial activities of its members.[14]

It is clear indeed that even the conduct of an association outside its rules may be sufficient to qualify as a decision under Article 85, The best example, however, comes from a case under Article 65 of the Treaty of Paris involving the German association of steel stockholders, the BDS,[15] but which the Tenth Annual Report confirms as having equal application under Article 85 of the Treaty of Rome.[16] BDS had prepared uniform price lists for rolled steel products and distributed them to members,

[12] [1975] 2 CMLR D20. An unusual feature of the case is that the 'Fair Trading Practice Rules' were written in such appalling English that the editor of Common Market Law Reports was moved (at p. 22) to disclaim any responsibility for it. An extract may give the flavour of this remarkable document (p. 26): 'Rebates which are granted openly for competitive reasons without any connections to quantities or functions are not unfair as such, but could easily lead to a claim of reactions boomeranging on the instigator.'

[13] [1974] 2 CMLR D50. It is a reasonably safe assumption that any rules of a trade association or statutory provisions which refer to 'fair trade' will involve a complex scheme for restricting price competition rather than increasing it. A prime example is the 'fair trade legislation' formerly in force in many states of the US, protecting the practice of resale price maintenance, though this State legislation was effectively nullified by the repeal in 1976 of the Federal Statutes (the Miller-Tydings and McGuire Acts) which had legitimated them.

[14] *Verband des Sachversicherer* v. *Commission* Case 45/85 (*German Property Insurers* case) [1987] ECR 405: [1988] 4 CMLR 264.

[15] [1980] 3 CMLR 193.

[16] pp. 75–6.

including also model calculations complete with specific values for each cost component and details of revised price lists drawn up by individual dealers for their own purposes. The justification raised by BDS was that the purpose of publication of all this information was 'market transparency', and that the price lists issued for the guidance of members merely reflected existing producer price lists together with allowances for freight. In practice, the Commission was satisfied that these publications had a powerful influence on the competitive process, since dealers tended to base their prices upon the lists, even though produced officially for guidance only. Furthermore, BDS pressed dealers hard to issue uniform prices based on their lists, encouraging them to negotiate regional prices and minimum prices which were often found to be above market levels. In finding the breach of Article 65, the Commission commented that no formal or even informal decision of BDS was required and that a 'decision' could be found on the basis of the actual conduct of the association or its members, it being assumed that no association of this kind acted without the express or tacit approval of its members.

As a result of its experience, the attitude of the Commission, however, has grown stricter over the years. In its early years it gave negative clearance to several cases involving trade associations which subsequently would have received a stricter review, especially in view of consequential effects on trading within the Community of what may have appeared relatively harmless restrictions applying on their face simply to exporting to non-EC countries. For example, in the *ACFM* case[17] (Commission 1968), an association of French manufacturers of machine tools had combined to share export opportunities. Members could join or leave the association at any time and were free to compete with manufacturers of competitive machines, although the association was appointed by each member to seek buyers and licensees for it outside France. The machine tool market at this time was in general both large and competitive, and the association did not have a particularly strong position within the market. Negative clearance was granted on the basis that the association acted simply as an intermediary and in no way restricted competition between manufacturers; there was indeed little competition in export markets between them in any case, since each manufacturer tended to specialize in its own type or range of machines for particular end uses, which were not in direct competition with machines from other manufacturers. Comparatively slight attention, however, seems to have been paid to the effect of this co-operation on the competitive process in the home markets of members.

Rather more restrictive was the agreement between the members of the

[17] [1968] CMLR D23.

Belgian nitrate fertilizer manufacturers' trade association known as *Cobelaz*[18] (Commission 1968). This association covered all the manufacturers of this product in Belgium in respect of export markets outside the Community. Manufacturers retained the right to sell freely in EC countries other than Belgium. The association, however, determined the total quantity of fertilizer required for both the Belgian market and for export markets other than the other five Member States of the Community and regularly informed members of such requirements. Sale prices were also determined by the association. Negative clearance was given because although there was a restraint on competition in that members could not themselves sell on the Belgian market, on the other hand the agreement did not affect trade between Member States, in the view of the Commission, because members remained free to allocate their outputs between the five other Member States of the Common Market, and Cobelaz (for all other territories). There was no indication from the evidence that prices of the products sold through Cobelaz itself were controlled in such a way as to discourage direct exports by individual members. On the other hand, these prices might well have been varied at a later time, and once negative clearance had been given, no control over its prices would be maintained by the Commission which would clearly find the task of having to supervise a large number of pricing arrangements of trade associations in a number of separate markets extremely time consuming and outside the range of its limited resources at that early date. Again, it is unlikely that a negative clearance would be given at the present time to such arrangements, and even if exemption were granted under Article 85(3), close supervision by the Commission would be a necessary obligation, so as to ensure that the valid purposes of the association did not act as a 'cover' for other activities.

A similar doubt about the level of prices which a trade association maintained for its members in export business arose in a case in the following year, 1969, the *Dutch Paint & Varnish Manufacturers* (*VVVF*).[19] Here, an old established Dutch trade association claimed that its object was principally the maintenance of quality control for exports and the maintenance of minimum prices for all export sales outside the Common Market. The Commission appeared satisfied that export prices were actually lower than the effective prices within the Common Market, and that therefore there was still room for competition between its members in those export markets. In the course of negotiation, the Commission had, however, required the removal of the obligation on the members of the association to observe minimum prices for sale within the Common

[18] [1968] CMLR D45. Originally there had been restrictions also on sales in EC countries, but these had been deleted.

[19] [1970] CMLR D1.

Market. With this amendment, the rules of the association were approved.

The Commission often found, however, that arguments put forward on technical grounds for the restriction in choice of customers were more concerned with reduction in competition than with ensuring high standards for the consumer. This formed one of the many issues raised in the well-known *VCH* case, concerning the Dutch Cement dealers' association whose rules were challenged by the Commission, whose decision went on appeal to the European Court in 1972.[20] At issue was the question of target prices imposed by the association, a form of control which without actually fixing prices had nevertheless a powerful effect upon price levels, because of the increased degree of certainty it brought to the trade of members. Their trade association had been founded as long ago as 1928, and its constitution embodied a 'general decision' equivalent in its broad sweep to legislative control over the Dutch cement industry. The effect was to give not only binding force to the decisions of the governing body of the association, but to control in great detail the activities of all the members, enabling through its supervision to ensure that all members complied with all its detailed rules. Members were required to notify the association of any change in the management, legal form, or object of their particular business. Their accounts were subject to strict supervision and they were, of course, required only to sell to other members, and to prevent the building up of stocks of cement by third parties. Expulsion was a remedy available to the association if a member failed to observe any of the rules of the association. It was argued by the association that control of the membership by this tight discipline enhanced the safe handling of the product as well as improving the quality of services provided to builders by members. These arguments carried little weight with either Advocate-General Mayras or the Court, which found that the discipline applied to members of the association and the strictness of the 'general division' were motivated not by a desire to improve the quality of service to customers, but by a need to restrict competition between members of the association to the greatest possible extent.

The combination of the imposition of a similar discipline with an extremely detailed set of trading rules was found in another Dutch case, the *Donck* v. *CBR* case, relating to the Dutch Bicycle Association.[21] Eighty-five per cent of the Dutch cycle manufacturers, wholesalers, retailers, repairers, and dealers belonged to the association whose rules restricted the domestic cycle business in a wide variety of ways. These included a ban on members dealing with non-member firms and a requirement that a firm do business only at its approved level, e.g. as a wholesaler or

[20] Case 8/72 [1972] ECR 977: [1973] CMLR 7.

[21] [1978] 2 CMLR 194.

retailer. There were prohibitions against wholesalers or retailers dealing with others at the same level of distribution, an obligation to impose resale price maintenance with a uniform calculation of profit margins on those goods not subject to it, and prohibitions on giving a guarantee for longer than a fixed period and restrictions on the granting of discounts. There were even restrictions on the limiting of price reductions for obsolete and damaged bicycles. Members were not allowed to resort to the Courts in connection with disputes with the association. Once again, the effect was to control the trade on a 'closed circuit' basis at the different trading levels, so that uniformly high profit margins were retained, and those members that might seek extra business by a reduction in margins, or by granting additional discounts, were unable to do so without rendering themselves liable to penalties under the rules of the association.

Similar rules were also found in the perfume business in the *Bomée Stichting*[22] case where the agreement covered dealers in cosmetics, toiletries, and perfumes who were allowed to deal only with other members in an appropriate category. It was described by the Commission as a 'coherent and carefully adjusted system which is collectively applied and imposed on wholesalers and retailers and which clearly has as its objects and effect the elimination of competition'. The effect of the restriction was to make the task of importing products into the local market considerably more difficult. Foreclosure of a great number of the normal channels of distribution meant that it was necessary for importers to offer extremely favourable terms before reasonable inroads could be made into the market, and trade between Member States was thereby substantially affected. In all these cases, the Commission had little difficulty in striking down the great majority of the restrictive provisions.

The cases of *VNP/Cobelpa*[23] raised some similar issues, where the Dutch and Belgian Paper Industry Federation had agreed respectively to observe each other's official channels of distribution; it also raised new issues relating to the exchange of financial information between associations. Important manufacturers in the two countries belonged to the relevant federations, and the information exchanged between them and their individual members included notification of relevant prices mentioning names of individual companies, their general terms of supply, sales, and payments including discount granted to particular customers, rebates, and changes in price levels. The original object of their co-ordination was claimed to have been a need to align prices with those of manufacturers in countries to which they exported but it gradually developed into a widespread collection, analysis, and distribution of virtually all the data

[22] [1976] 1 CMLR D1.   [23] [1977] CMLR D28.

that would have been required by undertakings within the two countries so as to anticipate the trading and pricing policies of their rivals.

The Commission's view was that the collection and analysis of figures from individual members of the association, with the object of preparing output and sales statistics, was itself a legitimate purpose for a trade association,[24] nor could there be objection to national associations or federations exchanging the same type of statistical information giving a general picture of the output and sales of the individual federation. Real objections, however, arise when individual undertakings within the body are identified, or where the simple exchange of aggregated information about the collective fortunes of the industry is replaced by a supply of individual information about the trading of members in the associations. In this case, the exchange was so detailed that it gave enough information to allow members to have an export sales policy fully planned, with exact knowledge of the prices being charged by the local manufacturers in that market in respect of various grades of paper. Even if some of the pricing information could have been obtained from other sources, the convenient basis on which it was provided saved the parties a considerable amount of time. Whilst there was some price legislation in effect, the exchange of information went far beyond mere exchange of those prices subject to control.

It is possible from the decided cases to identify those elements in an information exchange scheme operated by trade associations which are most likely to damage the competitive structure of individual markets. First, the extent to which the information provided is aggregated by taking a large number of individual companies, and producing global figures for individual regions or countries, or whether on the contrary the information is provided on the basis of individual undertakings so that their costs, turnover, and even profits can be calculated with some accuracy by rivals. The second relevant factor is the extent to which information as to relevant products are split into their individual grades or classes, or are again aggregated so as to make calculation of individual statistics more difficult. The third is the extent to which this information is made available to the public, and in particular to end-users and consumers; the more widely disseminated the information, the less likely that it will provide the

[24] Such a view had been cautiously expressed by the Commission in the Notice on Co-operation between Enterprises, issued in 1968, where the distinction was drawn between the provisions of aggregated data on the one hand and 'conclusions given in a form that they induce at least some of the participating enterprises to behave in an identical manner on the market' on the other. The same point is made in the decision of the Commission in the *Vegetable Parchment Association* case [1978] 1 CMLR 534. For a review of the Commission's attitude to such cases, see 7th Annual Report, pp. 18–21. For a recent case see *UK Agricultural Tractor Registration Exchange* 1992 OJ L 68/19, a decision dated 17 February 1992, now on appeal to the C.F.I.

basis for anti-competitive co-ordination. The fourth, and perhaps most important issue, is the time scale and the extent to which information is provided on a current basis or in arrears. The more current the basis upon which the information is provided, the more valuable it is as an aid in eliminating the uncertainty of competitive reaction in planning pricing policy. Other factors also relevant are whether restrictions are placed as to the identity of those to whom it may be communicated and the purpose to which it may be put. Significant also would be whether current conditions of sale were also notified together with lists of discounts, rebates, and other special deals given and whether the association included interpretative comments with the figures. Probably of minor importance only is whether the research has been carried out simply by the association itself through its own staff, on information provided by members, or whether a third party such as a market research organization or firm of independent chartered accountants carried out the research and provided it on an arm's length basis. An effective anti-competitive price information scheme could, however, be implemented on either basis, although members of the association will be able to place greater reliance on the accuracy of the figures if certified by such an independent agency.

Sometimes trade associations seek to regulate or reduce the degree of competition in a market which would otherwise be completely open to the competitive process, but where there is already substantial government involvement in the market, often for social or historical reasons. Thus, in a number of European countries, there are laws covering the retail price of tobacco and alcohol, and in particular price control has existed for many years for these products in both Holland and Belgium. The extent of government involvement and its effect on competition in the cigarettes and tobacco market is considered in Chapter 23 in the light of the *Fedetab*[25] and the *Stichting Sigaretten Industrie*[26] cases. The important conclusion to be drawn from both these cases (Court of Justice decisions in 1980 and 1985 respectively on appeal from the Commission) is that the main reduction in competition in the national cigarette markets came not from government intervention but from the ability of the relevant associations to place each trader within the market by reason of its staff, premises, financial resources, and other criteria into a particular category (retailer, wholesaler, or importer), and so prevent any competition in those categories *inter se*, as well as preventing traders in any category either from attempting to combine more than one function or from offering services additional to the norm in order to try and earn additional business or generate greater profit margins from existing customers.

[25] Also known as the *van Landewijck* Cases 209–215/78 [1980] ECR 3125: [1981] 3 CMLR 134.

[26] Cases 240–242/82 [1985] ECR 3381; [1987] 3 CMLR 661.

Rebates too were totally standardized regardless of whether orders were placed on members of the association or non-members, on home manufacturers or importers, so that there was no incentive for any retailer or wholesaler to try to seek to persuade customers to concentrate their own orders on one supplier. Moreover, the application of the rules of the association completely eliminated the possibility of extended credit, which both made entry more difficult for new undertakings and also prevented manufacturers or wholesalers from using the provision of credit as a competitive strategy.

The clearly anti-competitive purpose of the restrictions imposed at all levels of the associations in the *Fedetab* case meant that in the view of the Court it became unnecessary for the Commission to carry out a detailed analysis of the results of each individual restriction contained in the rules. This decision in its turn rendered the task of the Commission less onerous in the later *SSI* case where the combined effects of the restrictions on each level within the Dutch cigarette industry had reduced competition to the margins receivable, imposed restrictions on what price increases could be allowed, and banned the granting of special discounts to individual customers.

Where the membership of a trade association consists largely of producers of the same goods or services, its rules may seek to limit the extent of competition between them. The Danish Fur Breeders' Association provided a number of facilities for its members who sold fox and mink furs, including a production centre for pelts and an auction house. Use of the facilities was conditional upon the observation of a number of rules limiting the use by members of other auction houses for disposal of their pelts and also their ability to act as collecting agents for competing sales organizations. Each member moreover had to agree to a certain percentage of its production being sold at one of five auctions held annually by the Association. These rules not surprisingly were found to be in breach of Article 85 and a fine was imposed.[27]

Particular problems are also caused by members of trade associations combining in connection with exports. Often the law of individual states will permit export associations to make agreements that would normally be illegal if carried out purely for domestic markets. The oldest and best known of such laws is the United States legislation entitled the Webb-Pomerene Export Trade Act 1918 which provides that nothing in the Sherman Act shall make illegal 'an association entered into for the sole purpose of engaging in export trade and actually engaged solely in such export trade . . .'. The *Woodpulp*[28] decision of the European Court in 1988

[27] *Hudson's Bay and Annings* v. *Danish Fur Breeders' Association* (no. 1) [1989] 4 CMLR 340; on appeal some of the Commission's findings were overruled and the fine reduced.

[28] [1985] 3 CMLR 474: ECR 3831.

had to consider the status of such an agreement, namely that by the Kraft Export Association to which several pulp manufacturers in the United States belonged and who were accused by the Commission in that case of having agreed prices for sulphate pulp.

The Court's finding was that this Association could not be held responsible for any agreements on pricing made between its members since it made no separate recommendations to them but simply announced the terms of pricing agreements made by them in a number of separate meetings. The Association had always distanced itself too from the implementation of those agreements.

In considering the effect of restrictions that trade associations may seek to impose on their members, the degree of their particular influence and their power of exclusion from national or regional markets will be closely considered by DG IV. If, therefore, an association is able to control public markets or 'bottle neck' facilities, any restrictive provisions which cause detriment as a result either to member or non-member traders or to consumers will likely be in breach of Article 85(1). The *FRUBO*[29] case provides a clear example. Here, the associations of fruit importers and wholesalers between them controlled 75 per cent of the sales of apples, pears, and citrus fruits in The Netherlands, this fruit being sold exclusively through the Rotterdam auction. The associations were able to impose the sanction of total exclusion from such auction on any importer or wholesaler who dared to break the associations' rules, which with very limited exceptions provided that the Rotterdam auctions alone should be the method for importing fruit into The Netherlands. The object and the likely effect of these restrictions, if unchecked, would have been to give Rotterdam a virtual monopoly for the importing of fruit into The Netherlands and to prevent wholesalers from using alternative sources of supply of fruit originating from outside the Community. The rules of Aalsmeer flower auctions were similarly restrictive, which sought to compel wholesalers (who were tenants of the processing rooms that formed part of the market complex[30]) to buy their flowers for resale exclusively from the auction house run by the VBA, and by financial penalties imposed on tenants engaged in other forms of trading so that the likelihood of them obtaining supplies from other sources would be reduced.

Entry into foreign exchange dealing markets is also regarded by DG IV as the type of commercial activity which should be governed by objective and non-discriminatory criteria as was made clear in the case of *Sarabex*,[31] a case involving an association of London foreign exchange

[29] Case 71/74 [1975] ECR 563: 2 CMLR 123. The facts have already been considered in Ch. 6, p. 80.

[30] This was claimed by the owners, the VBA, as the building with the largest floor area in the world! *Florimex* v. *de Verenigde Bloemenveilingen Aalsmeer* [1989] 4 CMLR 500.

[31] [1979] 1 CMLR 262.

dealers referred to in the Eighth Annual Report;[32] here negotiations were successfully concluded with the Bank of England (without the need for a formal decision) to amend its rules so as to expand opportunity for entry into the association by brokers meeting required objective criteria. A number of other decisions have also been issued affecting associations of dealers in various London commodity markets, which provide organized facilities for trading in coffee, cocoa, sugar, and rubber. Negative clearance was granted, but only after deletion of fixed rates of commission and the introduction of revised membership rules based on objective and non-discriminatory criteria.[33]

Trade associations also frequently attempt to make rules for their members covering the holding of international trade fairs and exhibitions, since they claim that an absence of guidelines applied to the organization of and attendance by exhibitors at such occasions could seriously affect the reputation of individual major events by limiting the full range of products then displayed and thereby allegedly reducing the special impact which the exhibition may have on its market. The Commission has, therefore, attempted to apply Article 85(3) to such restrictions in a reasonable way, in order to provide in effect a 'rule of reason' for their assessment, distinguishing between those thought on the one hand ancillary to the need for some limitation on the 'free for all' that could exist if no restrictions were allowed, and on the other hand those restrictions which are unduly and unnecessarily onerous. Examples from a number of industries including motor cars, exhibitions, engineering, and dental products can be found in recent Annual Reports: a common thread running through the cases is the importance of preventing discrimination in the relevant rules on a national basis. Whilst examples from a number of industries can be found in recent Annual Reports, Commission decisions which illustrate the way in which it applies this principle include the cases of the *British Dental Trade Association*[34] and *CECIMO* (no. 3).[35]

Another potential method by which trade associations may seek to reduce competition and the possibility of new entry lies in their power to control standards and certification. These standards and certification rights are often originally linked with objective standards of health and safety whose enforcement is delegated to the associations by national government or authorities. Such privileges, however, can be abused.

In the case of *Community* v. *Anseau-Navewa*[36] (European Court 1982), fines here were imposed by the Commission and upheld by the Court of

[32] pp. 35–6. [33] 15th Report, pp. 72–3. [34] [1989] 4 CMLR 1021.

[35] [1990] 4 CMLR 231.

[36] Cases 96–102, 104, 105, 108, 110/82 [1983] ECR 3369: [1984] 3 CMLR 276, reported at the European Court level as *IAZ International Belgium* v. *Commission*. An earlier case decided by the Commission on similar issues is *Belgium Central Heating Agreement* [1972] CMLR D130.

Justice on associations of manufacturers, importers and distributors of washing machines and dishwashers in Belgium, as well as an association of water supply companies in that country which had statutory duties for the approval of appliances to be connected to the public water supply. Machines under these arrangements could only be installed if they had a certificate of conformity, and this certificate was only available from particular associations. These in turn restricted their membership to manufacturers and sole importers, so that general importers and dealers found themselves unable to obtain the necessary certificates. It appeared clearly that the intention of these arrangements was to make it more difficult for parallel imports of such appliances to occur, and exemption under Article 85(3) was therefore, refused. The Commission and Court both held that, where a trade association is a party to a restrictive agreement of this kind but does not have formal power to order its members to conform, nevertheless if its recommendations are in practice followed because of the grave practical problems which the manufacturers and dealers will encounter if they do not do so, then the agreement will be regarded as binding on the members and treated as a 'decision' under Article 85. The grave practical disadvantage referred to in this case was that foreign manufacturers of washing machines were forced to use sole agents to handle imports into Belgium; any other kind of importer, e.g. a distributor of a variety of such machines, would not be eligible to obtain the necessary certificate.

The reader may be surprised that the Article cited in the cases in this chapter is almost invariably Article 85 rather than Article 86, although Chapter 18 has considered the use made by the Commission of Article 86 against a special form of trade association, namely performing rights societies. The main reason is that the wide definition given by the Court and Commission to 'decision' as well as to 'concerted practices' makes the use of Article 86 unnecessary in the great majority of cases, since the structure of the association will usually ensure that any practices which it authorizes, or rules which it enforces or suggests, will come within one or more of these definitions. There will not be any need for the Commission to have to perform the more difficult task of showing first that the association is 'dominant' in some particular market and that it is abusing its dominance. Whilst in the special case of performing rights societies, it is often its members who are the subject of exploitation under Article 86, the burden of proof for the Commission in the case of the normal trade association is far simpler under the terms of Article 85.

This chapter has necessarily focused on the cases where trade associations have abused their powers; there are many trade associations which do not seek to regulate the affairs of their individual industries or markets in anti-competitive ways, especially when the associations represent only a single function within the market, e.g. manufacturers, and have no form of collective arrangements with any other associations or groups in that industry. Nevertheless, even the rules or constitutions of such trade associations remain subject both to Article 85 and 86 and must be clear, objective, and reasonable. The rules must be applied without discrimination of nationality or any other kind and without anti-competitive motives or effects. Moreover, it is clear from the cases decided by the Commission and the Court (especially those relating to performing rights societies mentioned in Chapter 18) that individual restrictions on members even if they can be justified under Article 85(3) must not go in their terms beyond what are strictly indispensable to the achievement of the legitimate objects of the association.

This review of some of the leading cases relating to trade associations illustrates how significant a vehicle they can become for the reduction of competition within a Member State in a particular market. Whilst it is probable that fewer cases relating to the almost complete control of individual industries within Member States by associations will need to be considered by the Commission in the future, the very facility which the structure of any trade association offers for the interchange of information and adjustment of conduct to that of one's competitors will in all probability mean that Adam Smith's famous words will remain as apt for the future as they have proved in the past.[37]

[37] Since this well-known quotation from Adam Smith has been included in the great majority of books on competition law ever published, there is an incentive to omit it. For the record, however, what was actually said in *The Wealth of Nations* (1776), bk. 1, ch. 10, was: 'People of the same trade seldom meet together even for merriment and diversion, but the conversation ends in a conspiracy against the public, or in some contrivance to raise prices. It is impossible indeed to prevent such meetings by any law which either could be executed, or would be consistent with liberty and justice. But though the law cannot hinder people of the same trade from sometimes assembling together, it ought to do nothing to facilitate such assemblies, much less to render them necessary.' For several decades now the trade association has represented the modern form of the 'assembly' to which Adam Smith refers as the most convenient forum for such activities, whether innocent or otherwise.

22

Community Law and National Law

1. The Issue of Supremacy

The entry into force on 1 January 1958 of both Articles 85 and 86 did not, of course, have the effect that Community competition authorities completely took over responsibility for the enforcement of competition policy within the Common Market. The inadequate resources of the Commission in general, and of DG IV in particular, especially in the early years, would have made such a step impractical, quite apart from any other considerations. In any case, however, the relevant provisions of the Treaty embodied the principle that the competition policy referred to in Article 3(*f*) needed joint implementation with Member States, with both national courts and competition authorities having an important role in its enforcement alongside the Commission. Whilst the primary role of such national competition authorities was the enforcement of their own domestic competition laws, their other responsibility would be to help the Commission to ensure that Community law on this subject was effectively and uniformly enforced.

That this was the original scheme of the Treaty is clearly shown by the terms of Article 87(2)(*e*). In referring to the regulations or directives to be adopted by the Council 'to give effect to the principles set out in Articles 85 and 86' this mentions the areas in which in particular such regulations or directives shall operate, including those necessary 'to determine the relationship between national laws and the provisions contained in this Section or adopted pursuant to this Article'.[1] The exact pattern of co-operation could not be foreseen at this early stage; the working out of the consequences for the Common Market of the terms of Article 85, let alone Article 86, were to take at least a decade, as we have already seen in Part I.

Since the provisions of the Treaty relating to competition applied only to activities which of themselves 'affect trade between Member States', Member States remained, of course, quite free to enact legislation to cover purely domestic issues.[2] In fact, at the time when the Treaty came into force in 1958 there was limited legislation in Member States in existence, ranging from the relatively sophisticated but untried (Germany) to the

[1] The 'Section' refers to Articles 85 to 90 inclusive, the 'Article' to Article 87 itself.

[2] Though the scope of such domestic legislation would clearly be restricted in time by the wide interpretation given by the European Court to that very phrase. See Ch. 7.

non-existent (Italy). Although all Member States now have some form of national legislation dealing with the control of restrictive agreements and abuse of monopoly its administration and enforcement still ranges from the advanced to the rudimentary. It has seemed convenient in a number (though not all) of Member States for such legislation to be enacted in a form which at least follows the main outlines of Articles 85 and 86, since this means that conflicts with the relevant terms of the Treaty are less likely.[3] One of the types of agreement most likely to be declared void under Article 85(1) is, of course, thc price-fixing cartel, and as we have seen in Chapter 14, the Commission has taken quite a severe approach towards vertical price-fixing, usually referred to as 'resale price maintenance'. Member States' domestic legislation dealing, therefore, with issues of pricing runs the risk of coming into conflict with these two Articles, if it makes it unnecessary for undertakings to enter into prohibited agreements, by providing an official method whereby the objectives of such agreements can be achieved without their utilization. Such legislation, however, also comes within the range of Articles 30 to 36 which, through a wealth of cases decided by the European Court of Justice over the last twenty years, has very considerably reduced the freedom of Member States to legislate in this area. Consideration, therefore, of the validity of national legislation under Articles 85 and 86 on the one hand and Articles 30 to 36 on the other may often be found combined in the same case as will be seen in Ch. 23.

All these developments, however, still lay in the future at the time of the signature of the Treaty. It was Articles 88 and 89 that laid down basic ground rules to govern the immediate future until the adoption of Regulations within the scope of Article 87 by the Council. During this interim period, Member States were free to apply both Articles 85 and 86, including the exemption provisions of 85(3), whilst the Commission for its part was entitled to apply the Treaty provisions in this area 'in co-operation with the competent authorities in the Member States, who shall give it their assistance'. Even Regulation 17 in 1962, although giving very extensive powers to the Commission in the administration of its competition policy, itself relied on a procedural structure that involved an important element of co-operation with national authorities. While under Article 9 of the Regulation the Commission reserved to itself powers to grant exemption under Article 85(3), so that Member States were not

[3] Although the United Kingdom published Green and White Papers in 1988 and 1989 respectively proposing to replace the Restrictive Trade Practices Act (with its long-established 'form-based' rules) by an 'effects-based' law with greater similarity to Article 85, no steps have yet been taken to implement these proposals in legislation. By contrast a number of members of EFTA and some East European countries including Poland, Romania, Czechoslovakia, and Hungary have already adopted national legislation based on the general approach of Articles 85 and 86.

entitled to take action in any case where the Commission had commenced its own proceedings (as in no other way in its view could the exemptions be guaranteed consistent application) the Regulation provided for the possibility of substantial involvement of Member States, both in the investigatory process and the adjudicatory proceedings.

Article 10 refers to the contact required between DG IV and the 'competent authorities' of Member States. They are entitled to receive copies of applications for notification and exemption and the main supporting documents. Article 10(2) sets out the principle that administrative procedures shall be carried out in close and constant liaison with these competent authorities and paragraph (3) provides for the creation of an Advisory Committee on restrictive practices and monopolies which is to provide the forum in which national authorities may express their views on the procedure. As mentioned in Chapter 4, this is the forum in which Member States become involved in the individual decisions, the views of the Advisory Committee being influential though not decisive. DG IV prefers naturally to receive the blessing of at least a majority of the Advisory Committee to a proposed decision. If there is substantial opposition within the Committee, the proposals may be taken back for reconsideration in either major or minor respects. On occasions, however, when DG IV feels convinced that it is correct in its approach to a particular case, it will be prepared to persevere in taking a proposed decision to the full Commission even if the vote or opinion of the Advisory Committee has been adverse.

The necessary consequence of such involvement for Member States in the decision-making process of the Commission is an obligation to cooperate with the administrative and investigatory procedures of the Commission. Articles 11 to 14 of the Regulation, therefore, contain a number of provisions setting out the way in which this help is to be provided. Thus:

(*a*) Under Article 11(1), the Commission may obtain 'all necessary information' from both Governments and competent authorities of Member States as well as from undertakings and associations within those Member States.

(*b*) Article 13(1) provides that the competent authorities of Member States may be asked to conduct investigations which the Commission itself could conduct, including the examination of books and business records, the making of extracts and the requiring of oral explanations 'on the spot' of the contents of these documents. This power has not yet been much used but may become more important in the future. It is significant that the Commission officials are entitled to help in such national investigations either at the Commission's own request or at the request of the national competent authority, under the terms of Article 13(2).

(*c*) Article 14, by contrast, gives the Commission necessary powers to carry out its own investigations, but with the help of the competent authority which is entitled to (and normally does) escort the Commission officials on their investigations. In the United Kingdom this authority is the Office of Fair Trading. Member States are required moreover under Article 14(6) to give all necessary assistance to the officials authorized by the Commission to make such investigations.

These investigations often involved physical inspection of the books and files of undertakings and can be arranged in advance on a voluntary basis (under Article 14(2)) or sprung as a surprise by way of a 'dawn raid' following a formal decision of the Commission to that effect under Article 14(3).

The exact nature of the Commission's powers under Article 14 have been the subject of a trio of cases in the European Court of Justice involving the chemical industry. In the *Hoechst*[4] and *Dow*[5] cases the rights of the Commission to carry out searches necessary for the particular stated objectives of its own decision were examined. If the national undertaking seeks physically to bar access to the Commission, its officials cannot use force to obtain entry to business premises and must obtain a national court order for entry. The national court must make an order in favour of the Commission if satisfied that the necessary causal link exists between the investigation proposed and its stated objective. It cannot enquire whether the Commission's objectives are themselves arbitrary or disproportionate. In the *Solvay* and *Orkem*[6] cases the Court drew a distinction between the making of factual and objective enquiries from undertakings in the course of such an investigation (which is allowed) and the asking of questions to which replies may force those undertakings to make statements which necessarily admit a breach of the law (not permitted). It is, therefore, inherent in Regulation 17 that the competent national authorities are to some extent involved in the administrative, fact finding, and investigatory processes as well as in the decision-making of the Commission. Member States are also involved, through the Advisory Committee, in reviewing proposed drafts of new Regulations, including those putting forward proposals for new group exemptions. They have a chance to express their views on these after entering into consultations over the terms of the draft documents with a wide variety of national interested parties and pressure groups. Views of Member States are, of course, also made known to the Commission through a number of other

[4] Cases 46/87 and 277/88 [1989] ECR 2859: [1991] 4 CMLR 410.
[5] Cases 85, 97 to 99/87 [1989] ECR 3137, 3165: [1991] 4 CMLR 410.
[6] Cases 374/87 and 27/88 [1989] ECR 3283, 3355: [1991] 4 CMLR 502.

formal and informal channels. As a result, although Member States do not customarily initiate individual policies or issue their own suggested drafts for Regulations or for group exemptions, they do have both an important role and an influence in the framing of those policies culminating in both actual decisions and new Regulations.

It would, therefore, be unfair to suggest that the Regulations which eventually are issued from the Commission, following this lengthy process of consultation, involve some kind of diktat on Member States imposed from above. The legislative process is slow-moving, and possibly too extended in its reaching out for a compromise solution, in which all the interested parties have participated and had their extensive opportunity to make their views known. The extent of the participation of Member States in the legislative process has relevance to the difficult issue of the extent of 'supremacy' of Community law over national law, to which we must now turn.

This issue, though difficult, can be simply stated. Is an agreement which has been exempted by the Commission under Article 85(3) capable of prohibition by national law, or does the grant of an exemption by reason of the primacy of Community Law preclude the possibility of prohibition by national law? In general it is clear that agreements must satisfy both Article 85 and national law (the 'double barrier'), but this may not be so where the Commission has granted an exemption. Because of the very wide interpretation placed by the early decisions of the European Court on the concept of 'trade between Member States', this question took on increasing importance; not many agreements of commercial significance, it seemed, would on the basis of such cases be outside the range of Community law, and it thus became important that the European Court of Justice provided an authoritative ruling.

Article 177 of the Treaty provides the appropriate means of obtaining a preliminary ruling on issues of Community law arising in cases being heard in national courts. Its purpose, and a very necessary one, is to ensure that Community law is interpreted in a like manner in all Member States. It was through such a preliminary ruling under this Article that a Berlin Court obtained in early 1969 guidance on the relationship of national law and Community law, in the context of parallel investigations being carried out by both the Commission and the German cartel authority (Bundeskartellamt).

The background to this case[7] was that certain major manufacturers of aniline dyes had entered into discussions over price increases for their product in circumstances which led to strong suspicion that an agreement had actually been concluded between them as to the timing and level of

[7] *Walt Wilhelm* v. *Bundeskartellamt* Case 14/68 [1969] ECR 1: CMLR 100.

future increases. Investigations carried out by the Commission led eventually to the bringing of the *Dyestuffs* case;[8] in the meantime, the BKA had also begun its own investigation of those agreements in which German companies had apparently been involved, and Wilhelm was one of the directors of Bayer involved in this investigation. His claim was that the BKA should desist from carrying out its own investigation whilst DG IV was simultaneously carrying out investigations for the Commission into the same matters. The Kammergericht in Berlin referred to the European Court of Justice, therefore, the issue of whether the BKA could continue with its investigation, or whether under Community law it was required to suspend them, pending completion of the investigation being carried out by DG IV.

The Court pointed out that Community and national laws considered cartels and restrictive agreements from a different point of view; whilst Article 85 regarded them in the light of the obstacles which they would present to trade between Member States, national legislation proceeded on the basis of its own individual priorities and considerations; though there might be some overlapping and interdependence of the two separate investigations, there was no reason in principle why the same agreements could not be the object of two separate parallel proceedings. To ensure, however, that the general aims of the Treaty were respected, such simultaneous investigation could only be allowed on the national level if three conditions were satisfied:

(*a*) that the national application of the law must not prejudice the full and uniform application of Community rules, or of the full effect of the measures adopted in the implementation of those rules ('the uniformity rule');

(*b*) that, so as to ensure in any conflict between Community and national rules the Community law takes precedence (to protect the effectiveness of the Treaty), Member States should not introduce or retain any measures which would prejudice this aim (usually called the 'practical effectiveness rule'); and

(*c*) that it was the responsibility of individual Member States to take appropriate measures to avoid any risk that a national decision on an issue of competition law might conflict with Commission proceedings still in progress (called the 'parallel proceedings' rule).

Whilst these principles sufficed to give a clear answer to the Berlin Court as to the proper procedure in the *Walt Wilhelm* case, their application to other circumstances of potential conflict is far from clear, as many

[8] *ICI* v. *Commission* Case 48/69 [1972] ECR 619: CMLR 557; several other related cases are also reported in [1972] ECR.

commentators have noted.[9] If the agreement or practice is forbidden by the provisions of Article 85(1), then the requirements of both uniformity and practical effectiveness underline the impossibility of allowing national law to provide an exemption for it.

In the converse case where the agreement or concerted practice is outside the scope of Article 85(1) or perhaps because it has insufficient effect on trade between Member States, the national court is clearly left free to intervene if it wishes to invoke national law with regard to the domestic effects of the agreement or practice. The intervention in such circumstances is in an area where the Commission either has no jurisdiction under the Treaty or has elected on policy grounds not to exercise any jurisdiction. With regard to the problem of conflict between parallel national community investigation and proceedings, the Commission stated in the Fourth Annual Report[10] that the onus was firmly placed on the national authorities to avoid the risk of conflict, an obligation that could be complied with so long as the national authorities had early consultations with the Commission or even suspended proceedings whilst the Commission investigation was completed.

Whilst acknowledging that the principles in the *Walt Wilhelm* decision provided an answer to the relatively simple situations so far described, the Report also acknowledged that the Court's decision failed to provide answers to the more difficult issue, the status of an agreement granted exemption by the Commission under Article 85(3) when a national authority nevertheless wished to be able to declare it illegal in whole or part. The Commission in the Fourth Annual Report purported to read into the *Wilhelm* decision a rejection by the Court of the argument that exemption by the Commission withdraws only the Community barrier to a restrictive agreement under Article 85(1), leaving unimpaired the national authority's power to prohibit such agreement under its own law (the 'double barrier' theory). Such rejection can, however, be found if at all only implicitly between the lines of the Court decision and not expressly stated in it. The Fourth Annual Report's review of the case concludes rather weakly that this judgment nevertheless left open the question of 'whether the primacy of Community exemptions constitutes a strict rule, or whether it should be regarded rather as a flexible principle in the application of which it is permissible to take account of the respective interests of the Community and of Member States'.

It is clear that there are two situations, therefore, where the *Wilhelm* case provides incomplete guidance. First, where the Commission has given an individual exemption to an agreement or concerted practice which a

[9] Including K. Markert, 'Some Legal and Administrative Problems of the Co-Existence of Community and National Competition Law in the EEC'. 11 *CML Rev* 92 (1974).

[10] pp. 27–31.

national authority in a Member State wishes later either to reject completely or to place under some restriction or condition. The second problem arises when there is a Regulation in force containing group exemption for particular forms of agreement, from which the Member State likewise wishes to derogate.

To provide an answer in these situations, it is necessary, whilst bearing in mind the terms of the judgment of the European Court in *Wilhelm* also to consider the inherent nature of the exemption process for individuals and groups respectively. Although the 'four conditions' to be applied by the Commission are in principle identical for both individual and group exemptions, they are applied in a different way. If the exemption flows on the one hand from an individual decision (which is properly described as administrative or possibly quasi-judicial) it is given on the basis of the particular facts disclosed in the course of the Commission's administrative procedure, rather than from a 'legislative' decision that a particular category or class of agreement both having certain 'plus' (or 'white') features and lacking certain defined anti-competitive 'minus' (or 'black') elements should be exempt. In the latter case, the facts of the individual cases benefiting from the group exemption will never need to be disclosed to DG IV, since the very concept of the group exemption, emphasized by the recitals to the enabling Council Regulation 19/65, is to provide a legislative solution to the 'mass problem' by excluding the need for the notification of a large number of separate agreements.[11]

In the *Walt Wilhelm* case, the Court refused to accept the argument of the Advocate-General that the application of a stricter national law to an individual agreement, thereby possibly overruling the effect of an exemption given by the Commission under Article 85(3), was not to the prejudice of Community law because each system was concerned essentially with the same objectives. Whilst the reasons for its refusal to follow him are not explicitly stated in its judgment, the essential difference in the approach of the Court is found in this key section:

Article 85 of the E.E.C. Treaty applies to all the undertakings in the Community whose behaviour it governs either by prohibitions or by means of exemptions granted—subject to conditions which it specifies—in favour of agreements which contribute to improving the production or distribution of products or to promoting technical or economic progress. While the Treaty's primary object is to eliminate by this means the obstacles to the free movement of goods within the Common Market and to confirm and safeguard the unity of that market, it also permits the Community authorities to carry out certain *positive* though indirect, action with a

[11] See also the important *Tetrapak* judgment of the Court of First Instance discussed in Ch. 19, pp. 383–5, for comment on the difference between group and individual exemptions under Article 85(3).

view to promoting a harmonious development of economic activities within the whole Community, in accordance with Article 2 of the Treaty.[12]

This passage clearly gives the Community both a positive and negative function in carrying out its responsibilities. If so, it is hard to see how the granting of an individual exemption by the Commission should not be treated as such a 'positive' act, grounded both on the investigation of the relevant circumstances affecting a particular agreement, and a finding that on balance its preservation is desirable in the public interest of the Community. To quote an early commentary[13] on this issue:

> . . . signature of the Treaty by the Member States implies their acceptance of competition policy (as formulated and implemented by the Community authorities) as one of the means of achieving the aims set out in Article 2 of the Treaty and that accordingly the positive aspects of the policy must be given the same weight as its negative aspects. Balance must be struck between too little competition at the expense of the consumer and too much competition leading in certain circumstances to inefficiency. If it is right that the Commission's actions should not be frustrated by the application of incompatible national legislation in relation to the former, it must also be right that the Commission should have an equally free hand in relation to the latter. The driver must have the use of the accelerator as well as the brake.

A separate distinction has been suggested between these exemptions granted purely for 'public policy reasons', when the Commission's decisions would have to be respected by national courts and authorities, whereas in all other cases a national authority or court would be free to impose further requirements. It is difficult to see how this distinction could be applied even to individual exemptions, where one has the advantage of a specific finding as to the relevant facts, as it will be hard to distinguish those cases where the public policy element is so great that the exemption should be respected by national courts. In every case where an exemption is granted, the fact that the agreement or practice must comply with all the four conditions contained in Article 85(3) makes it difficult to suggest that an element of public policy does not enter into this assessment. Public policy clearly must enter into an assessment as to whether, for example, an agreement contributes to improving the production or distribu-tion of goods or the promotion of technical or economic progress but is also likely to be relevant to the assessment of whether consumers receive a fair share of the resulting benefit. An important element in the opinion of Advocate-General Roemer is the statement that:

[12] [1969] ECR 1, 14: CMLR 100, 119.

[13] D. Barounos, D. Hall, and J. R. James, *EEC Antitrust Law: Principles and Practice* (London: Butterworths, 1975), 1–3.

If national authorities thwart the Community exemption through the application of a national rule of prohibition, they no more threaten the objectives of the Treaty than do the parties to that agreement when they refrain from applying it, which can occur at any time. This conclusion applies as a general rule because in principle cartels cannot be considered as instruments of the organisation of the Common Market.[14]

This, however, is surely misleading. For, if an individual agreement is no longer applied by the parties to it, this has no precedential value nor does it affect any other agreement between the same parties or between any other parties. On the other hand, an individual exemption has value as a precedent and may affect other persons and Courts in Member States which may take it into consideration in future cases, quite apart from those who are the parties to the agreement. It is difficult then to see how the status of a group exemption cannot equally be described as 'positive' and 'an act of policy' within the spirit of Articles 2 and 3(f) of the Treaty. Indeed, the involvement of Member States in the legislative process leading up to the enactment of group exemptions by way of Regulation is, as we have seen, substantial even though they do not control the final terms in which they are produced. National competition authorities, and Member States themselves, should not, therefore, be entitled to disregard their terms merely because they prefer a different and stricter solution for their national legislative purposes.

Nevertheless, this broad statement of principle requires some careful qualification. Detailed though group exemptions have now become (and this has been a gradual and continuing trend), they do not contain as much detail as by contrast a United Kingdom Act of Parliament. To their substantive provisions (including their recitals) there has been added in some cases a lengthy explanatory memorandum giving the Commission's view of their proper interpretation. Moreover, the Commission itself has on more than one occasion indicated an awareness of the fact that a delicate balance between the proper jurisdiction of the Commission and the Member States must still allow for a degree of flexibility in the way in which the group exemption can be dealt with by Member States. Thus Recital 19 of the group exemption on exclusive purchasing[15] reads:

To the extent that Member States provide, by law or administrative measures, for the same upper limit of duration for the exclusive purchasing obligation upon the reseller in service station agreements as laid down in this Regulation but provide for a permissible duration which varies in proportion to the consideration provided by the supplier or generally provide for a shorter duration than that permitted by this Regulation, such laws or measures are not contrary to the objectives of this Regulation which, in this respect, merely sets an upper limit for the duration of

[14] [1969] ECR 1, 23: [1969] CMLR 100, 110.

[15] Reg. 1984/83.

service station agreements. The application and enforcement of such national laws or measures must, therefore, be regarded as compatible with the provisions of this Regulation.

In this recital, we find the group exemption itself permitting national legislatures or competition authorities to have freedom to impose yet stricter limits upon the period for which service station exclusive purchasing agreements may be permitted; the Regulation itself merely sets an upper limit in time for which such agreements may be enforced. Moreover, the absence of any reference in the recital to similar arrangements applying either to provisions for general exclusive purchasing agreements or for the special kind of agreements covered in Articles 6 to 9 inclusive, (beer supply agreements) does not necessarily mean that such agreements might also not benefit from the provisions of the recital.[16]

Recital 19 is not unique. In the more recent Regulation 123/85, the group exemption for motor vehicle distribution agreements, Recital 29 preserves a degree of flexibility for Member States even after the adoption of the Regulation, stating the Regulation to be:

Without prejudice to laws and administrative measures of the Member States by which the latter, having regard to particular circumstances, prohibit or declare unenforceable particular restrictive obligations contained in an agreement exempted under this Regulation; the foregoing cannot, however, affect the primacy of Community law.

There is thus more inherent flexibility of application in group exemptions than in individual exemptions. Quite apart from the recitals referred to, all the group exemptions contain a 'safety valve clause'[17] under which they can be withdrawn if, after investigation, the Commission is satisfied that their application in a particular case has been incompatible with the conditions of Article 85(3), and in particular where they have had the effect of excluding competition, foreclosing alternative suppliers, or allowing undertakings to impose discriminatory terms on their customers. It

[16] Whilst the canon of statutory interpretation 'expressio unius est exclusio alterius' is familiar under English law (see *Craies on Statute Law*, ed. G. Edgar, 7th edn. (London: Sweet & Maxwell, 1971), 259; F. R. Bennion, *Statute Law* (London: Oyez Publishing Ltd., 1980), 84), it is not necessarily applicable to the interpretation of a recital to a group exemption (this is the view adopted by at least some Commission officials). In *R.* v. *Henn and Darby* [1980] 2 All ER 166 at 196, the House of Lords stressed the dangers of applying English canons of statutory interpretation to Community legal provisions, including regulations and directives.

[17] Examples are Article 6 of the original Exclusive Distribution Group Exemption Reg. 67/67, now replaced by 1983/83 (where the equivalent article is also Article 6) and 1984/83 (where the equivalent article is Article 14); Article 9 of the Patent Licensing Group Exemption Reg. 2349/84; Article 8 of the Specialization Agreement Group Exemption Reg. 417/85; Article 10 of the Research and Development Agreement Reg. 418/85; Article 8 of the Franchise Agreement Group Exemption Reg. 4087/88; and Article 7 of the Know-how Licensing Group Exemption Reg. 556/89.

appears, therefore, that the Commission retains rather greater freedom to review the grant of group exemption (whereas *Tetrapak* emphasizes it has not had the opportunity of considering its application in individual circumstances) than in practice it retains to review the operation of the exemption, individually granted, in cases other than those falling within the terms of Regulation 17, Article 8(3). This may in practice reduce the potential for conflict between such Community Competition exemptions and member States' desire to amend or extend them in domestic application.

2. The Use of National Courts to Enforce Community Law

The enforcement of Community competition law is clearly not, therefore, to be left solely to the institutions of the Community itself; the terms of the Treaty, the wide variety of relevant issues arising in different parts of the Community and the limited resources of the Commission in general and DG IV in particular all favour the involvement to the greatest degree possible of national courts in enforcing Articles 85 and 86. The Commission has sought from its earliest days to encourage Member States to utilize their courts to receive hospitably private actions of this kind hoping that, as in the United States, the burden of public enforcement can be lightened by the increasing use of private actions. The Court and Commission have not hesitated to state in their decisions that enforcement of competition policy is not a simple matter of public law but also of private law enforcement. Implementation, however, of these developments has taken a great deal longer than was originally hoped, even though in a number of Member States, including the United Kingdom, there have now been clear signs of gradual movements in this direction.

A major factor in slow growth of the development of private litigation in this area has been the easy availability of an alternative remedy, namely the complaint to DG IV which, at first sight, has appeared to have many attractive features. The Commission has shown an apparent willingness to give priority to the handling of complaints even in fairly trivial matters, over certain of its other responsibilities;[18] the wide powers of investigation and discovery under Regulation 17 and the apparent cheapness of involving its services (not to mention the relevant informality with which it could be done) have all seemed both to undertakings and their lawyers good reasons for approaching the Commission rather than instituting litigation in national courts when problems have arisen. The

[18] The Commission takes the view that it has a legal obligation to investigate complaints made to it, and that if it declined to do so, on the grounds of prior claims upon its limited resources, action could be brought against it under Article 175.

Commission also had the advantage that, if it offered an administrative remedy, this would be applicable throughout the Community rather than merely in the one particular Member State where court action had been started. If formal decisions were later given by the Commission as the result of its investigation of the complaint, then this would be accepted and respected by all national courts within the Community even if the position was less satisfactory with regard to informal 'comfort letters' to which the Commission has increasingly been resorting.

The Court of Appeal underlined the importance of the complaint procedure in a case[19] coming before it that arose from the *Hasselblad* dispute which we have already mentioned in Chapter 13. It ruled on a point of law, albeit only by a majority of two to one, that the public interest in ensuring that the Commission was not frustrated in its duty of enforcing the provisions of Articles 85 and 86 was sufficient to provide a defence of 'public interest' against a claim for defamation brought by a party which had been the subject of a complaint, namely Hasselblad. The Master of the Rolls referred to the risk that if an action in defamation could be brought in the United Kingdom in connection with the subject matter of a complaint, the substance of that complaint might have to be simultaneously considered by, on the one hand, the Commission and the European Court of Justice and, on the other hand, by the UK courts.

Gradually, however, it has been realized that some advantages could still be obtained by undertakings bringing the matter before the national courts for, after all, the Commission was unable to award an ex-parte injunction nor could it award compensation to the complainant (nor legal costs) nor did it have the flexibility of being able to consider claims under related heads, such as unfair competition or breach of copyright, at the same time as the main substantive complaint under Article 85 or 86. The foundation for private actions in national courts was laid by the important European Court case of *Belgian Radio and Television* v. *SABAM*[20] which established two important principles. The first was the direct applicability and, therefore, enforceability in national courts of both Articles 85 and 86. Whilst the direct application of the Treaty had been confirmed in a number of earlier cases, in particular the European Court case, *Van Gend en Loos*[21] in 1963, such application of Articles 85 and 86 was confirmed for the first time in unequivocal terms here, as well as the obligation of the national courts to enforce their provisions.

[19] *Hasselblad (GB)* v. *Orbinson* [1985] 1 AER 173: 2 WLR 1: QB 475. See Ch. 13, p. 248.

[20] Case 123/73 [1974] ECR 313: 2 CMLR 238.

[21] Case 26/62 [1963] ECR 1: CMLR 105. This case involved not a competition law point but the issue of whether the Dutch Government's imposition of an increased rate of import duty on ureaformaldehyde in breach of Article 12 of the Treaty of Rome could be challenged in the Dutch courts by an undertaking claiming to have been prejudiced by the tax.

The legal position, however, is also affected by the provisions of Article 9(3) of Regulation 17 which states that 'authorities' of Member States have jurisdiction and competence to apply Articles 85(1) and 86[22] so long as the Commission has not 'initiated any procedure' under that Regulation.[23] It was originally thought that this expression meant not only all legislative and executive agencies but also all courts and judicial bodies. This view, however, was later shown to be incorrect when in the *BRT* v. *SABAM* case the meaning of the latter words was then limited to courts 'especially entrusted to apply domestic competition law or to ensure its application by domestic authorities'.[24] This would include probably the Restrictive Practices Court and also the High Court to the extent that, through the procedural mechanism of judicial review, it ensured the application of appropriate administrative law principles under domestic legislation by bodies such as the Monopolies and Mergers Commission.[25]

National courts hearing disputes between litigants who choose to bring into issue the direct effect under national law of Articles 85 and 86 do not, however, have automatically to halt proceedings before them simply because of current proceedings before the Commission, although they have the discretion to suspend hearings of such a case, or to make a reference under Article 177, such a reference being voluntary for all courts below the House of Lords, but compulsory under the Article for the House of Lords itself as the court of last resort. If the ordinary courts have a case in which an issue arises under Article 85 or 86, the court should determine first if the agreement or practice referred to is either clearly within or outside the scope of the Articles. They clearly have jurisdiction to establish if a claim is outside the terms of the Article altogether or falls clearly within the terms of a group exemption.[26] What the court cannot

[22] But not Article 85(3).

[23] In some sectors not covered by Reg. 17 where the necessary implementing regulations permitting the application of Article 85(1) are not yet in force national courts may nevertheless be able to apply Article 86. See *Ahmed Saeed* Case 66/86 [1989] ECR 803: [1990] 4 CMLR 102.

[24] C. S. Kerse, *EEC Antitrust Procedure* (London: European Law Centre Ltd., 1987), 2nd edn., p. 140.

[25] As in the case of *R.* v. *Monopolies and Mergers Commission and Secretary of State for Trade and Industry ex parte Argyll Group Ltd.* [1986] 2 AER 257. The phrase, however, does not include criminal courts: Cases 209–213/84 *Ministère Public* v. *Asjes* (the *Nouvelles Frontières* case) [1986] 3 CMLR 173. The Department of Trade and Industry and the Office of Fair Trading jointly comprise the 'competent authority' for the United Kingdom. Kerse 142. For the purposes of the Merger Regulation the Monopolies and Mergers Commission has been additionally nominated as a 'competent authority'.

[26] As for example in the *Fonderies Roubaix* Case 63/75 [1976] ECR 111: 1 CMLR 538. The Court may, however, make a ruling if it falls within the scope of the *acte clair* doctrine, namely a point which is covered by an established body of case-law of the European Court or, more doubtfully, if the Community rule is regarded as clear by the national court even if there is no case decided by the European Court upon it. See T. C. Hartley, *The Foundations of European Community Law*, 2nd edn. (Oxford: Clarendon Press, 1988), 270–2.

do is to determine whether an individual agreement would obtain exemption under Article 85(3), though the terms of the judgment in the *Concordia*[27] case do suggest that it might be permissible for a court to go so far as to say, in the light of past precedents, that such exemption was unlikely to be available.

If a United Kingdom court finds in a private action that a breach of Article 85(1) or 86 can be established, the remedies available to the claimant can now be more clearly stated than before. The leading authority is *Garden Cottage Foods Ltd.* v. *Milk Marketing Board*[28] (House of Lords 1983). The Board had a statutory function to sell milk as well as butter and cheese. Garden Cottage had been buying butter from the Board since 1980 and had resold the greater part of it to a Dutch purchaser. In March 1982, however, a new distribution policy was introduced under which Garden Cottage ceased to be able to buy direct from the Board but would have to buy from one of four approved distributors chosen by the Board. The result of the introduction of these four preferred distributors was to reduce Garden Cottage's profits on the resale of butter, as it would have difficulties in competing with the prices that the four preferred distributors could quote to customers with their privileged position as direct purchasers from the Board. Proceedings were started against the Board on the ground that the refusal to supply Garden Cottage with butter in bulk was an abuse of the Board's dominant position. Parker J. at first instance refused to grant any interim relief, such as an injunction, pending the trial of the action, as he felt that there would be considerable difficulty in framing the terms of an injunction against the Board and that in any case damages might provide an adequate remedy for Garden Cottage.

The Court of Appeal, however, allowed the appeal by Garden Cottage and granted an injunction on an interim basis against the Board preventing it from confining its sale of bulk butter to any particular category of persons and also from discriminating between their different customers save than in accordance with normal practice. In addition to allowing the application for the injunction, two Lords Justices, Lord Denning and May L.J., indicated that they would be in favour of allowing a claim for damages in such circumstances, whilst the third judge, Shaw L.J., expressed doubts. Faced with so important a decision against it, the Board not surprisingly decided to appeal and achieved a degree of success. On the question of the injunction granted by the Court of Appeal, the House of Lords decided by four votes against one that the injunction should be discharged. Lord Diplock for the majority emphasized that the Court of Appeal should not in principle interfere with a trial judge's decision on an

[27] Case 47/76 [1977] ECR 65: 1 CMLR 378.

[28] [1983] 3 WLR 143: 2 AER 770: 3 CMLR 43: [1984] AC 130. The Court of Appeal hearing is reported at [1982] AER 292: QB 1114.

application for interlocutory relief unless the judge had reached a decision that was clearly wrong on the facts or law or where fresh evidence or change in circumstances had emerged. In the view of the majority, none of these situations applied to the present case so that the original order by Parker J. refusing the injunction should be restored. Lord Wilberforce disagreed, being in favour of allowing the injunction to maintain the status quo which he considered would be more just, and indicated in his view that damages would not be a complete remedy for Garden Cottage, as it would be out of the butter market for a considerable period of time.

In the course, however, of reaching these decisions, the House of Lords entered into a general discussion of the rights of parties in United Kingdom courts under Articles 85 and 86. None of the courts below nor the House of Lords found any difficulty about accepting the principle laid down in *BRT* v. *SABAM* that under these Articles individuals had direct enforceable rights in national courts. The plaintiffs' claim could be categorized in English law probably as a breach of a statutory duty 'imposed not only for the purpose of promoting the general economic prosperity of the Common Market but also for the benefit of private individuals to whom loss or damage is caused by a breach of that duty'. Hence the issue before the House of Lords was whether to interfere in an existing grant of an injunction on an interim application. The question of whether damages were payable to Garden Cottage if successful in their case strictly did not arise as an issue. Nevertheless, the majority of the House saw no reason why damages should not be granted in such circumstances for a breach of Article 86 as a breach of statutory duty, whilst not reaching a final decision, as this issue was not before the Court. Lord Wilberforce, however, had considerable doubts; whilst he agreed that Community rights had to be enforced in national courts, he felt that a complainant under these Articles was not necessarily entitled to be awarded compensation so that in effect the grant of damages by a national court might be to enlarge the rights of a claimant rather than merely enforce them.[29]

It seems likely nevertheless, on the basis of the majority verdict in the House of Lords, that in future cases claims to damages for breaches of Articles 85 and 86 in United Kingdom courts will be available, given suitable facts, though the likelihood of being able to obtain injunctive relief on an interim basis has thereby probably been reduced, save in exceptional circumstances where money payments would not adequately compensate the plaintiff. The argument raised by Lord Wilberforce seems in

[29] As to the availability of damages to private litigants following governmental action in alleged breach of Article 30 of the Treaty see the Court of Appeal decision in *Burgoin* v. *MAFF* (1986) QB 716: [1985] 3 AER 585. If damages would not be an adequate remedy for a breach of Article 85 or 86 an injunction may be granted. *Cutsforth* v. *Mansfield Inns* [1986] 1 AER 577.

principle incorrect; his dissent on the question of damages appears to stem from a belief that Articles 85 and 86 (as well as the various Regulations and Directives issued by the Council and Commission) comprise an exclusive range of remedies for complainants, and that national law is not entitled to award any remedy which the claimant could not himself obtain from the Commission or European Court. This overlooks the basis for the operation of competition policy generally, which is co-operation between the Community and national enforcement authorities, given that each has some advantages and some disadvantages in relevant resources, remedies, and range of jurisdiction. The function of the national courts is to enable adequate enforcement in private suits of the principles laid down in the Treaty (and Regulations made under it), and one of the most effective ways of ensuring this is by extending the remedies of a complainant in national courts beyond those set out in the Regulations relating to the making of decisions simply by Community procedures.[30]

[30] The analogy of the treble damage action in the US is often referred to by commentators on this issue, and perhaps some of its undesirable features have had the effect of casting an unfairly prejudicial blight on the use of the private action under UK law, without giving sufficient weight to the value that the private action (for single not necessarily treble damages) can play in assisting the objectives of competition law. See A. Neale and D. Goyder, *The Antitrust Laws of the United States*, 3rd edn. (Cambridge University Press, 1980), ch. 14, pp. 418–36.

23

Member States and EC Competition Rules

1. Articles 85 and 86: Rules Applicable to Undertakings

Although the provisions of the Treaty can have direct application to persons and undertakings, the majority of the Articles of the Treaty of Rome are directed towards Member States themselves, either by the stating of principles which they are required to observe, or by the laying down of specific tasks for performance by them. By contrast, Articles 85 and 86 laying down the main principles of competition law are not directed primarily towards the Member States themselves, but instead to individual undertakings and associations of undertakings.

If the activities of Member States were confined simply to the exercise of the traditional powers of governments,[1] such as the responsibility for the conduct of foreign affairs and defence, and the implementation of education, housing, and social policies, then their involvement with Articles 85 and 86 might indeed be limited. In fact, however, all Member States without exception are involved also in a wide variety of both industrial and commercial activities, often in direct or indirect competition with private undertakings; moreover, Member States also finance in part or whole the operations of many public undertakings, bodies, authorities, and utilities which have many dealings with other undertakings even if not directly in competition with them. These might include the grant of a monopoly by statute for *inter alia* the public supply of gas, water, electricity, or telephone services, the provision of pilotage services or lighthouses or the transmission of television or radio services. The framers of the Treaty anticipated the problems that would arise if public undertakings of Member States, and any commercial undertakings in which they might become involved, were not made subject to the common rules of competition. Thus, Articles 85 to 94 which comprise the first Chapter of Part 3 entitled the 'Policy of the Community' have a substantial effect on the activities of Governments of all Member States, and a number of leading European Court decisions have dealt with the extent to which the administrative and legislative activities of Member States are subject to them.

A useful starting point in considering the effectiveness of such

[1] Insofar, of course, as the Treaty of Rome has not itself placed them wholly or partly in the hands of institutions of the Community itself.

Community judicial control over national legislation in this area is the case of *Inno* v. *ATAB*[2] (European Court of Justice 1977). This was an Article 177 reference from the Belgian Court of Cassation on the validity of a Belgian statute which required retailers to sell cigarettes and tobacco only at the official price. The official price was the price notified to the authorities for purposes of the assessment of excise duty together with value added tax of an additional amount. The statute had the effect of preventing retailers from competing with each other on price and would have enabled their trade association to take advantage of its dominance in the Belgian market which, under the protection of such legislation, it would have had little difficulty in exploiting and abusing. The European Court of Justice referred back to the principles embodied in Articles 3(*f*) and 5 of the Treaty, and from their combined effect found no difficulty in reading into the Treaty an implied duty on Member States not to adopt or maintain in force any measure which could deprive Article 86 of its effectiveness. The Court also stated that Member States should not enact measures enabling private undertakings to escape from the constraints imposed by the competitive scheme of the Treaty; it added that any national measure which had the effect of facilitating the abuse of a dominant position capable of affecting trade between Member States would generally have been incompatible with Articles 30 and 34.

The freedom of Member States to enact legislation reducing or eliminating price competition arose as the central issue in a more recent case under Article 177, where a French court sought the ruling of the European Court of Justice on whether the Treaty prohibited the establishment of a system within a Member State to control the price at which books, whether locally published or imported, could be sold only at a price not more than 5 per cent below the official price named by the publisher or importer. We have already considered the facts of this case (the *Leclerc Books*[3] case) in Chapter 14 and noted the effect of the decision was that the French Government was entitled to enact resale price maintenance legislation for books published in France so long as there was no Community sectoral policy yet adopted relating to the pricing of books, and so long as the legislation did not conflict with other provisions of the Treaty, notably of course Articles 30 and 34. It was on the basis of these latter articles, however, that the Court ruled on the other hand that price restrictions imposed on those books imported into France would be illegal, since they would rank as being of 'equivalent effect' on imports; the only exception to that would be if it was established that the original exports of the books had been for the sole purpose of circumventing the legislation

[2] Case 13/77 [1977] ECR 2115: [1978] 1 CMLR 283.
[3] Case 229/83 [1985] ECR 1: 2 CMLR 286.

by subsequent reimportation.[4] The argument of the French Government that the legislation in its entirety was indispensable to protect both specialist book shops and books as a cultural medium against the likely reduction in variety and availability of books if unrestricted price competition was allowed failed to convince the Court, and the exceptions contained in Article 36 were held not to apply here so as to effect any qualification to the prohibitions in Article 30.

The issue of the enactment by Member States of legislation reducing or eliminating price competition also arose in the air transport sector. As already noted in Chapter 6 the European Court decided in the *Nouvelles Frontieres*[5] case that Member States and the Commission remained free, in the absence of any Community rules, to adjudicate under the terms of Articles 88 and 89 respectively on the admissibility of agreements and practices and the abuse of dominant position. In these circumstances it was uncertain whether Member States were entitled to continue to create and enforce domestic legislation covering such matters as tariffs, charter flights, discount fares and the general terms and conditions upon which air travel was provided to other countries within the Community. It could have been argued, by analogy with *Leclerc Books*, that until a Common Community policy on air transport had been adopted by the Council, that Member States retained jurisdiction in this area, even if the effect might be to permit agreements or practices that individual undertakings could not have engaged in consistently with the provisions of Articles 85 or 86.

In the last five years, however, this issue has ceased to have such importance since in 1987 the Council finally enacted two important regulations. Regulation 3975/87 is the equivalent for the air transport sector of Regulation 17/62 (which as already explained lays down the Commission's general powers and procedures for most sectors). Regulation 3976/87 authorizes the Commission to give group exemptions and is currently operative until the end of 1992. Exemptions have been provided under this Regulation for joint planning and co-ordination agreements between airlines, including consultation both on air fares and slot allocation at airports but without allowing airlines to make binding agreements on these subjects. Other exemptions granted relate to the development of computer reservation services and ground handling systems. More liberal arrangements have also been introduced for increasing capacity on certain routes within the Community and a system for introducing 'double disapproval' both for normal and for discounted fares is gradually being introduced. Under this system the proposals of an airline have to be

[4] The issue of whether the reimportation was genuine or artificial, i.e. simply to avoid the effects of the French law, would be for the national court to decide. Its companion case is the *Leclerc Petrol* Case 231/83 [1985] ECR 305: 2 CMLR 524, officially *Cullet* v. *Leclerc*.

[5] Cases 209–13/84 [1986] 3 CMLR 173. See pp. 83–4.

accepted by the two Member State national authorities involved unless both agree to disapprove. Eventually it is anticipated that airlines will have freedom to fix the great majority of fares.

It is well established that undertakings themselves cannot claim that Articles 85 and 86 do not apply to them, simply because national legislation or executive action has reduced the degree of competition within the market. The leading cases are also concerned (as was the *Inno/ATAB*[6] case) with the cigarette and tobacco markets, in Belgium and Holland respectively. In the *Fedetab* case[7] (European Court of Justice 1980), the Belgian legislation imposing strict price control of the retail level considerably reduced the scope for competition. Excise duty in fact comprised some 70 per cent of the price of an average packet of cigarettes. In this situation, the legislation laid down strict requirements for the affixing to every packet of the exact price upon which the relevant duty had been calculated in order to reduce the scope for fraud. These laws, of course, considerably reduced the scope also for competition but the European Court ruled that competition could still have remained in three important areas, namely profit margins for both wholesalers and retailers, rebates, and terms and conditions of supply including credit terms. These were all areas where in a competitive market variations would have been anticipated, as particular undertakings sought to increase their market share by improving the terms offered to their more favoured customers, but which under the rules of the relevant trade association had been suppressed so as to exclude the element of uncertainty which such competition would have brought. Fedetab's appeal against the decision of the Commission that its recommendation to members as to standard terms of business and prices to be charged contravened Article 85(1) was, therefore, rejected.

To similar effect was the Court's decision in the later *SSI*[8] case (1985) relating to Dutch legislation along similar lines but involving not only regulations relating to excise taxes but the permitted level of price increases by Dutch manufacturers and restraints on their profits. Once again, while the Court accepted that this legislation had considerably affected the climate of competition, it held that it did not by its terms prevent all aspects of competition between those engaged in the trade. Competition would still have been possible (had the relevant trade associations not prevented it) on such promotional matters as the giving of additional discounts or rebates or the provision of more favourable terms and conditions of sale. A decision finding breaches of Article 85(1) by the association of Dutch tobacco dealers was, therefore, also upheld.

It will, therefore, be comparatively rare for government involvement

[6] Case 13/77 [1977] ECR 2115: [1978] 1 CMLR 283.
[7] Case 209/78 [1980] ECR 3125: [1981] 3 CMLR 134.
[8] Cases 240–242, 260–262, 268–269/82 [1985] ECR 3381: [1987] 3 CMLR 661.

through legislation or administrative action to be so all pervasive as to eliminate the possibility of competitive activity to such an extent that Articles 85 and 86 can no longer have any application. The only case in which this has been found to occur was the *Sugar Cartel* case[9] (European Court of Justice 1975) where the Italian Government limited competition with regard to imports into Italy of sugar in no fewer than four major respects. The first was that the implementing regulations compelled the market for the import of sugar into Italy to take a particular form which it would not have adopted absent such regulations; second, by making the purpose of those regulations the bringing of supply and demand into an exact balance so that the normal incentive for traders to seek additional customers and additional volume was reduced, if not totally eliminated; third, by taking away through its regulations the freedom of sellers of imported sugar from negotiating freely with the buyers whom they selected; and fourth, by intervening directly through officials of the Italian Government to make clear on all possible occasions to undertakings that import negotiations should only be conducted in the way which that Government considered appropriate, rather than in any other way dictated or influenced by normal competitive requirements.

There is no other recorded case in which all these conditions have applied, and it seems unlikely that defendants will be allowed to rely on this precedent of the *Sugar Cartel* case save in the most unusual circumstances. Even though in the *SSI* case it was true the Dutch authorities had on some occasions held formal consultations with the cigarette industry on its agreed objectives, including the guarantee of substantial tax revenue for government and stable income for retailers, the Dutch Government had not on that occasion gone to the extent of suggesting to undertakings that they should support those objectives by anti-competitive arrangements.

It is comparatively rare, therefore, to find an example under the domestic law of a Member State under which the State has on a mandatory basis passed legislation or taken administrative steps to require undertakings subject to Articles 85 and 86 to enter into agreements or practices which, if adopted without such compulsion, would have necessarily involved them being in breach of these Articles.

The *Ahmed Saaed* case[10] may be a rare example of this principle where the German Government actually required travel agents to abide by fares which it had approved for particular air routes. National legislation has, however, also been considered in two cases by the European Court in which, if it did not actually require anti-competitive conduct in breach of these Articles, nevertheless served to provide considerable support for

[9] Case 40/73 [1975] ECR 1663: [1976] 1 CMLR 295.

[10] Case 66/86 [1989] ECR 803: [1990] 4 CMLR 102.

private agreements of such a kind. Both cases were Article 177 references from Belgian courts. The first was the *Flemish Travel Agents* case[11] where a Royal Decree required travel agents not to depart from official tariffs and not to share commissions with their customers. Standard forms of agreement between such travel agencies and tour organizers existed and were ruled by the Court to constitute a breach of Article 85. Under the Decree indeed a travel agent which complained of 'cheating' by other travel agents could obtain an injunction against them and could also ask for their operating licences to be withdrawn. The judgment of the Court was unequivocal, that Belgium was in breach of Article 5. The second case, *Van Eycke* v. *ASPA*,[12] led to a less clear-cut answer. Here the Royal Decree provided that the holders of deposits in savings accounts were only entitled to tax relief on the income if interest rates did not exceed 7 per cent. Mr Van Eycke alleged that the decree merely confirmed an existing restrictive agreement between the banks. The Court ruled that in such a case the central issue was whether the legislation merely confirmed the method of restricting interest rates which the interbank arrangements had set out or whether alternatively it went further. Finding ambiguity in the evidence on this issue the Court referred it back to the national court.

In the majority of cases, as we have seen, the defence when raised (as in *Fedetab*, for example) has been dismissed on the ground that the State action did still leave sufficient 'breathing space' in which the undertakings could operate without being required to act in contravention of Articles 85 or 86. If, however, the State action was sufficiently compelling and allowing no room for manœuvre, it would provide a defence to the undertaking, both against Commission proceedings and in response to any private action by an injured third party.[13] It may well be that the Court will regard *Leclerc Books* in the future as the 'high water mark' for allowing, national legislation to weaken the effectiveness and width of application of Articles 85 and 86. It is possible to detect from the language of the case (and also of its companion case *Leclerc Petrol*) an inference that the prohibitions of *Inno/ATAB* on the use of national legislation to provide an 'escape route' for undertakings from these articles will be more widely applied against Member States by proceedings brought by the Commission under the terms of Article 169 and in reliance on Article 5(2). Much will depend on the ability of the Commission and national courts to identify

[11] Its full title is considerably longer. *VZW Vereniging van Vlaamse Reisebureaus* v. *VZW Sociale dienst van de Plaatselijke en Gewestelijke Overheidsdiensten*. Case 311/85 [1987] ECR 3801: [1989] 4 CMLR 213.

[12] Case 267/86 [1988] ECR 4769: [1990] 4 CMLR 330.

[13] There is, however, an argument based on European Court decisions in other fields that an undertaking which seeks to rely on State action as a defence to a claim of anti-competitive behaviour must itself take the risk that the State was not authorized by Community law to compel such action.

suitable cases to bring to the Court of Justice under either Article 169 or 177 so that the control of the Community over national legislation in this area can be reinforced.[14]

It can moreover be said with confidence that the mere encouragement by Member States' governments that undertakings may establish prices and industry regulation by methods reducing or eliminating competition provides no defence if the legislation is permissive rather than strictly mandatory. Nor will it be any defence to show that there is governmental ratification for the implementation of schemes for the regulation of prices and other industry conditions that were originally agreed between the undertakings themselves. Justification for this statement can be found in the twin French cases of *Pabst and Richarz KG* v. *BNIA*[15] (Commission 1976) and *BNIC* v. *Clair*[16] (European Court 1985).

In the *Armagnac* case, this famous brandy was produced under the auspices of the relevant trade association, BNIA, whose function was to regulate the affairs of the industry and in particular to regulate the quality and control the terms upon which it was sold as well as providing technical and research assistance to the individual growers. The association itself had a chairman appointed by the French Ministry of Agriculture and twelve delegates representing producers and co-operatives whilst another twelve represented wholesalers, distillers, and brokers. All sales of Armagnac were graded from 0 to 5, the most expensive variety being grade 5, VSOP Napoleon at least five years old. Forty per cent of industry production was sold in France, and sixty per cent was exported either in bottles or in bulk. The majority of exports were in bulk, and the association had authority to issue certificates of age for these. The association effectively controlled all movements of bulk Armagnac and was fully informed of the stocks of various ages and grades held by the members of the association.

Concern had apparently been felt that some German importers had sold the brandy under false trade description, claiming that lower grades were in fact grade 5. The association, purporting to deal with this abuse,

[14] Unless, of course, as in *Leclerc Books*, the Government could show that there was as yet no Community policy adopted for the relevant industrial or trade sector as required by the terms of the Treaty so that Article 5 of the Treaty could have no application to the duties of the Government in respect of that sector. An influential article in French, emphasizing by contrast the alleged unwillingness of the Court to provide solutions in the area of national price-fixing legislation on the basis of Articles 85 and 86, is Y. Galmot and J. Biancarelli, 'Les réglementations nationales en matière de prix au regard du droit communautaire', 21(2) *Revenue trimestrielle de droit européan*, 269–311 (1985).

[15] [1976] 2 CMLR D63. The Commission's comment on this case in the 6th Annual Report, p. 68, was that it showed 'where associations or undertakings entrusted with certain statutory powers act beyond their powers and take measures to regulate the market, the object or effect of which is to promote the uniform conduct of its members, they cannot evade the rules of competition and hold the State responsible for their actions'.

[16] Case 123/83 [1985] ECR 391: 2 CMLR 430.

imposed a ban upon bulk deliveries in the future of grades 4 and 5.[17] German importers, who had become used to receiving bulk deliveries of the higher grade, complained to the Commission that this restriction was in breach of Article 85. It was claimed that, whatever the alleged justification for the interference by the association in trading practices, its effect would be artificially to reduce the supply and increase the price of the higher grades of brandy. The Commission ruled that the decision of the association was a breach of Article 85; it pointed out that the effect of the ban on bulk supplies was far in excess of the measures that would have been necessary to carry out all the quality protection obligations assigned to BNIA by the relevant French decrees. The objective of BNIA's decision appeared rather to be to restrict competition within the Common Market and to increase the price of the highest grades by reducing their availability on the German market. The Commission ruled that other administrative and legal means to stop any unfair practices or false trade designations were available, and those should have been applied in a way that was proportionate to the disclosed problem, rather than by the exercise of powers to prevent altogether bulk deliveries of the two highest grades.

In 1982, the Commission ruled that similar practices in fixing the price of Cognac by a similar professional association responsible for that industry was likewise in breach of Article 85, this decision being later confirmed by the 1985 decision of the European Court in the *Clair* case. Here, the association also had the function of acting as a joint trade organization between the various interests in the Cognac industry. It provided technical details and information as to the supply and demand for the products, it supervised maintenance of industry standards, and subject to government supervision was entitled to implement rules for the marketing of Cognac, including prices and terms of payment.

A committee representing the various interests in the trade, similar to that in the Armagnac industry, would decide on the relevant prices and conditions of sale. Once the committee had made a decision, this had to be approved by a three-quarters majority of the members of the association which included delegates of wine growers, dealers, brokers, workers, and technicians. Once the rules had thus been adopted, the association was then entitled to apply to the French Minister of Agriculture to extend the agreement compulsorily for a specified period to all members of the trade, whether or not they were participants in the joint trade organization. Once the ministerial decree had been made, any contract failing to comply with it would be void. In the relevant year, a minimum price of wines for distillation into Cognac had been fixed, but Clair had made con-

[17] The only exception was for deliveries of less than ten hectolitres.

tracts for sale to various growers at prices below those allowed. The BNIC then claimed that this agreement was void; Clair responded by arguing that the price-fixing itself was contrary to Articles 85 and 86. The French Court before whom the case came then made application to the European Court under Article 177 to determine whether this form of price-fixing was covered by Article 85(1).

The answer given by the European Court was in the affirmative.[18] The Court ruled that any agreement which had the object or effect of restricting competition by the fixing of minimum prices for the purchase of a semi-finished product (wine) could affect trade between Member States, even if the semi-finished product was not itself traded between Member States, as long as it formed the basis of a final product (Cognac) marketed elsewhere in the EC. The fact moreover that the agreement had been concluded within the framework of a public law, with the 'blessing' of government, could not affect the application of Article 85; nor would the fact that the persons who signed the agreement were themselves largely appointed by a government minister on the nomination of the relevant trade organization be considered relevant.

Another example of the fact that mere government encouragement provides no defence to a claim under Article 85 or 86 was provided in the *Aluminium Cartel*[19] case. The Commission decision, issued in 1984, concerned agreements between major Western producers of aluminium which had ensured that all Eastern Bloc source material during the years 1963 to 1976 had passed into the hands of the leading Western producers, thereby preventing their competitors from obtaining the benefits of metal at lower prices and removing a major source of price competition from Western markets. The original scheme for preventing the marketing of Eastern Bloc aluminium in the West, in a way which would reduce the overall level of prices, had been entered into with at least the tacit support of the British Government, which had been concerned with the risk that the market price for aluminium would be substantially reduced by Eastern Bloc material at a time when there was no UK domestic law limiting the agreements that could be made with regard to such foreign source material,[20] and, of course, at a time several years before the entry of the United Kingdom into the European Community and its acceptance of the Treaty

[18] In a further Article 177 case also involving the Cognac industry the Court predictably ruled that a levy on a French winegrower for exceeding his production quota was a private arrangement subject to Article 85. The French Government was held in breach of Article 5 by adopting and extending to non-members of the relevant trade association the burden of such production levies and restrictions. *BNIC* v. *Aubert* Case 136/86 [1987] ECR 4789: [1988] 4 CMLR 331.

[19] See Ch. 6, p. 159, for a fuller account of the facts of the case.

[20] The Restrictive Trade Practices Act would not have been applicable to such an agreement, even between UK companies, made abroad to cover the purchase of a commodity supplied from the Eastern Bloc.

including Articles 85 and 86. The Commission held that the encouragement given by the British Government did not serve to provide any defence to a charge under Article 85, although it might be a relevant factor in reducing the culpability of defendants and, therefore, in the assessment of the size of any fine or whether any fine at all should be imposed.[21]

These decisions of the Court have themselves merely served to confirm the distinctions drawn by the Commission itself in its decision on *Franco-Japanese Ballbearings*[22] in 1974, when it contrasted the agreements in that case between French and Japanese companies on price levels to be maintained for their products in the French market (which would be prohibited by the terms of Article 85(1)) with any measures that might be imposed by the Japanese Government itself (which would take any resultant arrangements or practices imposed thereby on the parties outside the scope of Article 85(1)). Even specific authorization or approval by government or governmental agencies alone, therefore, is always insufficient by way of defence to a claim made under Article 85 or 86, since the parties in such circumstances remain at liberty to make their own decision as to whether to enter into the relevant arrangements or practices.

2. Article 90: Public Undertakings under the Rules of Competition

It is not only by legislative or executive acts that Member States may become involved in interference with the normal working of competition. Member States are themselves involved in the provision of public services in many ways, either through nationalized industries or public corporations of many kinds. Their right to preserve their own particular balance of public and private enterprise is specifically retained under Article 222.[23] It is, however, undoubtedly true that if public undertakings are treated in a more favourable manner by Member States than other enterprises, it would be possible for considerable damage to be done to the process of competition in national markets in ways that would affect trade between Member States. For this reason, Article 90 was included in this part of the Treaty; unlike Articles 85 and 86, it is specifically aimed at Member States, but its function is to clarify the application of the competition rules to them rather than to restrict their application. This Article has a close connection with Article 5, which provides principles that

[21] In the event, no fines were imposed on the participants to the agreement.

[22] [1975] 1 CMLR D8.

[23] Article 222 reads 'This Treaty shall in no way prejudice the rules in Member States governing the system of property ownership.'

underlie many of the specific obligations set out in the remainder of the Treaty. Article 5 states that Member States are to take all appropriate measures to ensure that the obligations which they have accepted under the Treaty or which arise from action taken by the institutions of the Community are fulfilled and are to facilitate achievements of the tasks laid down for the Community and to abstain from any measures which could jeopardize their attainment.

These requirements apply particularly to the conduct of public undertakings. The relevant Article in the Treaty, Article 90, makes clear in its opening paragraph (1) that the rules of the Treaty do apply to such bodies and even proceeds (for the avoidance of doubt) to mention those most relevant to them, namely Article 7 and Articles 85 to 94 inclusive. Moreover, these rules apply not only to public undertakings in the strict sense but also to quasi-public or private undertakings to which 'special or exclusive rights' have been granted. The general rule applies, therefore, *inter alia* to all *public* bodies which possess an identity independent from that of the Member State itself; this appears to exclude only actual departments or Ministries of the Government itself forming an integral part of the executive or administrative function of the State.[24]

To this broad principle stated in paragraph (1), the following paragraph (2) establishes a narrow exemption, containing three cumulative elements which have all to be satisfied before it can operate. These are:

(i) The undertakings concerned must have been entrusted with the 'operation of a service of general economic interest' or 'having the character of a revenue-producing monopoly'.
(ii) The application of the Treaty rules would obstruct the performance in law or in fact of the particular tasks assigned to it.
(iii) In any case the development of trade must not be affected (if the exemption were otherwise available) to an extent contrary to the interests of the Community.

Help in interpretation of the meaning of 'public undertaking' itself in Article 90(1) is obtained from Article 2 of Commission Directive no. 80/723 of 25 June 1980 on 'transparency of relationships', introduced under the authority of Article 90(3), defining it as follows: 'Any undertaking over which the public authorities may exercise directly or indirectly a dominant influence by virtue of their ownership of it, their financial participation therein or the rules which govern it'.[25]

Such a definition covers a very wide range of bodies, including large-scale nationalized industry responsible for the production and distribution

[24] See A. Pappalardo, 'Measures of the States and Rules of Competition of the E.E.C. Treaty', *Eleventh Annual Proceedings of Fordham Corporate Law Institute* (1984) 515–41.

[25] OJ [1980] L195/35.

of gas and electricity, national railway authorities, and many other public bodies, large and small, engaged in some form of public economic activity. It includes the provision of services, even if these also involve the sale of goods.[26]

By contrast, the second category of undertaking mentioned in Article 90(1) is narrower since, although public undertakings often could themselves be described as enjoying 'special' or 'exclusive' rights, the phrase refers by contrast to undertakings not themselves actually public but which nevertheless have been granted such special privileges in order to carry out functions regarded as important by Member State governments. In this category, for example, are included bodies authorized to operate State television and broadcasting services.[27] It also covers bodies with rights to prescribe standards for particular sectors, such as agricultural produce, and those responsible for regulating those employed in particular trades or professions. The 1984 decision of the European Court of Justice in *IAZ International Belgium* v. *Commission*[28] concerned a trade association granted such statutory rights by the Belgian Government for approving appliances for connection to the public water supply, evidenced by the award of official certificates of conformity. The association, however, used its rights in a way designed to force all manufacturers and importers of washing machines to become members, by making difficulties in the granting of such certificates to parallel importers and concerns which preferred to import appliances without joining the association. The Court held that the association was subject to Article 85, notwithstanding its statutory functions.

Such an association would also come within the category of undertakings referred to in Article 90(2) as 'entrusted with the operation of services of general economic interest'. It is, however, clear in view of the cumulative triple test applied that the measure of exemption afforded by Article 90(2) to such bodies is markedly less generous than that conferred by Article 85(3); Article 90(2) moreover applies to a far more limited category of undertakings. The criteria for inclusion within this category were analysed in the early years of the Treaty by Professor Deringer[29] as follows:

(1) that the undertaking does not receive official authority merely to carry out a service, if it so chooses, but also has positive duties

[26] It is less certain if public bodies simply engaged in the production of natural resources such as coal or oil are included within the definition.

[27] See, e.g. *BRT* v. *SABAM* Case 127/73 [1974] ECR 313: 2 CMLR 238 and the *Sacchi* Case 155/73 [1974] ECR 409: 2 CMLR 177.

[28] Case 96/82 [1983] ECR 3369 [1984] 3 CMLR 276 (also known as *Anseau-Navewa* v. *Commission*). See Ch. 21, pp. 423–4).

[29] A. Deringer, *The Competition Law of the European Economic Community. A Commentary on the EEC Rules of Competition* (CCH Editions Ltd., 1968), 248–9.

imposed upon it to maintain the services which it may not arbitrarily change or discontinue;

(2) the nature of the services is such that if they were not carried out by the undertaking in question they would probably have to be supplied by the State, since they are essential for meeting a vital part of need of the general public or for enabling the economy to continue to function;

(3) public law will have endowed these undertakings with particular powers, or with certain exclusive rights relevant to their function;

(4) the undertaking will have reserved for itself not only general supervision of its function but also retains a broad power through regulations to organize its services and fix its prices;

(5) the prices which it charges are determined not strictly according to the laws of supply and demand but are influenced by social and political considerations, which may require that some disadvantaged sections of the population should receive free or reduced cost supplies, e.g. telephone or postal services for inhabitants of remote rural areas;

(6) the services supplied must be available to all, on a non-discriminatory basis.

The effect of these requirements is to limit the scope of the exemption to those activities of a public undertaking which have a direct relationship with its main statutory functions and are not, for example, simply commercial activities which are ancillary to them. To obtain the benefit of this restricted exemption involves the undertaking in showing there is an inherent conflict between the particular task which is imposed upon the undertaking and the application of the Treaty rules. That this is a substantial burden is clearly illustrated by the *Telespeed* (*British Telecom*) case decided by the European Court of Justice in 1985.[30] The Commission's decision was that British Telecom, at that time 'a public corporation established by statute holding a statutory monopoly on the running of telecommunication systems in the United Kingdom',[31] was abusing a dominant position under the meaning of Article 86. The particular way in which abuse of its dominant position had occurred was through the prohibition of private message forwarding agencies in the United Kingdom from transmitting messages, whose originators and recipients were resident in other countries, through the British network. This could be done at a cheaper rate than by sending the messages direct between two other countries. It is interesting that the appeal against the Commission's

[30] Case 41/83 [1985] ECR 873: 2 CMLR 368, reported under the name of *British Telecommunications: Italy* v. *Commission*.

[31] The quotation is taken from an account of the case at p. 96 of the 15th Annual Report. The case related to practices of British Telecom that occurred between 1975 and 1981.

decision was brought not by British Telecom itself, which accepted the Commission decision, but by the Italian Government which obviously regarded it as a precedent of relevance for its own public undertakings.

Amongst the arguments raised by the Italian Government was that Article 90(2) protected BT in acting to prohibit the activities of such independent agencies, so as to protect the services or general economic interest which BT was required by statute to carry out. The European Court rejected this argument on the grounds that in taking action against the private agencies BT was acting not in any official capacity, but simply responding in its commercial capacity, like any other undertaking engaged in business. The management of public telecommunications equipment and the placing of that equipment at the disposal of consumers upon payment of a fee was inherently a business, not a governmental, activity. The rules laid down under which telex messages were forwarded had been prepared by British Telecom alone and were not laid down by statute. They clearly formed an integral part of its business strategy and activity. Similar reasoning was applied in the case decided in the same year of *CBEM* v. *CLT and IPB.*[32] The Court ruled that this exception in Article 90(2) cannot apply when the activity to which Article 86 is being applied is a commercial activity which is ancillary to the main function of the undertaking. The enforcement of the prohibition on abuse of its dominant position (however obtained) will, therefore, apply both to such activities and also even to those activities more closely related to the performance of its statutory tasks, unless it can be shown in the latter case that the prohibition is incompatible with the proper carrying out of the responsibility. The Commission has taken a similar approach in decisions[33] relating to the Dutch and Spanish postal services where legislation favouring them against private courier services for certain smaller packages (up to 500 grams in Holland and up to 2 kilograms in Spain) has been held ineligible for exemption under Article 90(2). In both cases it ruled that the existing letter post monopoly was adequate to enable the postal services satisfactorily to carry out their public service functions. The European Court has confirmed the right of the Commission to issue decisions under Article 90(3) against Member States which have taken action in breach of that Article.

The Court is clearly not disposed, therefore, to give a broad interpretation to Article 90(2) even where the functions performed by the undertaking are central to the proper functioning of the economy of a Member

[32] Case 311/84 [1985] ECR 3261 [1986] 2 CMLR 558.

[33] *Dutch Courier Services* [1990] 4 CMLR 947: *Spanish Courier Services* [1991] 4 CMLR 560. In the *Dutch Courier Services* case decided 12 Feb. 1992 (Cases no. C-48 and 66/90) the Commission's decision was, however, quashed on procedural grounds by the European Court.

State. A further example is its important 1981 case, *Züchner* v. *Bayerische Vereinsbank*.[34] Züchner was a customer of this German bank and queried whether it was entitled to agree with all other banks in the Federal Republic of Germany a uniform service charge for transfers of sums of a similar amount to other Member States. The bank argued that by reason of the special nature of the services which banks provide and especially in regard to transfer of capital they should be considered as undertakings 'entrusted with the operation of services of general economic interest within the meaning of Article 90(2)'. Although the Court accepted that the transfer of funds from one Member State to another as normally carried out by banks is an operation which has a public element, it ruled that they do not come within the classification of Article 90(2), unless in performing such transfers they could be shown to operate a service of general economic interest with which they have been entrusted through measures adopted by government. The Court declined to reach such a finding.

The Commission keeps the treatment of public undertakings by Member States carefully under review, and an important section of each Annual Report is devoted to this subject. The Directive adopted in 1980 on transparency of relationship has been extended by a later Directive in 1985 (no. 85/413) to include a group of sectors previously excluded from it: public water authorities, energy undertakings, post and telecommunication authorities, and transport and credit bodies. The Fifteenth Annual Report[35] records the first Commission decision[36] under Article 90(3) declaring a Greek law requiring all public property to be insured with a particular state-owned insurance company to be incompatible with the provisions of the Treaty, the first occasion on which the Article has been applied to a Member State's grant of preferential treatment to one of its own public undertakings.

In conclusion, therefore, the extent to which Article 90(2) provides any relief from the requirements of Articles 85 and 86 is very limited, and future development of case-law by the Commission and Court is likely to move further in the direction of limitations of its scope rather than to permit it to be used to weaken the uniformity of application of those Articles to undertakings in both private and public sectors.

The Commission has also made effective use of Article 90(3), which gives it the obligation to ensure the application of Article 90 and empowers it, where necessary, to address appropriate directives or decisions

[34] Case 172/80 [1981] ECR 2021: [1982] 1 CMLR 313.

[35] pp. 201–2. The 16th Annual Report records (at p. 199) that the Greek Government had not yet complied with this decision, and the Commission indicated that it would have, therefore, to serve a reasoned opinion under Article 169.

[36] OJ [1985] L152/25.

to Member States. It had for some time viewed with concern the practices of Member States in granting special or exclusive rights to national telecommunications organizations in drawing up specification for the installation and operation of terminal equipment such as for example telephone exchanges. In order to remedy the situation it adopted Directive 88/301 which set out a number of principles for Member States in order to improve the competitive situation applying to this equipment and in order to bring to an end existing breaches of the Article. The Directive provided that

(i) Member States should withdraw such exclusive or special rights from their own favoured undertakings relating to telecommunications terminal equipment;
(ii) Member States should ensure that undertakings have the right to import, market, connect, and maintain terminal equipment;
(iii) type approval specifications should be drawn up by bodies independent of the operation of telecommunications network or terminal providers; and
(iv) all contracts for leasing or maintenance of such equipment should be terminable on not more than a year's notice, thus preventing the use of long-term contracts as a means of shutting out new competitors for a long period of time.

The validity of this Directive was challenged by France and some other Member States. Eventually the issue reached the European Court of Justice[37] which upheld the right of the Commission to deal with breaches of Article 90 by such Directive which specified in the general way the obligation which the Treaty imposed on Member States. It was not necessary, said the Court, for the Commission to utilize instead Article 169 which applied only when specific infringements of the Treaty had occurred. Nor was it necessary for the Commission to have to obtain from the Council separate measures under Article 87 or 100(A) of the Treaty. Dealing with the provisions of the Directives (i) and (ii) above, it upheld the requirement by the Commission that exclusivity should no longer be permitted for telecommunications operators since restrictions issued for this purpose by Member States would in any case be in breach of Article 30. Point (iii) relating to type approval was also approved. The Court did, however, annul the prohibition in the Directive on special rights (largely because of failure by the Commission to provide an adequate definition of this category) and also those relating to long-term contracts under (iv). The reason for this latter ruling was that the practice of long-term contracts was one adopted by independent undertakings rather than Member

[37] *France and Others* v. *Commission*, Case C-207/88, a decision dated 19 Mar. 1991.

States and should, therefore, be dealt with under Article 85 or 86 rather than Article 90.[38]

Perhaps the most important aspect of the judgment, however, was the clear statement that the reference in Article 90(1) of the Treaty to 'special' or 'exclusive' rights did not contain any inference that such rights were necessarily acceptable. The grant of any such rights commonly accorded by Member States in a variety of sectors is subject to all the rules of the Treaty, including the principle of proportionality as interpreted by the Court in many previous cases. Moreover, although the provisions of the Directive dealing with special rights were annulled because of the lack of clear explanation and the manner in which they infringed the Treaty, there seemed no reason why such rights should not remain subject to the Commission's powers under Article 90(3) so long as in future cases such a link can be clearly established. As a result of this case, new directives under Article 90(3) are likely to be issued by the Commission in other sectors, notably in respect of electrical supply.

[38] A later Directive (90/388) on telecommunications services contains a number of analogous provisions which have also been challenged by certain Member States before the European Court.

24

EC Competition Law in the Context of World Trade

Lawyers can become so immersed in the detailed study both of the theory and practice of competition law that they forget that it cannot sensibly be analysed or understood except in the context of the particular markets (whether local, regional, national, or international) within which it operates. Moreover, the competition law of the Community does not exist in a vacuum, separate from all the pressures of international trade, but on the contrary is substantially affected and influenced by them. Any review, therefore, of this law, and of the policies that underline it, needs also to take an outward look at the influence which such law and policies have on business undertakings in the wider context of international trade and on the patterns of business and commerce outside the territorial bounds of the Community. In this final chapter of Part Two, we shall, therefore, conclude our consideration of the substantive law of the Community relating to competition by looking at the jurisdictional problems that arise when undertakings against whom it is desired to enforce such laws operate from bases outside the territorial boundaries of the Community. It is also necessary to make a brief mention of the effect which Community trade policies have on the operation of its competition policies, and in particular, the effect on such policies of the measures taken to arrange voluntary restraints covering imports into the Community by third parties (notably Japan) and for the control both of dumping and of foreign government subsidies affecting such imports.

1. The Extraterritorial Application of Articles 85 and 86

As will already have become apparent from the preceding chapters in Part Two, many of the companies affected by these two Articles are foreign multinationals, operating either through subsidiaries or by way of distribution and licensing arrangements within the Market. The wide terms of the two Articles are interpreted by the Commission as applying to all undertakings whose operations or agreements meet the stated jurisdictional test, namely that they affect trade between Member States of the Community and have as their object or effect restraint upon competition within the Community. If United States or Japanese companies, for example, enter into distribution or licensing agreements direct with undertakings based in the Market or operate through a branch in one or more Member States, then normally by their actions those companies will have placed themselves within the territorial jurisdiction of the Community and the Treaty in accordance with the normal principles of international law. Such undertakings are as clearly bound to abide by the rules of the Treaty as they would be to abide by the rules of any individual Member State within which they chose to operate; no novel problem of international law arises.

Borderline cases can arise, involving questions of fact, when the precise degree of this direct involvement is disputed. A more difficult situation, however, involving issues of international law rather than simply of fact, arises when the foreign company operates through one or more subsidiary companies incorporated in one or more Member States, when the Commission seeks to enforce its rules not simply on those subsidiary companies but also on their parent company operating from a headquarters outside the Community.

The claim is often then raised by the parent company that the Community, through the Commission, is seeking to apply its law no longer on the basis simply of the parent company's own observable links and connection with activity within the geographical area of the Community, but simply on the basis of having performed actions outside the Community which have produced effects within it.[1] In an age of instant worldwide communication when the business operations of most large multinational groups are carried on both substantially within and without the Community, it is hardly surprising to find that the Commission seeks to enforce the competition rules against companies considered to have directly caused anti-competitive effects within it or to have entered into anti-competitive arrangements with or through subsidiaries which have such effects. This may occur even though the parent company itself may not have had any direct physical involvement with the Community, apart from normal two-way communication with its subsidiaries operating there. The question that arises naturally in these circumstances is whether the Commission can truly be said in such circumstances to be claiming jurisdiction on the basis of its own 'effects doctrine' against such parent companies.

The Commission itself has claimed a right to implement both Article 85 and 86 against undertakings which have none of the conventional links with the Community upon which jurisdiction is universally accepted,

[1] Such an extraterritorial claim to jurisdiction over undertakings situated outside the physical area of the authority claiming jurisdiction has been a feature of US antitrust law for many years, though both its jurisprudential pedigree and current application by US Courts is a matter of considerable controversy. See, e.g. J. Attwood and K. Brewster, *Antitrust and American Business Abroad*, 2nd edn., 2 vols. (New York: McGraw-Hill Book Co., 1981, 1991 Supplement).

under international law, namely nationality, place of business, domicile or residence. The Commission has done so, however, on two alternative bases both of which contain an extraterritorial element, but only one of which has received to date unqualified approval from the Court. This first ground, which has been approved by the Court, is that a subsidiary located within the Common Market which is under the effective control of other companies in its group is deemed, when itself acting within the Community, to be performing actions which are equally those of its controllers, by virtue of the corporate relationship between them. This is so regardless of whether the parent company is incorporated in or operates from Tokyo, New York, or Zurich or any other jurisdiction or location outside the Community. This presumption is referred to as the doctrine of the 'group economic unit' and has received support from a number of Court decisions.[2]

This doctrine, whilst causing no difficulty in establishing or refuting the existence of liability under Article 85 when all the undertakings concerned are within the Community, is not without its own difficulties if applied to parent companies outside the Community. The key to this problem is the emphasis which the Court has placed on the corporate structural relationship, i.e. the ability to control between a parent and subsidiary company regardless of whether this control has actually been exercised. It is surely more correct that analysis should focus on what has actually happened with regard to the operation of control by the parent over its subsidiary company. There are many borderline situations where it is far from clear whether the so-called ability to control in a legal sense (*de jure* control) actually confers a real commercial ability to control the policy of the subsidiary company (*de facto* control). In each case, all one can do is carry out a careful analysis of the facts.

If there were a company with shareholdings held by different parent companies of say 40, 30 and 30 per cent respectively, an analysis would clearly need to be done as to whether control is exercised by (a) the parent company holding 40 per cent or (b) by those collectively holding 60 per cent or (c) by a combination of one of the 40 per cent and one of the 30 per cent companies. If such an analysis would be appropriate, as even the Commission would presumably acknowledge, should it not be equally so for the subsidiary company whose share capital is allotted between different owners operating with say percentages of 55, 25, and 20 per cent? The company having an equity holding over the 50 per cent mark normally has the right and the ability to exercise control of its subsidiary, and in any individual case could perhaps reasonably be presumed to have done so. The burden of proof, however, should be rebuttable if the facts

[2] See Ch. 6, pp. 90–1.

justify it (see n. 7); if, therefore, a subsidiary in which a non-EC parent held 50 per cent or more was unusually left alone by its parent, in order to make its own policy, then no finding of presence or activity within the Community should be made against its parent company solely on the grounds of deemed control, i.e. as the result of the group economic unit. The emphasis in the use of this principle should be on actual exercise of control, which in most cases will be clearly illustrated by telexes, memoranda, and correspondence passing between parent and subsidiary, the status and function of persons operating the affairs of the subsidiary and evidence of the independence or otherwise of actual commercial policies adopted by it. Clearly, if the parent company can be shown itself to have determined such policy for its subsidiary, it will *by its own actions* have been engaging in business in the Community utilizing the subsidiary for this purpose, and in this situation the mere fact of separate corporate identity should not protect it from the jurisdiction the Community seeks to extend. Support for this case-by-case approach can be found in *Metro/Saba* (no. 2)[3] where the Court of Justice refused to assume for the purposes of Article 86 that the market share of Saba should have added to it the market share percentages of other companies in the Thomson-Brandt group to which it belonged, Metro having failed to show that Saba and the other Thomson-Brandt companies co-ordinated their marketing strategies.

Similar situations also arise with regard to Article 86 where the chief authority is the *Continental Can*[4] case in which the European Court followed the earlier judgment in the *Dyestuffs* case in its approach to the exercise of jurisdiction over foreign companies with subsidiaries actively trading in the Community. The facts of this case were considered in Chapter 20, dealing with mergers, and it will be remembered that Continental, a United States company, held 85 per cent of the shares in its German subsidiary which held a dominant position in the German market for the supply of tins for meat and fish products and for metal closures other than crown corks. The case arose when Continental wished also to acquire a Dutch company, which made all the tins for meat and fish products and half the metal closures in Holland. There was at all times active involvement by the parent company in the plans made for this proposal, it being agreed that the company to take over the Dutch company would be a Delaware subsidiary of Continental, to be known as Europemballage Corporation, and to which would be transferred Continental's interest in the German subsidiary. Continental then

[3] Case 74/84 [1986]ECR 3021: [1987] 1 CMLR 118.

[4] Case 6/72 [1973] ECR 215: CMLR 199. The important paragraph in the Court's decision is paragraph 16, where the Court states 'it is certain that Continental caused Europemballage to make a takeover bid . . . and made the necessary means available for this . . . this transaction . . . is to be attributed not only to Europemballage, but also to Continental.'

arranged that this new Delaware subsidiary made an offer to the Dutch shareholders, having provided it with the funds to make this offer, which went through successfully. The Delaware subsidiary could not have made the offer without the arrangements made and financial support given by its parent company.

The Court found no difficulty in attributing the conduct by its US and German subsidiaries within the Community to Continental as parent company, but the analysis of the action of Continental focused clearly on what it had actually done to promote the takeover rather than merely its legal rights of control. It is suggested that this is the correct approach and that the extent of the activity of Continental was sufficient to place it in the position of itself 'doing business' within the Community, either through its own actions or at least through the directed activities of its subsidiary companies.

The second ground for the extraterritorial claims of the Commission is that under the terms of Articles 85 and 86, any person, company, or undertaking responsible for actions coming within the terms of these Articles, is liable to have proceedings brought against it (regardless of its nationality, place of incorporation, domicile, residence, or other circumstances) provided only that its actions can be shown to have had the necessary effects required by those Articles on trade between Member States. As has been pointed out, the Commission in its published decisions and Annual Reports does not make a clear distinction between the two quite separate issues which this claim raises: first, the existence of the right of the Community itself to extend extraterritorial jurisdiction against such undertakings. This is a question of international law which cannot be determined simply by reference to the terms of Articles 85 and 86 or indeed any other provision of the Treaty. A subsequent and quite separate issue is whether the obligations of the Commission, as the executive arm of the Community, under the Treaty of Rome require it in applying and enforcing the Articles of the Treaty to take a particular form of action against undertakings that under international law are within its jurisdiction. The powers of the Community cannot, however, be expanded with respect to undertakings outside its jurisdiction under international law simply because of the internal allocation of duties between the institutions of the Community.[5]

The historical development of this important issue begins so far as the

[5] See e.g. D. Edward (appointed the United Kingdom judge on the European Court at the beginning of 1992). 'The Practice of the Community Institutions in Relation to the Extraterritorial Application of EC Competition Law', in R. Bieber (ed.), *The Dynamics of EC Law* (Nomos: Baden-Baden, 1987), pp. 355–74. On the other hand, the German influence in support of the 'effects doctrine' is strong: see the German Law against Restraint on Competition. Sec. 98(2).

Court is concerned with the *Dyestuffs*[6] case. ICI was alleged to have assisted in the concerted practices relating to price increases for various dyestuffs by giving appropriate price instructions to its wholly owned subsidiary incorporated in Belgium, although itself as a parent company demonstrably neither directly present in nor trading with the Community. The arguments relied on by the Commission in support of the imposition of a fine on ICI rested on three separate though related bases; first, that ICI by its actions, notably by the giving of such instructions for performance by its subsidiary had *actually been engaged* itself in the concerted practices, treating the subsidiary company as if it were simply the agent of the parent. The second basis claimed was that ICI was present within the Community *by reason of the corporate control* it was entitled to exercise over its subsidiary (the 'group economic unit' argument), and thirdly, because in any case the actions taken by ICI had *produced effects* within the Common Market.

The opinion of Avocate-General Mayras in this case interestingly was in favour of acceptance only of the third argument. He conceded that under international law the Community could exercise jurisdiction outside its territorial bounds, but based its right to take action simply on the effects doctrine for the application of which he laid down three conditions, namely that the effects must be (a) direct and immediate, (b) reasonably foreseeable, and (c) substantial. It did not extend to enable the Community to implement 'coercive measures or indeed any measure of enquiry, investigation or supervision outside the territorial jurisdiction of the authorities concerned where execution would inevitably infringe the internal sovereignty of the state on whose territory they claim to act'. The Court was careful to avoid either specific acceptance or rejection of the findings of the Advocate-General but based its decision against ICI on the second argument raised, namely that of the 'group economic unit' theory saying:

> The fact that a subsidiary has separate legal personality is not sufficient to exclude the possibility of imputing its conduct to the parent company. . . . In the circumstances, the formal separation between these companies, resulting from their separate legal personality, cannot outweigh the unity of their conduct on the market for the purposes of applying the rules on competition.[7]

The Fourteenth Annual Report reviews further the application of the effects doctrine within the context of the important recent cases of *Zinc*,

[6] Cases 48 to 57/69 [1972] ECR 619: CMLR 557. See Ch. 9, pp. 155–7.

[7] [1972] ECR 662–3: CMLR 629. The 'group economic unit' concept was further confirmed soon afterwards in both *Centrafarm BV* v. *Sterling Drug Inc.* Case 15/74 [1974] ECR 1147: 2 CMLR 480, and *Istituto Chemioterapico Italiano and Commercial Solvents Corporation* v. *Commission* Cases 6–7/73 [1974] ECR 223: 1 CMLR 309. Advocate-General Warner at ECR 264, 1 CMLR 321, stated that it is difficult but not impossible for a subsidiary to rebut the presumption of *de facto* control by its parent.

Aluminium, and *Woodpulp*.[8] The attitude of the Commission can be summarized as follows:

> The *Aluminium* and *Woodpulp* cases illustrate the Commission's assumption of jurisdiction over non-EEC undertakings when the activities of those undertakings have a direct and appreciable effect on competition and trade within the EEC. This reflects the policy, which is essential in view of the realities of modern world trade, that all undertakings doing business within the EEC must respect the rules of competition in the same way, regardless of their place of establishment.[9]

It is still, however, to be doubted whether the Commission's decisions in any of these cases need the support of the 'effects doctrine'. On analysis of the facts in these cases, it seems that the parties which were the subject of the decisions were all doing business within the Community itself in some way, either directly or through instructions given to subsidiary companies, branches, or agents.

The Commission's own decision on the issue of jurisdiction in the *Woodpulp* case left ambiguities as its wording was unfortunately vague. Much, therefore, was expected from the European Court on the outcome of the appeal by the large number of companies against whom the original decision had been rendered. The Court's decision on the jurisdictional aspects of the case alone was given in September 1988.[10] The appellant companies had argued that the Commission's decision was based on the 'effects' doctrine on the basis of the economic consequences within the Common Market of conduct restraining competition alleged to have occurred outside it.

The Court's judgment was surprisingly brief. It ruled that there were two elements of conduct relevant to the analysis of any agreement which infringed Article 85 by restriction of competition in the Common Market. The first is the formation of the agreement, the second its implementation. Assuming for the moment the Commission's decision on the facts to be correct,[11] the original agreements on the pricing of the pulp took place in most cases outside the Common Market; but the implementation of those agreements by the announcement of prices from time to time by the various producers and the subsequent sales and purchases within the Common Market at those prices took place within it. It did not matter whether such sales were made direct through branches or whether through subsidiary companies or agencies.

[8] Already discussed in Ch. 9, pp. 158–60.

[9] 14th Annual Report, p. 59. This paragraph is also unconvincing, since whereas the first sentence is referring specifically to the 'effects doctrine' the second refers to a quite separate (and far less contentious) point relating to undertakings merely doing business *within the EC*.

[10] Cases 89, 104, 114, 116, 117, 125–9/85 [1988] ECR 5193: 4 CMLR 901. Judgment on the substantive issues is likely to be given in 1992.

[11] [1988] 4 CMLR 932.

In its judgment the Court deliberately avoided specific reference to the 'effects' doctrine. The opinion of Darmon A.-G. had been that under international law it was permissible for the Commission to take jurisdiction over the agreements so long as they had effects within the Community that were substantial, direct, and foreseeable. It is clear that the Court did not accept this opinion in its entirety and did not refer to it in the judgment. What is unclear from the Court's judgment is the degree to which the analysis of 'formation' and 'implementation' reaches the same result as the 'effects' doctrine. It is likely that in the majority of the Commission's cases jurisdiction will not be affected by the difference in the approach between the Advocate-General and the Court, though it may still provide a fertile source of argument in the more borderline cases. The increasing likelihood of concurrent and potentially conflicting jurisdiction over cases involving economic activity in both the Community and the USA led in September 1991 to the signing of an agreement between them entitled a 'Competition Laws Co-operation Agreement'.[12] Under this notification can be given of enforcement activities which also affect the other and information exchanged to facilitate the effective application of both Community competition law and US antitrust law. Co-operation and co-ordination is thus to be encouraged, within the limits of their respective resources. By prior consultation the parties will seek to avoid the risk of conflict over enforcement activities. Although the terms of the Agreement are broad they are subject to important qualifications, in particular that neither party is required to alter its existing laws nor to provide the other with information if this would be prohibited by the law of the party possessing it.

2. Relationships of the Competition Policies of the EC with its International Trade Policies

The previous section has emphasized that the application of Articles 85 and 86 cannot realistically be considered as applying simply to undertakings incorporated or resident within the Community but may often affect undertakings incorporated, and with their main areas of operation, outside the Community. It is equally important to appreciate that competition policy, far from being applied in isolation, has at all times to co-exist with the other policies of the Community, including its trading policies towards the USA, Japan, and other trading blocs and nations.

For the most part, the creation and implementation of the trade policies

[12] [1991] 4 CMLR 823. France has challenged the *vires* of the Commission to make this agreement by a case taken to the European Court (Case C-327/91).

of the Twelve do not have direct effect on the work of DG IV. These trading policies are governed largely by the obligations of Member States under the General Agreement on Tariffs and Trade (GATT), in respect of many of whose activities the Community represents its Member States. Member States must conduct international trade on the basis of the general rules laid down in that Treaty; and they are permitted in particular to take all necessary steps to prevent unfair trading practices, notably dumping by way of export sales into the Community from outside of goods at prices which are, after adjustment for transport costs, found to be lower than those charged on the exporters' home market. Both the 1974 *IFTRA Virgin Aluminium*[13] case—and the 1985 *Aluminium*[14] case—emphasized the basic principle that Article 85 itself does not permit an agreement or concerted practice to be put into operation simply as a response to dumping by third parties, even if the dumping itself be proven rather than merely alleged. Such remedies against dumping have to be carried out by the Commission itself through DG I, DG IV having no responsibility for such measures.

Anti-dumping procedures have some elements in common with proceedings under Article 86, but in reality the resemblance is more superficial than real. In an Article 86 investigation of predatory pricing designed to strengthen the market position of an undertaking already with a position of dominance over competitors, the emphasis is on the relationship between costs and prices and on the margin between them. Only if the margin is so low or non-existent as to suggest that an undertaking has abandoned its normal commercial principle of making profits in favour of the purpose of elimination of competitors, by driving prices temporarily down to a level unprofitable for all, would an adverse decision under Article 86 seem likely. On the other hand, the scope of an anti-dumping enquiry is narrower, being limited to an assessment of the margin between the normal value at which the relevant goods are sold in their home market, against the actual export price for the goods sold into the Community. This actual export price, however, may be replaced by a 'constructed export price' in certain instances, e.g. where the sale is not made in the first instance to an individual buyer or where goods are not resold into the Community in the same condition as that in which they were exported.[15] Before an anti-dumping duty can be imposed to cancel the advantage obtained by sales at a figure below the domestic price level,

[13] [1975] 2 CMLR D20. [14] OJ [1985] L92/1 [1987] 3 CMLR 813.

[15] The constructed export price is calculated by taking production costs and adding a reasonable margin of profit so that there is in such cases some measure of similarity between this calculation and that to be made in the review of predatory pricing cases under Article 86.

a material injury must be shown to undertakings within the Community, a further requirement not found in Article 86 cases.[16]

Many anti-dumping enquiries are resolved without the need for formal decision by the Commission. It may instead obtain undertakings as to revision of prices, written statements filed with the Commission by the relevant exporters whose prices into the EC have been challenged. The objective of these undertakings is to satisfy the Commission that export markets are not being subsidized by home markets, and for this reason invariably the Commission is concerned as to whether the non-EC entity has adopted a uniform approach to pricing in both markets. At the same time, in deciding whether to accept these undertakings, the Commission has to satisfy itself that the price levels of the EC markets will not be immediately increased as a direct response to the price revision undertakings obtained. In this way, the Commission may find itself unwillingly performing a role as the arbiter of the 'reasonable price' in respect of a range of EC imports as well inevitably as for the competing domestic products. This can lead to a state of affairs not far different in its effect from a simple agreement as to price. Naturally, DG I in making its decisions attempts to avoid a situation where the effect of its order is to reduce competition between the foreign suppliers.

If, of course, EC manufacturers themselves export products to non-member countries, they are themselves at risk that equivalent action will be taken against them, both on the grounds of dumping as well as under the claim for 'countervailing duties', made when it is alleged that the price of goods exported had been reduced through the intervention of government subsidy. This leads naturally to a consideration of whether Article 85 would apply to any agreements or concerted practices entered into by such exporters in order to fix directly or indirectly prices or trading conditions or the allocations of territories or exports. The essential question for the Commission here is whether the arrangements will themselves have an effect on trade within the EC, whatever their effects may be on other markets outside the EC. It is true that the law in most Member States treats such export cartels more generously than the equivalent cartels in domestic markets since it is not the interests of such Member States which tend to suffer as a result of the arrangements, though of course the

[16] The Regulation governing anti-dumping proceedings by the Commission is 2423/88. For further details of the procedures adopted, see I. Van Bael and J. F. Bellis, *International Trade Law and Practice of the European Community: EEC Anti-dumping and other Trade Protection Laws* (CCH Editions Ltd., 1985). The Community has also adopted in 1984 Reg. 2641/84 to enable it to respond to any 'illicit commercial practices' by third countries causing injury to undertakings in the Community, and to obtain withdrawal or modification of them. For a spirited account of the tensions between competition and Trade policy in the Community, see J. H. J. Bourgeois, 'Antitrust and Trade Policy: A Peaceful Coexistence' (1989) 17 *International Business Lawyer* 58, 115.

attitude of the law of the importing country is unlikely to be as generous.[17] One does find examples in both Commission and Court decisions of agreements which have been largely designed to control prices in export markets and in a number of these, including the *Sugar Cartel*[18] case, the decision has been that, although the agreement related to export markets, its effect would not be limited to such markets and would inevitably 'overspill' into the EC.

The number of such export cartels has fallen since their peak was reached between the First and Second World Wars, and in particular since the Treaty of Rome, though the rate of reduction has apparently slowed up in the last ten years. This reduction has happened, although export cartels are often encouraged by individual Member States within the Community notwithstanding the competition policies either of that State or of the Community generally. The reasons why countries seek to strengthen export trade by allowing domestic firms to make these arrangements have been referred to in the recent review of competition and trade policies published by OECD.[19] They are stated as including:

> Interfirm co-operation to achieve scale efficiencies by pooling personnel, equipment or other resources or to spread high risks, capital outlays or demands on capacity imposed by export activities may be needed to enable medium and small firms to engage effectively in trade. Further, such arrangements may serve as a mechanism to collectively acquire and share the special skills and knowledge needed to successfully export to foreign markets such as where restrictions on imports distort export trade. In some case, export cartels may also be used to countervail joint buying practices in the importing country. To the extent that these policies enable firms to export that would otherwise not be able to do so, national export cartels may increase competition in foreign markets.[20]

In the United Kingdom, it is mandatory that export cartels are registered with the Office of Fair Trading, but provided registration is effectively completed they are exempt from review by the Restrictive Practices Court as to whether or not they are in the public interest; the information on the register moreover remains confidential.[21] In Germany, on the other hand, export cartels can be specifically permitted provided that the restraints within them are necessary to ensure the desired regulation of competition, do not violate principles of international law, do not lead in the domestic market to substantial restraints of competition, and do not

[17] Compare, for example, the respective attitudes of the USA and EC to export cartels alleged to have an effect on price levels in Western European markets in the *Woodpulp* case [1985] 3 CMLR 474.

[18] Cases 40–48, 50, 56, 111, 113, 114/73 [1971] ECR 1663: [1976] 1 CMLR 295.

[19] *Competition and Trade Policies: Their Interaction* (Paris: OECD, 1984).

[20] Ibid. 31.

[21] This is the combined effect of Sections 24 and 28 of the Restrictive Trade Practices Act 1976.

represent an abuse of a dominant position. Only few such export cartels have been approved and notification in full to the Federal Cartel Office is mandatory before they can be put into effect.[22]

In the United States, special legislative provision has also been made for the benefit of similar export activities. The Webb Pomarene Act of 1918 provides a limited exemption from the Sherman Act for export trade associations formed for the sole purpose of engaging in export trade provided that the acts of the association do not restrain trade within the United States or the trade of any domestic competitor of the association; in addition the trade association must not make any agreement or do any act which enhances or depresses prices within the United States or restrains trade in the United States. Moreover, the 1982 amendment to the Sherman Act widened the effect of the 1918 Act by narrowing the application of the Sherman Act to export cartels.[23]

The position, therefore, in general is that countries regard with a more tolerant eye restrictive agreements involving exports from them than they would view agreements applying similar restrictions to imports into them. An import cartel is an agreement between undertakings made for the express purpose of co-ordinating the importation into one country of particular goods or services. They are often the means whereby local undertakings seek to combat the market power of export cartels with their favoured position in the exporting country. It appears, however, that frequently import cartels within the Community do not benefit from the same range of exemptions afforded to export cartels, since the effect on the competitive structure of national markets is normally more obviously damaging. On some occasions, however, exemption may be possible in Member States for agreements where the importers can show that the agreements made are necessary in order to obtain a balance of bargaining on fair and reasonable terms with a dominant supplier of the product.[24] Import cartels appear to be found more often in connection with commodity and raw material markets than with manufactured products.

So far in discussing export and import cartels, we have assumed that

[22] *Competition and Trade Policies*, p. 33.

[23] The Sherman Act no longer applies to conduct involving trade or commerce (other than import trade or commerce) unless it has a direct, substantial, and reasonably foreseeable effect on domestic trade or commerce or on the export trade or commerce of a person engaged in it in the US.

[24] See for the position under UK law the Restrictive Trade Practices Act, Section 10(1) (*d*) where this argument may justify a 'gateway' for an exception to the normal presumption that registrable restrictions are against the public interest, and also the *National Sulphuric Acid Association Agreement* [1963] LR 4 RP 63; 3 AER 73; and [1966] LR 6 RP 210; also the Commission's decision in [1980] 3 CMLR 429. The German law against restraints of competition also protects import cartels in such situations (see Section 7). An English translation of the German law can be found in *German Antitrust Law*, by R. Mueller, M. Heiderheim, and H. Schneider, 3rd edn. (Frankfurt: Fritz Knapp Verlag, 1984).

they have been organized simply by commercial undertakings. It is, however, true that in many cases there is some element of government involvement although this does not necessarily mean that the agreements are thereby removed from the scope of Article 85. The export cartel is sometimes the means under which a *voluntary export restraint* is implemented, this being often in response to the need claimed by the potential importing country (whether an EC Member State or not) to protect domestic industries during difficult transition periods, although sometimes what are intended to be transitional arrangements tend to have an unexpected element of permanence.[25] If the voluntary export restraints are based on a formal understanding between governments, they are usually classified as *Orderly Marketing Agreements* (OMA). If the arrangements are entered into under a direct order or mandate from the government, neither Article 85 nor 86 can have application, since agreements between parties which have been imposed upon them do not fall within the reach of Article 85 and actions imposed on an undertaking by governmental authority cannot constitute an abuse under Article 86.

With some products, OMAs have been openly entered into between governments and such measures are outside the scope of competition rules. A number of recent examples may be found including those relating to steel and video recorders. Other restraint agreements, however, although receiving government encouragement, have been sought to be implemented by private arrangements. The Commission first publicly dealt with this issue in an opinion issued on 10 November 1972 headed 'Opinion relating to imports into the Community of Japanese products covered by the Treaty of Rome'.[26] The opinion warned individual undertakings against involvement in agreements relating to voluntary restraint, pointing out that these would be covered by Articles 85 and 86. If such agreements were entered into, application for exemption under Article 85(3) would be essential. Two years later, the Commission issued its first decision relating to this kind of agreement in the *Franco-Japanese Ballbearings*[27] case. The agreement here, based on the evidence both of letters and of minutes of various meetings, was that Japanese ballbearing manufacturers had agreed to increase their prices to be charged in France to match those charged for equivalent products by a French manufacturer. The Commission in its decision ruled that such an agreement was incapable of exemption under Article 85(3), owing to lack of any consequential benefit to consumers.

In order to remain outside the application of Articles 85 and 86, the

[25] For a detailed discussion of the long-standing VER restricting the import of Japanese cars into the UK, see the MMC Report on New Motor Cars (1992) Cmnd. 1808, ch. 6, pp. 73–8, and ch. 13, pp. 370–1 and 388–90.

[26] JO [1972] C111/13.

[27] [1975] 1 CMLR D8.

measures taken would either have to be taken in pursuance of specific trade agreements (OMAs) between the EC (exercising its right to make commercial policies) and Japan, or clearly and unilaterally imposed by mandatory sanction on Japanese undertakings by the Japanese authorities. Even then, any additional agreements or concerted practices designed to improve the working of the arrangement but going beyond its mandatory official terms would itself constitute a notifiable agreement. Mere authorization moreover or informal 'blessing' of arrangements would again be insufficient.[28]

A further Commission decision followed in 1975 in the *Preserved Mushrooms*[29] case where a written agreement executed in Taipei between five of the main French producers of preserved mushrooms and an export association of Taiwan mushroom packers, providing for the sharing of the German market between them as well as for the setting of annual quotas and price increases for exports of preserved mushrooms, was held in breach of Article 85(1).

In summary, therefore, DG IV remains aware that competition policies have to be enforced alongside the Community's trading policy as well as the other policies for which the Commission is responsible. It recognizes that on occasions when the adjustment process is proving difficult for industries within particular Member States, some slight 'safety valve' allowing some relaxation of the strict application of the rules of competition may have to be permitted, to allow adjustment on a temporary basis. This occurs in the context of voluntary export restraints just as with State aids, and it is ultimately the task of the Court to draw the line between, on the one hand, the type of aid or voluntary restraint which can be permitted on a limited basis and, on the other, those categories which offend the basic principles of competition too substantially to be permitted, or, alternatively, appear to be offering a permanent restraint on freedom of contract rather than the temporary improvement of a difficult situation. Particular suspicion will naturally fall on agreements involving both the foreign potential exporter and the local potential importer rather than on those agreements to which only firms from one category are parties.

The Commission nevertheless has shown itself sensitive to the political importance of not being too inflexible with regard to such problems. With the benefit of experience through the cases such as those discussed above, the effect of the competition rules on international trade seems not as clear cut as might have been originally expected; a softening of the rigid rules is available on a limited basis where the circumstances appear to be particularly difficult for individual undertakings within Member States.

[28] Compare the cases referred to in Ch. 23 such as *Cognac*, *Armagnac*, and *Aluminium*, which established this same principle in other circumstances.

[29] [1975] 1 CMLR D83.

The Commission has had to accept that its main concern may be the 'art of the possible', and that this may be the only way of ensuring the allocation of sufficient resources to DG IV in the long-term interests of the implementation of competition policy. On the other hand, the Commission is equally aware of the opposite danger, that of allowing exemption to be given (even on a temporary basis) in too many cases where political and trading considerations argue for it, so that the carefully limited basis for the exemptions under 85(3) are watered away to the point where it no longer commands sufficient respect. It is this tightrope that the Commission has to tread in its treatment of these difficult issues of international trade when they come into conflict with the central principles contained in Articles 85 and 86.

PART III

Conclusions and Prospects

To be in hell is to drift: to be in heaven is to steer.

George Bernard Shaw, Don Juan in *Man and Superman* (Act 3).

On ne gouverne bien que de loin, on n'administre bien que de près. (To govern properly, you need to be some distance away, to administer you have to be close at hand.)

attributed to Napoleon Bonaparte I.

Was du ererbt von deinen Vätern hast, Erwirb es, um es zu besitzen! (That which thy fathers have bequeathed to thee, earn it anew if thou would possess it.)

Goethe, *Faust* (Nacht).

25

Thirty-Five Years of the Treaty of Rome

Over the thirty-five years since the Treaty of Rome was signed Community competition law has developed into a mature system and it is, therefore, right in this final part of the work to look back at its development. There is, of course, in reality no perfect occasion for such a review since development of the law and its implementation proceeds on a gradual, unpredictable, and frequently interrupted basis. Nevertheless, after such a period, both the competition policy of the Community and the institutions which are responsible for it may be assumed both to have achieved a measure of maturity and to have reached a point of development when it is both necessary and valuable to provide a general assessment of its achievements. In this final part, therefore, such an assessment of the competition policy of the Community will be attempted, in order to see if the high hopes of the founders of the Community as described in Part I of the book have been fulfilled; in the following chapter, we shall look ahead to possible changes in the responsibilities and priorities of the Commission and DG IV over the next decade.

The framework of the Treaty of Rome sets out the main objectives and tasks of its institutions; these tasks comprise principally the promotion, throughout the Community, of economic growth and expansion, an increase of its stability and living standards, and the maintenance of close relationships between Member States. The means whereby the objectives were to be achieved were a number of common policies laid down in Article 3 including, of course, in paragraph (*f*) 'the institution of a system ensuring that competition in the Common Market is not distorted'. The characteristics of this system would necessarily have included:

(*a*) the establishment of primary rules, both substantive and procedural, of sufficient width, clarity, and certainty to ensure a reasonable degree of compliance in all economic sectors covered by the policy, which in turn could provide a stable foundation for business development, to apply equally to horizontal and to vertical relationships;

(*b*) the consistent application of competition policy throughout the geographical area of the Community, actively carried out not only by the institutions of the Community but by Member States themselves, to ensure that the implementation of the relevant provisions of the Treaty was both legally consistent and complete;

(*c*) the even-handed application to both Member States and undertakings within the Community of the rules comprised in this system, within the framework of the Treaty and regulations made under it; and

(*d*) acceptance by the Council of Ministers and Member States that the human and financial resources required for the effective implementation of the system by relevant Community institutions should be adequate for the considerable tasks involved.

Any attempt to make a balanced assessment in this way must, of course, take into account not only periods of success and forward movement but also of difficulties and problems, both internal and external. The Commission itself has certainly over this period faced considerable difficulties in both the economic and political sphere. In particular, the economic background against which it has had to operate has undergone disruption and substantial change. The early years of the Six, even up to the start of the 1970s, can be seen in retrospect as a comparatively calm period remarkable for steady economic growth throughout much of the territory of the Community. The continuous growth of both the internal and external trade of the Community and rising standards of living in its individual Member States (other than a few clearly identifiable regions such as the southern part of Italy) contributed to the willingness of Governments of Member States to give ready acceptance to the grant of extensive powers to the Commission in this area of competition policy, both those relating to procedures and to the application of substantive law. Even during the period of the mid-sixties when the Council remained largely inactive as the result of political deadlock, the powers already granted to the Commission and the decisions that began towards the end of this time to come from the European Court of Justice ensured that in this area at least progress continued. At the same time, a relatively relaxed approach was taken by the Community and its Member States to State aids which were not regarded initially as having much influence on the competitive structure of markets within the Community.

All this, however, changed during the early seventies. The sharp increase in oil prices in late 1973 led to nearly a decade characterized both by stagnation in economic growth and an acute increase in mutual sensitivity between Member States concerned as to the effect of government involvement in sectors such as steel, textiles, and shipbuilding, where the economic consequences of the new era were most keenly felt. It was, of course, also unfortunate, as many commentators have pointed out, for all Member States that the expansion of the Community first to nine and then to ten Member States took place against such a background, thereby accentuating problems as well as opportunities for the enlarged Common Market.

Nevertheless, the development of competition policy continued steadily over this and the following decade. Under active Competition Commissioners in Peter Sutherland (for the period 1985–8) and Sir Leon Brittan (from 1989) DG IV was able to place greater emphasis on the positive aspect of its responsibilities. Special attention was paid to liberalization of air transport and some progress made in the enactment of both procedural and substantive regulations within whose framework progress could be made (if slowly) towards a greater freedom for airlines to fix their own fares without having on all occasions to obtain Government approval. Maritime transport was also given a group exemption and a number of cases brought against price fixing within this sector. After their early neglect there was at last conspicuous activity in the banking, insurance, and financial sectors and a number of important decisions rendered.[1] Perhaps most important the Commission used its powers under Article 90(3) to put pressure on Member States to open up their telephone terminal equipment and services in order to establish a functioning market within the Community for large-scale information technology. The Court's decision in March 1991[2] largely upholding the terms of the Directive and confirming the use by the Commission of such powers was an important step in this direction. Greater pressure was also applied by the Commission on Member States under Article 90 in connection with a variety of other activities including postal services and employment agencies and appropriate support received from the European Court.

One method of following the administrative and policy-making progress of DG IV, through *inter alia* relevant Commission and Court decisions, is by reading the relevant Annual Reports on Competition Policy; these are normally published by the Commission in the latter part of each year to cover the previous calendar year's activities. Whilst these Annual Reports are valuable in their coverage of the events of that particular year, the calendar year is itself necessarily an artificial period; inevitably, therefore, they include reference to many developments that are themselves incomplete. To gain a better idea of the progress of DG IV, it is more helpful, therefore, to read the reports for several years consecutively. Moreover, the Annual Report on Competition Policy suffers from the inherent defect of having been written by the very officials whose administrative actions form the principal subject of the report. An objective assessment of their work has normally, therefore, to be left either to writers of textbooks or of articles in legal periodicals or to the more ephemeral comment of daily and weekly newspapers. Whilst Annual Reports, therefore, provide an invaluable record of tangible legislative and administrative achievements and decisions of a calendar year, they inevitably omit much that is necessary to a full

[1] See Ch. 6.
[2] *France and Others* v. *Commission* Case C-202/88 dated 19 Mar. 1991.

understanding of what has actually occurred in much the same way that the Annual Report of a public company, whilst truthfully recording the salient features of a single year's trading, may not necessarily give the outside reader (or even a shareholder) much information about vital changes in management style or boardroom policy.

Nevertheless, it may be best to begin our review with a reference to the aims expressed in the very first Annual Report published to cover the period to the end of 1971. In listing these, DG IV had, of course, already had the benefit of over a decade of experience in administering Articles 85 and 86, and special priority was accorded to the following:

(*a*) taking action with special vigour against restrictions on competition and practices jeopardizing the unity of the Common Market notably sharing markets, allocating customers and collective exclusive dealing agreements, and preventing agreements which indirectly resulted in concentrating demand on particular producers;
(*b*) with regard to systems of distribution, any degree of exclusivity permitted must not prevent distributors and consumers from obtaining goods in Member States 'on the terms customary in that State';
(*c*) heavy fines should be expected by those undertakings engaged in restrictions on competition causing serious damage to consumers' interests, and which were clearly forbidden;
(*d*) the Commission would reinforce the competitive position of undertakings by exempting, both by individual decisions and by means of a group regulation, positive forms of co-operation, particularly between small and medium-size businesses;
(*e*) the Commission did not intend to apply the prohibitions of Article 85 to restrictions on competition which had no appreciable effect on the Common Market.

The extent of the success of the Commission in carrying out these aims has been considered in detail in the various chapters of Part II, and in general within the limits of its resources the Commission has been reasonably effective. Experience does show that, while in applying Articles 85 and 86 many difficulties have arisen, the Articles themselves are not flawed nor inadequate as a foundation for the establishment of the necessary system referred to in Article 3(*f*) of the Treaty save in the area of merger control, for which a special Regulation has proved necessary. These two Articles, however, are merely the framework for the application of the detailed rules of the system and cannot themselves provide answers to many of the questions which inevitably would arise over the subsequent implementation of the detailed system of rules. It is, therefore, to the delegated legislation adopted by both Council and Commission, and to the case-law of the Commission and Court, that we must turn for these answers. It is clear that to have expected DG IV to have administered a competition law in the form contained simply in those two Articles, even as supplemented by Articles 87 to 90, would have been unreasonable given that responsibility for the development of competition policy was clearly assigned under the terms of the Treaty to the Commission and not simply to be enforced through national courts or the European Court of Justice.[3] The Commission was, of course, to be subject in making decisions to a right of review by the European Court under Article 173 for the benefit of affected undertakings. This right of review primarily applied to matters of procedure rather more than to substantive criteria, though the Court's decisions would soon make clear that in the course of such review important principles of substantive law would also have to be laid down.[4] Whether competition policy would be a question of 'steering' or merely 'drifting' was very much for the Commission itself; in particular, the key area of interpreting the basis of granting exemption under Article 85(3) whether by way of individual decision or by application to groups of agreement, rested exclusively with it, a Treaty requirement reinforced by the unequivocal terms of Regulation 17 of 1962.

Hence, only the implementation of Regulation 17 had brought about the early problem of mass notification, the major priority for DG IV became the enactment of detailed legislation to be contained in Regulations in order to lay down the exact criteria for group exemptions. This required analysis of the different kinds of agreement and the variety of clauses found within them which needed to be classified into the categories of 'harmless', 'borderline', and 'harmful', before any form of group exemption could be provided. The adoption of legislation embodying the terms of group exemptions, though sometimes a painfully protracted process, has ultimately enabled the Commission to dispense with the need for individual notification of a very large number of agreements, particularly in the area of vertical agreements such as distribution and licensing agreements. Particularly valuable have been the group exemptions of 1983/83 and 1984/83 on exclusive distribution and exclusive purchasing, which in turn built on the familiarity which the earlier and rather simpler Regulation 67/67 had gained as the result of its widespread application for the previous fifteen years. Group exemptions have also played a valuable part in enabling commercial interests to frame their agreements

[3] This policy differed from the choice made by the US when first enacting the Sherman Act in 1890. It is the decisions of Federal courts over subsequent years that have provided the corpus of legal principles and refinements by which the Act has been interpreted, without the assistance of delegated legislation in the nature of group exemptions that Article 85 for its part appears to have expressly contemplated (i.e. by its reference to exemption for 'categories' of agreement).

[4] From the decision in *Consten–Grundig* onwards, Cases 56 and 58/64 [1966] ECR 299: CMLR 418.

in a manner which allows them in effect to self-certify exemption by reason of compliance with the detailed categories of permitted clauses set out in the relevant regulation. It is noticeable, however, that group exemptions are tending to become more and more complex, not only because of the extension of the list of clauses in the relevant categories, but also by reference to the number of recitals.[5]

It seems clear, however, that the work of DG IV can in practice only be carried out if the terms of the Treaty are fully supported by such Regulations which enable the elimination from individual scrutiny of many thousands of commercial agreements falling within the approved pattern. Nevertheless, group exemptions also raise problems for DG IV. This is so, first because their original negotiation takes an extremely long time and requires a solid case-law base to provide data and justification for the content of such exemptions.[6] Only upon such a base can relevant 'white', 'grey' and 'black' lists be satisfactorily prepared. The preparation and publication of an individual draft regulation, however, then leads into an extended period of consultation and negotiation, when as we have seen relevant drafts are successively discussed with various interest groups from both individual Member States and the Community as a whole. The Advisory Committee referred to in Regulation 17[7] has to meet probably more than once to discuss such proposals and it is rare to find unanimity or even a clear majority support for the whole of a proposal, if one is to judge by the length of time that it takes for the final group exemption to emerge from the consultative process. Some changes made during this process may well be improvements or refinements for which the Commission may be grateful, but other changes may be little more than concessions to the special interests of pressure groups or requests from individual Member States.

Moreover, it is clear that once adopted the use of group exemptions also raises problems to which the Commission has not yet found a satisfactory answer. Opposition procedures have not been as utilized as much as had been anticipated and the process required for removing the benefit of a group exemption from individual agreements has proved too cumbersome to be put into formal operation save in a very occasional case. Nor is there apparently any satisfactory and published procedure, when the end

[5] Thus, the recitals in Reg. 67/67 were approximately 70 lines long, and that for Reg. 2821/71 on Research and Development only 40 lines; by comparison, 230 lines of recitals were required for the Patent Licence Group Exemption Reg. 2349/84, and even more (346 lines) in the special group exemption for car distribution (Reg. 123/85).

[6] It will be remembered that this is a point made in the recitals of Reg. 19/65. Whilst in the case of some group exemptions, e.g. those for exclusive purchasing and exclusive distribution, this requirement has been met, in other cases, e.g. that of the Research and Development Group Exemption Reg. 418/85, it appears doubtful.

[7] Established under Article 10(3) of the Regulation.

of the finite life of a group exemption approaches, for the systematic review of its past operation and future potential in the light of all relevant evidence.

These problems have been a particular feature of Regulations concerned with individual sectors, including Parts II and III dealing with beer and petrol respectively in the exclusive purchase Regulation 1984/83, and Regulation 123/85 dealing with automobile distribution. In particular, the latter Regulation in an attempt to meet the requirements of both manufacturers and distributors provides an almost complete code or model agreement, to the extent that a French court has even referred to the European Court of Justice the issue of whether the parties to such agreements are free if they wish to substitute other forms of agreement even if it involves the trouble and expense of having to seek an individual exemption.[8] The tendency of such a process in individual sectors is to force all relevant agreements in that sector into a similar mould. This result may not be totally in the interests of a competitive market. The difficulty for undertakings is that to enter into agreements that do not fit this pattern is to risk unforeseeable days in approval, especially with regard to the automobile distribution Regulation where there is no provision for the opposition procedure.

Moreover, it is apparent that the main thrust of group exemption relates to vertical rather than horizontal agreements[9] since it is generally those group exemptions dealing with vertical agreements which go into far greater detail in their 'white' lists[10] and enable a large proportion of the normal range of agreements to benefit from their existence, quite apart from the possibility of using the opposition procedure to obtain a reaction from the Commission on clauses in the doubtful 'grey' area. It is notoriously far harder for horizontal agreements, in particular those concerning joint ventures relating to both research and development and specialization, to benefit from the limited scope of those group exemptions that have been issued. This may be argued to arise mainly from the inherently greater danger to competition posed by such agreements (often made between companies which are more likely to be in actual or potential competition with each other) that must, therefore, be examined with special care by DG IV. But at least some of the difficulties and delays

[8] The answer given by the European Court in *VAG France* v. *Etablissements Magne* Case 10/86 [1986] ECR 4071 [1988] 4 CMLR 98, was, not surprisingly, that the parties were quite free to adopt their own form of agreement with the possible consequence of having to seek individual exemption.

[9] In the transport sector, however, group exemptions do relate mainly to horizontal arrangements.

[10] White lists in early Group exemption regulations are extremely brief, but have become considerably more extensive in later examples, e.g. the Franchising Agreement and know-how Licensing Group Exemptions.

which DG IV has encountered in providing criteria for such agreements arise not simply from the inherent nature of the subject matter, but from the past policy decisions of the Directorate-General to devote an unduly high proportion of its resources and energies to consideration and investigation of vertical relationships whilst too little by comparison has been devoted to the problems of horizontal agreements. The Commission has undoubtedly tried, however, in recent years to redress the balance and has brought a number of cases against large groups (notably in the chemical industries) which have led to substantial fines being imposed under both Article 85 and 86.[11]

The Commission has also found in recent years that it has become the victim of rising expectations, in that it is being required to give greater weight at the same time both to the well-established 'negative' aspect of competition policy and also now in more cases to the 'positive' requirements and aims of such policy. This has become a central issue not only for the Commission but for all national competition authorities. Legislative aims in framing competition policy are usually directed at the prevention of agreements or practices or mergers or the growth of monopoly power regarded in themselves as undesirable, and the early targets for such legislation are usually obvious. Moreover, there is substantial public support normally found for this control from which direct consumer benefit is assumed. The Sherman Act in the United States is the best example of such legislation and other examples may be found in post-war French legislation[12] covering refusals to deal as well as in the restrictive trade practice legislation of the United Kingdom. Such legislation provides sanctions which may be simply civil, rendering void and unenforceable agreements that fall within the prohibition, or involve alternatively criminal penalties. The enforcement of such negative aims by competition authorities is comparatively straightforward so long as the authorities themselves have been given adequate powers of investigation and enforcement.

Once, however, a competition authority has achieved a reasonable degree of success in implementing such forms of negative control, it is often then realized that such control alone may not in the long run be

[11] On appeal to the Court of First Instance the Commission's experience has been mixed. For example in the *Polypropylene* case (Cases T1 to 15/89, decision of 10 Mar. 1992) fines totalling 23.5 million ecus on a large number of companies were upheld or only slightly reduced. On the other hand the decision and fines in the *PVC* cases (Cases T 84, 85, 86, 89, 91, 92, 94, 95, 98, 102, and 104/89, decision of 27 Feb. 1992) were struck down because of procedural defects in the Commission's decision. It is possible that other existing decisions of the Commission in such cases may be open to challenge on similar grounds.

[12] The French Government has now enacted a new Competition Act (ordinance no. 86–1243 dated 1 Dec. 1986) which replaces the earlier legislation on, *inter alia*, refusals to deal, though retaining much of its content.

sufficient to satisfy public demand for the achievement of tangible results. It then has to become the aim of competition authorities to combine their negative responsibilities with the positive need to provide a framework within which competition itself in all its many forms and varieties can be encouraged. Competition has to be encouraged not only in structural terms, by ensuring that there are sufficient undertakings capable of competing with each other within individual product and geographic markets, but also by giving support to the agreements and practices that undertakings may desire to enter into either in the avowed interests of fostering competition in development of new products (especially in the area of advanced technology), and also in adopting new methods of distribution, e.g. franchising. This pressure for positive competition policy is, of course, particularly acute if the authority itself has the jurisdiction to grant exemptions from the basic prohibition contained in the competition law.

The authority will soon find (as has the Commission) that it is being constantly urged by the Press, as well as by academic and professional sources, to adopt an approach that will allow those agreements and practices to continue which appear to offer long-term economic benefit or technical progress, even if in the short term it produces a reduction in competitive activity or in the number of competing undertakings. Moreover, even if powers of legal control are available over mergers that may prima facie appear anti-competitive because of the size of the participating enterprises and their potential ability to compete with each other, pressure may be applied similarly on the authority to permit an individual merger on the grounds that the merged company may itself provide the size and capacity to compete with other powerful undertakings in important international markets. The Commission has also been criticized for failing to give sufficient weight to the potential competition in Community Markets from groups based in Japan and elsewhere outside the Community. The pressure, therefore, to combine the positive and the negative elements in competition control poses major problems, as DG IV has discovered for itself.[13]

At the same time, however, the competition authority remains under existing pressure to continue to apply effective negative control in the traditional way; the maintenance of an appropriate balance between the two policies may be hard. These considerations apply with particular emphasis to the Commission, especially in view of its extensive powers to adopt

[13] The difficulty of combining them has been expressed by the Director General of Fair Trading in the UK when he said: 'To obtain the perfect policy which manages to combine the rigorous impact of market discipline with the best interests of the public—in terms of a commercial and industrial structure in the U.K. which is competitive both internally and externally—is possibly akin to the search for the Holy Grail' (Sir Gordon Borrie, 164 *Lloyds Bank Review* 8 (Apr. 1987).

Regulations and Directives to enquire into whole sectors as well as individual undertakings throughout the Community's wide geographical area, and to administer the law over a wide variety of different market sectors, combined with a wide range of procedural and investigatory powers and ultimately the authority to issue a decision binding on the undertaking subject only to right of appeal under Article 173 to the Court.

The Fifteenth Annual Report, looking ahead to the target for completion of the internal market by the end of 1992, spelt out the pressures to which the Commission is subject.

> It is one of the Commission's top priorities to strengthen the economy of the Community, so as to give a decisive push to growth and to improve international competitiveness. Directly linked to this goal are the consolidation of the internal market by 1992 and the creation of a 'technology Community'. These will open out new perspectives and improve the longer-term outlook for our industry. Within this economic strategy competition policy has a crucial role to play. Dynamic innovative competition, led by entrepreneurs, is the life-blood of the economy. . . . Innovation must hence be encouraged, by making it worthwhile. For this purpose, on the one hand, co-operation between enterprises needs to be facilitated and the influence of innovation should be given a certain degree of protection. On the other hand, a satisfactory level of effective competition must be maintained between innovators and between centres of research and development. In this respect, evaluation of the intensity of competition at Community level must also take into account the impact of international competition. . . . This approach guided the Commission in its block exemption regulations on research and development and on technology transfer (patent licences) which went into effect in 1985. . . . Aid for research and development appears to be an appropriate method for fostering dynamic competition. For this reason, the Commission has generally taken a favourable view of it. But there are limits which cannot be exceeded.[14]

So far the Commission and DG IV have proved more adept at handling the negative rather than the positive aspects of competition, though clearly the group exemptions have made an important contribution to establishing a pattern for distribution and licensing which should encourage inter-brand competition. More difficult for the Commission is to implement the positive aspects of policy through individual decisions, in spite of these claims of the Fifteenth Annual Report. It may be a mistake to expect too much from the Commission and DG IV in the way of positive encouragement of particular pro-competitive market structures or agreements purely as the result of such individual decisions. There will indeed always be a certain natural tension between administrators anxious about giving too liberal an interpretation of the four exemption conditions when considering individual agreements or practices and, on the other hand, the policy makers (perhaps within other Directorates-General of the Commission as well as on

[14] 15th Annual Report, pp. 11–13.

the Council) who, whilst not wishing the negative role of DG IV to be ignored, are concerned that the positive role of the Commission be also emphasized again if this means taking a more relaxed attitude to individual cases.

Nevertheless, there are clear signs in certain more recent Commission decisions noted in Part II of the book that it is well aware of the problem and a more liberal approach is now being adopted in granting individual exemptions in cases involving both distribution and intellectual property rights such as *UIP* and *Moosehead*.[15] A new approach can also be detected in recent joint venture decisions. A more relaxed approach to Article 85(1) has been adopted in the recent *Odin-Elopak* case.[16] Instead of claiming that Article 85(1) applied almost as a matter of course to the joint venture itself because of the possibility that the parent companies could one day conceivably become interested in competing with each other, and then going on to grant exemption to the joint venture under Article 85(3), the Commission gave a negative clearance to the proposal on the ground that the parent companies neither competed nor were likely to do so in the foreseeable future.

The need to ensure those joint ventures which are essentially 'concentrations' came within the jurisdiction of the Merger Regulation has meant that considerable thought has gone into the Notice issued in September 1990 which indicates the distinction between that category of joint venture and those which by contrast are merely 'co-operative'. Where the object or effect is the co-ordination of the competitive behaviour of undertakings which remain independent, then the joint venture falls into this latter category and is not subject to the Merger Regulation.

Whether the highly subtle distinctions drawn between the two forms of joint venture in the Notice are actually the most appropriate is a matter on which opinion differs; both the procedural and substantive advantages of having a classification of concentrative joint venture provides too great an incentive to the parties to seek to emphasize the concentrative elements even when this may not reflect their primary commercial interests. The Commission has indicated its awareness of the need for prompt review of co-operative joint ventures by its issue of the draft Guidelines in January 1992, setting out the detailed considerations behind its policy. Sympathetic consideration is to be given in particular to those joint ventures which create substantial new capacity or which substantially increase their parents' existing capacity provided that these do not provide a network nor strengthen an existing oligopoly. The desirability of rendering decisions within the same periods as those applicable to concentrative

[15] *UIP* [1990] 4 CMLR 749; *Moosehead* [1991] 4 CMLR 391.
[16] [1991] 4 CMLR 832.

ventures is acknowledged, at least when clearance can be given by way of comfort letter, though not if a formal decision is required.

For many years, of course, it was the absence of merger control which was the chief weakness in the Commission's armoury. Whilst it was limited to application of Article 86 in the situation covered by the *Continental Can* decision and in respect of Article 85 within the bounds of the *Philip Morris* case, the ability of DG IV to exercise any real form of control or influence over market structures was limited. The final adoption, however, in 1989 of the Merger Regulation after seventeen years of negotiation between Member States was a most significant step, even if the jurisdiction requirements of world and Community turnover were set at a level regarded as too high by some Member States as well as the Commission itself. Its successful operation to date during its early life has meant that the system of merger control is now beginning to become familiar. The number of cases coming within the Regulation has been sufficient to provide continuous activity for the Task Force operating it and adequate case-law experience for the Commission, without at the same time creating demands upon it that its limited resources could not deal with.

In general, however, in spite of this past weakness in control over mergers, it is clear that in terms of those administrative and legislative powers accorded to it by both primary and secondary legislation, DG IV has achieved much during its life so far. The question, however, still arises whether the achievements of DG IV would have been greater had the resources allocated been more generous and also whether the actual resources available to DG IV were appropriately allocated between its various responsibilities.

It always seems the case that resources allocated for expenditure on public purposes will be regarded as inadequate by those responsible for administering them as well as by those for whose benefit they are administered; whether the public purpose be health, social security, education or defence, the same principle holds good. By comparison with these major areas of public expenditure, the sums available and expended for the purposes of enforcing competition policy are small indeed. Nevertheless, problems of inadequate resources still remain for such authorities even though their direct call upon public expenditure may be relatively insignificant. Governments in Member States tend to take an ambivalent view towards their own competition authorities. On the other hand there is clear political and economic need for the existence of such authorities to deal with those clearly anti-competitive agreements, practices, mergers, and monopolistic activities whose termination or prevention can easily be shown to benefit both the consumer and the general economy. On the other hand, Governments are also concerned by the problems which competition authorities may cause by threatening to challenge or investigate activities which individual ministries may regard as their own responsibility, and where the link between the authorities' intervention on the side of competition and consequent benefit is less obvious. This ambivalence of view is likewise to be found within the European Community.

The early history of the Community, notwithstanding its small initial allocation of manpower and resources to DG IV, showed the important place which Articles 85 to 90 held within the overall aims and structure of the Treaty, as reflected also by the extent of the powers given to DG IV by Regulation 17. As, however, the Community has grown to the present number of twelve Member States, and against a background of recurring budget crises as well as the financial pressures within Member States for specific State aids and assistance for crisis cartels, the willingness of the Council to authorize increased finance or manpower resources for the work of DG IV has conspicuously failed to match its increasing importance within the extended Community. Any assessment, therefore, of the activities of DG IV must take into account that, although it has received increases in personnel over this period from its initial small beginnings, it has remained at all times understaffed and underfinanced in relation to its increasingly heavy responsibilities, including now those of implementing the Merger Regulation.

It is this under-resourcing which has been largely responsible for many of the inadequacies in its procedures to which attention is regularly drawn in the Press. In many respects, its procedures have been greatly improved since the early days, and the Annual Reports show a continuing consciousness of, and preoccupation with, the importance of minimizing the difficulties and delays that are inevitably imposed by institutional structure and bureaucracy of such an organization, not to mention the difficulties caused by the existence of nine official languages. With regard to those aspects of procedure which DG IV does control, it has simplified them to a considerable degree. Amongst these we may list the introduction of the group exemption incorporating on some occasions an opposition procedure, the streamlining of the notification procedure, the introduction of a hearing officer to enable a greater degree of formality at the oral hearing,[17] the greater availability of interim measures in urgent cases, and the continually increasing use of devices to short-circuit the need for formal decisions. These devices include informal letters of assurance known as 'comfort letters', short-form decisions in suitable cases, and a general willingness to try to settle cases on an informal basis when agreement can be reached on the deletion of particular clauses objected to in notified agreements, whose removal can then enable the remaining terms to be approved. In many areas, the extent of case-law has been

[17] Another aim of the creation of this post was to allow an objective report to be made on the substance of individual cases to be given to the Director-General of DG IV.

sufficient to establish a set of compliable rules on whose basis legal advisers can give practical advice on a large number of issues, even if as already indicated there are still areas where such advice remains considerably more difficult owing to lack of specific authority from case-law, let alone from legislation. Cases that are settled (even of some significance) may receive no more publicity than a short Press release, which is often too brief to serve as a reliable precedent. The Commission is also more prepared now to accept undertakings from individual companies rather than insisting on the making of formal interim decisions. From the viewpoint of practitioners, the main difficulty is the unpredictability of the time-scale for obtaining either an informal or a formal decision since there does not appear to be any clearly stated policy of the priorities to be accorded to particular kinds of cases. On occasion also, the quantity of decisions, and in particular the number of decisions issued in a particular calendar year, seem to be regarded as more significant by the Commission than the quality of those decisions in terms of their importance to the sectors affected and the size of the undertakings involved, or even their value as precedents.

Unfortunately, none of these procedural changes, valuable though each may be in its own way, can compensate for the unavoidable weakness of the administrative system which is operated for the Commission by DG IV. The fundamental problem is that at any one time there will be several thousand notifications of agreements sitting in files awaiting decision by the Commission, and whose status may remain in doubt for some years whilst the parties will probably continue to operate them without any degree of certainty as to the ultimate outcome of the notification (though doubtless as the decades pass with a greater feeling of freedom). This of course is not necessarily of concern to all companies which have made a notification and indeed some may well prefer not to hear anything from the Commission for a considerable period (if ever) on the assumption that they will receive a 'comfort letter' in due course and in the meantime can enjoy the benefit of freedom from fines on an indefinite basis: they may even prefer this uncertain status to being granted an exemption limited in time, possibly subject to reporting and other conditions. Other notifying companies may feel that if they later require a ruling, formal or informal, they can then apply sufficient pressure to the Commission to ensure a response. The actual number of pending cases at any one time therefore does not represent an equivalent number of dissatisfied companies unable to place reliance on commercial arrangements they have made and notified. Even so the size of the backlog of pending cases remains a problem for both DG IV and many of its industrial and commercial clients. With so large a backlog, whose number is annually confirmed by the figures of outstanding cases given in the Annual Report, the criteria for priority remain important, but in practice are rarely disclosed. Clearly, however, the strictly chronological order of notification is not all important.

Apart from the responsibilities of preparing draft regulations, notices, and explanatory memoranda, examples of which are referred to in the chapters in Part II, the workload of DG IV arises mainly from three sources: notification of agreements, the handling of complaints, and 'own initiative' investigations. One of its unsolved problems has always been the difficulty in the allocation of resources devoted respectively to complaints and notifications. Although the backlog of notifications has now been substantially reduced from the levels prevalent in the 1960s, largely as a result of the use of group exemptions, the figures published in recent Annual Reports show the numbers of agreements notified, but undealt with, still obstinately and undesirably high. It is likely that some of these agreements date back many years and may indeed have now been abandoned by the parties or overtaken by changes in circumstances. Resources should ideally have been concentrated on eliminating or at least very substantially reducing this backlog so that the Commission could operate under a predictable time-scale for its remaining and current notifications. Clearly, however, resources that should have been applied in this way have had to be devoted to the receiving and considering of a wide range of complaints that regularly reach DG IV from all parts of the Community about alleged agreements or practices, or abuses of dominant position, alleged to be in breach of Article 85 or 86. The tradition of DG IV is that all such complaints, even those apparently trivial, are dealt with as promptly as possible and inevitably this detracts from its ability at the same time to handle notifications to any kind of firm time-scale. Judgments of the Court have indicated that the Commission is required to give priority to such complaints and much time is spent in investigation into matters that may turn out in the end far more appropriate for investigation by the competition authorities in Member States, since they often involve activities in a single Member State and mainly breaches of national rather than Community law. The problem continues even in spite of the issue of regularly updated Commission Notices stating that agreements of minor importance will not be held to fall within Article 85(1) and, therefore, will not require investigation by the Commission. The difficulty for DG IV is, of course, that it is not always apparent (at least until after substantial investigation) whether the complaint relates to an agreement or practice below the *de minimis* level or above it.[18]

[18] DG IV often tries to deal with complaints by forwarding them quickly to the company concerned and making on its reply a prompt assessment of whether the complainant has a solid basis of fact and law in its support. It is also more inclined in appropriate cases now to suggest that complaints are better dealt with at national level by Member State competition authorities.

Regardless also of the resources made available to DG IV as a competition authority, its achievements can be assessed in another way. To every institution such as the Commission, opportunities are presented often by a random conjunction of events, quite separate from its own acts or decisions, enabling unexpected progress to be made in the implementation of its policies. These may metaphorically be described as 'windows of opportunity'. The structure of the Treaty and of the organization of the Commission have meant that from time to time such opportunity has arisen, mainly as the result of rulings by the Court of Justice. These cases have arisen either as a result of requests for preliminary rulings by national courts under Article 177 or, alternatively, as a result of requests for review by undertakings of the legality of decisions by the Commission, under Article 173. The chronological table at the front of this book illustrates that during the early years of the operation of the Treaty cases came only slowly, so that the first decisions of the Commission itself are not found until 1964 and the first decisions of the Court under Article 173 do not occur until the *Grundig* case in 1966. Looking at the cases decided in the years immediately after 1966, the Commission's preoccupation with vertical relationship (so often the subject of complaint by the distributor or licensee) is noticeable, especially those dealing with cases of exclusive distribution, whilst only a minority of the decisions relate to cartels, trade associations, and other mainly horizontal relationships. This preoccupation with the vertical arose directly from the early policy choice made in the course of negotiations leading up to the adoption of Regulation 17, that the range of vertical agreements for which in practice notification would be needed would be very wide. Had the mesh of the net for notification been less fine and fewer notifications thus required in the early stages of the Commission's work, more attention could have been paid at that vital time to horizontal agreements and in particular to cartels allocating national markets. As it was, it was not until the end of the 1960s that we find the Commission and the Court coming to grips with substantial horizontal cartels of the type found in the *Dyestuffs* and *Sugar* cases,[19] which of course present much greater evidentiary problems.

The assumption has often been made that this policy choice became inevitable as a result of the *Grundig* case and that the importance attached by the Court to the principle of the unity of the market made it inevitable that DG IV would choose to concentrate first on vertical rather than horizontal relationships. This viewpoint, however, can be challenged since, notwithstanding the importance of the *Grundig* case, the policy choice to concentrate on vertical agreements initially was made long before the

[19] *ICI* v. *Commission (Dyestuffs)* Case 48/69 [1972] ECR 619: CMLR 557. *Suiker Unie (Sugar Cartel)* Case 40/73 [1975] ECR 1663: 1 CMLR 295. See Ch. 9 generally on these cases.

adoption of the *Grundig* decision by the Commission in 1964. The critical policy choice was inherent in the structure of Regulation 17 itself which, coupled with the wide interpretation initially given by the Commission and Court to Article 85(1), gave only very limited measures of relief[20] of agreements from notification. The chief strength of Regulation 17 lay in the powers of investigation and decision which it conferred upon the Commission and the fact that it gave Member States both an opportunity and an obligation to co-operate in the system without giving them, however, more than an advisory status in the course of the actual decision making. The Regulation also confirmed the 'prohibition' rather than the 'abuse' theory, a decision that removed any doubts that might have remained about the initial interpretation of Article 85 but itself contributed substantially to the width of the jurisdiction taken by the Commission. Much of the Regulation, therefore, represented an opportunity presented at a crucial early time in the development of policy which gave DG IV a strong procedural base of its future activities. The weakness, however, of the Regulation was that it encouraged such a burden of notifications to DG IV that it was left unbalanced and overburdened in a way that would affect its approach not for just the rest of that decade but for the remainder of its history to date.

Nevertheless, during its first fifteen years of existence the Commission's chief windows of opportunity came undoubtedly from the Court which went out of its way to emphasize the importance that both Article 85 and 86 played in the overall scheme of the Treaty by giving broad interpretation in a number of leading cases to the essential elements in those Articles. The extent to which the Court had taken trouble to do this has been pointed out from an authoritative source.[21] After stressing that the application of Article 177 always posed particular difficulties for the Court because of the risk that the findings of the national court were wrong or did not arise from the evidence established or that the questions put to the European Court might have been inadequately phrased. Sir Gordon Slynn (then Advocate-General) pointed out that the Court nevertheless has in a number of ways given great assistance to the Commission in its task. It has treated as the priority from the outset the creation of a single market for Member States rather than focusing on issues of efficiency or even of consumer protection. To quote his words:

> It is wholly consistent with this approach that the Court took a broad view of what constitutes a concerted practice . . . that it should treat as part of contractual arrangements what on the face of it could appear to be background or contextual material . . . that it should see one individual contract as part of a network

[20] Reg. 17. Article 4(2).

[21] In an address 'E.E.C. Competition Law from the Perspective of the Court of Justice', *Twelfth Annual Proceedings of Fordham Corporate Law Institute* 383 (1985).

> rather than in isolation . . . that it should construe 'undertaking' broadly, 'relevant market' narrowly, often by reference to the nature of the product, of the end uses, or in relation to intraband rather than broader market considerations, that it should accept a narrow geographical base such as that of one or several Member States rather than by taking a broader region as the area appropriate for the relevant market; that it should define abuse in terms of its effect on competition rather than by balancing whether the allegedly abusive conduct might contribute to economic efficiency.[22]

The Court also made it clear that it would not intervene on applicants' behalf in the course of Commission investigations unless there were cogent grounds for doing so nor would it annul Commission decisions on procedural grounds unless there was either a cardinal flaw in the process or a total inadequacy of economic analysis as, for example, in the *Belgian Wallpapers*[23] case.

One must give credit to DG IV for taking full advantage of the support which it had received from the Court over the twenty-five years since *Grundig* was decided. As can be seen by a study of its individual decisions year by year, it has fully utilized the width of interpretation accorded by the Court to the principal elements of Articles 85 and 86. It has not found itself handicapped, as have some national courts, by technical or narrow interpretation of its jurisdiction and powers. Moreover, the Commission has done its best to pay attention not only to the words of the Court's decision but also to the rather broader 'signals' which the Court has apparently been trying to send to it as to the general conduct of its procedures, the *Belgian Wallpapers* case being the best example of a warning that superficial analysis would not be tolerated even if the actual decision on the facts appeared to be from the Commission's viewpoint an obvious one.

Since the end of 1989 the creation of a Court of First Instance at Luxembourg with jurisdiction over appeals in competition cases from the Commission has brought a major new factor into the situation. The Court[24] was created in order to try and reduce the burden of cases considered by the European Court and in particular it was hoped that it might be able more fully to deal with factual as well as legal aspects of the cases brought, in a way that the very nature and heavy responsibilities of the European Court in other sectors made difficult. After two years of experience, it is agreed that the standard of judgments from the Court of First Instance in competition cases has in general been of a high order and the judgments, for example, in *Tetrapak* and *Magill* and other early cases illustrate both thoroughness of approach and clear articulation of principle. Moreover, both Commission officials and private lawyers have

[22] Ibid. 394.

[23] Case 73/74 [1975] ECR 1491: [1976] 1 CMLR 589.

[24] Known in French as the 'Tribunal', so that cases in its list have a prefix 'T' whilst those in the list of the Court of Justice now have the prefix 'C'.

been impressed by the quality of the detailed analysis to which Commission decisions have been subject. Indeed the decision of 27 February 1992 in the *PVC* appeal that it had failed to comply with its own rules of procedure came as a considerable shock to the Commission. It is too early to know whether this decision will prove an isolated one or whether it presages a large number of other past or pending Commission decisions being declared void and fines quashed. The creation of the Court has without doubt enabled a more rigorous examination of the Commission's procedures and substantial reasoning to be introduced. Article 177 cases, of course, still have to be referred to the European Court which, by the end of 1991, still found itself burdened with so many cases that judgment was still taking some 18 to 24 months from first being set down (and considerably longer in the largest cases). It may be, therefore, that the Court of First Instance will be required in the future to take a further range of cases. Appeals lie to the European Court from the Court of First Instance on matters of law.[25]

The work of DG IV has been carried on against a background of changing economic circumstances which have deteriorated, in particular since late 1973. DG IV operates as one of more than twenty Directorates-General administering the whole range of Community policies and co-ordinating with them as required when matters are of overlapping interest. They have in practice been well aware of the stress imposed by these new circumstances and the tendency for requests for exemption to be made which in normal economic circumstances would undoubtedly have stood no chance of acceptance with the framework of Article 85(3). In this area, DG IV seems to have taken a realistic view of the importance of its own continued existence. In the same way that a well designed bridge will be able to move a few inches under the force of a hurricane so as to preserve its structural integrity, and thereby prevent total destruction in the eye of the storm, so in the same way a competition authority may find it better to show itself reasonably sensitive to the problems of particular sectors and undertakings in times of economic difficulty, in order to enable it to retain its major principles intact for the future. It may thus avoid the risk of being totally reconstructed and suffering wide changes to its functions and jurisdiction on the alleged grounds of crisis which may not be restored even when the economic climate reverts to a greater state of normality. A number of the decisions in the late 1970s and early 1980s seem to show a more flexible stance in this respect, notably the *Synthetic Fibres, URG/KEWA* and *International Energy Authority*[26] cases.

[25] Among the first such appeals are two of the *Magill* cases.

[26] *Synthetic Fibres* [1985] 1 CMLR 787, *KEWA* [1976] 2 CMLR D15: *URG* [1976] 2 CMLR D1. *International Energy Authority* [1984] 2 CMLR 186.

A key element in the work of the Commission is its handling of the conditions for exemption in Article 85(3). The way in which it has interpreted these conditions has been considered in detail in Chapter 8. Whilst generalizations in this area are dangerous, it appears that the Commission has handled the positive conditions satisfactorily, but has encountered more difficulties in the interpretation of its negative conditions dealing with 'indispensability' and 'protection of competition' respectively. It appears that on many occasions it has been the issue of indispensability which has posed the major problem[27] and frequently the solution adopted has been to approve the agreement after the deletion of certain restraints felt to go beyond the reasonable needs of the parties in preserving the relevant technical progress or improvement in manufacture or distribution which has been the original justification for the claim to exemption. In some of the judgments, the element of consumer benefit appears to have received mechanistic treatment rather than to have been individually considered, particularly as in many cases the parties appeared to be in difficulties in proving actual or potential specific consumer benefit above and beyond that likely to flow from an improvement in the manufacturing or distribution methods, or in technical or economic progress within a competitive market.

In assessing, therefore, the progress achieved to date, the greatest progress has been made in those areas where the allocation of resources has been most generous. Thus, in both exclusive distribution and exclusive purchasing the degree of attention paid to these problems has enabled a clear pattern to emerge of the form of agreement which will be permitted and has substantially reduced the need for individual clearance. Whilst the application of these general principles has been one of the major successes of DG IV, the application of group exemptions to special sectors while administratively convenient has been less satisfactory, notably to beer and petrol under Reg. 1984/83 and to automobile dealers under Reg. 123/85. As already noted, the effect of such group exemptions in these fields has been to make the pattern of distribution more rigid, and the history of the negotiations leading up to the adoption of the terms of the group exemptions has been apparently more influenced by the competing strengths of national and commercial interests than by an objectively logical analysis of the sector's particular needs and circumstances. It is doubtful if the multiplication of such sectoral exemptions is really in the ultimate interests of competition, although it may be for the initial convenience of the particular trade sector and of the Commission itself in that the number of future individual notifications will be substantially reduced. The fact that detailed explanatory memoranda setting out the Commis-

[27] See e.g. the Commission's decision in *Rennet* on appeal to the Court as *Cöoperative Stremsel-en Kleurselfabriek* v. *Commission* Case 61/80 [1981] ECR 851: [1982] 1 CMLR 240.

sion's viewpoint have been required for these special sectors illustrates how difficult it is to provide legislative solutions which adequately cover all aspects of the individual industries.

The Commission has also devoted considerable resources to selective distribution and, as has already been indicated in Chapter 13, may well have expanded the definition too extensively. It has accorded this status to a number of sectors where it is logically difficult to understand, notwithstanding the support given to its approach by the Court in both the *Metro/Saba*[28] cases. The Commission does appear to have been in difficulties in finding broad and consistent criteria to characterize its policies in the closely related areas of selective distribution, exclusive distribution, and franchising. It has, however, taken a progressive attitude in recent years to the preparation and issue in 1988 of new group exemptions for franchising agreements and know-how licences, and has been prepared, with increasing understanding of their commercial value, to accept that its original restrictive approach to both could be substantially relaxed. This has been a welcome development, although the time taken for the adoption of the know-how licence group exemption still seems rather protracted.

With regard to horizontal agreements, the difficulties have been less of establishing legal principles than of providing resources to handle what are often extremely large numbers of defendants and complex factual situations, the *Woodpulp* case being an example with over forty defendants and many countries involved, some within and some outside the Community. The sheer amount of labour involved in a case as heavy as say the *Sugar Cartel* or the *Dyestuffs* case leads to a hesitation to commit the resources of DG IV to such cases unless the evidence is strong. In suitable cases, however, the Commission has shown itself willing to challenge large numbers of substantial undertakings and in recent years has achieved notable success in having the great majority of its challenged decisions upheld by the Court with fines being quashed or reduced substantially in few cases until the *PVC* appeals referred to above. Moreover as illustrated by its run of successful cases against cartels operating in the guise of trade associations it has shown itself willing and able to prevent the abuse of such organizations for anti-competitive purposes.

The problem in the horizontal area the Commission finds most difficult to handle is that of the joint venture; it is notorious that many joint ventures which should technically have been notified have not been so notified both because of fear as to the changes which the Commission may require as well as of the delays apparently inevitable before a decision will be issued. Moreover, the group exemptions issued covering

[28] *Metro/Saba* (no. 1) Case 26/76 [1977] ECR 1875: [1978] 2 CMLR 1. *Metro/Saba* (no. 2) Case 75/84 [1986] ECR 3021 [1987] 1 CMLR 118.

research and development and specialization are far too restrictive to permit more than a handful of the proposals put foward within these broad categories to fall within them. It is clearly in this area that some fresh thinking will be required if the effective control of DG IV over its area of jurisdiction is to be assured. The publication in January 1992 of the draft Notice on guide-lines for co-operative joint ventures shows the emphasis which is now being given to this area, though its concern with a more speedy consideration of straightforward cases is not accompanied by any major policy initiative. By contrast, the use of Article 86 has in the leading cases been generally successful.[29] This Article remains a dynamic and flexible instrument in the hands of DG IV, with scope for further development as indicated in Chapters 17 to 19.

Overall, therefore, DG IV has achieved much given the fact that in its modest beginnings progress was necessarily slow and later on limited by the restricted resources made available to it. Moreover, it has had in the second half of this period to contend with adverse economic circumstances which have rendered its task far more difficult than would otherwise have been the case. It has also had to contend with the extension of the Community to six additional Member States, some of whom had only limited experience, if any, of national law relating to the control of competition. Less difficulty has actually been encountered as a result of all these factors than might have been expected. The Commission and the Court have produced some notable decisions which have had wide influence in changing the basis upon which economic activity within the Community has been carried on, and it can safely be said that the degree to which the competition rules and policies are now familiar within the Community means that businessmen even in Member States with less intrusive national competition laws are aware in general terms of the prohibitions and exemptions which are available.

The Commission, whatever its past achievements, has, however, no chance of resting on its laurels since dramatic changes are now taking place in the rest of Europe which will necessarily have their effect on its operations. With the enlargement of Germany to include the whole of former East Germany, this is itself a large new territorial area whose undertakings are fully subject to the rules of competition as well as the other Treaty provisions. It is also likely[30] that during 1993 the countries in the European Free Trade Area (EFTA) will have moved into a closer relationship with the Community by the creation of the European Economic Area (EEA) in which *inter alia* the competition rules of the Treaty will have full

[29] Though the strictures of the Court of First Instance's on the Commission's *Italian Flat Glass* decision on Art. 86 need to be carefully considered by the Commission.

[30] Subject to resolution of some legal difficulties of a constitutional nature raised by the European Court opinion on the proposed treaty establishing the EEA.

application to all undertakings operating within it. It is likely that these rules will be administered by another authority in parallel with DG IV, whose place of operation is yet to be determined, and rules for the allocation of jurisdiction will have to be drawn up to the satisfaction of the European Court of Justice.

The process of achievement of all these objectives will no doubt encounter some of the same problems which the Community itself has faced over its history as already described. This, however, is not the end of the story since some of the newly liberated countries of Eastern Europe, having entered into treaties of association, are anxious to obtain membership of the Community in the course of the next few years. Many of them have already adopted national laws embodying the main principles of Articles 85 and 86. This issue of how the Community can and should maintain its competition policies in this new world will be considered in the final chapter.

26

The Future of European Competition Law

Those who have the power of creating institutions to implement new ideas often themselves have little idea how, with the passage of time, such institutions will later develop. It is, therefore, both an interesting and legitimate question to ask whether those who drew up the Treaty of Rome and in particular its rules of competition policy, as contained in Articles 85 to 90 of the Treaty, would have been satisfied with the progress which Community institutions have made in this area so far.

They would have found cause for both encouragement and disappointment. The growth of the Community to include a number of new states far beyond its original membership, and the acceptance even by many countries not within it of the basic principles that underlie the Treaty would have surprised them. This acceptance has included that of Articles 85 and 86, which have in some countries of Eastern Europe been taken as a model for their new national competition laws. The adoption after many years of merger control of the new Regulation giving the Commission at last an opportunity of preventing the creation of dominance in particular markets and thus influencing competitive structures, in a way that Articles 85 and 86 alone could not easily accomplish, would also have pleased them. In other respects, however, the lack of progress would have caused them grave disappointment for the existing state of affairs still falls short in some important respects, though by no means in all, of the circumstances which they might well have hoped and expected would have prevailed after no less than thirty-five years of continuous administrative implementation of the rules contained in the Treaty, aided by the legislative intervention of the Council and important judicial pronouncements in the many cases decided by the European Court of Justice.

In the previous chapter, an attempt was made to draw up a balance sheet containing both relevant debits and credit entries, with a brief account of some of the reasons for them. It may now be helpful in this final chapter to suggest some steps that could be taken that might improve the workings of the system in the future. Nor is it occasionally a bad idea to stand back from detailed consideration of Community institutions and their current policies, and rather than just counting the individual trees, to look for a while at the nature of the forest that has been created; only then will we be able to know if it is likely to grow in the way which will be most in the interests of future generations.

At the time when this Chapter was written for the first edition of this book in 1988 inadequate resources (both financial and of personnel) were identified as the chief continuing weakness. This weakness had prevented the Commission from ever completely digesting the overload of notified agreements which it received immediately after the implementation thirty years ago of Regulation 17/62. This problem still remains but in the meantime the whole landscape has been changing, as the territorial application of the rules of competition is being required to spread far beyond the Twelve to the members of the European Economic Area and on a more distant timescale also to the countries of Eastern Europe. How far DG IV, already stretched by existing commitments including the operation of the Merger Regulation, can adjust over the coming years to the additional areas of responsibility is of major concern. Notwithstanding the use made of group exemptions and other procedural changes in order to reduce the number of agreements not yet disposed of, the backlog has remained and is still substantial. This has in turn led to an uncertainty about the value of notification in the case of new agreements (except where the consequences of failure to notify appear too serious to risk) and a widespread cynicism in business circles about the efficacy of the exemption process for all individual agreements unable for any reason to benefit from group exemption. The only way forward in present circumstances is by giving priority in the immediate future to clear at least the greater part of this backlog even at the sacrifice of some 'own initiative' investigation, even too at the expense of a temporary reduction in the resources allotted to the investigation of current complaints, though some of these complaints could be referred to Member States for investigation. Once the backlog has been substantially cleared, an extended version of the already familiar opposition procedure should be introduced for all individual notifications so that some form of timetable would apply to all of them. A feature of the new system could continue to involve making greater use of the resources of competition authorities in Member States to assist DG IV in appropriate cases, thereby reducing some of the burden of the current caseload on DG IV. It is important for both practical and symbolic reasons that the administration of competition policy is not conducted simply at a distance by competition authorities 'over there' (in Brussels) whose intentions and activities can easily be represented (or misrepresented) in individual Member States as the imposition of foreign rules,[1] rather than the implementation of common policies which the Treaty always envisaged would be on a co-operative basis for the mutual benefit of Member States as well as of the Community itself.

[1] To the extent that national competition rules come closely to resemble Articles 85 and 86 it seems less likely that undertakings will regard the latter as 'foreign rules'.

Turning now to consider these proposals in more detail, the prime requirement for any mature competition authority is, of course, a set of substantive rules which effectively cover the whole range of subject matter for which the authority is responsible, clear and comprehensive enough to provide clear guidelines for compliance on the basis of legal advice in the vast majority of cases. These rules should also contain sufficient flexibility to accommodate the wide variety of circumstances and special situations which will always arise in individual cases. In some areas, as we have seen in the previous chapter, the Commission does possess such rules and, in its application of Article 85(3), the ability within reasonable parameters to respond to individual situations. In other areas, as we have seen, there is still a lack of clear rules and guidelines and although these may have been discussed at length and may exist in draft form, they need to be decided upon and implemented as soon as possible. In those difficult areas where it may be impossible either for legal or political reasons to provide regulations, there should at least always be guidelines or notices published to indicate the main lines of Commission thinking, even if these cannot provide the degree of certainty or detail normal in a regulation. Moreover, if Notices cease to represent Commission policy (as has quite often occurred), they should be promptly withdrawn and not left in place whilst having a diminished authority, for example as in the case of the Agency Notice originally issued in 1962. It is fair to mention that Annual Reports have been used in the past to indicate new thinking, for example in the determination of potential competition in joint venture cases and in regard to information exchanges. The absence, however, of any even informal indication of Commission policy in any important area of the law which it has to administer under Articles 85 and 86 is a serious weakness at this stage in its development and needs rectification as soon as possible. The existence of a more complete set of guidelines will not, however, prevent the continuing growth of case-law arising out of individual decisions both of the Commission and the two Courts in Luxembourg; that growth will, nevertheless, be based on more solid foundations if the structure of legal rules and guidelines operated by the Commission is more complete. It is not possible always for the Commission to wait upon a ruling in a leading case from the Court before taking policy initiatives.[2]

Given its limited resources and the financial constraints of the Community, DG IV should make its future plans on the assumption that it will have no more than its existing resources to cover the whole range of its present activities and should consider, therefore, whether and how its priorities may need adjustment. From the early stages of development, DG

[2] The Court in effect criticized the Commission on this ground in the *Leclerc Books* case (see n. 13).

IV has taken a broad view of the scope of Article 85(1) in order to bring as many as possible of all agreements containing any form of restriction on trade or effect on competition within its scope, thereby ensuring that it was its own judgment alone that could be exercised under Article 85(3) to determine the extent to which exemption could be applied to agreements, decisions, and practices, rather than limiting the scope of Article 85(1) initially so that only a more limited selection of the agreements relating to restraint of trade reached it for consideration under 85(3). This wide assumption of jurisdiction has produced both good and bad results for the Community. The benefit of the policy has been to ensure that very few types of agreement containing restraints on trade have been outside the scope of investigation of the Commission, on the grounds normally either that they affect matters with only trivial effect on competition or that they only relate to individual national markets; the effect of this wide interpretation of Article 85(1) has been that the interdependence of national markets has been underlined and the unity of the Common Market thereby strengthened. The detriment, however, flowing from that policy has been that DG IV has actually been unable through lack of resources, in particular in view of the increasing size and geographical area of the Community, to handle the enormous caseload resulting from so wide a claim within a reasonable time-scale or on a predictable basis in terms of its own priorities. While such difficulties may have been understandable and even acceptable in the early days of DG IV, it is much less so now that DG IV is a well established part of the administrative machinery of the Commission with so many cases already determined (whether by formal or informal process).

The Commission regularly indicates in its Annual Reports its continuing uneasy awareness of the problem of the backlog of cases, caused in its view mainly by staff shortages. The number of pending cases under Articles 85 and 86 at the end of 1991 were 2,287, a reduction of 447 over the year and the third successive year in which a reduction had been achieved. This is a welcome new development since during the 1980s the number of pending cases tended to increase steadily. On the other hand, in 1991, although the Commission disposed of 835 cases, of these only 13 were formal decisions. The rest were made up of 146 cases disposed of by comfort letters and 676 by other informal means (including withdrawal by the parties). Of the pending cases at the end of 1991, 1,732 were applications or notifications (of which 282 had been submitted during the year), 328 were complaints from undertakings (83 in the year), and finally 227 were 'own initiative' investigations (23 started in the year). It is the large figure of applications and notifications which gives the most cause for concern, especially as in neither the 20th nor 21st Annual Reports is there a more detailed breakdown of the type of cases of which they consist and for how many years they have been awaiting decision.

These figures do suggest that DG IV should both be able to handle the number of new cases now arising and still reduce its backlog of outstanding cases to manageable proportions. It is significant that figures disclosed in earlier Annual Reports and referred to in the first edition indicate that a large proportion of these relate to intellectual property rights cases or exclusive distribution agreements, categories for which group exemptions exist and where there is a substantial body of existing case-law. This leaves a much smaller residue of possibly some 20 per cent in the area of horizontal agreements, many of which are presumably joint ventures.

It is true, as already mentioned, that joint ventures present special problems of analysis and that it is difficult to provide general criteria for their assessment just as it is to provide such criteria for mergers generally. It is not clear, however, why the Commission should not at least be able to eliminate that part of the backlog consisting of the vertical agreements representing over four-fifths of its pending cases and which presumably would give in the majority of cases less difficulty in disposal. Alternatively, even if it were not possible to eliminate all the backlog, it would surely be possible, perhaps by the appointment of a task force, to analyse the cases remaining on the 'pending' list so that the obsolete ones could be promptly disposed of and the significant ones given a degree of greater priority in their disposal.

A number of procedural steps being taken to reduce the backlog are referred to by this and previous Annual Reports, including internal reorganization and accelerated procedures. No official claim has been made, however, that these steps will do more than make minor inroads into the problem. Conspicuous by its absence is any published medium- or long-term plan for eliminating the backlog or at least reducing it to a manageable level so that DG IV could reach the point in the relatively near future where it could accept, as a basic operating principle, a mandatory time limit for handling all notifications. It is in fact noticeable that the Commission operates under two quite different regimes. In the area of merger control DG IV operates under tight time limits imposed by Regulation 4064/89 of one month for the preliminary stage and four months thereafter for a full investigation. On the other hand, in the departments dealing with applications, notifications, and complaints under Articles 85 and 86, the position is that agreements submitted several years ago have simply not been considered in any real sense, owing to lack of time or resources to review them; no time limits nor official principles of priority whether imposed externally by way of regulation or self-imposed by way of internal organization seem, however, to operate in this area.

In fact the co-existence within DG IV of a separate and highly motivated Merger Task Force operating under rigorous time limits, alongside existing officials continuing their regular work on Articles 85 and 86, is already imposing new pressures on the organization. It is entirely to the credit of the officials of the Merger Task Force that, by careful advance planning and continuing efficient handling of cases, their procedures have won respect and co-operation from the undertakings involved and their professional advisers, whatever problems may have arisen on the final substantive resolution of cases. It has, however, been at the expense of raiding the rest of DG IV for some of its more experienced staff members and this may have left the remainder of DG IV less well resourced to cope with its own heavy case load. Moreover, it led to unfavourable comparisons being made by outside commentators between the 'user friendly' and sensitive implementation of the Merger Regulation on the one hand as compared with the often relatively unpredictable and protracted handling of both major and minor cases under Articles 85 and 86, as the terms of the recent draft Guidelines on Co-operative Joint Ventures themselves seem to acknowledge.

It would, of course, be unfair to make such comments an occasion or general ground for criticism of DG IV, which in general carries out the implementation of Articles 85 and 86 in a competent fashion. The absence, however, of any time limits on its work and the consequent effects on both the undertakings concerned and possibly also to the morale of DG IV has become evident. The following suggestions made for introducing time limits and additional procedures to speed up the sectoral divisions of DG IV in their review of individual notification applications remain relevant and perhaps more so than when originally made. The effect of the implementation of Merger Task Force procedures has been particularly felt in the area of joint ventures. An immediate consequence has been to emphasize the advantages of presenting joint ventures to the Commission as concentrative joint ventures rather than co-operative. This approach has an advantage not only in terms of the much shorter time limits that prevail under the Merger Regulation, but also involves the benefit of less demanding criteria under those Regulations than operate under Article 85. Moreover, in cases where a co-operative joint venture has been closely linked to a concentration (whether a merger proper or a concentrative joint venture) as for example in the *Volvo/Renault* and *Appollinaris/Schweppes*[3] cases, the Merger Task Force has applied pressure that the investigation into the co-operative joint venture be speedily carried out so that a decision can be rendered at the same time as on the concentration. Obviously such an outcome cannot be guaranteed in every

[3] Cases IV/M.004, decision dated 7 Nov. 1990, and IV/M.093, decision dated 24 June 1991.

case, particularly when the facts are potentially complex, but the situation is not altogether satisfactory since it introduces a built-in inducement to structure joint ventures in the concentrative form wherever possible.

It is impossible, of course, to impose rigid time limits realistically on a competition authority whose function is investigatory, making inquiries of its own initiative or reacting to complaints, since these may vary from the trivial to the far-reaching, from cases where substantial information is initially provided to those where many months of enquiry are needed to ascertain even the basic facts. It is, however, both possible and desirable to impose reasonable time limits when the authority is considering the claim for exemption of notifications which have been lodged in a set form against a well-established background of existing case-law. Once the backlog has been cleared or substantially reduced, such time limits should be introduced for all future formal notifications, operating from the date of receipt of a fully completed form A/B. Under the latest version of this form, the parties are required to provide a great deal of information about their product market and undertakings at time of first notification, which should itself enable DG IV to give a speedier response. As with those existing opposition procedures, found in current group exemptions, the time limit would operate on the basis of a presumption that a duly notified agreement would be deemed to have been exempted unless a preliminary reaction to the contrary had been issued within the required period. This preliminary reaction would not necessarily constitute a formal decision but would place the parties on notice of the likely decision of the Commission, a form of preliminary announcement already provided for under Article 19(3) of Regulation 17. The target period for the issuing of such preliminary decisions should not be longer than six to twelve months from notification. The issuing of such a preliminary decision would place the parties to the agreement on notice that it was likely that the agreement would either be exempted (either with or without amendments) or, alternatively, declared as prohibited by Article 85(1) or given negative clearance. A period of negotiation would then normally follow in the majority of cases, and all notifications could then be subject to a formal decision within say a further maximum period of six months. In this way, the parties would know where they stood as to the likely validity of agreements at a much earlier stage than is now possible in many cases, and the interests of legal certainty considerably enhanced.[4]

[4] Even these improved time limits would not necessarily be satisfactory for national courts asked to consider the legality under Article 85(1) of particular agreements, and once the new procedures were well established, consideration could be given to establishing also a special procedure (with shortened time limits) for the consideration of those agreements referred by national courts to DG IV in connection with pending cases. Another possibility would be a system for providing from the Commission preliminary opinions for national courts in such cases. If DG IV is concerned to encourage national courts and competition

The introduction, therefore, of reasonable time limits for notification decisions, whilst putting considerable pressure upon the administration of DG IV, should make possible an improvement in its relationships with both commercial and industrial sectors throughout the Community and also with Member States. No longer would undertakings be faced with the unsatisfactory prospect referred to by Forrester and Norall[5] whereby they had no idea when a particular joint venture or distribution or licensing agreement was to be considered, with the consequence that either notification would not occur or would occur as a precaution but without any reasonable expectation that a decision would appear at any date early enough to have commercial value to the parties.[6] Time limits imposed on bodies responsible for the issuing of decisions in specific cases are found in the legislation of some Member States dealing with competition, and are regarded both as providing an incentive for the efficient administration of national competition authorities and as a means for ensuring public acceptance of their function notwithstanding the business inconveniences or actual detriment which such procedures may impose on the parties concerned. A system of time limits for dealing with notifications does, however, depend on the assumption that resources allocated for this purpose within DG IV should not be diverted to the handling of complaints or the Commission's 'own-initiative' procedures. Given the traditional attitude of DG IV that complaints should receive priority, this may require further internal reorganization as well as a change in the relationship with Member States' competition authorities to be referred to at a later stage in this chapter.

The introduction of time limits is a procedural reform that would be desirable in its own right. It would, however, also be assisted by a change of substantive law, that of a more restricted interpretation of Article 85(1). The European Court of Justice has over the last twenty years given a number of signals starting with the *STM*[7] case in 1966 and continuing through cases on selective distribution and open exclusive licences up to *Pronuptia*[8] in 1986 and *Erauw Jacquery*[9] in 1988 that a broader view should be taken, encouraging the concept of what might loosely, if

authorities to take greater responsibility for enforcing Articles 85 and 86 some assistance of this kind will be desirable (but see n. 19).

[5] I. Forrester and C. Norall, 'The Laicization of Community Law: Self-help and the Rule of Reason', 21 *CML Rev* 11 (1984).

[6] The service probably most valued by undertakings and their advisers would be an informal but expedited procedure to give them guidance as to whether any actual proposed agreement required notification and if so whether it would qualify for an individual exemption with or without amendment. DG IV will often do its best to provide such a service but normally only on an informal basis.

[7] Case 56/65 [1966] ECR 235: CMLR 357.

[8] Case 161/84 [1986] ECR 353: 1 CMLR 414.

[9] Case 27/87 [1988] ECR 1919: 4 CMLR 576.

inaccurately, be called the 'rule of reason' within Article 85(1) thereby reducing the width of the Commission's jurisdiction in this area. The issuing of further group exemption regulations, on outstanding issues including copyright licences, and even joint ventures may also help in this way. It may be that DG IV would serve its own proper objectives under the Treaty best by reducing the scope of its claimed jurisdiction in order to ensure that it was able to carry out adequately those functions that lie at the heart of its responsibilities under Articles 85 and 86.

The major institutional problem facing the Commission, however, is the manner in which merger control is administered. The euphoria which has accompanied the smooth introduction of these procedures from September 1990 onwards has been accompanied by a growing realization voiced by many, including the heads of leading national competition authorities,[10] that the machinery established by the Regulation, though working well for the uncontentious case, has important potential weaknesses in the handling of major cases. Here the competition issues which dominate the content of Article 2(1) of the Regulation may conflict with the desires of certain Member States for greater weight to be attached to the 'economic and technical progress' also there referred to. These factors were sharply brought out by the decision in October 1991 on a bid by France's state-owned Aerospatiale and Italy's state-owned Alenia for de Havilland of Canada.[11] This transaction would have given a joint venture owned by the French and Italian companies some 50 per cent of the world market and 65 per cent of the Community market, for commuter planes with between 20 and 70 seats. Clearly the merger would create a dominant position for the joint venture within the Community. The argument of France and Italy, however, was that the joint venture should have been allowed in order to enable the Community itself to support a sufficiently powerful company in the world commuter aircraft market.

Whatever the merits of the individual decision (which would seem to be appropriate and consistent in terms of normal competition criteria), the case illustrates the inherent weakness of any system of merger control where the inevitable 'political' element operates not openly but behind closed doors, before the relevant facts and legal criteria have been arrived at by objective assessment and then published. A decision which lacks a sufficient degree of transparency has substantially less value as a precedent and reduces the confidence of undertakings and their advisers in the objective operation of the system and in its basic integrity. It illustrates

[10] Including the heads of the German Bundes-Kartellamt (BKA) and of the United Kingdom Office of Fair Trading and Monopolies and Mergers Commission. For a more detailed review of these issues see my article 'The Implementation of the EC Merger Regulation: New Wine in Old Bottles'; *Current Legal Problems 1992*, Volume 45, Part II, p. 117, (Oxford University Press, 1992).

[11] Case IV/M.053 decision dated 2 Oct. 1991: [1991] O JL 334/42: [1992] 4 CMLR M2.

by contrast the value of the merger control system in France, Germany, and the United Kingdom, which in their different ways ensure that the 'legal' and 'political' stages are kept separate, and that any final decision to permit a merger on political or social grounds is transparently seen as such, and is not confusingly presented as the 'legal' outcome of the case.

The Commission's dual role both as investigator and final decision maker (subject only to review by the European Court) makes such transparency, however, very difficult to achieve. Moreover, much of the essential information needed in order to form an objective assessment of the individual case is not included in the published decision, even in the important cases subject to the four-month 'second stage' investigation under Article 6(1)(*c*) of the Merger Regulation. It might, therefore, be better if the investigative process in merger control at least were carried out by a separate office or body which could perform the investigation and publish in full its findings of law and fact. As a final stage, however, but after publication of the decision political arguments could still be brought before the Commission, if this were regarded as sufficiently important in an individual case, by Member States, and the outcome of the case even varied if it were regarded as generally desirable in the interests of the Community rather than simply of one or two Member States. Clearly, however, such an outcome would, and should be, very rare and limited to highly unusual circumstances.

Earlier chapters have shown that the application of competition rules to Member States themselves began very slowly and cautiously, and that it is only in recent years that these rules have been applied more often by the Commission to Governments and governmental authorities. As was pointed out by Judge Pescatore,[12] himself for many years a member of the Court, there has been a surprising asymmetry in the application of the relevant legal principles under the Treaty. Under Articles 30 to 36, the Court and Commission have gone far in restricting the ability of Member States to place obstacles in the way of free movements of goods and services between them, and though the Articles are directed specifically towards the actions and omissions of Member States, the effect of the decisions (particularly in the field of intellectual property) has been substantially felt by undertakings as well. By contrast, Articles 85 to 86 have been applied strictly to undertakings, which have been substantially affected by the numerous rulings, but had not been applied in such an extensive way to Member States.

In some cases where the European Court has had an opportunity to apply the competition rules in their full vigour to Member States, it has adopted an easier solution by relying on Article 30 as justification for the

[12] *Thirteenth Annual Proceedings of Fordham Corporate Law Institute*, pp. 381, 389–95 (1986).

decision even where this Article provides only a partial answer. Notwithstanding the provisions of Article 5 of the Treaty which require co-operation in ensuring the fulfilment of the obligations arising under the Treaty and abstention from measures which could jeopardize the attainment of those objectives, the Court in the *Leclerc Books*[13] case appeared to indicate that Member States could in principle validly produce effects under their legislation which were comparable to those to be produced by an agreement that would itself be prohibited by Article 85. The only circumstances of an exceptional nature in which such measures were prohibited would in the Court's view be if the Member State was shown to be favouring or facilitating the conclusion of agreements between undertakings or extending the effect of such agreement to third parties or permitting undertakings to escape the obligations of Article 85 or 86. The rather surprising justification given by the Court for allowing the French Government to issue legislation fixing the price of books was that the Community itself had not yet adopted a policy in this sector. Nevertheless the recent cases discussed in Chapter 23, especially *Ahmed Saeed, Flemish Travel Agents*, and *Van Eycke* v. *Aspa*, illustrate a greater willingness to apply Article 5 and Article 90 to restrain the actions of Member States, which is welcome.

The more uniform application of competition laws, however, would also be assisted by a greater level of co-operation with the Commission by national authorities in their enforcement of Community competition rules. It has already been made clear that Article 87(2)(*e*) and Regulation 17 (in particular Articles 10 and 14) envisage a fuller basis for co-operation than has so far normally occurred.[14] It is true that Member States participate regularly in the Advisory Committee set up under Article 10 of Regulation 17 to discuss the draft decisions in pending cases and proposed new Regulations. Likewise under the Merger Regulation the Advisory Committee is consulted on the draft decision and moreover here their opinion can be published if the Committee wishes, though its opinions so far published have been disappointingly brief. On the other hand, it is notable that Article 13 of Regulation 17/62 has been little utilized; this authorizes the competent authorities in Member States at the request of the Commission to carry out investigations which the Commission has either ordered by decision should be carried out or otherwise considers necessary, and provides for mutual help between the relevant Member States and the Commission in the carrying out of such investigations. It seems that the failure to utilize Article 13 or take over some investigative tasks from DG IV has not been caused mainly by opposition or lack of goodwill on the

[13] Case 229/83 [1985] ECR 1: 2 CMLR 286.

[14] For details of the current involvement of the OFT in Community cases see the 1990 Annual Report of the Director General (published by HMSO), pp. 47–8.

part of national authorities, but is more attributable to simple lack of resources of manpower in the national authorities and a consequent reticence on the part of DG IV to make adequate use of them.

Two changes are necessary in national law if this co-operation is to work effectively and if the principle of 'subsidiarity' is to be successfully implemented so that necessary enforcement action is taken at the appropriate level in each case. First, national competition law of all members of the Community needs to be brought generally into line with the Community law, particularly Articles 85 and 86, so as to ensure that it is based entirely on the objects and effects of agreements and practices on competition rather than on other principles.[15] The second change needed is that national competition authorities should be authorized, as some already are, to apply not only national law but also Articles 85 and 86 of the Treaty so that, by arrangement with the Commission, those cases more suitably dealt with in a Member State can be handled there subject to the same legal rules as those applied in Brussels by DG IV. This could even extend in time to the giving of 85(3) individual exemptions in straightforward cases, though a change in Regulation 17 would be required to make this possible.[16]

The larger the Community grows the more important it would be that DG IV reserves its resources for those important cases under both Articles which deserve its attention, leaving other cases to the appropriate national authority, once these show themselves capable of handling them. The Commission will, however, find it easier to do this if such national authorities are applying laws closely based on Articles 85 and 86, whether or not their own national laws totally reflect the wording of those Articles and even if they depart from it in some minor respects. Such reforms would not however necessarily make the provision of documentary assistance and exchange of information easier between the Commission and national authorities as under Regulation 17 national authorities and the Commission can only use information acquired as a result of the application of Articles 11 to 14 inclusive if it is to be used for the 'relevant request or investigation'. Whether this restriction is given a narrow or liberal interpretation may be known after the hearing of the pending *Spanish Banks* case by the European Court.[17]

[15] Such, in particular, as the 'form-based' principles underlying the Restrictive Trade Practices Act in the United Kingdom.

[16] Problems of consistency would of course arise.

[17] Case C67/91 *Dirección General de Defensa de la Competencia* v. *Associación Española de Banca Privada*. The recent decision in this case dated 16 July 1992 has made it clear that in applying both national and Community competition law national authorities cannot treat as evidence information provided by parties under Articles 2, 4, or 5 of Reg. 17/62 or in response to Commission inquiries under Article 11. National authorities may however use such information as the starting point for their own investigations.

The Commission has made greater efforts in the area of encouraging private actions through national courts, although with only a limited degree of success. Decisions have been rendered in a number of courts in Member States accepting the principle that damages can be obtained for breaches of Articles 85 and 86,[18] but there appears to be a general reluctance for litigants in such courts to attempt to obtain such remedies as opposed to utilizing the institutions of the Community, namely the Commission and Court.[19] It is possible that if assistance could be obtained from DG IV in providing guidance to such courts, again possibly with co-operation of the national competition authorities, then national courts might find it easier to apply the provisions of Articles 85 and 86 in individual cases. Once a single well publicized award of damages had been gained by one undertaking, the fashion for utilizing this law might grow as has occurred in other contexts.[20] *Hasselblad* v. *Orbinson*[21] is a precedent under United Kingdom law for the intervention of the Commission in cases before English Courts and further applications and extensions of the principle are highly desirable.

It is true that DG IV itself has always taken a very cautious view of possible involvement in cases in national courts (again usually because of its limited resources) but it may well be that the time for this caution has now largely passed, especially as a new generation of lawyers has now been educated in Member States who are far more familiar than their predecessors with the rules made under the Treaty and with the jurisprudence of the Court of Justice. National judges themselves have adapted, albeit slowly, to this new situation where national courts have the responsibility of applying Community law. New Member States in particular are normally anxious to show themselves 'Community minded' and might be willing to accept the formal intervention by the Commission in individual private suits involving competition issues. In practice, although it is the occasional refusal by a Member State to co-operate with the Commission which receives headline treatment (as for example in the *Hoechst*[22] case in 1987), the normal pattern of relationships between DG IV and Member States' competition authorities operates on a co-operative if low-key basis.

If the Community and national institutions are to work more closely together, officials of national competition authorities should be seconded for reasonable periods to Brussels to gain experience of the workings of

[18] See Ch. 22, pp. 440–2.

[19] The question of cost and speed of action may well be important factors. The Commission has now published a draft Notice on the application of Articles 85 and 86 by national courts, set out at [1992] 4 CMLR 524.

[20] Thus the practice of including a retention of title clause in conditions of sale became almost universal once the consequences of *Aluminium Industrie Vaassen BV* v. *Romalpa Aluminium* [1976] 1 WLR 676 were appreciated, whereas previously it had been very rare.

[21] *Hasselblad (GB)* v. *Orbinson* [1985] 1 AER 173: 2 WLR 1: QB 475.

[22] See Ch. 22.

DG IV, and likewise members of the staff at DG IV should receive regular secondment to competition authorities in Member States in order both to obtain familiarity with national authorities' operating procedures and problems, as well as to ensure that Community law is being properly enforced alongside the national law. It is noteworthy that national experts have been seconded from Member States competition authorities to the Merger Task Force with considerable success. Once such exchange and cross-fertilization becomes well established, it should become considerably easier for the Commission to delegate to national competition authorities under Article 13 some of the enquiries into situations which are essentially national rather than Community-wide, upon which at present the valuable resources of the Community are themselves spent. Mutual assistance might also be particularly helpful in major cases involving visits to a very large number of undertakings, possibly involving the need to inspect many documents found within particular national jurisdictions. It is of course important that national competition authorities are themselves adequately resourced for these tasks.

An additional problem for any competition authority and certainly for DG IV is ensuring that its staff have a wide enough experience of commerce and industry to apply the relevant rules in a way which is in touch with current realities of industrial commercial life without sacrificing, or compromising on, issues of basic principle. The confidence that is required to achieve this delicate balance does require some up-to-date knowledge of the way in which business operates, and although there are difficulties relating to confidentiality and career prospects, some means should be found for short exchanges also between senior officials of DG IV with people in responsible positions in industry and commerce. For European competition law to be effectively enforced, it needs to be administered by officials who are fully aware of the problems and special circumstances of individual sectors of industry within the Member States, even though they must still be able to stand back from them and assess their activities objectively.

The 1990s will be a difficult decade for the Commission as it seeks to continue to implement competition policy in the extended Community where the internal market for the original Twelve will have been completed by the end of 1992. The demands to be made on DG IV and the Commission as a result of the formation of the European Economic Area (EEA) and the likely subsequent enlargement of the Community (with the need for fuller harmonization of both substantive and procedural law with Member States) will be substantial even if the principle of 'subsidiarity' is fully applied and benefited from. The overall framework within which DG IV should operate should be as disciplined and responsive as that of the Merger Task Force, taking into account the different types of investigation which it will be undertaking and the necessarily more extended period within which these can be carried out, but without any sacrifice of independence or of adherence to legal principle.

It is also important for the future of the Commission in this area that it maintains a close and genuine relationship with two other institutions of the Community which are involved in the procedures for adopting new regulations and the general review of policy. These are, of course, the Economic and Social Committee and the European Parliament. The Parliament in particular has played a crucial role at some important moments in the history of Community competition law, usually to the long-term advantage of the Commission. It is interested in the technical details of new legislation, but even more in the general oversight of policy development through its regular review of the Commission's own Annual Reports. The support of the Parliament may become of greater importance in years to come if the Commission is to obtain both the jurisdiction and the resources that it seeks. Annual Reports[23] pay at least lip service to the value which the Commission places upon help from both these institutions and it is important this is translated into closer working relationships from which the Commission could derive both ideas and political support as well as, on occasion, even some constructive and useful criticism.

Competition policy is only one of the common policies which the European Community has to implement, if the vision of the founding fathers is to be realized over the next decade. In an enlarged Community its requirements will inevitably on occasion have to give priority to the imperatives of other Community policies, as indeed it has already done on occasions in the past. Nevertheless, in a broad sense, it is difficult to conceive of a fully effective internal market that does not have, as one of its central and most important objectives, the maintenance of such a policy in all its different aspects, including merger control. The institutions of the Community have so far been able to co-operate in this area to produce results of great commercial and also social significance, and over the years they have been largely fortunate in the commitment and technical abilities of the individuals who have held responsible positions within them, and have played their part in the many different developments here recorded.

There is, however, no guarantee that the achievements of the past will necessarily lead to progress in the future. Indeed, if the necessary policy decisions are not taken and the prevailing atmosphere were to become one of 'drift' rather than 'steer', it would not take long for much of the good work so far done to be rendered abortive. Goethe's famous lines have their application even in the context of competition policy; for such policy

[23] See e.g. the 16th Annual Report, pp. 2–3.

is not self-executing but one which without constant renewal and reinvigoration quickly becomes a dry husk, rather than the dynamic force responsive to the needs of the Community which it should always remain.

Sources of Law and Bibliography

SOURCES OF LAW

Whilst legal practitioners and officials will be well aware of the sources from which detailed information about cases and competition policy can be obtained, students are often unsure as to where they should best look for this information. The Treaty of Rome is, of course, the essential starting point. It should be studied in one of the annotated editions published by the leading law publishers.[1] The official source of information in the Community is the Official Journal published in all official Community languages but in English only since 1973 when the UK became a Member State.[2] The series marked 'L' includes details of actual competition cases decided by the Commission, whilst series 'C' contains details of proposed exemptions for specific cases, negative clearances, and other relevant items relating to the work of DG IV. The complete text of the Regulations issued by the Commission or by the Council is also essential, and there are several collections of this material.[3] Careful attention should always be paid to the recitals of Regulations, since they can be of major help in the interpretation of the individual substantive clauses and are often referred to for this purpose by the Court.

Other important sources of official information are the Annual Reports of DG IV published since 1971 in a variety of languages, including English, which usually provide a useful short summary of the facts and decisions of complex cases in the course of the review of major decisions issued both by the Commission and by the Court of Justice. The decisions of the Court of Justice often have, however, to be read in full detail, and this can be done either in the official European Court Reports (ECR) or alternatively in the Common Market Law Reports (CMLR).[4]

[1] It is, of course, important to ensure that the copy purchased is fully up to date, since the original treaties have been subject to amendment on several occasions, principally by the Single European Act signed in Feb. 1986 and ratified by the Member States during 1986 and 1987, and the Maastricht Treaties of Monetary and Political Union signed in December 1991 but not yet ratified.

[2] Some special editions of the Official Journal containing English texts of important regulations, directives, and other official notices and publications prior to 1973 were published at the time when the UK joined.

[3] The Regulations and official notes published by DG IV, including official explanatory memoranda relating to group exemptions, can be found in a number of handbooks published by the main law publishers.

[4] The ECR series is published only very slowly, normally at least a year or two after the case has actually been decided, but with a full summary of the arguments raised by the parties, and often also by Member States or the Commission intervening in the proceedings, as well as the officially approved text of the Advocate-General's opinion and of the Court's judgment. By contrast, the reports of CMLR are normally produced with commendable speed on a weekly basis, though they do not normally include the detailed summary of the parties' arguments. For most practical purposes, however, they are the reports which the student will find most convenient to use in respect of recent cases, and especially when the original language of a case is not English, with the consequential delay in the publication of the official English translation. The most recent decisions of the Commission can be found initially only in the Official Journal, prior to any report becoming available in ECR or CMLR.

Students will also find in their law libraries a variety of different textbooks on the topic of European competition law, often of formidable length and scholarship. Unfortunately some of the best are not available in English and a knowledge of French, German, and other Common Market languages would be of considerable help, especially to the more advanced student. Those to whom English is the only language available need, however, not despair. The Bibliography in this book lists a number of textbooks available on the subject (the number of which has substantially increased in the period since 1988) though the date of publication needs always to be carefully considered in view of the continuing development of competition policy and the regular appearance of important new Commission or Court judgments changing the law in major or minor respects. This renders any book liable to become quickly outdated, whatever the original value of its analysis. Good, though simplified, general descriptions of the application of the law and Commission procedures are found in the publications of the Commission itself. These include a guide to the procedures and investigations of the Commission and also a guide to the rules applicable to small and medium-sized companies, and explanatory memoranda on some of the more recent group exemptions.

There are also a number of excellent specialist periodicals on the subject of European law, and these often contain articles on competition law; other periodicals exist which devote themselves entirely to the subject of EEC competition law. A list of articles is referred to at the end of this section.

The Bibliography which follows contains details of books which I have found useful in the course of preparation of this work. It does not purport to be exhaustive, and in particular I have not included a number of books which, whilst useful in their time, have now become virtually obsolete owing to changes in the relevant law or practice. On the other hand, some books are listed with publication dates in the 1960s and 1970s which have not been updated, but which still remain valuable, especially on historical aspects of the subject.

BIBLIOGRAPHY

1. General Books on EC Law

BROWN, L., *The Court of Justice of the European Communities*, 3rd edn. (Sweet & Maxwell, 1989).

COLLINS, L. J., *European Community Law in the United Kingdom*, 4th edn. (Butterworths, 1990).

GREEN, N., HARTLEY, T., and USHER, J., *Legal Foundation of the Single European Market* (Oxford University Press, 1991).

HARTLEY, T. C., *The Foundations of European Community Law*, 2nd edn. (Clarendon Press, 1988).

HUNNINGS, N., *Gazetteer of European Law*, 2 vols. (European Law Centre Ltd., 1983).

KAPTEYN, P., and VERLOREN VAN THEMAAT, P., *Introduction to the Law of the European Communities* (Sweet & Maxwell, 1973).

From the beginning of 1988 a special monthly series of the CMLR reports, entitled 4 CMLR, has been devoted exclusively to competition law cases and developments. From the beginning of 1992 important decisions of the Commission made under the Merger Regulation are included in a separate section of 4 CMLR prefixed M.

LIPSTEIN, K., *The Law of the European Economic Community* (Butterworths, 1974).

MATHIJSEN, P., *A Guide to European Community Law*, 3rd edn. (Sweet & Maxwell, 1980).

MILLETT, T., *The Court of First Instance* (Butterworths, 1990).

SMITH, H., and HERZOG, P., *The Law of the EEC: A Commentary on the Treaty*, 6 vols. (Matthew Bender, looseleaf, 1976).

WYATT, D., and DASHWOOD, A., *The Substantive Law of the EEC*, 2nd edn. (Sweet & Maxwell, 1987).

2. Books on EC Competition Law (General)

BAROUNOS, D., HALL, D., and JAMES, J. R., *EEC Antitrust Law: Principles and Practice* (Butterworths, 1975).

BELLAMY, C., and CHILD, G. (eds.), *Common Market Law of Competition*, 3rd edn. with 1991 Supplement (Sweet & Maxwell, 1987).

CAMPBELL, A., *EEC Competition Law: A Practitioner's Textbook* (North-Holland, 1980).

DERINGER, A., *The Competition Law of the European Economic Community: A Commentary on the EEC Rules on Competition* (CCH Editions Ltd., 1968).

GIJLSTRA, D., and MURPHY, D., *Leading Cases and Materials on the Competition Law of the EEC*, 3rd edn. (Kluwer, 1984).

GLEISS, A., *A Common Market Cartel Law*, 3rd edn., trans. from the German (BNA Inc., 1981).

HAWK, B. E., *United States, Common Market and International Antitrust: A Comparative Guide*, 2nd edn., 4 vols., vol. ii (1990 Supplement) (Prentice-Hall Law and Business, 1990).

JACOBS, D. M., and STEWART-CLARK, J., *Competition Law in the European Community*, 2nd edn. (Kogan Page, 1991).

JACQUEMIN, A. (ed.), *Merger and Competition Policy in the European Community* (Basil Blackwell, 1990).

JONES, C., LEWIS, X., and VAN DE WOUDE, M., *EEC Competition Law Handbook*, 2nd edn. (Sweet & Maxwell, 1991).

KORAH, V. L., *An Introductory Guide to EEC Competition Law and Practice*, 4th edn. (ESC Publishing Ltd., 1990).

MONTAGNON, P. (ed.), *European Competition Policy* (RIIA/Pinter Publishers, 1990).

RAYBOULD, D. M., and FIRTH, A., *Law of Monopolies, Competition Law and Practice in the USA, EEC, Germany and UK* (Graham & Trotman, 1991).

ROWE, F., et al., *Enterprise Laws of the 1980s* (American Bar Association, 1980).

VAN BAEL, I., and BELLIS, J. F., *Competition Law of the EEC*, 2nd edn. (CCH Editions Ltd., 1990).

3. (a) Books on Special Aspects of EC Competition Law

BALFOUR, J., *Air Law and the European Community* (Current EEC Legal Development Series, Butterworths, 1991).

CAWTHRA, B., *Patent Licensing in Europe*, 2nd edn. (Butterworths, 1986).

CLOUGH, M., and RANDOLPH, F., *Shipping and Competition Law* (Current EEC Legal

Development Series, Butterworths, 1991).

COOK, J., and KERSE, C., *E.E.C. Merger Control* (Sweet & Maxwell, 1991).

DOWNES, T. A., and ELLISON, J., *The Legal Control of Mergers in the European Communities* (Blackstone Press, 1991).

ELLAND, W., *Beer Supply Agreements* (Current EEC Legal Development Series, Butterworths, 1991).

FINE, F. L., *Mergers and Joint Ventures in Europe* (Graham & Trotman, 1989).

GUY, D., and LEIGH, G., *The EEC and Intellectual Property* (Sweet & Maxwell, 1981).

HARDING, C., *Notices and Group Exemptions in EEC Competition Law* (ESC Publishing Ltd., looseleaf, 1980).

KERSE, C. S., *EEC Antitrust Procedure*, 2nd edn. (European Law Centre Ltd., 1987).

KORAH, V. L., and ROTHNIE, W. A., *Exclusive Distribution and the EEC Competition Rules* (Sweet & Maxwell, 1992).

—— *Patent Licensing and EEC Competition Rules: Regulation 2349/84* (ESC Publishing Ltd., 1985).

—— *Research and Development: Joint Ventures and EEC Competition Rules; Regulation 418/85* (ESC Publishing Ltd., 1986).

—— *Know-how Licensing Agreements and EEC Competition Rules: Regulation 556/89* (ESC Publishing, 1989).

—— *Franchising Agreements and the EEC Rules: Regulation 4087/88* (ESC Publishing, 1989).

MENDELSSOHN, M., and HARRIS, B., *Franchising of the Block Exemption Regulation* (Longman, 1991).

(b) Books on Special Aspects of Subjects Related to EC Competition Law

GORMLEY, L., *Prohibiting Restrictions on Trade within the EEC* (North-Holland, 1985).

MENDES, M. M., *Antitrust in a World of Inter-related Economics: The Interplay between Antitrust and Trade Policies in the US and EEC* (Brussels University Editions, 1991).

VAN BAEL, I., and BELLIS, J. F., *International Trade Law and Practice of the European Community: EEC Anti-dumping and other Trade Protection Laws* (CCH Editions Ltd., 1985).

4. Books on UK and EC Competition Law

FREEMAN, P., and WHISH, R. (eds.), *Encyclopaedia of Competition Law*, 3 vols. (Butterworths, 1991).

GARDINER, N., *A Guide to UK and European Community Competition Policy* (Macmillan, 1990).

GREEN, N., *Commercial Agreements and Competition Law* (Graham and Trotman Ltd., 1986).

MERKIN, R., and WILLIAMS, K., *Competition Law: Antitrust Policy in the UK and EEC* (Sweet & Maxwell, 1984).

WHISH, R., *Competition Law*, 2nd edn. (Butterworths, 1989).

5. Other Works

DIEBOLD, W., *The Schuman Plan: A Study in Economic Co-operation* (Praeger, 1959).
MAYNE, R., *The Recovery of Europe* (Weidenfeld & Nicholson, 1970).
MONNET, J., *Memoirs*, trans. Richard Mayne (Collins, 1978).
Essays *In Memoriam J. D. B. Mitchell* (Sweet & Maxwell, 1983).
NEALE, A. D., and STEPHENS, M. L., *International Business and National Jurisdiction* (Oxford University Press, 1988).
SWANN, D., *The Economics of the Common Market*, 5th edn. (Penguin Books, 1984).

PERIODICAL ARTICLES FOR FURTHER READING

General Note

This list is selective and includes only articles in the English language in periodicals published in the Community. It does not generally include individual case notes in such periodicals nor articles in the United States law reviews, though both sources often contain valuable material relevant to the subject matter of this book. Three periodicals with the largest number of relevant articles are the *European Competition Law Review*, the *European Law Review*, and the *Common Market Law Review*. Many of the articles listed below are also referred to in footnotes in the relevant chapters of the book. In addition to excellent full-length articles, the *Yearbook of European Law* also contains a full annual review of competition law developments for the preceding calendar year. The *Proceedings and Papers of the Annual Fordham Corporate Law Institute* held in New York each autumn (and published during the summer of the following year) are also of particular value and relevance because of the quality of the speakers invited to participate, and the *Fordham International Law Journal* maintains a particularly high standard.

Chapters 6 to 8

EVANS, A. C., 'European Competition Law and Consumers, the Article 85(3) Exemption', [1981] 2 *ECLR* 425.
GIJLSTRA, D. J., and MURPHY, D. F., 'Some Observations on the *Sugar* Cases', [1977] 14 *CML Rev* 45.
GREEN, N., 'Article 85 in Perspective', [1988] 9 *ECLR* 190.
HAWK, B. E., 'The American (Antitrust) Revolution: Lessons for the EEC', [1988] 9 *ECLR* 53.
KON, S., 'Article 85, paragraph 3: A Case for Application by National Courts', [1982] 19 *CML Rev* 541.
KORAH, V. L., 'Concerted Practices', [1973] 36 *MLR* 220.
MANN, F. A., 'The *Dyestuffs* Case' in the Court of Justice of the European Communities, [1973] 22 *ICLQ* 35.
STEINDORFF, E., 'Article 85 and the Rule of Reason', [1984] 21 *CML Rev* 639.
WHISH, R., and SUFRIN, B., 'Article 85 and the Rule of Reason', [1987] 7 *YEL* 1.

Chapters 9 to 11

FAULL, J., 'Joint Ventures under the EEC Competition Rules', [1984] 5 *ECLR* 358.

KORAH, V. L., 'Critical Comments on Commission's Recent Decisions Exempting Joint Ventures to Exploit Research that Needs Further Development', [1987] 12 *EL Rev* 18.

LANG, J. TEMPLE, 'Joint Ventures under the EEC Treaty Rules on Competition', [1977] vol. xii (new series) *Irish Jurist* 15.

SHARPE, T., 'The Commission's Proposals on Crisis Cartels', [1980] 17 *CML Rev.* 75.

VENIT, J. S., 'The Research and Development Block Exemption Regulation', [1985] 10 *EL Rev* 151.

Chapters 12 to 14

CHARD, J. S., 'The Economics of the Application of Article 85 to Selective Distribution Systems', [1982] 7 *EL Rev* 83.

DOWNES, T. A., 'Exclusive Dealing Agreements: A Change for the Worse?', [1979] 4 *EL Rev* 166.

FERRY, J., 'Selective Distribution and Other Post-sales Restrictions', [1981] 2 *ECLR* 209.

GOEBEL, R. J., 'The Uneasy Fate of Franchising under EEC Antitrust Laws', [1985] 10 *EL Rev* 87.

GROVES, P., 'Motor Vehicle Distribution: The Block Exemption', [1987] 12 *ECLR* 77.

GROVES, S. G., 'Refusal to Supply and Article 85', [1983] 4 *ECLR* 65.

GYSELEN, L., 'Vertical Restraints in the Distribution Process', [1984] 21 *CML Rev* 647.

HORNSBY, S., 'Public and Private Resale Price Maintenance Systems in the Publishing Sector: The Need for Equal Treatment in European Law', [1985] 10 *EL Rev* 381.

KORAH, V. L., 'Group Exemptions for Exclusive Distribution and Purchasing', [1984] 21 *CML Rev* 53.

—— '*Pronuptia*: Franchising the Marriage of Reason and EEC Competition Rules', [1986] 8 *EIPR* 99.

LASOK, K. P., 'Assessing the Economic Consequences of Restrictive Agreements', [1991] 12 *ECLR* 194.

LUKOFF, F. L., 'European Competition Law and Distribution in the Motor Vehicle Sector', [1986] 23 *CML Rev* 841.

SHARPE, T., 'The *Distillers* Decision', [1978] 15 *CML Rev* 447.

Chapters 15 to 16

BARENTS, R., 'New Developments in Measures Having Equivalent Effects', [1981] 18 *CML Rev* 271.

FRAZER, T., 'The Commission's Policy on Know-how Agreements', [1989] 9 *YEL* 1.

HOFFMAN, D., and O'FARRELL, O., 'The Open Exclusive Licence—Scope and Consequences', [1984] 6 *EIPR* 104. KORAH, V. L., A Reply, [1984] 6 *EIPR* 206.

Joliet, R., Marenco, G., and Banks, K., 'Intellectual Property and the Community Rules on Free Movement: Discrimination Unearthed', [1990] 15 *EL Rev* 224.

Pickard, S., 'The Commission's Patent Licensing Regulation—A History', [1984] 5 *ECLR* 158.

—— The Commission's Patent Licensing Regulation—A Guide', [1984], 5 *ECLR* 384.

Rothnie, W. A., 'EC Competition Policy, The Commission and Trademarks', [1991] 16 *IBL* 495.

Venit, J. S., 'EEC Patent Licensing Revisited: The Commission's Patent Licensing Regulation', [1985] 30 *Anti Trust Bulletin* 457.

Winn, D. B., 'The Commission's Know-how Regulation', [1990] 11 *ECLR* 135.

Zeyen, C., and Odle, A., 'The EC Know-how Regulation: Practical Difficulties and Legal Uncertainties', [1991] 12 *ECLR* 231.

Chapters 17 to 19

de Jong, H. W., 'Unfair and Discriminatory Pricing under Article 86', [1980] 1 *ECLR* 297.

Fuller, C. Baden, 'Economic Analysis of the Existence of a Dominant Position: Article 86 of the Treaty', [1979] 4 *EL Rev* 423.

Gyselen, L., and Kyriakis, N., 'Article 86: The Monopoly Power Issue Revisited', [1986] 11 *EL Rev* 134.

Korah, V. L., 'Concept of a Dominant Position within the Meaning of Article 86', [1980] 17 *CML Rev* 395.

Schödermeier, M., 'Collective Dominance Revisited', [1990] 11 *ECLR* 28.

Schulte-Braucks, R., 'European Telecommunications Law in the Light of the *British Telecom* Judgment', [1986] 23 *CML Rev* 39.

Sharpe, T., 'Predation', [1987] 8 *ECLR* 53.

Siragusa, M., 'The Application of Article 86 to the Pricing Policy of Dominant Companies: Discriminatory and Unfair Pricing', [1979] 16 *CML Rev* 179.

Smith, P., 'The Wolf in Wolf's Clothing: The Problem with Predatory Pricing', [1989] 14 *EL Rev* 209.

Temple Lang, J., 'Monopolisation and the Definition of Abuse of a Dominant Position under Article 86', [1979] 16 *CMLR* 345.

Vajda, C., 'Article 86 and a Refusal to Supply', [1981] 2 *ECLR* 97.

Vogelenzang, P., 'Abuse of a Dominant Position in Article 86: The Problem of Causality and Some Applications', [1976] 13 *CMLR Rev* 61.

Chapter 20

Bright, C., 'The European Merger Control Regulation: Do Member States still have an Independent Role in Merger Control', [1991] 12 *ECLR* 139, 184.

Brittan, L., 'The Law and Policy of Merger Control in the EEC', [1990] 15 *EL Rev* 351.

Elland, W., 'The Mergers Control Regulation', [1990] 11 *ECLR* 111.

Elland, W., 'The Merger Control Regulation and its Effect on National Merger

Controls', [1991] 12 *ECLR* 19.

KOVAR, R., 'EC Merger Control Revisited', [1990] 10 *YEL* 71.

SIRAGUSA, M., and SUBIOTTO, R., 'The EEC Merger Control Regulation: The Commission's Evolving Case Law', [1991] 28 *CMLR* 877.

VENIT, J. S., 'The Merger Control Regulation: Europe Comes of Age or Caliban's Dinner', [1990] 27 *CMLR* 7.

Chapter 21

CORONES, S. G., 'The Application of Article 85 to the Exchange of Market Information between Members of a Trade Association', [1982] 3 *ECLR* 67.

REYNOLDS, M. J., 'Trade Associations and the EEC Competition Rules', [1985] 23 *Revue Suisse de Droit International de la Concurrence* 49.

WATSON, P., and WILLIAMS, K., 'Application of EEC Competition Rules to Trade Associations', [1988] 8 *YEL* 121.

Chapter 22

BARAV, A., 'Enforcement of Community Rights in the National Courts', [1989] 26 *CMLR* 369.

BARAV, A., and GREEN, N., 'Damages in National Courts for Breach of Community Law', [1986] 6 *YEL* 55.

DAVIDSON, J. S., 'Actions for Damages in the English Court for Breach of EEC Competition Law', [1985] 34 *ICLQ* 178.

MARKERT, K., 'Some Legal and Administrative Problems of the Co-existence of Community and National Competition Law in the EEC', [1974] 11 *CML Rev* 92.

VAJDA, C., 'The *Perfume* Cases: The Role of National Courts in Applying National and Community Competition Law', [1980] 2 *EIPR* 397.

Chapter 23

DAGTOGLOU, P., 'Air Transport and the European Community', [1981] 6 *EL Rev* 335.

HOFFMANN, A. B., 'Anti-Competitive State Legislation Condemned under Articles 5, 85 and 86', [1990] 11 *ECLR* 11.

KUYPER, P. J., 'Airline Fare Fixing and Competition: An English Lord, Commission Proposals, and US Parallels', [1983] 20 *CML Rev* 203.

MARENCO, G., 'Public Sector and Community Law', [1983] 20 *CML Rev* 495.

PAPPALARDO, A., 'State Measures and Public Undertakings: Article 90 Revisited', [1991] 12 *ECLR* 29.

SLOT, P. J., 'The Application of Articles 3(*f*), 5, and 85 to 94', [1987] 12 *EL Rev* 179.

WAELBROECK, M., 'The Extent to which Government Interference can constitute Justification under Article 85 or Article 86 of the Treaty of Rome', [1980] 8 *Int Business Lawyer* 113.

Chapter 24

BELLIS, J. F., 'International Trade and the Competition Law of the EEC', [1979] 16 *CML Rev* 647.

BOURGEOIS, J. H. J., 'EEC Control over International Mergers', [1990] 10 *YEL* 103.

—— 'Antitrust and Trade Policy: A Peaceful Co-existence? European Community Perspective', [1989] 15 *IBL* 58, 115.

FERRY, J. E., 'Towards Completing the Charm: The *Woodpulp* Judgment', [1989] 10 *ECLR* 58.

KULMS, R., Competition, Trade Policy and Competition Policy in the EEC', [1990] 27 *CMR Rev* 285.

LANGE, D. G., and SANDAGE, J. B., 'The *Woodpulp* Judgment and its Implication for the Scope of EEC Competition Law', [1989] 26 *CML Rev* 17.

REYNOLDS, M. J., 'Extraterritorial Aspects of Mergers and Joint Ventures: The EEC Position', [1985] 6 *ECLR* 165.

SLOT, P. J., and GRABANDT, E., 'Extraterritoriality and Jurisdiction', [1986] 23 *CML Rev* 545.

Chapters 25 and 26

FORRESTER, I., and NORRALL, C., 'The Laicization of Community Law: Self-help and the Rule of Reason', [1984] 21 *CML Rev* 11.

FRAZER, T. M., 'Competition Policy after 1992: The Next Step', [1990] 53 *MLR* 609.

HORNSBY, S., 'Competition Policy in the Eighties: More Policy, Less Competition', [1987] *ECLR* 22 79.

SUTHERLAND, P., 'Towards a Competition Policy: Future Trends and Actions', [1986] 6 *EIPR* 283.

VAN BAEL, I., 'The Antitrust Settlement Practice of the EEC Commission', [1986] 23 *CML Rev* 61.

Index